Y0-BYA-119

Ultraclean Semiconductor Processing Technology and Surface Chemical Cleaning and Passivation

MATERIALS RESEARCH SOCIETY
SYMPOSIUM PROCEEDINGS VOLUME 386

Ultraclean Semiconductor Processing Technology and Surface Chemical Cleaning and Passivation

Symposium held April 17-19, 1995, San Francisco, California, U.S.A.

EDITORS:

Michael Liehr
IBM T.J. Watson Research Center
Yorktown Heights, New York, U.S.A.

Marc Heyns
IMEC
Leuven, Belgium

Masataka Hirose
Hiroshima University
Higashi-Hiroshima, Japan

Harold Parks
University of Arizona
Tucson, Arizona, U.S.A.

PITTSBURGH, PENNSYLVANIA

Single article reprints from this publication are available through
University Microfilms Inc., 300 North Zeeb Road, Ann Arbor, Michigan 48106

CODEN: MRSPDH

Published by:

Materials Research Society
9800 McKnight Road
Pittsburgh, Pennsylvania 15237
Telephone (412) 367-3003
Fax (412) 367-4373
Homepage http://www.mrs.org/

Library of Congress Cataloging in Publication Data

Ultraclean semiconductor processing technology and surface chemical cleaning and passivation / editors, Michael Liehr, Marc Heyns, Masataka Hirose, Harold Parks
p. cm.—(Materials Research Society symposium proceedings, ISSN 0272-9172; vol. 386
Includes bibliographical references and index.
ISBN 1-55899-289-8 (alk. paper)
1. Integrated circuits—Design and construction—Congresses.
2. Manufacturing Processes—Cleaning—Congresses. I. Liehr, Michael II. Heyns, Marc III. Hirose, Masataka IV. Parks, Harold V. Series: Materials Research Society Symposium Proceedings; Vol. 386
TK7874.U373 1995 95-31842
621.3815'2—dc20 CIP

Manufactured in the United States of America

CONTENTS

*Invited Paper

PART II: CHEMICAL MECHANICAL POLISHING AND POST-CMP CLEANING

*Invited Paper

*Invited Paper

PART V: GAS-PHASE CLEANING

*Invited Paper

*Invited Paper

*Invited Paper

PREFACE

This volume contains most of the papers presented at the 1995 MRS Spring Meeting, Symposium O, entitled "Ultraclean Semiconductor Processing Technology and Surface Chemical Cleaning and Passivation."

The symposium is the third in a series held on this general subject area. Attendance at the symposium, held April 17-19 in San Francisco, well exceeded expectations and is testimony to the continued—and increased—significance of this area of science and technology. In fact, for most of our symposium we occupied about twice the seats we anticipated we would need! Wafer cleaning, microcontamination, and surface passivation remained the key focus of the symposium. The symposium's move to include related technologies, such as chemo-mechanical polishing and the associated cleaning steps, proved successful and outlines future opportunities and challenges in this field.

The sessions on aqueous wafer cleaning focussed on control of surface morphology, as well as particle- and molecular-contamination removal. It was dominated by work aimed at improving the standard RCA SC-1 cleaning step, mainly through the use of elaborated design-of-experiments. It has become clear at this point that reduced-chemical-consumption cleaning is feasible without sacrificing cleaning efficiency. Such cleans have already been implemented by several semiconductor manufacturers. The session on aqueous wafer cleaning also emphasized TCAD modeling efforts to optimize equipment flow pattern and thus, e.g., rinsing efficiency to reduce process time and cost. Surprisingly, we learned that the effects of megasonics, in manufacturing use for several years, are not well understood. Significantly more effort is needed to elucidate the effects of megasonics on boundary layer diffusion, cavitation, and streaming potentials.

Chemo-mechanical polishing (CMP) was reviewed as an emerging technology to which this community can contribute significantly. Invited presentations made it clear that CMP has been in manufacturing use both at wafer suppliers and microelectronics manufacturers for many years. However, the field offers substantial opportunity for research as well as innovation, e.g., in the area of slurry development. Our understanding of the mechanisms of CMP, as well as post-CMP cleaning—i.e., particle adhesion and its dependence on pH and surfactant use—was seen to be rudimentary at best.

Gas-phase cleaning seems to emerge finally as a serious process option. Several presentations made clear that the state of the art in this area is progressing steadily, and that manufacturing implementation at a larger scale can be expected around the 1Gb generation.

A session on process integration highlighted the level of sophistication that present-day manufacturing—e.g., of the 64Mb generation—has to take into account to achieve acceptable yield levels. This session was particularly useful to put in context the field of wafer surface preparation with respect to preceding and subsequent process steps and the associated process interactions. It also highlighted the multiple options for process trade-offs. Several papers in this session dealt with the issues of cluster integration and device characterization that become critical in future technologies.

Acceptable microcontamination levels and techniques to characterize these ever-decreasing levels were subject to a significant number of symposium contributions. To be highlighted here is the emergence of organic contamination as a serious microcontamination issue. Mostly due to the complexity and scarcity of the contaminants, this presents a tremendous challenge to this community. The other area that emerges as a key focus is the application of scanning probe techniques *in-situ* (i.e., in solution) to enhance our understanding of cleaning processes while they occur at an atomic level.

We are indebted to Judy Kuneth, Rose Avitia, and Becky Hagenston for help in preparing this proceedings.

Michael Liehr
Marc Heyns
Masataka Hirose
Harold Parks

June 1995

ACKNOWLEDGMENTS

Sponsors

Balzers
CFM Technologies Incorporated
Dainippon Screen/Prizm
Fujikin Incorporated
Fujitsu Ltd.
Hitachi Ltd.
IBM Corporation
Legacy Systems Incorporated
Matsushita Electric Industrial Co., Ltd.
Mitsubishi Electric Corporation
NEC Corporation
Nomura Micro Science Co., Ltd.
OnTrak Systems Incorporated
Santa Clara Plastics
Sharp Corporation
Solid State Equipment Corporation
SubMicron Systems, Incorporated
Texas Instruments Japan Ltd.
Toshiba Corporation
ULSI R&D Laboratories—Sony Corporation

MATERIALS RESEARCH SOCIETY SYMPOSIUM PROCEEDINGS

Volume 352—Materials Issues in Art and Archaeology IV, P.B. Vandiver, J.R. Druzik, J.L. Galvan Madrid, I.C. Freestone, G.S. Wheeler, 1995, ISBN: 1-55899-252-9

Volume 353—Scientific Basis for Nuclear Waste Management XVIII, T. Murakami, R.C. Ewing, 1995, ISBN: 1-55899-253-7

Volume 354—Beam-Solid Interactions for Materials Synthesis and Characterization, D.E. Luzzi, T.F. Heinz, M. Iwaki, D.C. Jacobson, 1995, ISBN: 1-55899-255-3

Volume 355—Evolution of Thin-Film and Surface Structure and Morphology, B.G. Demczyk, E.D. Williams, E. Garfunkel, B.M. Clemens, J.E. Cuomo, 1995, ISBN: 1-55899-256-1

Volume 356—Thin Films: Stresses and Mechanical Properties V, S.P. Baker, P. Børgesen, P.H. Townsend, C.A. Ross, C.A. Volkert, 1995, ISBN: 1-55899-257-X

Volume 357—Structure and Properties of Interfaces in Ceramics, D.A. Bonnell, U. Chowdhry, M. Rühle, 1995, ISBN: 1-55899-258-8

Volume 358—Microcrystalline and Nanocrystalline Semiconductors, R.W. Collins, C.C. Tsai, M. Hirose, F. Koch, L. Brus, 1995, ISBN: 1-55899-259-6

Volume 359—Science and Technology of Fullerene Materials, P. Bernier, D.S. Bethune, L.Y. Chiang, T.W. Ebbesen, R.M. Metzger, J.W. Mintmire, 1995, ISBN: 1-55899-260-X

Volume 360—Materials for Smart Systems, E.P. George, S. Takahashi, S. Trolier-McKinstry, K. Uchino, M. Wun-Fogle, 1995, ISBN: 1-55899-261-8

Volume 361—Ferroelectric Thin Films IV, S.B. Desu, B.A. Tuttle, R. Ramesh, T. Shiosaki, 1995, ISBN: 1-55899-262-6

Volume 362—Grain-Size and Mechanical Properties—Fundamentals and Applications, N.J. Grant, R.W. Armstrong, M.A. Otooni, T.N. Baker, K. Ishizaki, 1995, ISBN: 1-55899-263-4

Volume 363—Chemical Vapor Deposition of Refractory Metals and Ceramics III, W.Y. Lee, B.M. Gallois, M.A. Pickering, 1995, ISBN: 1-55899-264-2

Volume 364—High-Temperature Ordered Intermetallic Alloys VI, J. Horton, I. Baker, S. Hanada, R.D. Noebe, D. Schwartz, 1995, ISBN: 1-55899-265-0

Volume 365—Ceramic Matrix Composites—Advanced High-Temperature Structural Materials, R.A. Lowden, J.R. Hellmann, M.K. Ferber, S.G. DiPietro, K.K. Chawla, 1995, ISBN: 1-55899-266-9

Volume 366—Dynamics in Small Confining Systems II, J.M. Drake, S.M. Troian, J. Klafter, R. Kopelman, 1995, ISBN: 1-55899-267-7

Volume 367—Fractal Aspects of Materials, F. Family, B. Sapoval, P. Meakin, R. Wool, 1995, ISBN: 1-55899-268-5

Volume 368—Synthesis and Properties of Advanced Catalytic Materials, E. Iglesia, P. Lednor, D. Nagaki, L. Thompson, 1995, ISBN: 1-55899-270-7

Volume 369—Solid State Ionics IV, G-A. Nazri, J-M. Tarascon, M. Schreiber, 1995, ISBN: 1-55899-271-5

Volume 370—Microstructure of Cement Based Systems/Bonding and Interfaces in Cementitious Materials, S. Diamond, S. Mindess, F.P. Glasser, L.W. Roberts, J.P. Skalny, L.D. Wakeley, 1995, ISBN: 1-55899-272-3

Volume 371—Advances in Porous Materials, S. Komarneni, D.M. Smith, J.S. Beck, 1995, ISBN: 1-55899-273-1

Volume 372—Hollow and Solid Spheres and Microspheres—Science and Technology Associated with their Fabrication and Application, M. Berg, T. Bernat, D.L. Wilcox, Sr., J.K. Cochran, Jr., D. Kellerman, 1995, ISBN: 1-55899-274-X

Volume 373—Microstructure of Irradiated Materials, I.M. Robertson, L.E. Rehn, S.J. Zinkle, W.J. Phythian, 1995, ISBN: 1-55899-275-8

MATERIALS RESEARCH SOCIETY SYMPOSIUM PROCEEDINGS

Volume 374—Materials for Optical Limiting, R. Crane, K. Lewis, E.V. Stryland, M. Khoshnevisan, 1995, ISBN: 1-55899-276-6
Volume 375—Applications of Synchrotron Radiation Techniques to Materials Science II, L.J. Terminello, N.D. Shinn, G.E. Ice, K.L. D'Amico, D.L. Perry, 1995, ISBN: 1-55899-277-4
Volume 376—Neutron Scattering in Materials Science II, D.A. Neumann, T.P. Russell, B.J. Wuensch, 1995, ISBN: 1-55899-278-2
Volume 377—Amorphous Silicon Technology—1995, M. Hack, E.A. Schiff, M. Powell, A. Matsuda, A. Madan, 1995, ISBN: 1-55899-280-4
Volume 378—Defect- and Impurity-Engineered Semiconductors and Devices, S. Ashok, J. Chevallier, I. Akasaki, N.M. Johnson, B.L. Sopori, 1995, ISBN: 1-55899-281-2
Volume 379—Strained Layer Epitaxy—Materials, Processing, and Device Applications, J. Bean, E. Fitzgerald, J. Hoyt, K-Y. Cheng, 1995, ISBN: 1-55899-282-0
Volume 380—Materials—Fabrication and Patterning at the Nanoscale, C.R.K. Marrian, K. Kash, F. Cerrina, M. Lagally, 1995, ISBN: 1-55899-283-9
Volume 381—Low-Dielectric Constant Materials—Synthesis and Applications in Microelectronics, T-M. Lu, S.P. Murarka, T.S. Kuan, C.H. Ting, 1995, ISBN: 1-55899-284-7
Volume 382—Structure and Properties of Multilayered Thin Films, T.D. Nguyen, B.M. Lairson, B.M. Clemens, K. Sato, S-C. Shin, 1995, ISBN: 1-55899-285-5
Volume 383—Mechanical Behavior of Diamond and Other Forms of Carbon, M.D. Drory, M.S. Donley, D. Bogy, J.E. Field, 1995, ISBN: 1-55899-286-3
Volume 384—Magnetic Ultrathin Films, Multilayers and Surfaces, A. Fert, H. Fujimori, G. Guntherodt, B. Heinrich, W.F. Egelhoff, Jr., E.E. Marinero, R.L. White, 1995, ISBN: 1-55899-287-1
Volume 385—Polymer/Inorganic Interfaces II, L. Drzal, N.A. Peppas, R.L. Opila, C. Schutte, 1995, ISBN: 1-55899-288-X
Volume 386—Ultraclean Semiconductor Processing Technology and Surface Chemical Cleaning and Passivation, M. Liehr, M. Hirose, M. Heyns, H. Parks, 1995, ISBN: 1-55899-289-8
Volume 387—Rapid Thermal and Integrated Processing IV, J.C. Sturm, J.C. Gelpey, S.R.J. Brueck, A. Kermani, J.L. Regolini, 1995, ISBN: 1-55899-290-1
Volume 388—Film Synthesis and Growth Using Energetic Beams, H.A. Atwater, J.T. Dickinson, D.H. Lowndes, A. Polman, 1995, ISBN: 1-55899-291-X
Volume 389—Modeling and Simulation of Thin-Film Processing, C.A. Volkert, R.J. Kee, D.J. Srolovitz, M.J. Fluss, 1995, ISBN: 1-55899-292-8
Volume 390—Electronic Packaging Materials Science VIII, R.C. Sundahl, K.A. Jackson, K-N. Tu, P. Børgesen, 1995, ISBN: 1-55899-293-6
Volume 391—Materials Reliability in Microelectronics V, A.S. Oates, K. Gadepally, R. Rosenberg, W.F. Filter, L. Greer, 1995, ISBN: 1-55899-294-4
Volume 392—Thin Films for Integrated Optics Applications, B.W. Wessels, D.M. Walba, 1995, ISBN: 1-55899-295-2
Volume 393—Materials for Electrochemical Energy Storage and Conversion—Batteries, Capacitors and Fuel Cells, D.H. Doughty, B. Vyas, J.R. Huff, T. Takamura, 1995, ISBN: 1-55899-296-0
Volume 394—Polymers in Medicine and Pharmacy, A.G. Mikos, K.W. Leong, M.L. Radomsky, J.A. Tamada, M.J. Yaszemski, 1995, ISBN: 1-55899-297-9

Prior Materials Research Society Symposium Proceedings available by contacting Materials Research Society

Part I

Aqueous Si Surface Cleaning

Recent Advances in Wet Processing Technology and Science

Steven Verhaverbeke*, Rochdi Messoussi, Hitoshi Morinaga and Tadahiro Ohmi

Tohoku University, Sendai, Japan

*Presently at CFM Technologies, West Chester, PA

1. INTRODUCTION

For the cleaning of silicon substrates, a mixture of an oxidizing and an etching agent is often used. One of the first successful mixtures (1970) is the well known APM or SC-1 cleaning ($NH_4OH/H_2O_2/H_2O$) (1). Since then, other mixtures (oxidizing agent/etching agent) such as HNO_3/HF (2), H_2O_2/HF (3) were developed. For the cleaning of silicon substrates, a mixture of an oxidizing agent and an etching agent always results in good particle performance, since the Si under the particles is continuously oxidized and etched simultaneously (2,3).
Oxidizing agents are not only used for oxidizing the Si surface, but also to oxidize the metallic impurities in order to get them dissolved in the solution. The most well know mixture of this kind is the HPM or the SC-2 cleaning solution ($HCl/H_2O_2/H_2O$) (1).
For oxidizing agents, either H_2O_2 or HNO_3 is used commonly. Recently, also O_3 has been used as an oxidizer. In this paper, the different solutions which contain H_2O_2 will be reviewed.

2. APM solution

The APM (Ammonia Peroxide Mixture) solution was developed by W. Kern in 1970 (1). It consists of a mixture of $NH_3/H_2O_2/H_2O$. This mixture was developed to remove organics and particles. The organics and particle removal works by the solvating action of the NH_3, the oxidizing action of the H_2O_2 to break down the organics and the undercutting of the particles by the etching of SiO_2 by H_2O at high pH. Moreover, the oxidizing action of the H_2O_2 forms a protective layer on the Si. NH_3 was chosen because of its volatility and thus, it would not leave any residue on the surface.
One of the fundamental properties of the APM solution is the double layer repulsion which prevents the particles from redepositing (4).
Since 1970, the only fundamental change has been the adding of megasonic energy. This has provided us with an additional particle removal mechanism. Without megasonic energy the particles are removed by undercutting. However, the megasonic energy provides enough energy for physical dislodging and thus the undercutting is not necessary any more. This has led to a substantial decrease in temperatures needed to perform the APM cleaning.
Since 1970, the only process advances which have been made are the choice of concentrations for the APM cleaning. It was found that the APM solution had some negative effects such as Si surface roughening (5), metallic contamination deposition, evaporation of the NH_3 and decomposition of the H_2O_2 (1) and Si etching which leads to bird's beak problems and irreproducible V_T's (threshold voltage). Therefore, the early work was mainly focused at limiting the Si etching and roughening of the surface (6). The main outcome of this work was the reduction in NH_3 concentration. However, recently the attention was shifted to increasing the particle removal. Initially, it was thought that higher concentrations would yield better particle removal efficiencies. It was shown already early on however that there was an optimum NH_3 concentration for particle

Mat. Res. Soc. Symp. Proc. Vol. 386 © 1995 Materials Research Society

removal and that higher NH_3 concentrations resulted in decreased particle removal efficiencies (7). This was confirmed recently by S. Cohen *et al.* (8). There are a number of mechanisms which are responsible for the drop in particle removal efficiency at higher NH_3 concentrations. At higher NH_3 concentrations, the double layer repulsion decreases because the ionic strength of the solution increases. Secondly, H_2O is the purest chemical. So, evidently the more dilute the cleaning solution, the more pure. Finally, at higher pH, the hydrogen peroxide decomposes faster and will form O_2 bubbles because of the low solubility of O_2 in aqueous solutions. The bubbles will also lead to increased particle deposition (9). A newer development is the reduction in the hydrogen peroxide concentration. Also a reduction in the hydrogen peroxide concentration leads to improved particle removal efficiencies, mainly through the same mechanisms. At first a lower hydrogen peroxide concentration leads to lower bubble densities, since the bubble density is directly proportional to the hydrogen peroxide concentration and secondly the cleanliness of the bath improves with lower hydrogen peroxide concentration since H_2O is the purest chemical. It has been found by T. Futatsuki *et al.* (10) that the hydrogen peroxide can be reduced by a factor of 10 before any surface roughening starts.

3. HPM solution

Also, the HPM (Hydrogen chloride hydrogen Peroxide Mixture) solution was developed by W. Kern in 1970 (1). It consists of a mixture of $HCl/H_2O_2/H_2O$. According to W. Kern this mixture could remove gold at 90^oC. In order to remove Fe or alkali metals, HCl/H_2O at room temperature was enough (11). However, since gold was an important impurity at that time, W.Kern advised to use $HCl/H_2O_2/H_2O$ as a general purpose mixture. W. Kern had chosen HCl again because of its volatility so that it wouldn't leave any residues. Recently these results were confirmed by J.S. Glick (12). He found, similar to W. Kern, that for Fe, Zn and Al removal, hydrogen peroxide is not necessary. This can be understood very easily by looking at the solubility of metals as a function of pH and oxidation potential. This can be represented in a Potential-pH diagram as e.g. shown in fig. 1 for the case of Cu. This is a so called Pourbaix diagram. In fig. 2 is shown the position in this diagram of the different cleaning solutions. From comparison of fig. 2 and fig. 1 we can see that all of the cleaning solutions which contain an acid with or without an additional oxidizer such as peroxide fall in the dissolution area of Cu and thus will remove Cu impurities. For every metal such a diagram exists and thus we can easily determine whether a metal impurity is soluble in a certain solution or not.

Since most cleaning solutions for metals are close to pH=0, we can also use the standard reduction potentials for determining if a metal can be dissolved in its oxidized state or not in the solution. Some relevant standard reduction potentials are shown in table 1. From this table it is clear that dissolved O_2 can oxidize most of the metals, but that H_2O_2 is necessary to oxidize Au. Therefore, W. Kern had to use a $HCl/H_2O_2/H_2O$ mixture in order to remove Au impurities from the surface. However, the amount of H_2O_2 is not critical. The oxidation potential of solutions containing H_2O_2 is determined by the oxidation potential of the following reaction :

$$H_2O_2 + 2H^+ + 2e^- = 2H_2O$$

The oxidation potential of this reaction is :

$$E = E^o + 0.0347 \log[H_2O_2] = 1.78 - RT/F\ pH + 0.0347 \log[H_2O_2]$$

For SC-2 solutions, which have a pH around 0, this gives for a concentration of
$[H_2O_2] = 1$ mol/l $\quad E = 1.78$ V and for a concentration of
$[H_2O_2] = 0.26*10^{-3}$ mol/l (=10 ppm) $\quad E = 1.66$ V.
So, even at 10 ppm H_2O_2 concentration the oxidizing power is strong enough to keep Au into the solution. This value is important, since 10 ppm is around the solubility limit of O_2 in aqueous

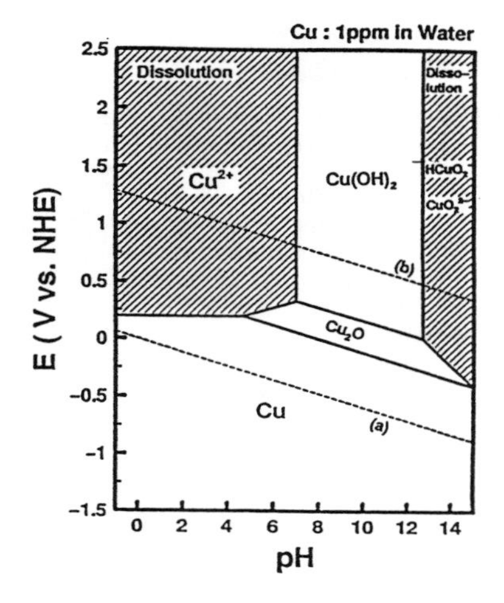

Fig. 1. Potential-pH diagram (Pourbaix diagram) of the Cu-water system calculated from the equilibrium constant at 25°C.

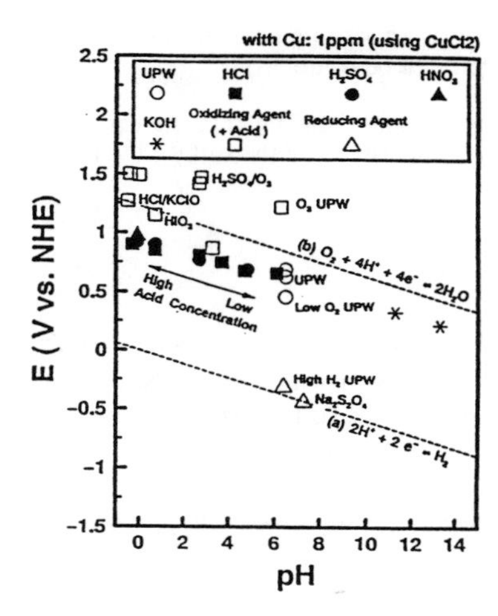

Fig. 2. The pH and redox potential of various solutions (measured value).

Table 1 Standard Reduction Potentials of some relevant reactions

	E^o (V vs. NHE)
$O_3 + 2H^+ + 2e^- = O_2 + H_2O$	2.07
$H_2O_2 + 2H^+ + 2e^- = 2H_2O$	1.776
$Au^{3+} + 3e^- = Au$	1.50
$O_2 + 4H^+ + 4e^- = 2H_2O$	1.229
$Ag^+ + e^- = Ag$	0.799
$Cu^+ + e^- = Cu$	0.521
$Cu^{2+} + 2e^- = Cu$	0.337
$2H^+ + 2e^- = H_2$	0.000
$Pb^{2+} + 2e^- = Pb$	-0.129
$Ni^{2+} + 2e^- = Ni$	-0.228
$Fe^{2+} + 2e^- = Fe$	-0.440
$SiO_2 + 4H^+ + 4e^- = Si + 2H_2O$	-0.857
$Al^{3+} + 3e^- = Al$	-1.662
$Mg^{2+} + 2e^- = Mg$	-2.37
$Na^+ + e^- = Na$	-2.714

solutions. So, at this concentration of H_2O_2, almost no O_2 bubbles will be formed since all the oxygen will be soluble.

The temperature effect can be understood by considering the reactions involved. In order to dissolve a metal impurity from the surface the following reactions are going on (from right to left)

$xM^+ \leftarrow M_xO_y$
or
$M^+ \leftarrow M^o$

These reactions are characterized by a Gibbs free energy ΔG^o.
The equilibrium which exists between the metals dissolved in the liquid (M^+) and the metals deposited on the surface as oxides, hydroxides or metallic (M_xO_y and M^o) can be expressed by the following equation :

$$C_{deposited} = C_{dissolved} * \exp(-\Delta G^o/RT)$$

where $C_{deposited}$ and $C_{dissolved}$ are the activities of the metal on the surface and in the solution. For the very dilute case, these are directly proportional to the concentration. Thus, the amount of metallic impurities on the surface is always directly proportional to the amount of impurities in the liquid, regardless of pH. Of course, pH will influence the exponential factor. Concerning temperature, we can conclude that when $-\Delta G^o<0$, a low temperature will result in more dissolved impurities, if $-\Delta G^o>0$, a high temperature will result in more dissolved impurities. Therefore, high temperature will not necessarily result in higher cleaning efficiency, as is expected intuitively.

Finally, HPM solutions remove metallic impurities by dissolving them. Dissolution can only take place when metallic impurities are on top of the surface. W. Kern (11) found that most of the metallic impurities are on top of the native oxide surface. However, a small fraction is inside the oxide and those cannot be removed by dissolution. These impurities have to be removed by etching such as in a dilute HF solution. However, HPM solutions are often used to remove metallic impurities deposited from APM solutions. When we disregard diffusion at the temperatures of the APM solution, which are close to room temperature, then all impurities have to end up on top of the chemically grown oxide, since the oxide grows on at the SiO_2/Si interface. This is shown in fig. 3. The initial impurities (x) and the during the APM cleaning deposited impurities (z) are remaining on top of the surface, since the oxide grows at the Si/SiO_2 interface. However, metals from other sources, especially from sources with a high energy, such as ion implantation or plasma processes can be beneath the surface and these impurities cannot be removed by HPM cleaning. Also, after thermal processing, metallic impurities can be driven in by diffusion and those impurities cannot be removed by HPM cleaning.
So, as a conclusion, in the future, HPM cleaning will require much lower concentrations of H_2O_2 and if no noble metals such as Au or Ag have to be removed, then the H_2O_2 can be completely omitted.

4. FPM solution

In 1970, W. Kern (11) already reported that Au deposits on the surface when it is in contact with the bare Si surface. Nowadays, there is not so much concern about Gold any more. There has been a report recently about Ag contamination (14), but generally the only noble metal which can sometimes be found in chemicals or DI-water is Cu. Cu^{++} deposits on bare Si surface regardless of the solution, so this is true for Cu^{++} in HF solutions, but it is as well going on in DI-water solutions. In fig. 4, this is shown for Cu^{++} deposition in ultra pure water depending on the substrate.

Recently, an extensive review of the Cu^{++} deposition was published by Morinaga *et al.* (15).

However, the Cu^{++} deposition on bare Si surfaces can be prevented by giving the solution a high oxidation potential. This is shown in fig. 5. From this figure, we can conclude that the Cu^{++} deposition can be prevented by an oxidation potential higher than 0.75 V. There is a very obvious way to give a solution a high oxidation potential. This is to add H_2O_2. This was proposed by T. Shimono and M. Tsuji (16) and is called the FPM (hydrogen Fluoride hydrogen Peroxide Mixture). However, there is a potential of surface roughening in these solutions as an etchant and an oxidizer are mixed. There exist a number of alternatives to avoid the Cu^{++} deposition. The first is to add an acid to the solution. The adding of an acid will by itself raise the oxidation potential of the solution, since the oxidation potential is pH dependent and increases at low pH values. This can be seen in fig. 2. The effect of adding an acid is shown in fig. 6. The mechanism of reducing the deposition is the same as that for the addition of H_2O_2. This can also be concluded from fig. 5, where it is clear that when lowering the pH and when keeping the dissolved O_2 concentration constant, the area of non-Cu^{++} deposition is entered. Finally, there are also some alternatives which prevent the Cu^{++} deposition by a different means, namely by neutralizing the Cu^{++} so that it can't react any more with the Si surface. This can be done by complexing, as is shown in fig. 7 or it can be done by adding a surfactant as is shown in fig. 8.

As a conclusion there are many ways by which the Cu^{++} deposition on the bare Si surface can be prevented : H_2O_2 addition, acid addition, complexing agents and surfactants.

5. SPM Solution

The SPM (Sulfuric acid hydrogen Peroxide Mixture) solution consists of a H_2SO_4/H_2O_2 mixture and was originally developed for removing photoresist. AFM measurements showed that this oxide is one of the thickest chemical oxides obtained (17). It also exhibits the slowest etching rate in HF solutions (18). However, because of the viscous character of the sulfuric acid, sulfur containing residues are difficult to remove completely from the surface even after the standard 10 minutes rinse in deionized water. Recently, an etching study and an electrical evaluation study were performed using an $H_2SO_4/H_2O_2/HF$ mixture by T. Ohnishi et al. (19). Therefore, we have investigated this mixture that we call SPFM from Sulfuric acid Peroxide hydrogen Fluoride Mixture according to the terminology used by T. Ohnishi et al. (19).

At first the hydrophobicity/hydrophilicity after the cleaning and before the DI water rinsing was investigated. Unlike the conventional SPM cleaning after which the treated surfaces are hydrophilic, the SPFM cleaning leads to hydrophobic surfaces as shown by contact angle measurements for various HF concentrations in fig. 9 and 10 on bare Si and on oxide surfaces after the cleaning itself. The SPFM treatment time was held constant at 1 min. If longer processing times are used, then the curves shift even to the left, i.e. higher contact angles for lower HF concentrations. After the following DI water rinse, the surfaces become hydrophilic again (fig. 9 and fig. 10). After the SPFM treatment, there is still a chemical oxide on top of the Si wafer. This is shown in fig. 11 as a function of the HF concentration. In this figure the thickness of the native oxide as after SPFM and followed by a DI water rinse is shown as a function of HF concentration. The SPM treatment without HF yielded an oxide thickness of 0.9 nm. Up to a concentration of 0.01 vol % of HF the same thickness is obtained. Only for higher HF concentrations, the obtained oxide thickness is thinner. XPS measurements show that the surface is F passivated after SPFM, but the F passivation is removed by rinsing in DI water. In fig. 12 the F1s peak is shown after SPFM treatments for different HF concentrations and before DI water rinsing. After DI water rinsing the surface is completely hydrophilic again as can be seen from fig. 9 and fig. 10. Fluorine cannot be detected any more after DI water rinsing. This shows that the fluorine is being replaced during rinsing and thus, after rinsing the same surface is obtained as for a SPM treatment where no HF is present. Since the surface after SPFM is hydrophobic however, both the bare Si and the oxide parts of the wafer,

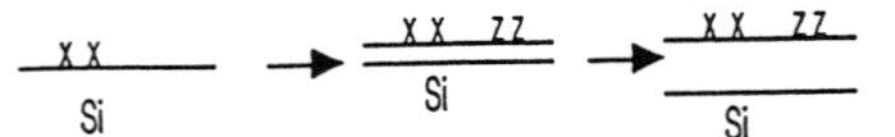

Fig. 3 Behavior of metallic impurities (initial x and during APM cleaning deposited z) during APM cleaning.

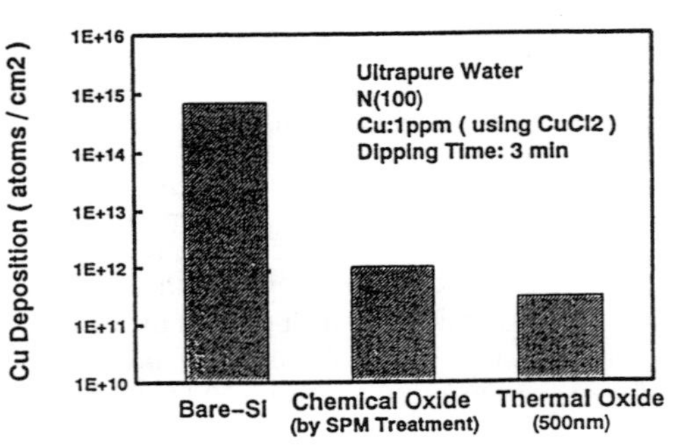

Fig. 4. Dependence of substrate on the Cu deposition in ultra pure water.

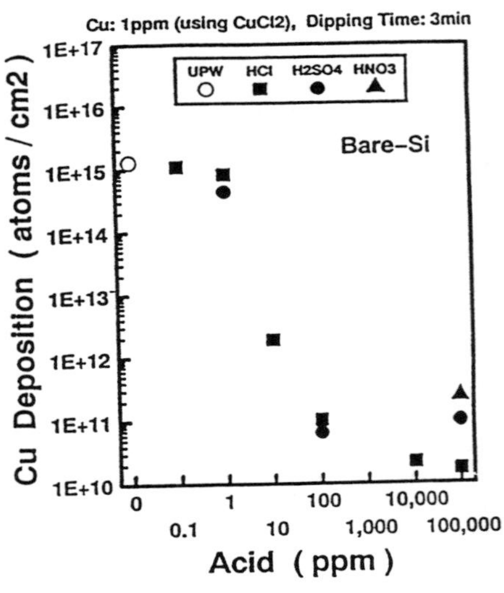

Fig. 6 Effect of the acid concentration on the Cu deposition onto the Si surface.

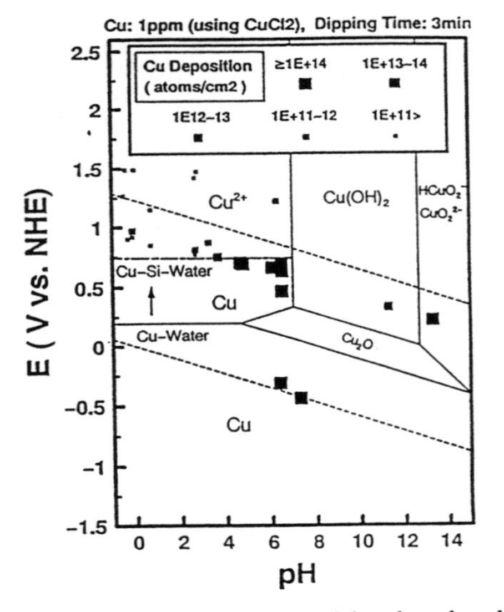

Fig. 5. The effect of pH level and redox potential of solutions (ultra pure water with various chemicals added) on the Cu deposition onto the Si surface.

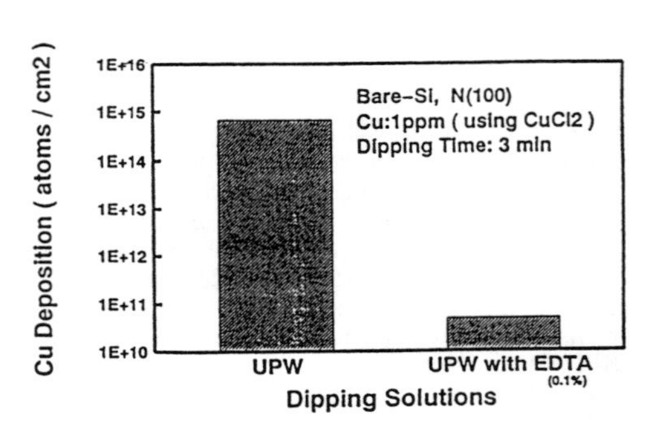

Fig. 7. The effect of a chelating agent on the Cu deposition onto the Si surface in ultra pure water

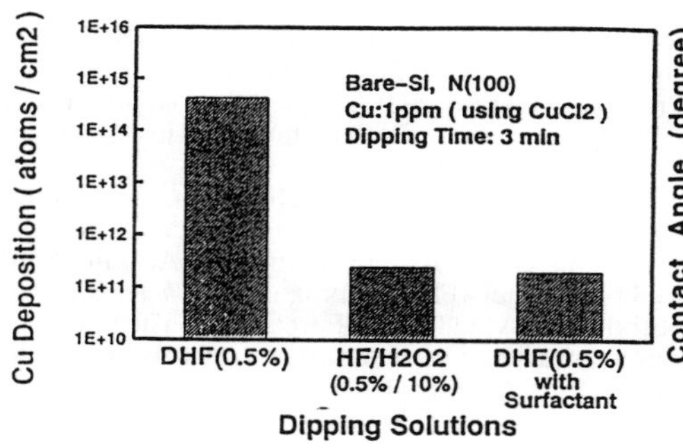

Fig. 8. The effect of injecting 10% H2O2 and compared to 0.1 % anion-type surfactant on the Cu deposition onto the Si surface in DHF solution.

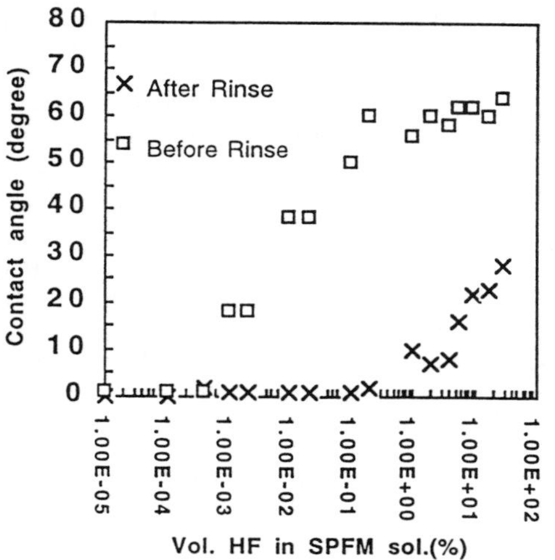

Figure 9. Evolution of the contact angle, before and after rinse, on bare Si surface with the HF concentration varied in different SPFM solutions.

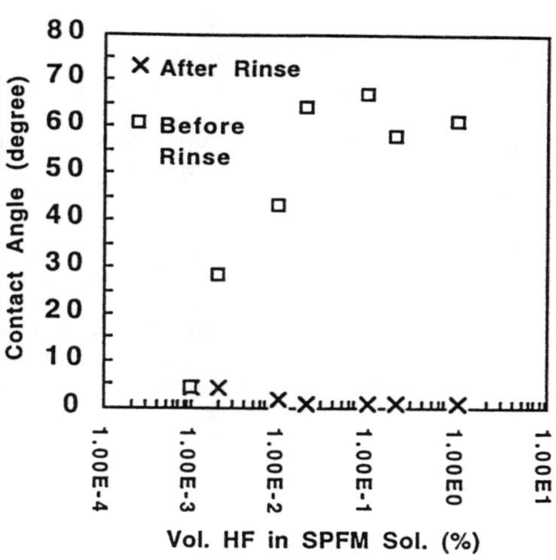

Figure 10. Evolution of the contact angle, before and after rinse on thermal oxide with the HF concentration varied in different SPFM solutions.

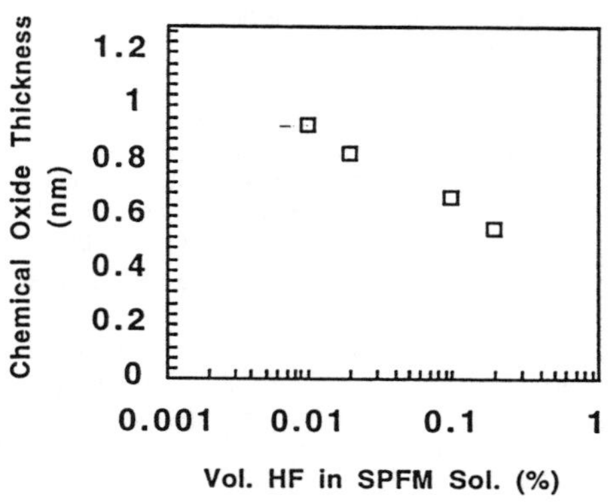

Figure 11. Evolution of the chemical oxide thickness after a 10 min rinse and following an SPFM treatment with different HF concentrations.

there is no chemical carry-over in the DI water bath and rinsing is performed much faster. The DI water rinsing efficiency after the cleaning in various SPFM solutions (different HF concentrations) compared to the standard SPM solution was investigated by measuring the time dependence of overflow rinse DI water resistivity as is shown in fig. 13. The resistivity of the DI water reaches 18 MΩ.cm much faster in the case of the SPFM solutions. In this experiment, the flow was maintained at a very low flow of 250 cc/min. In the case of the conventional SPM treatment, often, as in this case is shown, the resistivity shows an irregular behavior. Of course, this is not always reproducible, but this behavior can be observed regularly after SPM treatments. This is due to a randomly removal of sulfuric acid contamination.

Whenever an etching agent is mixed with an oxidizing agent, a possibility for surface roughening exists. Therefore, we investigated the surface roughening as a function of HF concentration. The result is shown in fig. 14. The roughness was measured on epi-wafers by Atomic Force Microscopy. The chemical oxide is first removed before the AFM scanning in a 0.5% HF solution. So, every cycle consists of an SPFM treatment followed by an 0.5% HF treatment. The maximum number of iterations was 14. From these measurements it can be seen that the roughness does not increase even after 6 cleaning cycles for HF concentrations of up to 0.01 vol. %. For a concentration of 0.02 %, we can detect a small increase in roughness after the 6th iteration. For a concentration of 0.1%, we can detect an increase in roughness already after the first SPFM treatment. However, after 3 iterations, the roughness is not increased with respect to only 1 SPFM treatment.

6. CONCLUSIONS

In this paper, the different solutions which contain H_2O_2 have been reviewed.

The APM solution consists of a mixture of $NH_3/H_2O_2/H_2O$ and was developed to remove organics and particles. Previously, the work was mainly focused at limiting the Si etching and roughening of the surface by a reduction of the NH_3 concentration. However, recently the attention was shifted to increasing the particle removal. It was found that particle removal efficiency drops at higher NH_3 concentrations. Also a reduction in the hydrogen peroxide concentration leads to improved particle removal efficiencies. The hydrogen peroxide can be reduced by a factor of 10 before any surface roughening starts.

The HPM solution consists of a mixture of $HCl/H_2O_2/H_2O$. This mixture can remove gold at 90°C. In order to remove Fe or alkali metals, HCl/H_2O at room temperature is enough. Although this has been reported for more than 30 years, people only now start to pay attention to this, since only now people have been able to completely eliminate Au from the processing. The future HPM cleaning will require much lower concentrations of H_2O_2 and if no noble metals such as Au or Ag have to be removed, then the H_2O_2 can be completely omitted.

When a solution such as HF or water is in contact with the bare Si surface, Cu^{++} , if present as an impurity will deposit on the surface. However, there are many ways by which the Cu^{++} deposition on the bare Si surface can be prevented : H_2O_2 addition, acid addition, complexing agents and surfactants. When H_2O_2 is added to HF, this is called the FPM solution.

The SPM solution, which is commonly used to remove photoresist from the wafer surface, leaves a lot of S residues on the surface. This can be prevented by adding minute amounts of HF to the surface.

We have shown that SPFM solutions result in hydrophobic surfaces, as well oxide as bare Si surfaces. In spite of the possible etching character of the SPFM solutions, chemical oxide still remains on top of treated bare Si surfaces. For the useful HF concentrations, the same chemical oxide thickness is obtained after SPFM as after the conventional SPM solution. This suggests that the HF, at the concentrations used, is only changing the termination of the surface of the oxide, but is not etching the oxide. This fluorine passivation results in better sulfur removal and shorter rinsing times, in addition to the power of particle removal. Concurrently a chemical oxide still can be formed on the treated surfaces and the fluorine passivation layer is removed during the DI water

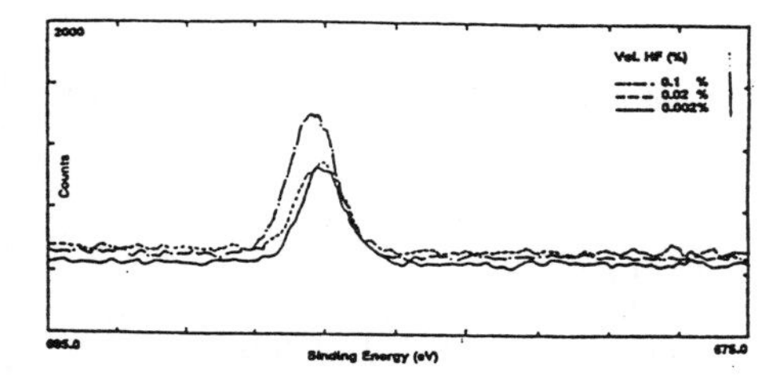

Figure 12 . F1s XPS spectra of Fluorine coverage after various SPFM treatments.

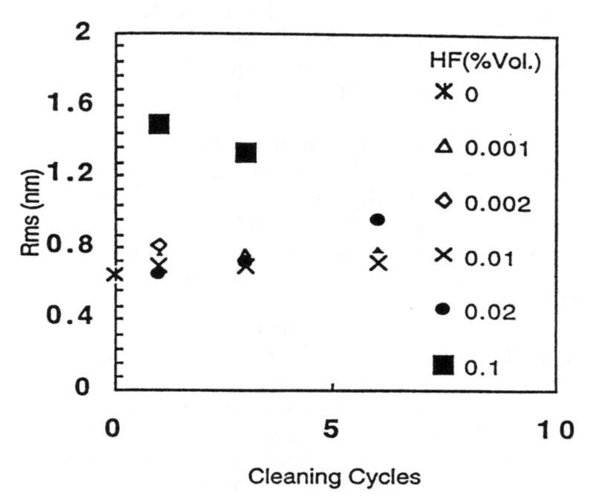

Figure 14. Evolution of the roughness as expressed in the Rms value in different SPFM solutions evaluated by AFM.

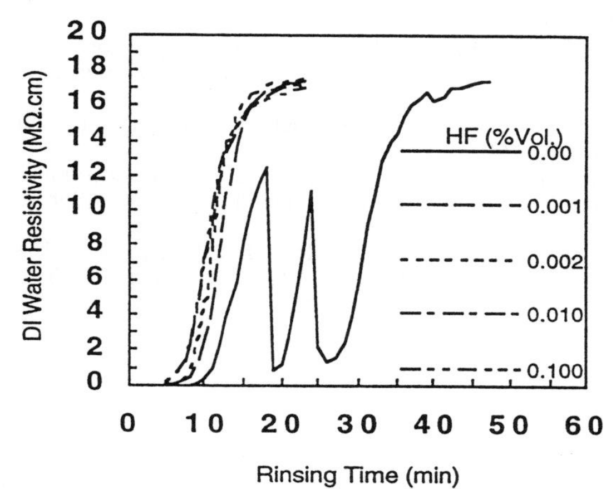

Figure 13. Variation of rinsing DI water resistivity as a function of rinsing time for various SPFM solutions and compared to the conventional SPM solution.

rinsing. Furthermore, the surface micro roughness is not affected by this cleaning for the HF concentrations which are useful for establishing this fluorine passivation.

REFERENCES

1. W. A. Kern and D.A. Puotinen, RCA Rev., 31, p. 187 (1970).
2. A. Ohsawa, K. Honda, R. Takizawa, T. Nakanishi, M. Aoki and N. Toyokura, in *Semiconductor Silicon 1990* (The Electrochemical Society, Pennington, N.J., 1990), H.R. Huff, K.G. Barraclough and J. Chikawa eds., p. 601.
3. T. Shimono and M. Tsuji, Abstract 200, p. 278, The Electrochemical Society Extended Abstracts, Vol. 91-1, Washington DC, May 5-10, 1991.
4. M. Itano and T. Ohmi, "Particle removal from Si wafer surface in wet cleaning process", *IEEE Tans. Semicond. Manuf.*
5. M. Miyashita, M. Makihara, T. Tsuga and T. Ohmi, "Dependence of surface microroughness of CZ, FZ and EPI wafers on wet chemical process," *J. Electrochem. Soc.*
6. T. Ohmi, M. Miyashita, M. Itano, T. Imaoka and I. Kawanabe, IEEE Trans. on El Dev., Vol. 39, No. 3, March 1992, p.537.
7. M. Itano, F.W. Kern,Jr., M. Miyashita, I. Kawanabe, R.W. Rosenberg and T. Ohmi, "Particle adhesion and removal in wet cleaning processes for ULSI manufacturing," *IEEE Trans. Semicond. Manuf.*
8. S.L. Cohen, W. Syverson, S. Basiliere, M.J. Fleming, B. Furman, C.Gow, K. Pope, R. Tsai and M. Liehr, in *Proceedings of the Second International Symposium on Ultra-Clean Processing of Silicon Surfaces*, p. 35, 1994.
9. G. N. DiBello, S.T. Bay, C.F. McConnel, J.W.Parker, and E.A. Cheney, in *Proceedings of the Second International Symposium on Ultra-Clean Processing of Silicon Surfaces*, p. 23, 1994.
10. T. Futatsuki, K. Ohmi, K. Nakamura and T. Ohmi, in *Proceedings of the International Conference on Advanced Microelectronic Devices and Processing*, p. 425, 994.
11. W. A. Kern, RCA Rev., 31, p. 256, (1970).
12. J.S. Glick, SPWCC 992.
13. M. Pourbaix, "Atlas of Electrochemical Equilibria in Aqueous Solutions," Pergamon Press, London (1966).
14. D. Levy, P. Patruno, L. Mouche and F. Tardif, in *Proceedings of the Second International Symposium on Ultra-Clean Processing of Silicon Surfaces*, p. 293, 1994.
15. H. Morinaga, M. Suyama, M. Nose, S. Verhaverbeke and T. Ohmi, "A model for electrochemical deposition and removal of metallic impurities on Si surfaces," submitted to J. Electrochem. Soc.
16. T. Shimono and M. Tsuji, Abstract 200, p. 278, The Electrochemical Society Extended Abstracts, Vol. 91-1, Washington DC, May 5-10, 1991.
17. S. Aoyama, Y. Nakagawa and T. Ohmi, in *Extended Abstracts of the 1992 International Conference on Soid State Devices and Materials* (Japan Society of Applied Physics, Tokyo, Japan, 1992), Tsukuba, 1992, p. 126.
18. K. Nakamura, T. Futatsuki, K. Makihara and T. Ohmi, Abstract no. 343, p. 563, The Electrochemical Society Extended Abstracts, Vol. 93-2, New Orleans, LA, October 10-15, 1993.
19. T. Ohnishi et al., SSDM 1993, Ext. Abst. p. 627, Makuhari, Japan.

STUDIES OF THE RELATIONSHIP BETWEEN MEGASONICS, SURFACE ETCHING, AND PARTICLE REMOVAL IN SC-1 SOLUTIONS

S. L. Cohen [a], D. Rath[a], G. Lee[c], B. Furman[a], K.R. Pope[a], R. Tsai[a], W. Syverson[b], C. Gow[b] and M. Liehr[a]

[a] IBM Research Division, T. J. Watson Research Center, Yorktown Heights, NY,
[b]IBM Microelectronics Division, Essex Junction, VT
[c]IBM Microelectronics Division, Hopewell Junction, NY

ABSTRACT

Wafer cleaning studies have been performed so as to understand the influence of acoustic (megasonic) energy on particle removal in dilute SC-1 solutions. Surface etching alone (up to 60Å) has been found to be insufficient to completely remove silicon nitride surface particles from native oxide surfaces in the absence of megasonics. For megasonic cleaning processes the minimum surface etching required for complete nitride particle removal is significantly lower (between 3-12Å) than for a non-megasonic process. The exact 'threshold' for surface etching will depend on the chemical nature of the particle/surface and the megasonics power. Megasonics energy does not appear to enhance chemical etching of the substrate, at least for silicon oxide substrates, however, it significantly improves particle removal. This data suggests that the particle removal process can benefit from both a thermally activated component (etching) as well as an acoustic component (cavitation/ acoustic streaming).

INTRODUCTION

The science and technology of semiconductor wafer cleaning has attracted considerable attention in recent years due to the stringent requirements for steadlily decreasing defect density and demands for lower cost, environmentally sound manufacturing practices. The traditional high concentration RCA cleaning sequence is increasingly being replaced with more dilute chemical processes (often still based on the H_2O_2/NH_4OH SC-1 and H_2O_2/HCl SC-2) so as to minimize surface roughening and chemical consumption [1,2,3]

Increasingly, the use of acoustic energy in the 600-900kHz regime (Megasonics) is being used to aid in particle removal in the SC-1 solution. Recent studies have demonstrated a dramatic improvement in cleaning ability of dilute SC-1 solutions with the addition of megasonics energy [3,4], however, a fundamental

Mat. Res. Soc. Symp. Proc. Vol. 386

understanding of the mechanisms of the megasonic cleaning process is still evolving. Attraction of particles such as silicon nitride from aqueous (neutral pH) solutions to silicon oxide surfaces is dominated by an electrostative zeta potential interaction since the positively charged nitride particles are attracted to the negatively charged oxide surface. [5] Subsequent detachment of these particles from the surface requires that strong adhesion forces (electrostatic as well as capillary, VanderWaals [6]) be overcome.

One mechanism to 'loosen' particles is surface etching. Previous studies of particle removal on thick oxides have shown that in the absence of megasonics energy, >20**Å** of etching is required for complete particle removal.[7] Addition of megasonics energy can further aid particle removal through cavitation (bubble formation and collapse) and/ or acoustic streaming (fluid motion). [8,9,10] The connection between megasonics energy and surface etching has not been fully explored. While it is anticipated that any change in the solution boundary layer due to bubble formation or fluid motion could contribute to changes in surface etching, the relative contribution of etching and acoustic energy have not been measured. There is clearly some contribution from a thermally activated process during particle removal since previous work has clearly shown a significant increase in particle removal efficiency at higher temperatures [3,4] In this paper, the relative contributions of surface etching and megasonics ('physical') energy on particle removal are discussed.

2. EXPERIMENTAL

Most of the megasonic SC-1 cleaning experiments were performed in a direct displacement cleaning tool as part of a large design-of- experiment based study. Using BestDesign, (an IBM proprietary statistics package), 14 parameters of the RCA clean were varied and response functions were generated for oxide and nitride particle removal and surface roughening (among others). Standard 8-inch diameter Si(100) wafers were precleaned with an RCA-based clean in a spray processor to create a smooth, thin chemical oxide on the surface. The wafers were subsequently contaminated in a controlled way by a dip into DI water containing either the nitride or oxide particles. Initial particle counts prior to deliberate contamination were less than 50/wafer; 1500-3500 particles/wafer were deposited using this method as measured on a Tencor 5500 at 0.3μm sensitivity. Wafer cleaning conditions varied from a 'standard' clean (8:1:1; H_2O: H_2O_2: NH_4OH) to dilutions of 20X for the chemically active components; furthermore HF etches were added at times before the SC-1 step. Surface roughness was measured by AFM for one wafer in each cleaning cassette and the oxide etch rate of each cleaning run was determined from a thermal oxide etch rate monitor. Complementary cleaning experiments were also performed for nitride particles deposited on thermal oxide films (80Å, 1000Å) so as to compare the cleaning results on bare silicon to that on thick oxides. These experiments were conducted in a traditional multi-tank wet bench equipped with a chain drive megasonics unit.

3. RESULTS

Our results suggest that surface etching alone (for etch depths of 50-60Å) , in the absence of megasonics energy, is not sufficient for high efficiency silicon nitride particle removal. Conversely, in the presence of megasonic energy, the amount of surface etching required is significantly less than in the absence of acoustic energy. This is demonstrated in Figure 1 for both nitride and oxide particles on native oxide covered silicon wafers. In this figure, the particle removal efficiency is plotted as a function of silicon substrate etch depth. The etch depth of the silicon substrate was inferred from the height of pillar structures on the surface of a clean wafer (as measured by AFM) in the same cleaning cassette as the particle standards [11,12]. The origin of these pillars is believed to be due to micromasking of the surface by, for example, a trapped surface bubble; the silicon surface directly underneath the bubble is not etched, while the surrounding area is. These pillars are nominally 1000Å- 3000Å in diameter and their height depends upon the etch rate of the process.

Figure 1 suggests that for both nitride and oxide particles, etch depths of 50-60Å are not always sufficient to provide complete particle removal in the absence of megasonics energy (squares). It also appears that the requirement for substrate etching depends on the chemical nature of the particle; oxide particles are generally easier to remove than nitride particles as previously discussed [3]. The shape of these two different particles may also affect their adhesion strength [3] . Thus, the cleaning requirements for 'real' process contaminants will vary significantly depending upon the chemical nature of the surface and the particle. Figure 1 suggests that as long as megasonics energy is used in the cleaning process, etch depths as low as 15Å are effective for removal of these specific particles. No pillars were observed for etch depths below 15Å.

It has previously been shown [3,4] that the SC-1 temperature can be a significant factor in silicon nitride particle removal from native oxide surfaces; for a given megasonic power, the particle removal efficiency improves at elevated temperatures. The variable temperature/megasonic response surface [4] suggests, however, that above a certain megasonic power, the effect of temperature become less important. The role of temperature in particle removal is illustrated in Figures 2 and 3. Figure 2 is a plot of the etch depth of a thermal oxide substrate as a function of SC-1 Temperature and NH_4OH concentration (all other process conditions remain fixed). Figure 3 illustrates the particle removal efficiency in the same temperature/ concentration range for nitride particles on a native oxide surface.

For a given NH_4OH/ H_2O_2 ratio, the etch rate of the thermal oxide increases substantially with temperature. At the highest temperatures, the etch rate increases strongly with increasing NH_4OH concentration. Particle removal efficiency also increases towards the higher temperatures (Figure 3) and clearly there is an optimum NH_4OH/ H_2O_2 ratio which provides the best particle removal. At the lowest NH_4OH concentrations, particle removal efficiency decreases due to lack of etching, while above that optimum ratio, surface etching increases leading to LPD addition (i.e. added defects in the Tencor light scattering map) and therefore an apparent decrease in cleaning efficiency.

The influence of etching on particle removal is also illustrated in Figure 4 where a comparison of nitride paritcles deposited on thermal vs. native oxide surfaces is shown. The high temperature 65C process provides 99% removal on all wafers as long as megasonics energy is used, but the lower temperature 25C procsss shows a significant loss of cleaning efficiency for the same megasonic power. Most interestingly, as indicated in Figure 4, the addition of the megasonics energy significantly improves cleaning efficiency without influencing the substrate etch rate. Conversely, Figure 4 also shows that thermal oxide removal of 12-14**Å** is not sufficient for particle removal in the absence of the megasonics energy consistent with the results in Figure 1 for native oxide surfaces. A clear reduction in cleaning efficiency is observed at room temperature where only 2-3**Å** of thermal oxide are removed. This suggests that for silicon nitride particles under these specific dilute SC-1 concentrations and at this particular megasonic power, complete removal requires >2-3**Å** surface etching.

Recent work confirms that for various types of annealed and unannealed oxides there is no observable enhancement in silicon dioxide etch rate with addition of megasonics energy. [13] This suggests that for silicon surface etching, while ion transport to/from the surface may be influenced by acoustic effects such as cavitation and acoustic streaming, silicon oxide surface etching is not limited by mass transport to or from the surface, but rather by the kinetics of the etching process itself. Thus the megasonic cleaning process appears to be a combined

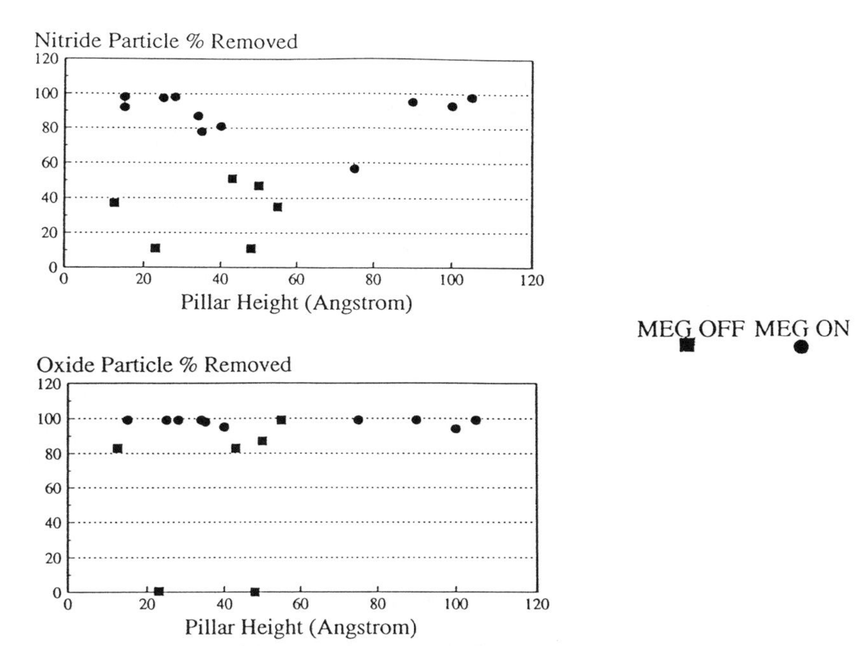

Figure 1. Particle removal vs. silicon substrate etch depth for particles deposited on native oxide covered silicon. Etch depth is determined from AFM measurements of pillars. Surface etching alone is not sufficient to remove particles in the absence of megasonics energy.

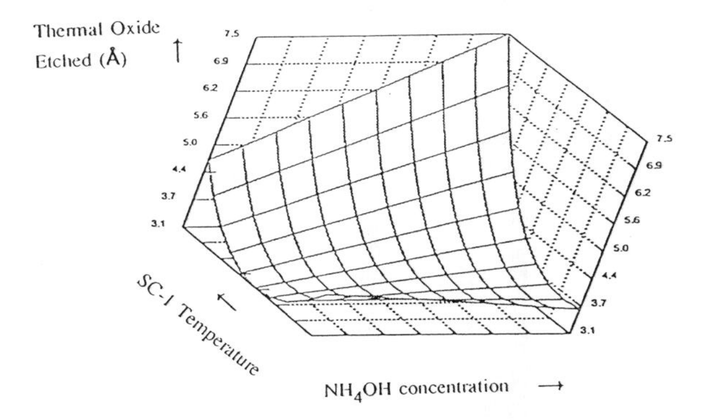

Figure 2. BestDesign response surface for thermal oxide etch depth versus SC-1 Temperature and NH_4OH concentration (same scale as Figure 3). All other parameters are fixed. The etch rate of thermal oxide is a strong function of temperature; at the highest temperatures, the NH_4OH concentration has the strongest influence on etch depth.

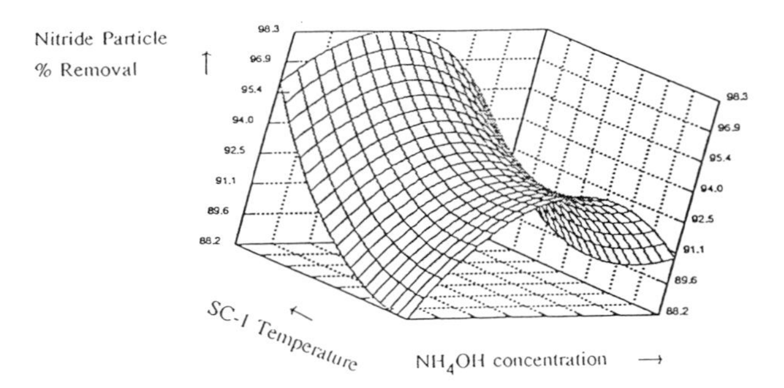

Figure 3. BestDesign response surface for nitride particle removal versus SC-1 Temperature and NH_4OH concentration (same scale as Figure 2). For a fixed megasonic power, there is a strong temperature dependence of particle removal on both variables.

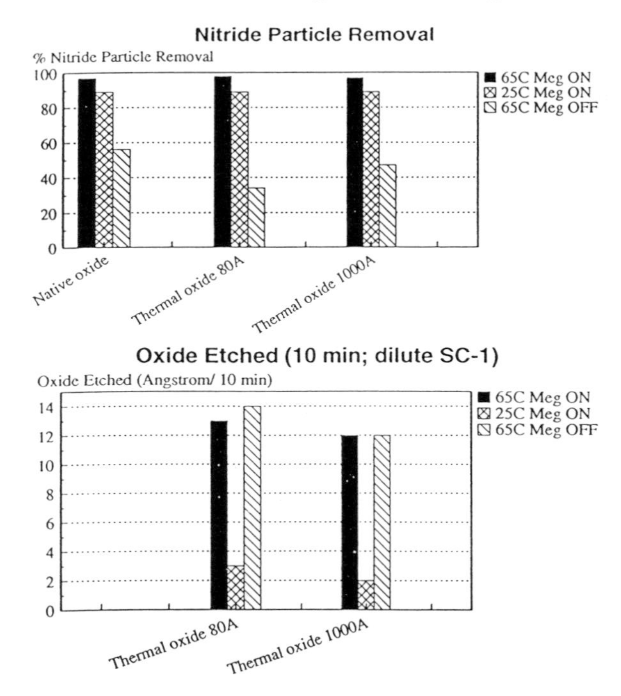

Figure 4. Megasonics energy significantly increases particle removal efficiency but does not increase surface etching for particles deposited on thermal oxide.

chemical and physical process. The chemical etching of the surface, which is dominated by temperature and chemical concentration aids in 'loosening' surface particles, while the megasonic energy provides significant physical energy to dislodge particles. Once the particles are removed from the surface, however, they remain in solution due to the high pH and zeta potential repulsion. [3,5]

Acknowledgments: The authors gratefully acknowledge support from CFM Technologies for portions of this study as part of Sematech project J80.

1 T. Ohmi, Proceedings of the Inst. of Environ. Sci. 287-296 (1992).
2 M. Meuris, M. Heyns, P.W. Mertens, S. Verhaverbeke and A. Philipossian, Microcontamination 10(2), 31 (1992).
3 S.L. Cohen, W. Syverson, S. Basiliere, M.J. Fleming, B. Furman, C. Gow, K. Pope, R. Tsai, M. Liehr, Proceedings of the Ultraclean Processing of Silicon Surfaces Conference , Sept. 19-21, 1994, Bruges, Belgium
4 P.J. Resnick, C.L.J. Adkins, P.J. Clews, E.V. Thomas and S.T. Cannaday, Symp. on Cleaning Tech. in Semi. Dev. Mfg., 184th ECS Mtg (New Orleans, LA, Oct 10-15 1993).
5 D. Jan and S. Raghavan, Symp. on Cleaning Tech.in Semi. Device Mfg., 184th ECS Mtg (New Orleans, LA, Oct 10-15, 1993).
6 **a**) D. J. Riley, R.G. Carbonell, J. Colloid Interface Sci, (1993) **b**) D. J. Riley, R.G. Carbonell, Journal of the IES, 28, Nov/Dec 1991 **c**) R.P. Donovan, Spring MRS meeting procedings (San Francisco 1993).
7 M.M. Heyns, S. Verhaverbeke, M. Meuris, P.W. Mertens, H.F.Schmidt, M. Kubota, A. Philipossian, K. Dillenbeck, D. Graf, A. Schnegg and R. deBlank, Materials Research Society Meeting (San Francisco CA, Apr. 12-16, 1993).
8 A.A. Busnaina, G.W. Gale, I.I. Kashoush, Precision Cleaning, Vol II, (1994) p 13.
9 S.J. Putterman, Scientific American, Feb 1995, p 46.
10 D. Zhang, D.B. Kittelson, B. Liu, Microcontamination Conf. Proceedings, San Jose, CA Oct 4-6 1994 p. 215.
11 H.F.Schmidt, M. Meuris, P.W. Mertens, S. Verhaverbeke, M.M. Heyns, L. Hellemans, J. Snauwaert and K. Dillenbeck, Symp. on Cleaning Technology in Semi. Device Mfg., 184th ECS Mtg (New Orleans, LA, Oct 10-15, 1993).
12 B.K. Furman, et.al. paper presented in this symposium, MRS San Francisco 1995
13 D.L. Rath, F. Abramovich, S. L. Cohen, C. Gow, M. Hevey, G. Ouimet, unpublished results

A STUDY OF CLEANING PERFORMANCE AND MECHANISMS IN DILUTE SC-1 PROCESSING*

P. J. Resnick, C.L.J. Adkins, P.J. Clews, E.V. Thomas, and N.C. Korbe, Sandia National Laboratories, Albuquerque, NM 87185

Abstract

A statistical design of experiments (DOE) approach has been employed to evaluate the effects of megasonic input power, solution chemistry, bath temperature, and immersion time on particle removal in SC-1 chemistries. Megasonic input power was the dominant factor in the response surface model. Substantially diluted chemistries, performed with high megasonic input power and moderate-to-elevated temperatures, were shown to be very effective for small particle removal. Follow-on studies to the original DOE have led to an investigation of ultradilute SC-1 chemistries with megasonic power. These chemistries ranged from 0 to 1000 ppm of NH_4OH and H_2O_2. Post processing light point defect (LPD) counts differ substantially between bare n and p-type Si<100>, and ambient lighting conditions are shown to influence LPD counts on p-type Si<100>. Solution properties such as pH and oxidation potential have been studied, and an investigation of post processing silicon surface properties is underway.

Introduction

An empirical response surface model was previously generated for particle removal in SC-1 chemistries with applied megasonic power [1]. Power was the dominant factor for these cleans, with temperature and chemical ratio (NH_4OH:H_2O_2) modifying the effect of power. High cleaning efficiencies were observed for the removal of silicon nitride particles (0.10-0.30 μm) from silicon wafers using substantially diluted SC-1 chemistries. Although the chemistry of the SC-1 solution was varied during this experimental matrix, no experiments were performed in deionized (DI) water alone. As a continuation to the original study, a small matrix was constructed and performed using DI water without any SC-1 chemistry at various megasonic power and temperature set points. The particle removal response surface from the DI water experiments was substantially different than the response surfaces generated when dilute SC-1 solutions were used. The disparate nature of these two matrices suggested that additional experimentation is warranted in the regime between DI water and the previously studied dilute chemistries. The objective of performing experiments in this "ultradilute" regime is not necessarily to optimize an existing clean, but rather to develop insight into the chemical action involved in the SC-1 megasonic clean. By developing a more fundamental understanding of the way in which the cleaning chemistry interacts with the wafer surface, optimized SC-1 chemistries may be used with more confidence for future generation device fabrication.

*This work was performed at Sandia National Laboratories, which is operated for the U.S. Department of Energy under contract no. DE-AC04-94AL85000. This work was funded through a cooperative research and development agreement with SEMATECH.

Experimental

A design of experiments (DOE) matrix was constructed to study particle removal efficacy with various SC-1 chemistries in a megasonic bath. An additional DOE was constructed to study particle removal using DI water in a megasonic bath. The particle source was 0.10 - 0.30 μm Si_3N_4, deposited from an aerosol. The cleans were performed in a Verteq focused beam megasonic, and particle metrology was performed on a Tencor Surfscan 6200. Blank correction wafers (clean, uncontaminated wafers) were included to help suppress confounding of the cleaning results with light scattering events other than particles (*e.g.*, increase microroughness).

Following the SC-1 and DI water experimental matrices, a third matrix was constructed in which NH_4OH and H_2O_2 concentration were varied from 0 to 5000 ppm at fixed megasonic power and temperature (150 or 200 W, 25°C). The intent of this matrix was not for response surface modeling, but instead to provide a framework for performing the ultradilute SC-1 chemistry experiments and to bridge the experimental space between dilute chemistry and DI water. Initially, silicon nitride particles were again used as the particle challenge. However, the number of light point defects (LPDs) added to blank wafers which were processed in the same bath with the contaminated wafers was highly variable and often quite large. Since the blank correction data can heavily leverage the particle removal efficiency calculations, the contaminated wafers were omitted from the studies, and LPDs added to clean wafers became the response of interest. Wafer type (n vs. p) and doping level were also considered during these experiments, as was the ambient lighting conditions. Additional measurements, including contact angle and electron paramagnetic resonance spectroscopy (EPR), were performed on wafers processed under both light and dark conditions.

Results

A contour plot of the response surface from the first SC-1 DOE is presented in Figure 1. The data presented in Figure 1 are from chemistries in which $NH_4OH:H_2O_2 = 0.01$ (volumetric ratio of standard concentration chemicals). Included in these data would be chemistries as dilute as 1:100:6900 ($NH_4OH:H_2O_2:H_2O$). Contour lines in this plot represent constant cleaning efficiencies. A quadratic response in power is observed and a linear response in temperature can be seen. Contour plots at various chemical ratios show similar response, although with higher chemical concentration a more robust process window is observed. Factors such as immersion time, chain speed, and absolute hydrogen peroxide concentration were deemed statistically insignificant (within the experimental space) in preliminary screening experiments, and thus were not included in the response surface. The DI water experimental results are shown in Figure 2. The difference in the contour plots of Figure 1 and Figure 2 are seen to be quite different; most notably, the power dependence is linear (not quadratic) when DI water is used. It is apparent that relatively dilute chemistries provide cleaning action that is not seen with DI water. An additional matrix was constructed in which the SC-1 chemistry (NH_4OH and H_2O_2) was the only factor. Megasonic power and temperature were held constant. The number of defects added to clean wafers is the response variable. The wafers used in these experiments

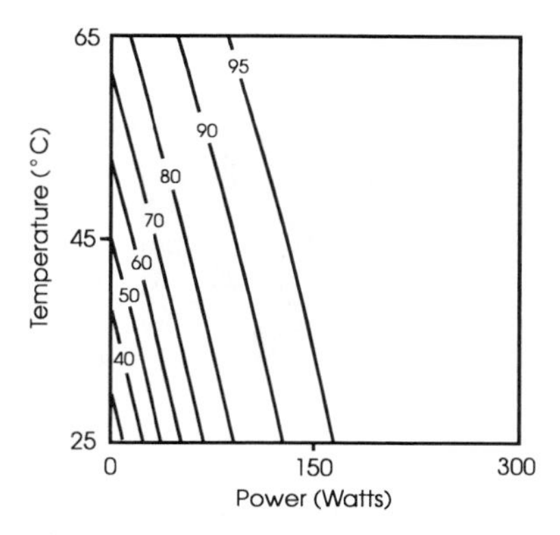

Figure 1. Si_3N_4 particle (0.10-0.30 μm) removal efficiency in dilute SC-1.

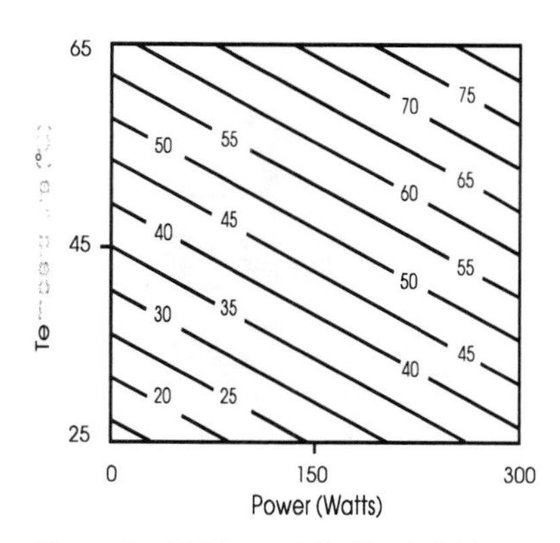

Figure 2. Si_3N_4 particle (0.10-0.30 μm) removal efficiency in DI water

(and all previous experiments) are n-type Si<100>, 2-20 ohm-cm. The number of defects added as a function of NH_4OH and H_2O_2 concentration is shown in Figure 3. Light point defect addition increases as hydrogen peroxide is increased and ammonium hydroxide is reduced.

The pH values of the system were calculated assuming a steady-state concentration of CO_2 based on the measured equilibrium pH of DI water exposed to cleanroom air. For these calculations, the ammonia concentration ranged from 0 to 35 ppm, and the hydrogen peroxide concentration ranged from 0 to 5000 ppm. pH is strongly dependent on ammonia concentration, and has a very weak dependence on hydrogen peroxide concentration. It appears that the addition of LPDs to the blank wafers cannot be explained entirely by alkaline attack and subsequent faceting of Si<100> material since LPD count was seen to be a function of both ammonia and hydrogen peroxide concentration. According to van den Meerakker and ven der Straaten [2], the etch rate of silicon in dilute ammonia solutions can be enhanced by the addition of hydrogen peroxide (again, suggesting a more complex interaction than a simple alkaline etch).

The oxidation potentials of this system were also calculated, using the method of Pourbaix [3]. The oxidation potential can be expressed using equation (1).

$$E = E^0 - \frac{2.3RT}{F} pH + \frac{RT}{2F} \ln([H_2O_2]) \quad (1)$$

Where R is the ideal gas constant, T is absolute temperature, F is the Faraday constant, and E^0 is the potential relative to the standard hydrogen electrode (SHE) for reaction (2)

$$H_2O_2 + 2H^+ + 2e^- = 2H_2O \quad (2)$$

The oxidation potentials were calculated based on calculated pH values, and are shown in Figure 4. The trend in oxidation potential is increasing with increasing hydrogen peroxide

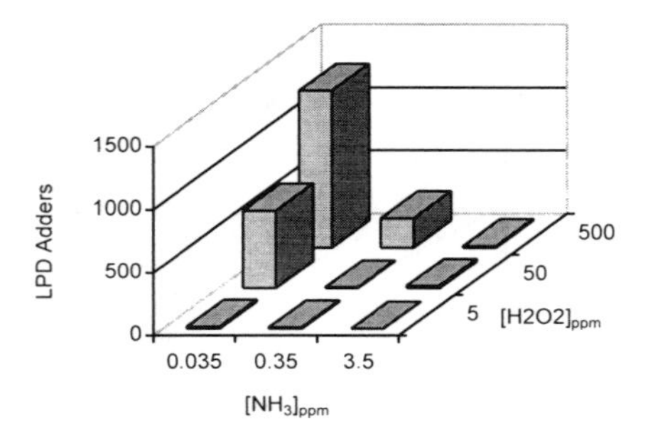

Figure 3. LPD (0.10 - 0.30 μm) addition in SC-1 @ 20°C, 150 W

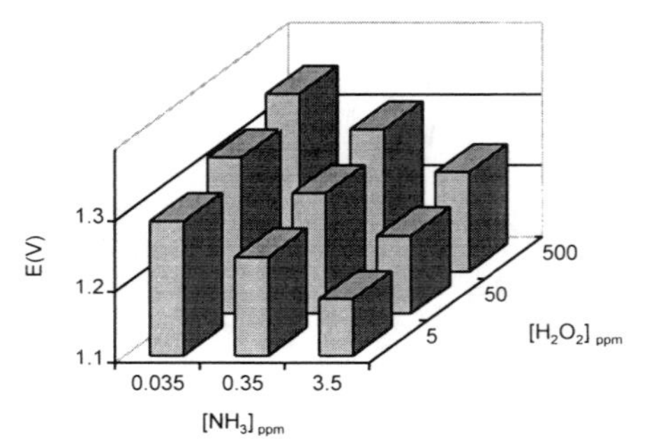

Figure 4. Oxidation potential, based on calculated pH values

concentration and decreasing ammonium hydroxide concentration. This general trend is similar to the trend observed for the number of LPDs added as a function of chemistry (Figure 3). A qualitative correlation can be seen between solution oxidation potential and added LPDs, but no causal relationship can yet be deduced.

A small selection of previously performed experiments was repeated with both n and p type wafers. These experiments were performed under normal cleanroom lighting conditions, using 2-20 ohm-cm Si<100>. Four wafers of each type were included in each run. The average number of LPDs added in each experimental run is given in order of increasing oxidation potential in Table 1. The wafers in the last two experimental runs clearly gained more LPDs than the others. The largest disparity between n and p type silicon was seen in the first run, in which the number of LPDs was reduced for the p-type wafers. Similar experiments were performed in which the lighting conditions and wafer type was varied. Dark processing conditions were created by covering openings in the bench with black plastic, and turning the lights in the bay off. Bright conditions were created by shining a tungsten halogen inspection lamp into the process bath. The cleans were performed in a 35 ppm NH_3, 5 ppm H_2O_2 solution at 25°C with 200 Watts megasonic power. Data from these experiments are given in Table 2. A substantial reduction in LPDs was seen for p-type wafers processed under dark conditions, especially for the more heavily doped material. Under bright conditions, p-type silicon LPD counts were similar to n-type under either bright or dark conditions.

Wafers which were processed under both light and dark conditions, using SC-1 solutions of both high and low oxidation potential, were selected for contact angle measurements and surface state measurements using electron paramagnetic resonance spectroscopy (EPR). The contact angles on all wafers (both p and n type) were too small to measure (< 5°). If subtle changes are occurring at the silicon surface (*e.g.*, a change in hydrophobicity resulting from a change in surface termination), such changes were not detectable by goniometry. In addition, the EPR measurements showed no detectable change in surface state concentration between wafers which were processed under bright conditions versus dark conditions. The samples on which EPR was performed had only a native oxide; there may be better chances of exploiting differences in surface state concentration by repeating EPR measurements using silicon on which a thermal oxide is grown.

				Added LPDs (0.10 - 0.30 μm)	
[H2O2]ppm	[NH3]ppm	E(V)	Meg Power (W)	n-Type	p-Type
5	35	1.13	150	33	-169
5000	35	1.22	150	-27	15
500	3.5	1.24	200	12	-55
5	0.35	1.24	150	-52	-40
50	0.35	1.27	200	14	-5
50	0.35	1.27	150	0	-7
500	0.35	1.30	150	232	207
5000	0.035	1.38	150	812	911

Table 1. Light point defects (0.10 - 0.30 μm) added to n-type Si<100>, 2-20 Ω-cm processed under normal cleanroom lighting conditions. Values presented are averages of 4-wafer data.

		Added LPDs (0.10 - 0.30 μm)	
Wafer Type	Resistivity (Ω-cm)	Light	Dark
n-type	0.10-0.40	36	24
n-type	2-20	5	15
p-type	2-20	5	-960
p-type	30-60	-27	-93

Table 2. Light point defects added to n and p type Si<100> under light and dark conditions at 20°C, 200 W, 5 ppm H_2O_2, 35 ppm NH_3. Values presented are averages of 4-wafer data.

Discussion and Conclusions

LPD addition during silicon processing cannot be related solely to pH values of the cleaning solution. A trend was observed in which the LPD addition increased with reduced NH_3 concentration as well as increase H_2O_2 concentration. If LPD addition depends only on pH, then the addition of H_2O_2 should have little or no influence on defect counts, since H_2O_2 is a very weak acid and thus has very little effect on the pH. The increase in LPD addition does coincide with an increase in oxidation potential.

The wafer type and ambient lighting conditions influenced light point defect counts at these dilute concentrations as well. Although n-type wafers of different doping levels appeared insensitive to lighting conditions, p-type wafers exhibited significantly different behavior depending on lighting conditions and doping levels. The population of electrons in the conduction band of p-type silicon is increased by illumination, while in n-type silicon, illumination would not have a significant impact on the majority charge carrier concentration. Therefore, these experiments indicate that reduced LPD counts could be a manifestation of reducing the electron concentration of the conduction band. According Memming [4], the reduction of hydrogen peroxide at a semiconductor surface occurs via a two step reaction, the first of which involves electron transfer from the conduction band to the electroactive species in solution. In addition, van den Meerakker [5] has noted that the reduction of hydrogen peroxide at an n-type semiconductor occurs in the dark, while illumination is necessary for reduction at p-type material.

Preliminary data indicate that the reduction of peroxide at the silicon surface (which is promoted by populating the conduction band with electrons) is undesirable from a light point defect standpoint. The best LPD performance was observed for p-type silicon processed under dark conditions (low electron concentration in the conduction band). Increased oxidation potential also appears undesirable with respect to LPD counts. This may be related to the ability of the electroactive species to capture electrons from the semiconductor conduction band for reduction of the oxidizing agent. It is not clear to what extent the LPD measurements performed thus far have been confounded with light scattering from surface roughness. It is also not known how significant ambient lighting conditions will be when traditional or more concentrated chemistries are used. Clearly, much work remains in order to understand the interaction of SC-1 chemistries with silicon surfaces, and the impact of these interactions on light point defect counts.

References

1. P.J. Resnick, C.L.J. Adkins, P.J. Clews, N.C. Korbe, S.T. Cannaday, E.V. Thomas, *SEMATECH Technology Transfer Document 94012177A-ENG* (1994).
2. J.E.A.M. van den Meerakker and M.H.M. van der Straaten, *J. Electrochem. Soc.*, **137**, 1239 (1990)
3. Pourbaix, M., *Atlas of Electrochemical Equilibria in Aqueous Solutions*, Oxford: Pergamon (1966).
4. R. Memming, *J. Electrochem. Soc.*, **116**, 785, (1969).
5. J.E.A.M. van den Meerakker, *Electrochemica Acta*, **35(8)**, 1267 (1990).

SELECTIVE ONE-STEP CHEMICAL ETCHING OF THE SILICON NITRIDE/SILICON PBL STACK FOR 0.5μm DEVICE FABRICATION

DAVID ZIGER*, Susan Vitkavage*, Charles Oberdorfer*, Juli Eisenberg** and Michael Hughes**
*ATT Bell Laboratories, 9333 S John Young Parkway, Orlando, FL 32819
**ATT Bell Laboratories, 555 Union Blvd, Allentown, PA 18103

ABSTRACT
Wet chemical removal of a silicon nitride/silicon Poly Buffer LOCUS (PBL) film stack with a phosphoric/nitric acid solution was characterized. Though silicon nitride etch rates remain constant, silicon etch rates decrease as a function of loading which severely limits the usable lifetime of the bath. This is caused by buildup of etching products which limits the amount of silicon that can be dissolved in the solution. Addition of HF to the phosphoric/nitric acid solution enhances the dissolution chemistry presumably by decomposing the silica reaction products and shifting the equilibrium. Characterization of this process was done to determine whether it could be applied towards PBL removal. Depending on the relative amounts of HF and HNO_3 added, silicon etch rates could be enhanced by two orders of magnitude (200-400A/min) and silicon nitride etch rates by a factor of two. Selectivities for etching silicon to oxide and silicon nitride to oxide were typically between 8-12:1 and 8-20:1 respectively.

INTRODUCTION
The Poly Buffer LOCOS (PBL) process is an enhanced isolation technique in which a thin polysilicon film is deposited under the silicon nitride. Consequently, a silicon nitride/silicon stack must be removed for formation of the field oxidation (FOX) isolation structure. Fig. 1 shows a schematic of this film stack before and after removal.

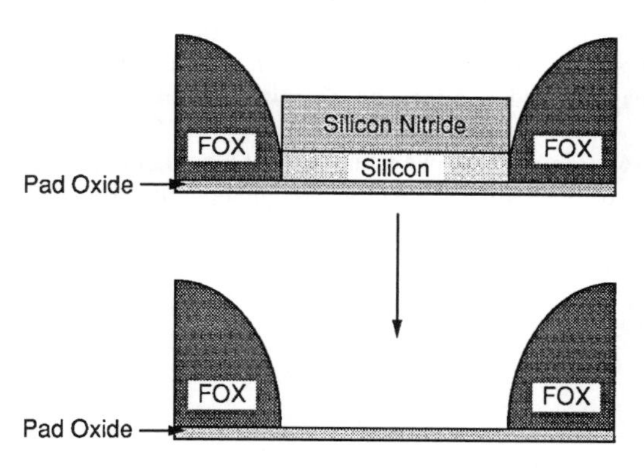

Fig. 1 Poly Buffer Locus Stack Structure

Phosphoric acid is a well known etchant for silicon nitride with selectivity to silicon oxide of about 30:1 and is used industry wide in the LOCOS (Local Oxidation of Silicon) process[1]. Typically, cassettes of wafers are immersed in a boiling bath of phosphoric acid for a set time. Water is injected into the solution to control the boiling temperature between 160-180C. Bath lifetime is usually limited not by degradation of the chemistry but by buildup of particles.

Chung and Fetcho[2] have studied the bath chemistry during silicon nitride etch. In particular, they were interested in characterizing the selectivity of this bath towards etching silicon nitride preferentially to silicon oxide and

Mat. Res. Soc. Symp. Proc. Vol. 386

extending bath lifetime by decreasing particle buildup with loading. The overall chemistry proposed by Chung and Fetcho for etching silicon nitride and silicon oxide is[a] :

$$Si_3N_4 + 4H_3PO_4 + 12H_2O \rightarrow 3Si(OH)_4 + 4NH_4H_2PO_4 \quad (1)$$

$$SiO_2 + 2H_2O \rightleftharpoons Si(OH)_4 \quad (2)$$

We have investigated the use of a phosphoric acid bath doped with ppm levels of nitric acid to remove both the silicon nitride and polysilicon films. The purpose of the nitric acid is to oxidize the silicon presumably via the reaction shown in Eq. 3:

$$3Si + 4HNO_3 + 4H_2O \rightarrow 3Si(OH)_4 + 4NO \quad (3)$$

In a fresh bath, the nitric acid increases the silicon etch rate from about 2 to 20 A/min without affecting either the silicon nitride (65A/min) or silicon oxide (2A/min) etch rates. Two main advantages of this approach are that process steps are minimized and that the etching is done by a metal ion free etchant. Other wet/wet approaches invariably use the phosphoric acid bath to strip off the silicon nitride and another bath chemistry to remove the silicon. The objective of this work was to determine whether the loading effect could be overcome.

THEORY

Though PBL stack removal using phosphoric acid with ppm traces of nitric acid has been used in AT&T for 0.5μm technology development, the process has been erratic; often leaving residual silicon on the wafers. Additional etching time sometimes not only failed to remove the remaining polysilicon, but selectively bored tunnels into the underlying pad oxide which affects final device characteristics.

Inspection of Eq. 1-3 suggests removal of the PBL stack necessarily leads to the formation of silicates not only from the silicon nitride/silicon stack but also etching of the surrounding FOX and underlying PAD oxide. Silicate buildup will not only decrease silicon oxide etching but could interfere with silicon dissolution either by redepositing oxide, consuming the available nitrate or perhaps limiting the amount silicon solubilized by the bath due to equilibrium considerations. Any of the above would explain loading effects on PBL stack removal.

Consequently, if we consume the silicate reaction product, we would expect loading effects to be minimized and perhaps reaction rates to be increased. An effective method therefore is to chemically destroy silicates with trace amounts of hydrofluoric acid[3] via the nominal reaction:

$$Si(OH)_4 + 6HF \longrightarrow H_2SiF_6 + 4H_2O \quad (4)$$

EXPERIMENTAL

Equipment and Materials

All etch studies were done with a manual Santa Clara Plastic (Boise, Idaho) wet bench station (17 liter volume) with automatic water injection for temperature control. Bath samples of 10-25 mls of etchant were manually removed as needed using a Teflon beaker. Polysilicon, silicon nitride and silicon oxide etch rate monitors were fabricated on 5 inch silicon wafers. All acid solutions were semiconductor reagents.

---

[a] Note that Eq. 1 is oversimplified since thermal decomposition of phosphoric acid into pyrophosphoric acid is ignored.

Fluoride and nitrate activities were determined using the proper ion selective electrodes while silicon content was done by atomic spectroscopy. Etch rates were done by measuring before and after film thicknesses after 10 minute etches at the same 19 sites on a Prometrix SM200E (Santa Clara, CA). Solution concentrations given in this paper refer to the volume of reagent diluted in water. Consequently, a 15% HF/15% HNO_3 was made by proportionately mixing 15 cc of concentrated HF, 15 cc of HNO_3 and diluting to 100cc.

Characterization of the Phosphoric/Nitric Acid Process

Bath loading effects with the phosphoric acid/trace nitric acid bath were characterized by measuring polysilicon, silicon nitride and silicon oxide etch rates before and after etching lots of 50 PBL wafers and after spiking with 100cc of 3% nitric acid solution. Fig. 2 shows the results of this study. Results from the bath loading show that the polysilicon etch rate decreases dramatically with wafer loading.

Characterization of the Phosphoric/Nitric/Hydrofluoric Process

An 8 run fractional factorial experiment was run to characterize the effect of HF and HNO_3 spiking on polysilicon, silicon nitride and silicon oxide etch rates. At each condition, etch rates were measured before and after spiking and before and after etching a load. Results are shown in Table 1. Note that Runs 8-9 were not completed. This is because Runs 4-5 demonstrated that there was no enhancement of polysilicon etch rates in the absence of HNO_3 which was later verified at other conditions.

Table 1 Effect of Spiking on Initial Etch Rates

					After Spiking			After 1st Si Load			After 2nd Spiking		
Run	Temp	Vol	%HF	%HNO3	Poly-silicon	Silicon Nitride	Silicon Oxide	Poly-silicon	Silicon Nitride	Silicon Oxide	Poly-silicon	Silicon Nitride	Silicon Oxide
1	167.5	200	3	3	66±2	103±1	6.9±0.2	128±12	98±1	3.9±0.2	103±3	95	8±1
2	175	300	5	6	195±4	158±1	17.9±0.3	105±5	149±1	10.4±0.3	313±4	193±1	20±0
3	160	100	5	6	42±1	85±1	5.3±0.2	38±2	81±1	3.6±0.2	67±1	100±1	6±3
4	160	100	1	0	2±0	62±5	2.8±0.3	2±0	65±4	2.2±0.3	2±0	68±0	3±0
5	175	100	5	0	3±0	117±0	8.0±0.2	2±1	117±1	7.7±0.1	2±1	117±1	8±0
6	160	300	1	6	27±1	72±1	3.4±0.5	25±1	69±1	3.8±0.6	38±1	75±1	4±0
7	175	100	1	6	47±2	103±1	3.9±0.1	42±1	103±1	3.3±0.1	50±1	108±1	3±0
8	175	300	1	0	NA	NA	NA	NA	NA	NA	NA	NA	NA
9	160	300	5	0	NA	NA	NA	NA	NA	NA	NA	NA	NA

Data in Table 1 indicate that initial etch rate selectivity to silicon oxide increases with temperature and at low HF spikings. Initial polysilicon, silicon nitride and silicon oxide etch rates as a function of spiking content were regressed from the data shown in Table 1 and augmented by subsequent experiments. For example, Fig. 2 shows that initial polysilicon etch rate is very sensitive to HF content only in the presence of HNO_3. Table 2 summarizes the normalized factor effects on these etch rates over the range of process conditions shown in Table 1. Note that the effect of nitric acid on the etching rates of polysilicon and silicon nitride is best modeled with a quadratic term and that the HF/HNO_3 interaction was significant.

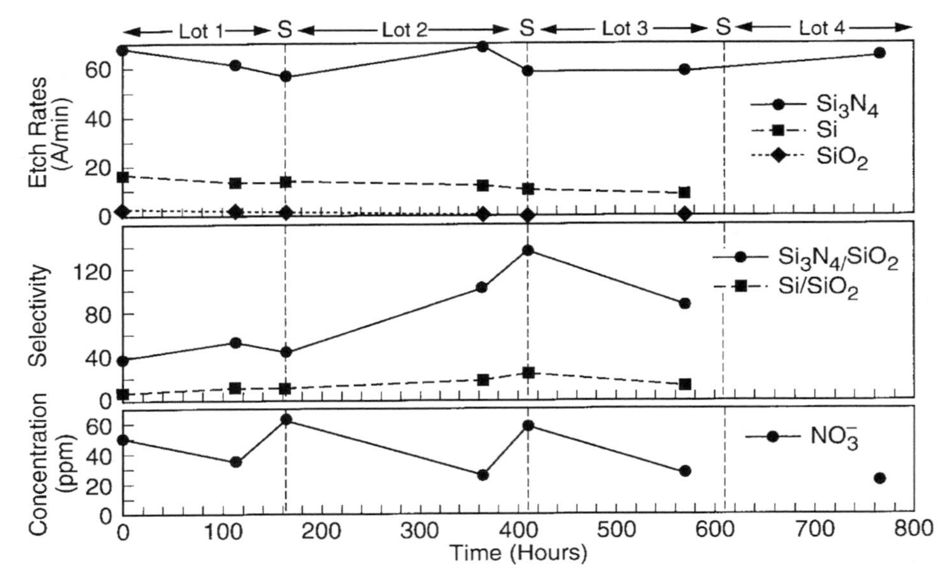

Fig. 2 Loading Effects in H_3PO_4/HNO_3 Baths

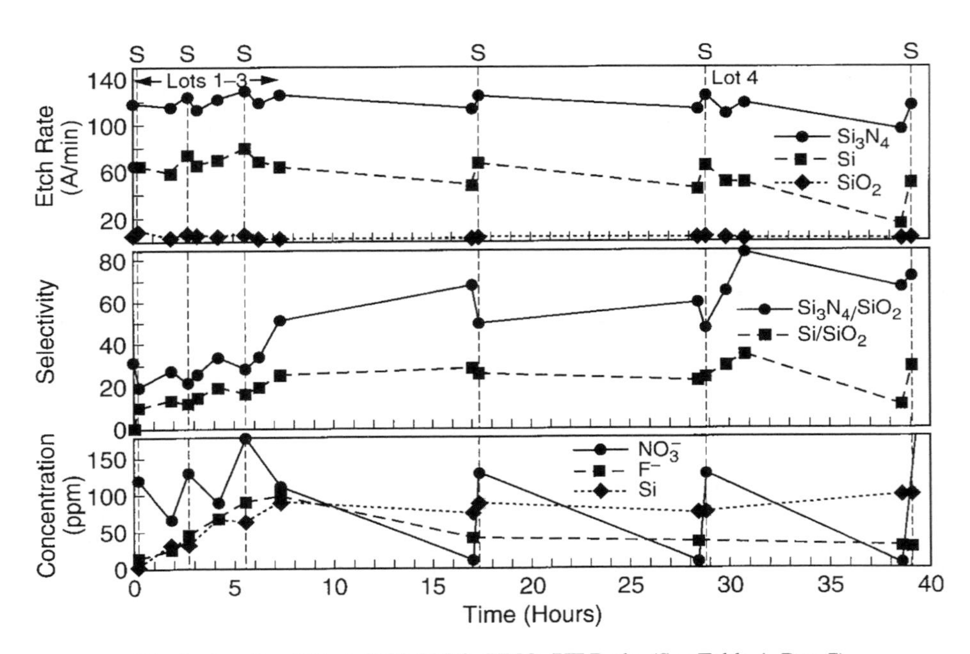

Fig. 3 Loading Effects With $H_3PO_4/HNO_3/HF$ Baths (See Table 1, Run 7)

Table 2 Regressed Factors Affecting Initial Etch Rate Data[b]

Factor	Polysilicon	Silicon Nitride	Silicon Oxide
Constant	74±3.4	107.0±1.6	8.1±.4
NO_3	49.7±4.1	11.3±2	3.1±.4
NO_3^2	-19.1±2.5	-4.7±1	0
HF	37.9±4.7	18.7±2.2	3.7±.5
HF/NO_3	38.5±4.7	8.4±2.3	3.7±.5
Temp	14.5±3.5	19.9±1.7	1.2±.4
Temp\NO_3	13.7±3.4	0	0
Temp\HF	0	0	0
No. of Pts	22	22	22
$\bar{s}$	10.7	5.7	1.2
r^2	.99	.97	.98

Loading studies were done to determine the stability with spiking of the PBL stack removal in the presence of dissolved reaction byproducts. Fig. 3 shows a loading study done at conditions similar to Run #7 in Table 1. Note that selectivity of etch rates to silicon oxide increased with the number of lots processed.

Discussion

Though we have learned enough to engineer a process to remove the PBL stack using phosphoric acid with trace amounts of nitric and hydrofluoric, our fundamental mechanistic understanding is quite immature. The biggest barrier towards understanding the mechanism is analyzing trace level reaction products in phosphoric acid solutions. We were able to quantify only the total silicon contents and the fluoride and nitrate activities. Due to low concentrations and phosphate interference, we were unable to measure the total fluoride content or structure of the silicon-fluoride reaction product. Consequently, we can only postulate about the mechanism from knowledge of the silicon content and fluoride and nitrate activities until better analytical techniques are found.

Equations 1-4 summarize the expected stoichiometry for the dissolution of silicon in nitric acid (Eq. 3) and the destruction of silicates by hydrofluoric acid (Eq. 4). Combining these equations yields the theorized stoichiometry for the dissolution of silicon:

---

[b] Etch Rate=Constant+$(F-F_-)/((F_+-F_-)$. where F_- and F_+ are the following low and high variable settings:

Factor	F_-	F_+
Temp (C)	160	175
NO_3 Concentrate Addition (cc)	0	18
HF Concentrate Addition (cc)	1	15

$$3Si + 4HNO_3 + 18HF \rightarrow 3H_2SiF_6 + 4NO + 8H_2O \qquad (5)$$

The most bewildering observation is that the analyzed fluoride activity increases with the amount of dissolved silicon independent of the spiking strategy (Fig. 4). This is contrary to expectations for the dissolution of silicon suggested by Eq. 5 which indicates that fluoride concentration should decrease as silicon is dissolved. Furthermore, since samples were diluted twentyfold before analysis, hydrofluoric acid dissociation cannot be the explanation. Rather a possible reason is that addition of hydrofluoric acid to the hot phosphoric acid produces fluorophosphoric acid:

$$H_3PO_4 + HF \rightleftharpoons H_2PO_3F + H_2O \qquad (6a)$$

$$H_2PO_3F + HF \rightleftharpoons HPO_2F_2 + H_2O \qquad (6b)$$

$$HPO_2F_2 + HF \rightleftharpoons POF_3 + H_2O \qquad (6c)$$

With this reasoning, it is the fluorophosphoric acid that enhances the dissolution chemistry in the presence of nitric acid by donating fluoride atoms in a manner consistent with Eq. 5. The fluoride measured fluoride activity would increase with silicon content if the fluorosilicic acid product hydrolyzed by the time the sample was cooled, diluted and neutralized.

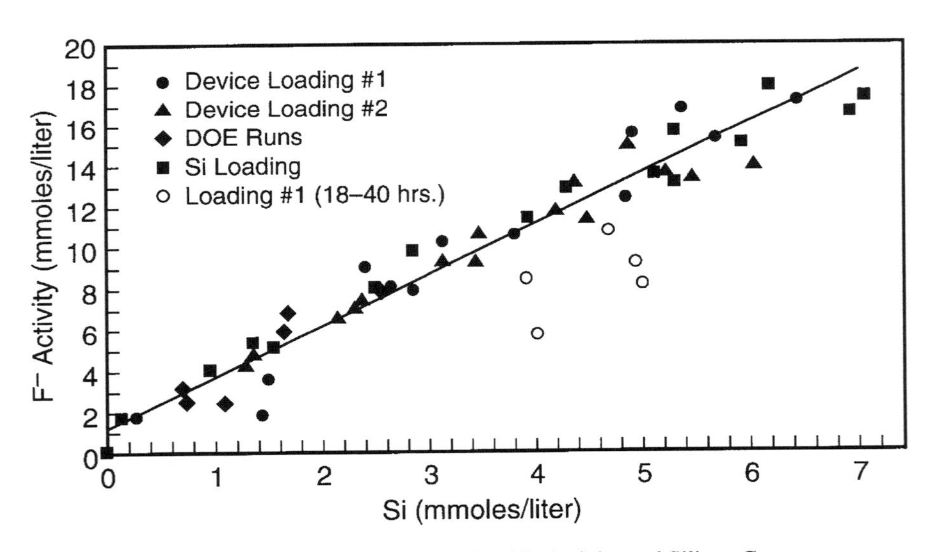

Fig. 4 Correlation Between Bath Fluoride Activity and Silicon Content

If we assume that the analyzed fluoride activity is really the hydrolyzed byproduct of fluorosilicic acid, than the slope of Fig. 4 is a measure of the stoichiometry and the intercept is the equilibrium concentration of unreacted HF which is presumed to dissociate during dilution. The measured slope is 2.5±.2 which indicates that perhaps there is a mixture of fluorosilicic products after dissolution.

Finally, we should note that the points deviating most from the line shown in Fig. 4 were those from a loading study in which a used bath was intentionally left unused for 9 hours. On a molar basis, the fluoride activity diminished at about 7 times the rate of silicon loss. This loss mechanism is assumed to be evaporation.

SUMMARY

The performance of the phosphoric/ trace nitric acid bath for removal of a silicon nitride/silicon film stack was characterized. Silicon loading effects were responsible for diminishing silicon etch rates perhaps via oxide deposition. Addition of hydrofluoric acid significantly alters the bath chemistry and can alleviate loading effects and can decrease process times by a factor of 5.

ACKNOWLEDGMENTS

The authors want to acknowledge the following individuals for their contributions: Dr. Brian Chung (ATT Bell Laboratories, Princeton, NJ), Drs. John Cuthbert and Lewis Katz (ATT Bell Laboratories, Orlando, FL) and Gladys Felton (ATT, Allentown, PA) for discussions. In addition, we want to thank Cindy Eckert for doing chemical analysis and Craig Bredbenner and Michael Gourniak for processing support.

---

[1] van Gelder, W., Hauser, V.E. *J Electro. Chem. Soc., 144*, 869 (1967).

[2] Chung, B, Fetcho R.F. "Silicon Nitride Etch Process in Phosphoric Acid", *unpublished AT&T Technical Memorandum*, 1987.

[3] Ziger, D., "Method Of Integrated Circuit Fabrication Including Selective Etching Of Silicon and Silicon Compounds", *US Patent 5,310,457.*

COMPUTATIONAL FLUID DYNAMIC MODELING AND FLOW VISUALIZATION OF AN ENCLOSED WET PROCESSING SYSTEM

STEVEN T. BAY*, CHRISTOPHER F. MCCONNELL*, HUW K. THOMAS*, MICHAEL G. IZENSON**, AND JAYATHI MURTHI***
*CFM Technologies, Inc., 1381 Enterprise Dive, West Chester, PA 19380
**Creare Inc., Etna Road, P.O. Box 71, Hanover, NH 03755
***Fluent Inc., Centerra Resource Park, 10 Cavendish Court, Lebanon, NH 03766

ABSTRACT

With the goal of optimizing point-to-point etch uniformity in a continuous flow wet processing system, a comprehensive fluid dynamics study was undertaken. Aspects of this work included developing a computational, finite-element, fluid dynamic model; performing a series of empirical studies based on dye-injection with photo detection and video analysis; and observing platinum-wire bubble generation with transparent wafers.

Various flow modification designs were tested including the base case of no inserted device, a stationary "spinner" insert, and a stationary "showerhead" insert. Unsteady, non-uniform, flow distribution and zones of substantial flow recirculation were observed in the case of using no flow modification device. These observations were confirmed with the computational fluid dynamic model.

Used in conjunction with the two different types of flow inserts, a new supplementary Teflon screen was designed and tested. The combined optimum configuration using this screen with an insert was then identified. Significantly enhanced point-to-point etch uniformity, with minimal recirculation and prompt fluid displacement, resulted from this design.

OBJECTIVE AND METHODOLOGY

Several prior studies have examined fluid flow patterns in wet processing systems with the objective of maximizing rinse efficiency[1,2,3]. The objective of this research program was to optimize the point-to-point etch uniformity in a 50-wafer, 200mm Full-Flow™ wet processing system. One way to accomplish tight uniformity is to minimize any flow maldistributions throughout the process chamber so that all the wafer surfaces are exposed to etchant for the same duration. The ideal behavior is a plug flow through the chamber with a sharp interface between the fluid entering the chamber and that already in the chamber. The sharp interface would be maintained to the exit of the chamber, and the rate at which the interface progressed would be governed by the flow rate. Our objective was to achieve this behavior as closely as possible.

APPARATUS DESCRIPTION

The test apparatus utilized in this project was an actual Full-Flow vessel fitted with various transparent components. Flow diagnostic instrumentation was added to the apparatus as shown schematically in Figure 1. The diagnostic techniques included hydrogen bubble tracers, pulsed dye injection, and light transmissivity measurements during dye injection.

Hydrogen Bubbles

To obtain a qualitative assessment of the local flow velocity at the inlet plane to the wafer carrier, we installed a series of 0.002" platinum wires which spanned the wafer carrier at the inlet plane. Three wires were parallel to the wafer stack and three wires were perpendicular to the wafer stack. By applying approximately 80 V of potential to these wires relative to the rest of the process

Mat. Res. Soc. Symp. Proc. Vol. 386 © 1995 Materials Research Society

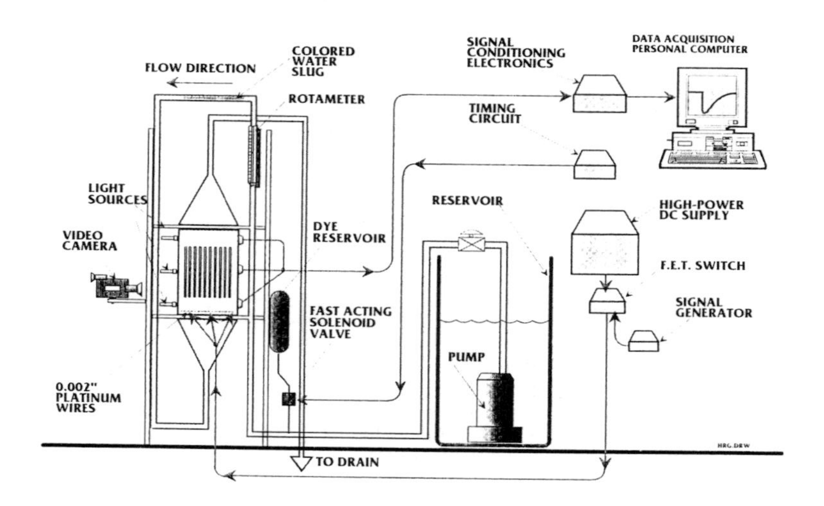

Figure 1. Flow Visualization Test Apparatus

chamber, a very fine stream of hydrogen bubbles was produced along the length of the wire. These bubbles, which act as flow tracers, have a diameter of the same order as the wire which produced them. Consequently, the buoyant forces on the bubbles are quite small relative to the dynamic forces, and the bubbles follow the flow with little effect due to buoyancy. The direction in which the bubbles travel is then a direct indicator of the local flow direction. Additionally, a qualitative assessment of the turbulence intensity in the vicinity of the wire could be made from the level of fluctuation in the flow direction.

Dye Injection

To provide a qualitative assessment of the gross flow behavior in the wafer carrier, a dye injection system was installed upstream of the etching station inlet. The dye injection system was designed to be pulsed to provide a slug of colored water of a consistent length from test to test. When using the dye injection system, the flow loop was operated in an open circuit manner so that the pump reservoir would not be contaminated with colored water. To aid viewing of the flow field in the central portion of the chamber, we replaced the silicon wafers with glass wafers of nominally the same thickness as the silicon wafers. The pulsed dye injection proved to be a very effective means of observing the gross flow behavior throughout the chamber. Video records of the dye transients for each test condition were taken and analyzed.

Light Sensitive Detectors

Although simple visual observation of the flow during dye transients is illuminating, it does not provide a firm basis for comparison between tests with different flow conditions and different flow distributors. To provide a basis for comparison, five phototransistors were mounted on one of the walls perpendicular to the plane of the wafers, and five light sources were mounted on the wall directly opposite the phototransistors, as shown in Figure 2. The output of the phototransistor is proportional to the intensity of the light incident on its face. Consequently, as the average dye

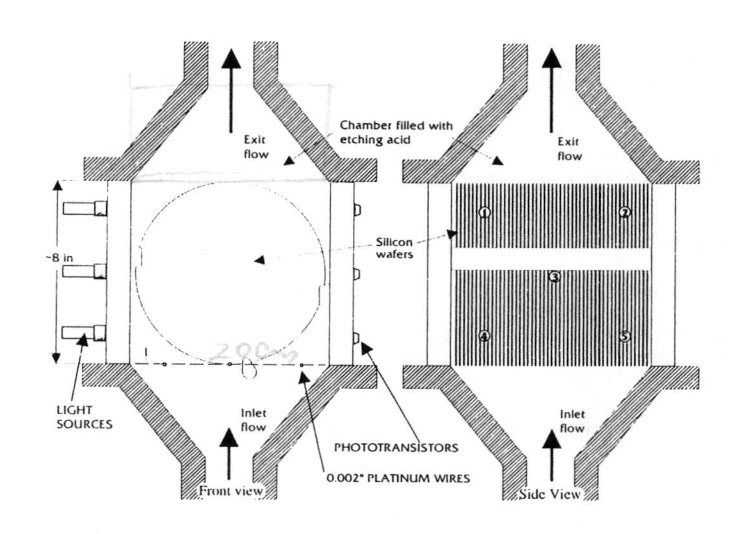

Figure 2. Location of Light Sources and Phototransistors

concentration along the transmission path between the light source and the phototransistor increases, the phototransistor output will decrease because of absorption and scattering. For plug flow, the ideal phototransistor output would be as shown in Figure 3 where the output remains constant at the initial value and then drops sharply to another level determined by the dye concentration. The output would then remain at this level for a length of time determined by the length of the colored slug of water. At the end of the dye transient, each phototransistor output would very sharply return to its initial value. It is important to remember that the phototransistor output will be related to the integral of the dye concentration along the light path.

In our experiments, it was not possible to produce a dye slug with a sharp gradient in dye concentration at the trailing edge because of diffusion and weeping from the dye injection line.

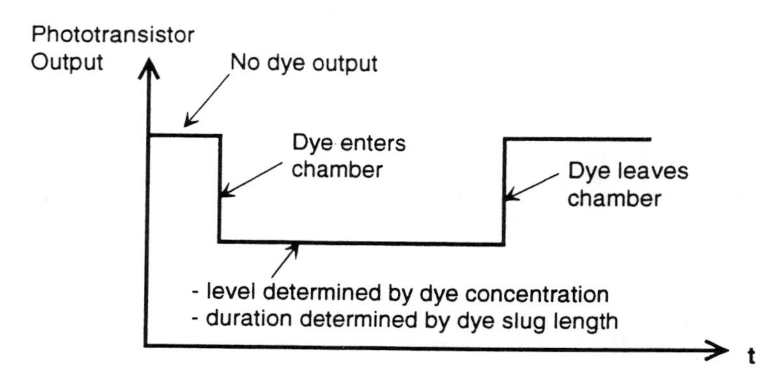

Figure 3. Theoretical Phototransistor Output for Ideal Plug Flow

Consequently, even if ideal plug flow conditions were achieved, we would not observe the ideal phototransistor output. However, the colored water slug was consistent from test to test, so comparisons of the phototransistor outputs between tests are certainly valid. The results are presented in nondimensional form where the phototransistor outputs have been normalized by their initial value prior to dye injection, and the time axis has been nondimensionalized by multiplying by the flow rate and dividing by the chamber volume. The time axis can then be thought of as the number of chamber turnovers.

SOURCES OF NONUNIFORM FLOW

Several sources of nonuniform flow are apparent in an unmodified plug-flow chamber:

- Jets and recirculation zones caused by the inertia of the liquid,
- Separated flow and recirculation regions in the inlet diffuser,
- Bypass of flow through the wafer-stack end gaps,
- Nonuniform spacing between the wafers, and
- Pressure losses due to flow between the wafers.

Flow distributors at the wafer carrier inlet can address the first two of these effects, but the introduction of a flow distributor at the inlet to the wafer carrier does not eliminate the recirculating regions beneath the distributor in the lower transition piece. This effect must be addressed by improving the diffuser design and/or adding flow distributors in the inlet diffuser. The third effect must be addressed by properly sizing the end gaps. The last two effects listed above may give rise to a background level of nonuniformity but they are very difficult to completely solve given the constraints of semiconductor wafer processing.

Separated Flow in the Inlet Diffuser

Even though flow though the wafer carrier can be made relatively uniform by an appropriate pressure drop at the inlet plane of the wafer carrier, there will remain regions of nonuniform, separated flow beneath the flow distributor. As a result, etchant can and will enter these recirculating regions during acid injection and slowly bleed out of these regions when water is introduced for rinsing. This behavior could lead to nonuniform etching and increased rinsing time. A major portion of this study focused on characterizing the flow patterns which exist in this region and then optimizing the hardware configuration to reduce or completely eliminate the recirculating regions. These methods included using a stationary "showerhead" flow dispersion device and using a stationary "spinner" flow dispersion device as disclosed in US and international patents owned by CFM Technologies[4].

Wafer Stack End Gaps

One source of flow maldistribution in the chamber results from the fact that the spacing between each wafer is approximately 0.090" and the spacing between the wafers at the end of the stack and the walls of the etch vessel is approximately 0.330". Since the pressure drop through a channel is related to the square of the channel width, the difference between the end gap spacing and the interwafer spacing results in a significant difference in pressure drop between the two. Consequently, the flow will tend to bypass the wafer stack and pass through the two end gaps since they offer less resistance to the flow. Previous work[3] has shown that in a conventional wet bench, a large majority of the flow bypasses the wafers completely. This situation is much improved in the standard Full-Flow vessel, but it is not eliminated entirely. The CFD calculations conducted in this study indicate that approximately 17% of the flow passes through the end gaps. If the end gaps were sized equal to the gaps between the wafers, then the end gap flow area would account for 4% of the flow instead of 17% of the flow.

Wafer Spacing Nonuniformity

The current 200mm Full-Flow vessel has a capacity for 50 wafers, half-spaced. As such, wafer spacing is on the same order of magnitude as wafer thickness, and small variations in wafer thickness can significantly alter interwafer spacing. In turn, these small variations in spacing can result in significant variations in flow. For this reason the thickness of each wafer used in the study was measured. The results (Figure 4) show a bimodal distribution of wafer thicknesses. This observed variation results in roughly a 5% variation in wafer spacing which results in a 10% variation in gap flow rate. The predicted flow variation was confirmed in visual observations during the dye injection studies. Video tape analysis clearly confirmed the minor variations in flow velocities between wafers. Unfortunately, no conceivable flow distribution device at the inlet plane of the wafer carrier can correct for this source of non-uniformity.

RESULTS OF CFD MODELING STUDY

To better understand the impact of these sources of non-uniform flow, a computational fluid dynamic model was developed for the unmodified etch vessel, and the inlet and outlet sections of this vessel, with no flow distribution devices. This simplified model also assumed symmetrical flow behavior so that only one quarter of the flow vessel needed to be modeled. A nonuniform 25x21x81 grid was chosen to simulate one quarter of the physical space inside the vessel. The wafers were simulated by using a porous-media assumption. A standard κ-ε turbulence model was used to provide closure on the equations of motion.

Figures 5a through 5c show the unmodified velocity distribution in the plane of three wafers in the stack. Figure 5a shows the velocity distribution at the end of the wafer stack, and Figure 5c shows the velocity distribution at the center of the wafer stack. Figure 5b illustrates the velocity distribution for a station halfway between the end and the center of the stack. In general, the CFD results agreed with the results observed in flow visualization experiments using no flow distribution techniques. Two large regions of reversed flow are evident in the velocity vector plots and there is a jet of fluid near the center of the wafer stack. Additionally, the variation in the flow between the different positions shown in Figures 5a through 5c is evident.

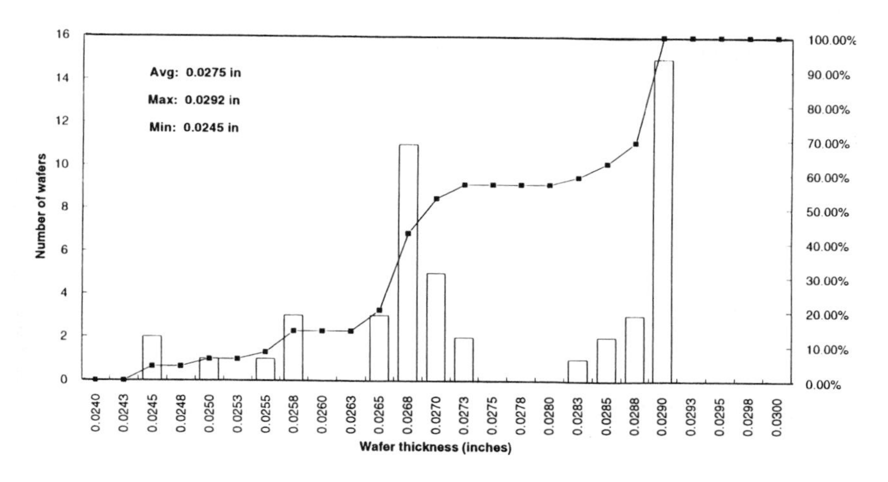

Figure 4. Measured Wafer Thickness

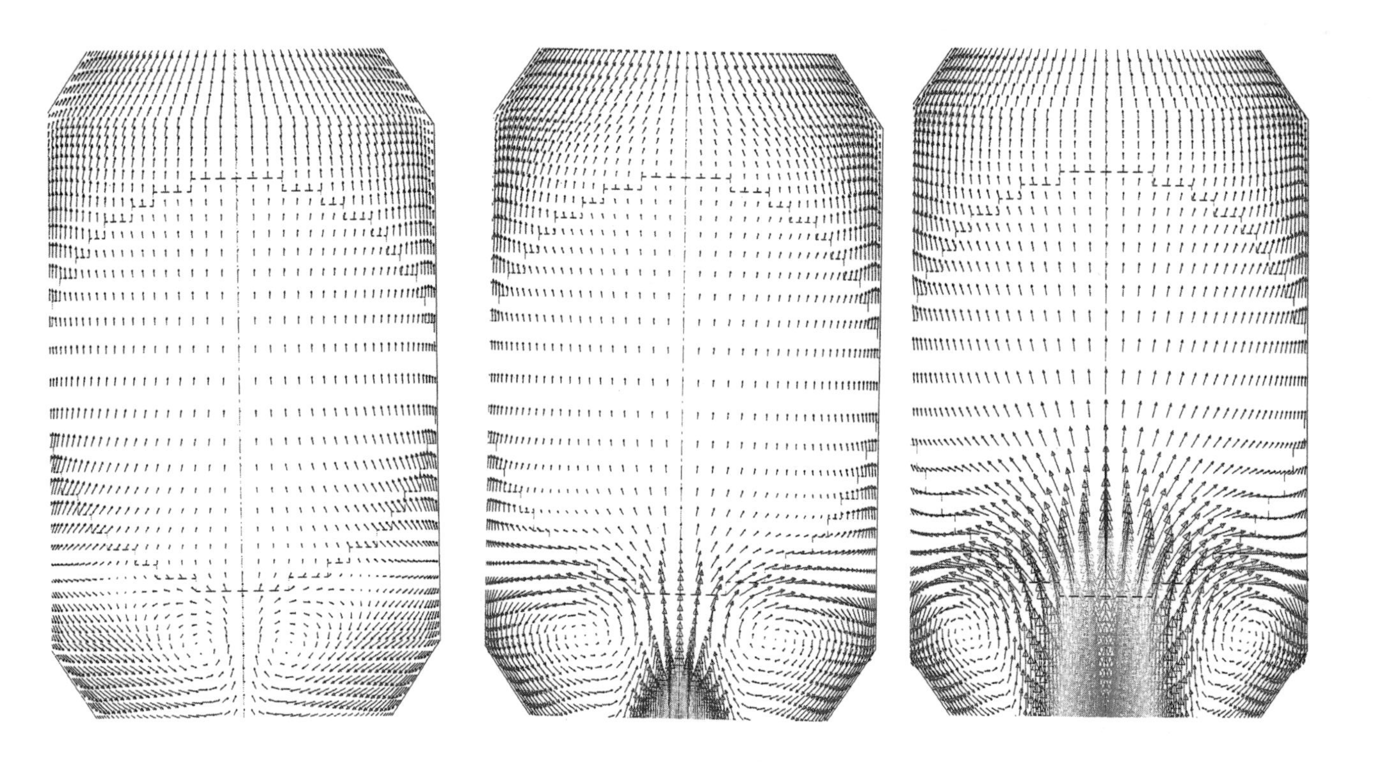

Figure 5a. End Velocity Profile of Simulated (Unmodified) Flow

Figure 5b. Mid Velocity Profile of Simulated (Unmodified) Flow

Figure 5c. Center Velocity Profile of Simulated (Unmodified) Flow

Figure 6 shows the streamwise velocity contours for the cross-sectional plane at the center of the wafer carrier. This plot illustrates the disproportionate amount of the fluid that passes through the end gaps as well as the central jet. Figure 7 shows the turbulent kinetic energy contours for both the in-plane and out-of-plane directions. This diagram represents an oblique view, looking up and across, from one lower corner of the Full-Flow vessel. It illustrates high turbulence near the chamber inlet, decay through the wafers, and increasing turbulence again near the chamber exit.

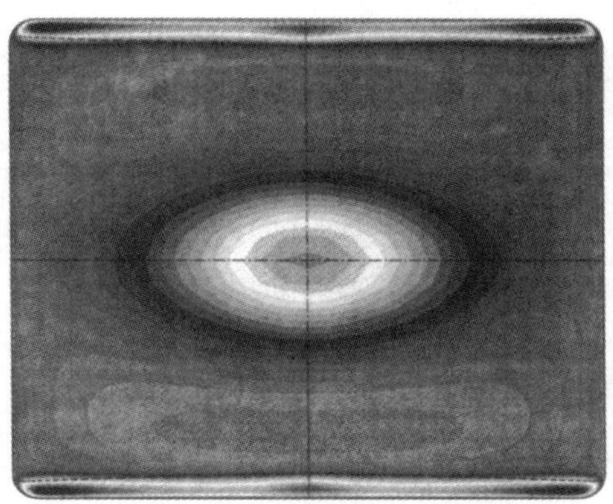

Figure 6. Cross-sectional Velocity Profile of Simulated (Unmodified) Flow

Figure 7. Kinetic Energy Contours for Simulated (Unmodified) Flow

ASSESSMENT OF FLOW DISTRIBUTORS

No Flow Distributor

To provide a baseline for comparison with the various flow distributors, the first experiments were conducted with no flow distributors. The hydrogen bubble tracers clearly indicate that there is a large, turbulent, recirculating region below the wafer stack in each of the corners. This results from flow separation in the inlet diffuser. Also, a jet of fluid passes through the center of the wafer stack. As mentioned previously, CFD results indicate that approximately 17% of the flow passes through the end gaps, and this was qualitatively confirmed. A set of representative phototransistor outputs during a dye transient is shown in Figure 8. The unsteadiness in the traces is indicative of the turbulence intensity in the chamber. Some of the unsteadiness is a result of dye passing through the end gaps and recirculating at the top of the wafer stack. As would be shown in later experiments, flow distributors reduce this unsteadiness. Although not apparent to the naked eye, the phototransistors show that there is evidence of dye in the chamber even after 16 chamber turnovers when the experiment was terminated.

Flow Distributor Test Results

The showerhead distributor design was tested with hydrogen bubble and dye injection techniques. The flow qualitatively appeared to be more well-distributed than the case with no flow distributor, but there was clearly room for significant improvement. Careful observation revealed that individual flows from each nozzle were not uniform across the showerhead array. Also, reversed flow regions were still noted above the showerhead and beneath the wafer stack; however, these recirculating regions were steadier than in the case above with no flow distributor. Figure 9 shows a representative phototransistor output for the showerhead flow distributor case. Similar experiments with CFM's spinner design showed significantly improved results. There was less recirculation, and less preferential flow. Nevertheless, the flow patterns were not ideal.

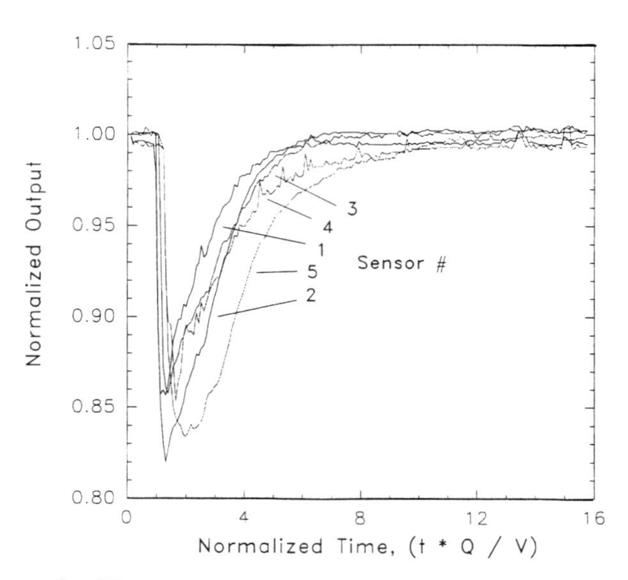

Figure 8. Phototransistor Outputs for No Flow Distributor

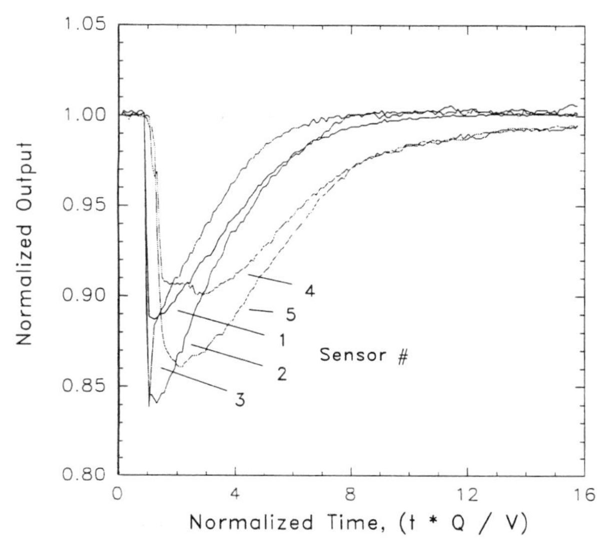

Figure 9. Phototransistor Outputs for "Showerhead" Distributor

Teflon Screen Flow Distributor Test Results

The effects of the flow separation in the inlet diffuser can be minimized by providing a pressure drop at the outlet plane of the diffuser comparable to the dynamic head of the incoming stream. The desired pressure drop can be obtained with an appropriately sized mesh which spans the entire inlet plane of the wafer carrier. The etching chamber offers two interesting challenges with regard to mesh selection:

- To achieve the desired pressure drop, the mesh must be relatively fine, and
- To withstand the hydrofluoric acid, the mesh must be made entirely of Teflon.

We were able to locate a Teflon mesh produced by Norton Performance Plastics (Wayne, NJ) which nearly meets all the requirements for the flow distributor. They supplied several samples of various mesh porosity. The sample selected for preliminary testing was Zitex SK TP High-Pressure mesh for which the manufacturer quoted a pressure drop of 0.01 bar through a 1 cm^2 sample at a flow rate of 12-14 l/min of water. Using these data, the approximate pressure drop across the etching chamber flow distributor is estimated as 0.025 inches H_2O at a 7 gpm flow rate which is lower pressure drop than would normally be expected to be required. However, since it was easily obtained, several simple tests were performed.

In combination with CFM Technologies' stationary spinner flow distributor, the Teflon mesh at the exit of the diffuser provided a dramatic improvement in the flow distribution throughout the chamber. The hydrogen bubble tracers show that the reversed flow regions beneath the wafer stack are completely eliminated and the turbulence intensity is reduced everywhere. Visual observation of the flow during dye injection transients demonstrates the improvement in flow uniformity across the chamber both in the plane of the wafers and perpendicular to the plane of the wafers. A representative set of phototransistor outputs are shown in Figure 10.

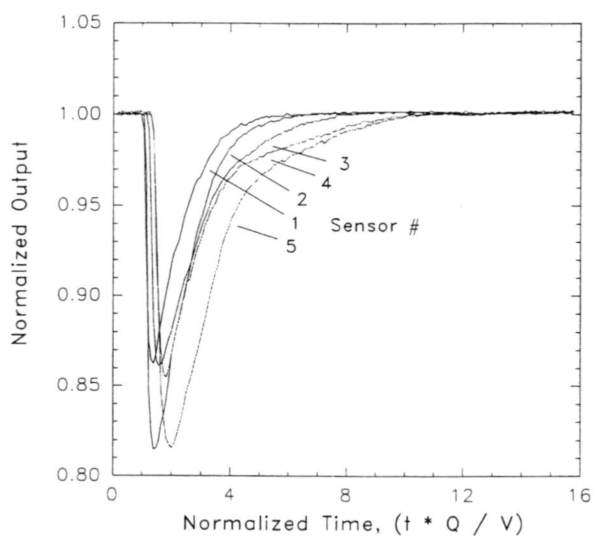

Figure 10. Phototransistor Outputs for "Spinner+Mesh" Distributor

Comparison of Figures 8 and 9, with Figure 10, shows the marked improvement in flow steadiness and the substantial reduction in the time required to completely remove the dye from the chamber. After 10 chamber turnovers, in the case of the spinner and Teflon mesh combination, there is no evidence of dye in the chamber from any of the phototransistor outputs. For a standard 50-wafer, 200mm Full-Flow system, this displacement volume corresponds to approximately 25 gallons of rinse water.

ASSESSMENT OF FLOW DISTRIBUTORS (ETCH DATA)

After obtaining such encouraging flow visualization results with the Teflon mesh flow distributor, a sample of the mesh was sent to a CFM customer for actual etch tests. Table I summarizes the results. Flow rates during injection and displacement ranged from 5 to 8 gpm. Nominal etch depth for was 100 Å. Before and after film thickness measurements were taken at 49 points. All tests were performed using identical inject, soak, and displace times. (Inject and displace times were 1 minute for all tests).

There were promising results at high flow rates, but also some severely non-uniform results at lower flows. This negative outcome was partially expected due to the reduced number of chamber turnovers at low flow. At the higher flow rates (8 gpm or 3.2 chamber turnovers) the etch was relatively uniform (standard deviations of 1 to 4%), but at lower flow rates (5 gpm or 2 chamber turnovers) the performance was worse than in tests with the showerhead distributor alone (standard deviations up to 15%). Insufficient pressure drop across the Teflon screen, and resultant sub-optimal flow dispersion, also may have contributed to the worse results at lower flow rates.

Further optimization work using the spinner-mesh combination revealed that somewhat higher etch flow rates generate significantly better etch results than those shown in Table I. For light etches (75Å to 250Å), point-to-point etch uniformity on Full-Flow systems operating in the range of 9 to 10 gpm generally lie between 0.5% and 1.5% in terms of standard deviation.

Table I. Summary of Initial Silicon Wafer Etch Test Results

Flow rate (gpm)	Wafer number	Points measured	Avg etch (Å)	Min etch (Å)	Max etch (Å)	Std. Dev. (%)
5	1	13	103.5	74.9	110.8	9.85
5	13	13	96.93	69.1	110.8	13.91
5	25	13	91.315	68.5	110.7	14.85
7	1	49	107.3	74.6	128.5	13.6
7	13	49	118.24	111.5	123.3	2.81
7	25	49	109.44	99.5	117.0	4.10
8	1	13	104.32	98.7	106.9	2.84
8	13	13	104.81	96.3	107.9	2.99
8	25	13	105.23	103.5	107.0	1.00
8	1	49	97.14	92.2	100.2	1.70
8	13	49	97.51	89.1	105.0	3.70
8	25	49	97.97	95.5	100.2	1.03

SUMMARY AND CONCLUSIONS

Computational fluid dynamic modeling results have been obtained for flow through a 50-wafer, 200mm Full-Flow chamber without any flow distributor at the inlet and with a complete wafer stack. These results were in good agreement with the general flow patterns observed in the chamber, reproducing the central jet through the wafer stack, recirculation regions beneath the wafers, and bypass flow around either end of the wafer stack.

We conducted a series of baseline tests using various flow visualization techniques, and have identified the main causes of nonuniform flow. We also determined a combination of flow distributors which produces very uniform flow at higher flow rates. The important points are:

- The gaps between the chamber walls and the end wafers are a significant source of nonuniform flow (17% of flow versus the ideal 4% of flow).
- The test series with the open inlet (no spinner and no showerhead) show a large amount of recirculation in the lower transition. Dye removal times were approximately twice those measured with the spinner or showerhead.
- The showerhead flow distributor does not produce an ideal plug-like flow. (We observed a central jet and recirculation regions beneath the wafers.)
- The spinner shows improved flow distribution characteristics over the showerhead, but still displays non-ideal behavior (some recirculation and non-uniformity).
- An additional Teflon mesh flow distributor contributed substantial improvements in the flow distribution. Optimum results were obtained when the mesh was used in conjunction with the stationary spinner device.

All of these results were confirmed quantitatively using phototransistor output charts that characterized rinse profiles during dye injection experiments. Further confirmation of optimum flow was obtained by actual etch experiments performed using various flow rates and selected hardware configurations. With the optimal flow distributor design and optimal flow rate conditions, point-to-point etch uniformities of 0.5% to 1.5% (standard deviation) are normally obtained in light etches (75Å to 250Å).

ACKNOWLEDGEMENTS

The authors are indebted to IBM Corporation in Burlington, VT, for their assistance in testing the etch uniformity of various hardware configurations, and for providing wafer and facilities for these activities. The following individuals were especially helpful and offered many valuable suggestions: Ted Barrett, Cathy Gow, Tom Harrigan, Fred Kern, and Bill Syverson.

REFERENCES

1. A. Tonti, "Contamination by Impurities in Chemicals During Wet Processing," in Proceedings of the Second International Symposium on Cleaning Technology in Semiconductor Device Manufacturing, The ElectroChemical Society, p. 409, 1992.

2. J.J. Rosato, R.N. Walters, R.M. Hall, R.G. Lindquist, R.G. Spearow, and C.R. Helms, "Studies of Rinse Efficiencies in Wet Cleaning Tools," in Proceedings of the Third International Symposium on Cleaning Technology in Semiconductor Device Manufacturing, The ElectroChemical Society, p. 94, 1993.

3. S.N. Kempka, J.R. Torczynski, A.S. Geller, J.J. Rosato, R.N. Walters, and S.S. Sibbett, "Evaluation of Overflow Wet Rinsing Efficiency," in Conference Proceedings of MICRO '94, MICROCONTAMINATION, p. 225, 1994.

4. C.F. McConnell and A.E. Walter, U.S. Patent No. 4 633 893 (1987), European Patent No. 0 233 184 (1988), Japanese Patent No. 1 592 557 (1988), and Korean Patent No. 028 929 (1989).

SURFACE PHOTOVOLTAGE MONITORING OF SILICON SURFACE NATIVE AND CHEMICAL OXIDES FOLLOWING WAFER CLEANING AND RINSING OPERATIONS

John J. Rosato, R. Mark Hall, Thad B. Parry, Paul G. Lindquist, Taura D. Jarvis
Santa Clara Plastics, 400 Benjamin Lane, Boise, ID 83704

ABSTRACT

We report on the use of the Surface PhotoVoltage (SPV) technique to monitor the Si surface bonding arrangement, and the impurity metallic contamination level prior to critical diffusion processes via the indirect measurement of surface charge and diffusion length, respectively. We show that the effectiveness of the pre-diffusion wet chemical cleaning and rinsing sequences can be accurately monitored via the real-time, nondestructive SPV measurement. In particular the nature of the surface passivation/chemical oxide formed during the cleaning and rinsing operations can be monitored by quantitative surface charge measurements. The importance of the prior wafer history is highlighted, as is the role of the Si starting material and measurement parameters.

INTRODUCTION

The chemical bonding arrangement of the silicon wafer surface is of critical importance to the manufacture of semiconductor integrated circuits. The silicon surface chemical bonding arrangement prior to a high temperature diffusion or Chemical Vapor Deposition (CVD) process ultimately affects the electronic properties of the devices fabricated on the wafer. The prediffusion silicon surface state, and contamination levels can impact the stoichiometry and therefore the electrical integrity of dielectric films used in the fabrication of electronic devices. Because of the sensitivity of electronic device properties to the silicon surface state, most chip manufacturing processes employ extensive use of wet chemical cleaning and rinsing operations. These cleans remove unwanted impurity contaminants(atomic, ionic, and particulate), and frequently result in the formation of a chemical or native oxide on the Si surface. This oxide can act as either a desirable passivation layer, which may result in controllable and predictable electronic properties of devices, or it may act as a major contributor of contaminant metal atoms to the Si wafer and deposited thin films. Unfortunately, until recently, it was difficult to quantitatively measure the Si surface bonding, passivation, and contamination properties in-line prior to critical diffusion processes.

The advent of optical measurement techniques such as the Surface PhotoVoltage (SPV) technique [1,2] has allowed for real-time, non-destructive measurement of the Si surface bonding arrangement indirectly via the quantitative measurement of surface charge, and minority carrier diffusion length.

Silicon Oxides and Surface Passivation

With the ever-shrinking device geometries associated with the advance of integrated circuit processing, the thickness of the dielectric films used in fabricating these devices has

Mat. Res. Soc. Symp. Proc. Vol. 386 © 1995 Materials Research Society

decreased proportionally. With gate insulators approaching only a few monolayers in thickness for state-of-the-art MOS devices, it becomes increasingly important to have an ultraclean Si surface prior to the gate oxide deposition. Since the Si surface is inherently unstable, it is difficult to avoid the uncontrolled spontaneous regrowth of a native oxide upon exposure to an oxygenated environment (e.g. a cleanroom environment or H_2O-based cleaning solutions). While some researchers have maintained that it is vital to completely suppress this native oxide growth prior to high temperature diffusion processes[3], others have maintained that the native oxide can be beneficial, as long as it is free of impurity contaminants. In this paper we propose the use of surface photovoltage measurements to monitor the state of the Si surface and the quality of the native oxide formed after wet chemical cleaning and rinsing sequences. Hydrogen peroxide-based cleans such as SC-1 (NH_4OH:H_2O_2:H_2O), typically result in an oxidized Si surface via the etching and subsequent reoxidation of the Si surface which occurs during the chemical cleaning process. This oxide is sometimes referred to as a *chemical oxide*, owing to its origin in the chemical solution, and frequently contains unwanted impurity metallic species such as Fe, Al, and Ca (all which have been shown to degrade electronic device performance). In this paper, we show examples of how the wafer cleaning sequence affects the surface properties as measured by SPV.

The charging effects of metallic impurities and other defects in Silicon oxides have been studied extensively, owing to their importance in MOS device fabrication[2, 3, 4]. A major source of metallic impurities is often the pre-diffusion wet chemical cleaning sequences, which often represent a compromise between particle removal efficiency and impurity metallic contamination. This is especially true of the alkaline-based solutions, such as SC-1, which are most effective at removing particles, but frequently deposit metals on the surface and within the chemical oxide[7, 10].

SPV Surface Charge Measurements

The electric charge on the Si surface is an important indicator of the wafer surface state. The charge on the surface of a bare Si wafer can be modeled analogous to that of a thermally oxidized Si surface, with the native or chemical oxide behaving similarly to a thermal oxide in terms of charge properties. The charge distribution characteristics of the Si/SiO_2 interface have been studied extensively [4,5] due to it's important role in the operation of Metal-Oxide-Semiconductor(MOS) devices, which are the fundamental building blocks of most modern integrated circuits [5, 6].

There are four major components to the total electric surface charge, $\mathbf{Q_{ST}}$. These include: 1) the *Si/SiO_2 Interface Trapped Charge*, $\mathbf{Q_{it}}$, which is due to electrons or holes trapped at interface defect sites; 2) the *Oxide Fixed Charge* $\mathbf{Q_f}$, which is caused by either positively ionized Si^+ at dangling bond sites, or by negative charge centers within the oxide interface region (typically due to impurity defects caused by contaminants such as Al or Ca incorporated within the oxide); 3) the *Oxide Trapped Charge*, $\mathbf{Q_{OT}}$, due to electrons or holes trapped at defect sites within the bulk of the oxide(often trapped as the result of electric field-induced tunneling currents through the oxide); and 4) the *Oxide Mobile Ionic Charge*, $\mathbf{Q_m}$, due to ionic impurities (such as Na^+, K^+, Li^+, H^+), which are mobile under the combined actions of thermal and electric field stresses. This total oxide surface charge, $\mathbf{Q_{ST}=(Q_{it} + Q_f + Q_{OT} + Q_m)}$ induces an image charge of opposite polarity near the Si surface in the form of a space charge region (either a depletion or accumulation region, depending on the polarity of the Si starting material). The image charge has a magnitude equal to the sum of the individual oxide charge components, but is of opposite polarity. This space charge region in the Si results in a bending of the Si energy bands near the surface and, thus, the formation of a surface potential barrier, **qV**.

The surface potential barrier, **qV**, can be quantitatively measured by the constant photon flux SPV technique [1,2]. In this technique, the surface potential barrier is measured by shining a high intensity light source(photon flux $>10^{19}$ photons/cm^2) of sufficient wavelength (800nm) to generate an abundance of electron-hole pairs in the bulk Si. These excess carriers then neutralize the surface potential barrier and create an approximate flat band condition. The surface charge is calculated from the change in the measured surface photovoltage with illumination. It is this quantity, **ΔqV** which is actually measured during a surface charge measurement. As implied from the earlier discussion, this quantity gives an indication of the state of the Si surface oxide and surface bonding arrangement. Therefore, the surface charge can be an important indicator of the Si surface quality following a wet chemical cleaning and rinsing process. Any chemical residues or impurities remaining on the surface following the cleaning process will directly alter the measured surface charge.

SPV Diffusion Length Measurements

Measurement of the minority carrier diffusion length is accomplished via the constant flux photon technique[2]. A diffusion length is calculated based on the change in measured surface photovoltage with varying wavelength of optical excitation. This diffusion length gives an indication of the overall bulk Si contamination levels(both lattice defect and impurity metallic). In order to monitor contamination introduced during wafer cleaning steps, surface contaminants must first be diffused into the Si bulk via a high temperature RTP treatment[2,7]. Following this heat treatment an optical measurement of the minority carrier lifetime can be used to quantify the bulk impurity contamination level [2,7]. More specifically the impurity iron level can be measured directly via the change in measured diffusion length with the dissociation of FeB pairs to interstitial lattice Fe in p-type Si via an optical excitation of a specific wavelength.

Because metallic impurity atoms have a different bonding arrangement and valence state than the Si atoms in the wafer crystal lattice, they introduce either: 1) electrical charges which can be detected by surface charge measurements; or 2) recombination centers which degrade the minority carrier diffusion length. Previous studies have correlated the impurity contamination with the surface electric charge magnitude and polarity [1,8].

EXPERIMENTAL

Bare Silicon wafers were processed through typical chemical clean/rinse/dry sequences in a SPS 9400 automated wafer cleaning system manufactured by Santa Clara Plastics. All chemicals used were of high purity with all impurity contaminants guaranteed to have a concentration of <5 ppb in solution. All wafers used were 200mm diameter with Cz-grown <100>crystal-orientation and p-type doping, unless otherwise stated. Rinsing operations following the chemical cleans were performed using Ultra High Purity(UHP) Deionized(DI) water, and all drying was accomplished with a SCP IPA Vapor Jet Dryer. SPV measurements were carried out using a Surface Charge Imager manufactured by Semiconductor Diagnostics, Inc.(Model CMS IIIA). Except where noted, all measurements were performed immediately following the final dry process. Chemical cleaning sequences investigated included those typically used in wafer manufacturing plants, which, in some cases, are known to introduce metallic contaminants to the wafer surface.

RESULTS

Figure 1 shows the effect of the chemical cleaning sequence on the SPV-measured surface Fe concentration and diffusion length. It is clearly evident that the HF process results in the lowest Fe concentration and the highest diffusion length, indicative of low overall metallic contamination levels. In contrast the SC-1 clean shows the highest metallic contamination levels. However it is well known that the SC-1 clean is the most effective at removing particles from the wafer surface[11], whereas the HF clean typically deposits particles on the wafer. The SPV data are corroborated with the TXRF results shown in Fig. 2, which also show the highest metallic contamination levels for SC-1 cleans.

Figure 3 shows the effect of cleaning sequence on surface charge for both n- and p-type wafers. Aside from the difference in charge polarity similar trends are observed for the different cleans, suggesting identical contaminant species.

In addition to monitoring metallic impurities, the surface charge can be used to indicate the state of surface passivation following an oxide-removal etch in HF. The Si surface state following an HF etch has been the subject of much study owing to the importance of understanding the Si surface passivation state. It is generally accepted that the HF solution passivates the unsatisfied Si surface bonds with either hydrogen or fluorine ions during the HF native oxide removal treatment[9], with the F ions collecting preferentially at chemically reactive sites such as lattice steps or defects. The extent to which this surface passivation occurs has been shown to depend upon both the strength and the temperature of the HF solution[9]. Indeed, a well passivated Si surface may dramatically suppress the formation of a native oxide upon subsequent exposure to an oxidizing ambient.

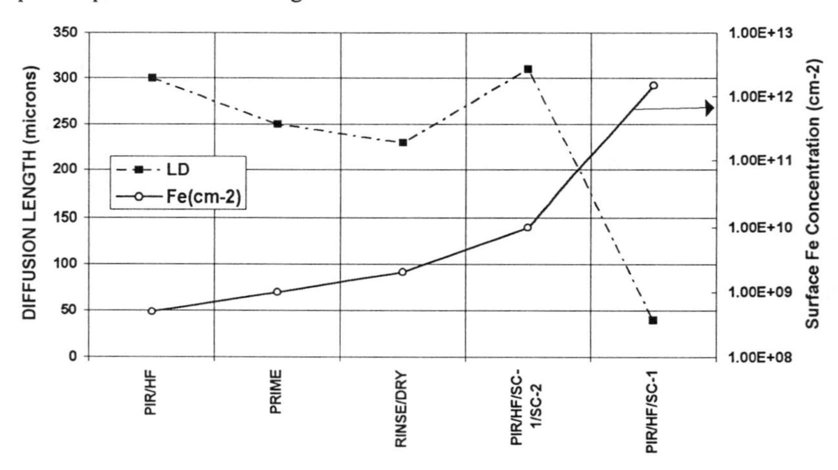

Figure 1. Effect of wafer cleaning process on SPV-measured diffusion length and iron concentration.

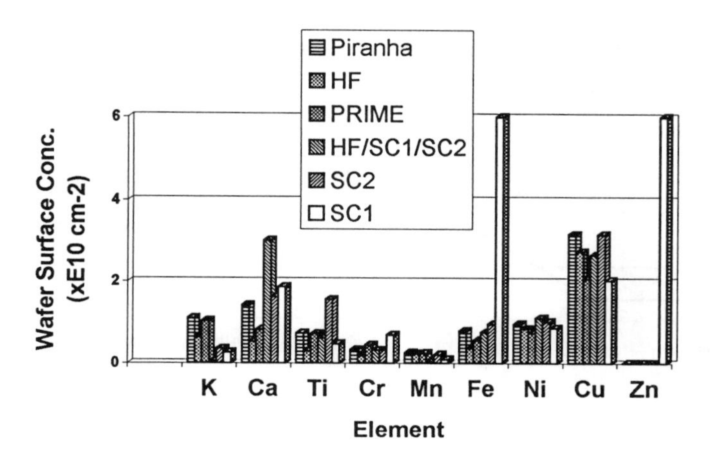

Figure 2. Effect of wafer cleaning process on TXRF-measured surface metallic contamination levels.

The results of Fig. 4 demonstrate that the surface charge can be used to monitor the state of the native oxide regrowth on HF-treated surfaces. All wafers shown in the figure were processed in dilute 50:1 H_2O: HF for 2 minutes at 25C followed by a DI rinse and IPA dry process. A characteristic feature of all the wafers is a delay time in the onset of surface charge increase, and an asymptotic approach to a saturation value. This type of behavior has also been demonstrated with ellipsometric data of native oxide regrowth following HF processes[3]. The similarities between the surface charge and ellipsometry curves suggest that the same phenomena is being measured. Further substantiation of this is given by the wafer of Run #6 in Fig. 4. This wafer was parked for over 3 hours in an inert nitrogen ambient immediately following the HF/rinse/dry process, and then later exposed to the clean room oxidizing ambient while surface charge was measured over the next 15 hours. The increase in surface charge is both delayed in time and suppressed in magnitude in comparison to the other wafers which were immediately exposed to the oxidizing ambient. This indicates a higher degree of surface passivation was achieved in the inert ambient which inhibited the subsequent native oxide regrowth. This suggests that there may be some value to storing wafers in inert ambients prior to processes which are highly sensitive to the presence of a native oxide (e.g. for silicide formation or metal contact formation).

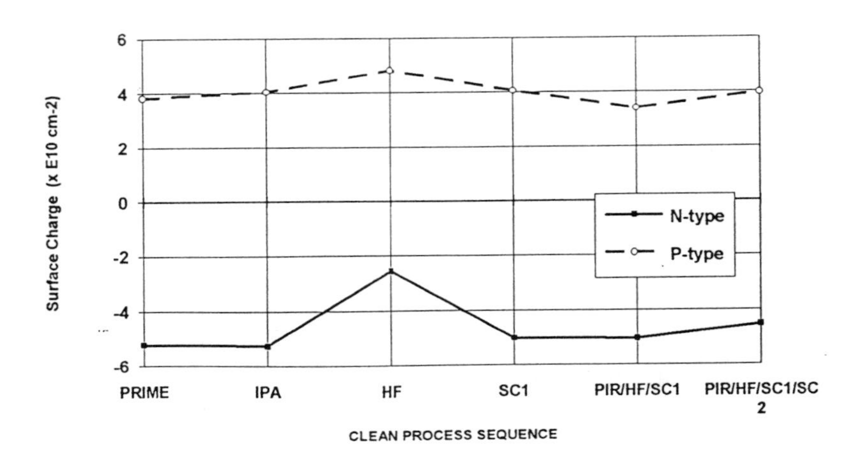

Figure 3. Effect of wafer cleaning process on SPV-measured surface charge for n-type and p-type wafers.

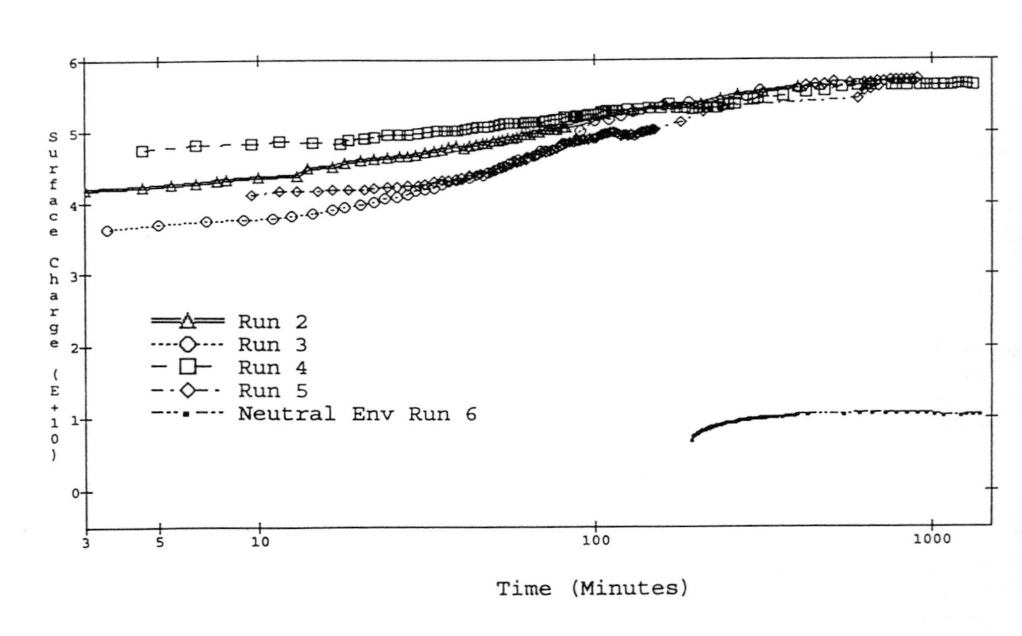

Figure 4. SPV-measured surface charge vs. cleanroom exposure time following an HF oxide etch process.

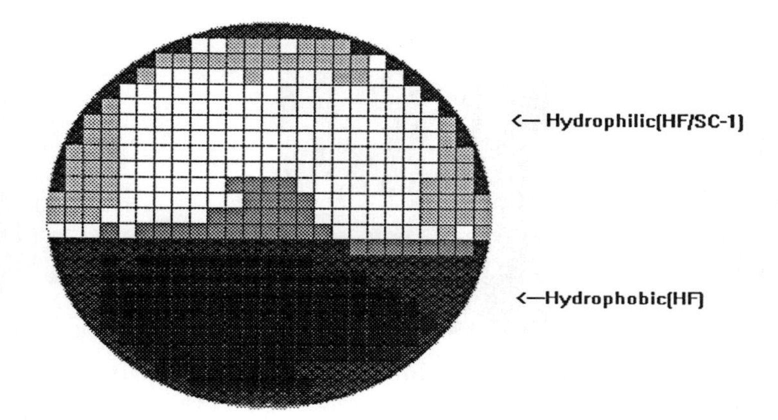

Figure 5. SPV-measured wafer surface charge distribution for a wafer which received an HF etch (entire wafer) and subsequent SC-1 dip (upper half only).

The surface charge wafer map shown in Fig. 5 further validates the role of surface charge measurements in detecting differences in Si surface oxides. This wafer was processed for 2 minutes in a 10:1 H_2O: HF process bath followed by a rinse and dry procedure. At a later time the upper half of the wafer was then dipped in a SC-1 alkaline bath. This resulted in the formation of a hydrophilic chemical oxide in the upper half region and a bare Si hydrophobic surface in the lower half region. The delineation is clearly visible in the surface charge map shown. Also evident from the figure is that the surface charge values measured in the 10:1 HF region are higher than those of the 50:1 HF processed wafers shown in Fig. 4. The difference may be the result of differing surface passivation properties with HF mixture strength as indicated by the FTIR-ATR studies of Hirose[9]. It is important in interpreting surface charge data to recognize the importance of the previous wafer history in influencing surface charge values. Surface charge measurements should always be of a comparative nature between wafers of the same vendor type, and similar processing histories.

Another application for surface charge monitoring is illustrated in Fig. 6, which shows a surface charge map for a wafer following a piranha(H_2SO_4:H_2O_2) clean/rinse/dry. Due to the viscous nature of the piranha solution, it is notoriously difficult to rinse the sulfate residues from the wafer surface with a simple DI rinse[10]. The end result is typically a sulfate residue on the wafer surface after the rinse/dry process. Upon exposure to the cleanroom ambient, this residue often reacts to produce large crystallites which are thought to be an ammonia-sulfate compound[8, 10]. These crystallites can represent a major source of particulate contamination. Recent studies have shown that this residue can be effectively removed if the piranha clean is followed by a rinse in hot DI water[10].Figure 6 shows the surface charge and wafer surface particle scan maps vs. exposure time following the rinse/dry process for a wafer which was processed in a piranha solution followed by a 6-cycle quick dump rinse with cold DI water. The bottom half of the wafer received an additional rinse in hot DI water, which has been shown to be effective at removing the sulfate residue[10]. Both the particle and surface charge maps show the change in wafer surface state with time as the crystallites form, and the efficacy of the hot DI rinse treatment.

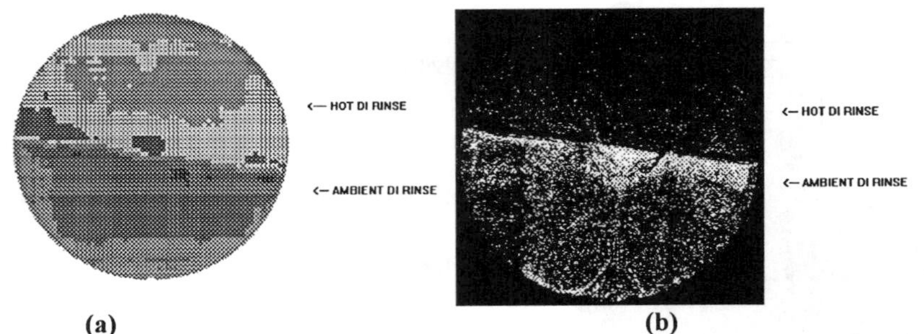

Figure 6. a) SPV-measured wafer surface charge distribution, and b) surface particle scan for a wafer which received a piranha clean followed by a cold DI rinse (entire wafer) and subsequent hot DI dip (upper half only). Note: wafer scans were performed after 6 days of exposure to cleanroom environment.

CONCLUSIONS

Surface PhotoVoltage measurements have been shown to be an effective indicator of the Si surface state and contamination level following wet chemical prediffusion cleaning processes. In particular the surface charge is shown to accurately monitor the native oxide regrowth following an HF etch, and also can be used to monitor sulfate residues following a piranha clean.

REFERENCES

1) P. Edelman, J. Lagowski, L. Jastrzebski, *1992 Proceedings of the MRS*, San Francisco, CA, April 1992.
2) J. Lagowski et al.,*Semicond Sci Technol*, 7, p. A185 (1992).
3) T. Ohmi et al.,*J.Electrochem Soc.* 139,11p.3317(1992).
4) S.M. Sze, *Physics of Semicond Dev*, chap7-8, J.Wiley & Sons (1981).
5) J. P. Uyemura, *Fund of MOS Digital Int. Cir.*, Addison-Wesley(1988).
6) Y.P. Tsividis, *Oper & Model of the MOS Transistor*, McGraw-Hill (1987).
7) J.J. Rosato et al, *1994 Proc of Electrochem Soc., 186th Fall Mtg, Diagn Tech for Semi Mtls & Dev*, Miami, FLA(Oct. 1994)
8) H. Shimizu et al., *1994 Proc of Inst of Env. Sci*, p.297 (1994).
9)M. Hirose et al., *Sol St. Technol.* p.43, (Dec. 1991).
10)R.M. Hall et al, *Proc.1994 SPWCC*, San Jose, CA (1994).
11)R.M. Hall et al., *1994 Proc of Electrochem Soc., 186th Fall Mtg*, Miami, (1994)

DETERMINATION OF RINSING PARAMETERS USING A WAFER GAP CONDUCTIVITY CELL IN WET CLEANING TOOLS

P. G. LINDQUIST, R. N. WALTERS, J. O. THRONGARD AND J. J. ROSATO
Santa Clara Plastics, Ultra Clean Research and Development, 400 Benjamin Lane, Boise, ID 83704

ABSTRACT

A wafer gap conductivity cell determined which rinsing parameters control the removal of wafer cleaning and etching solutions. The first part of this study focused on set up and calibration of the conductivity cell. Sodium chloride solutions with known ionic conductivities are used as model fluids. Wafer rinsing experiments showing the concentration as a function of time are analyzed. The sensitivity of the wafer gap conductivity cell is compared to a wall mounted probe, the typical method used to measure rinsing efficiency. The conductivity results are explained using a fluid dynamics model of the wafer gap.

The second part of this study focused on using the wafer gap conductivity cell to study the removal of chemicals typically used in wafer cleaning processes. Experimental results show the effect of tank geometry and flow parameters on the time required for rinsing. Analysis shows the rinse process can be modeled by assuming convective fluid flow and ideal mixing. These results provide critical insight into the most important wafer rinsing parameters.

INTRODUCTION

Improvements in rinse tank design and processes are driven by the need to reduce rinse water consumption and increase throughput as the number of process steps and wafer size increase. Methods employed in current products to improve rinse efficiency include minimizing the tank volume, reducing the wetted surface area by use of a reduced cassette, utilizing dilute chemistries, and by increasing the temperature of the rinse water. [1,2,3] Rinse tanks can be optimized through the use of computational fluid dynamics, followed by experimental verification.[4] To date, most data published to characterize and model the rinse process are based on the use of wall mounted conductivity probes.[1,2,5] There are many drawbacks to the wall mounted probe: 1) The probe is not a direct measurement of the ion concentration between the wafers; 2) Probe performance is affected by changes in tank configuration and flow field; 3) The probe response to different carry over chemistries do not correlate quantitatively to the amount of carry over on the wafer during rinsing.

This work is motivated by the need to develop an instrument that measures chemical concentrations in the wafer gap. Despite the drawbacks of conductivity probes, this technique is appealing because the conductivity of a well mixed solution is proportional to the ion concentration[6]. Recognizing that the ideal conductivity cell is two flat electrodes separated by a fixed gap, a conductivity cell fabricated from a pair of wafers is logical, see Figure 1. The cell used in the reduced cassette quantifies the rinsing behavior of wafer surfaces when the carry over chemistry is varied and it is a useful tool to examine what the effect of changing tank geometry and fluid flow has on rinse efficiency.

Mat. Res. Soc. Symp. Proc. Vol. 386 © 1995 Materials Research Society

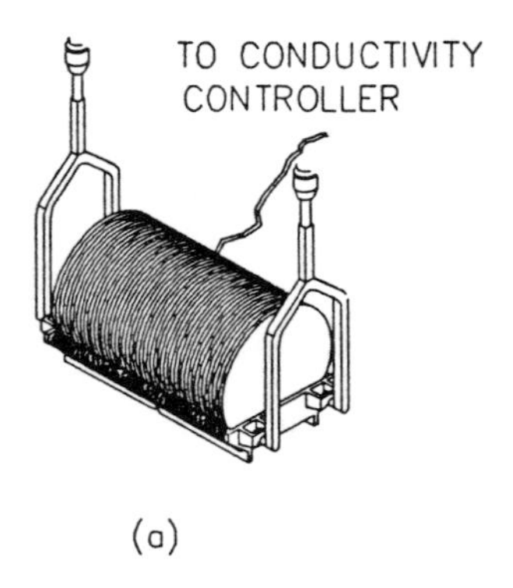

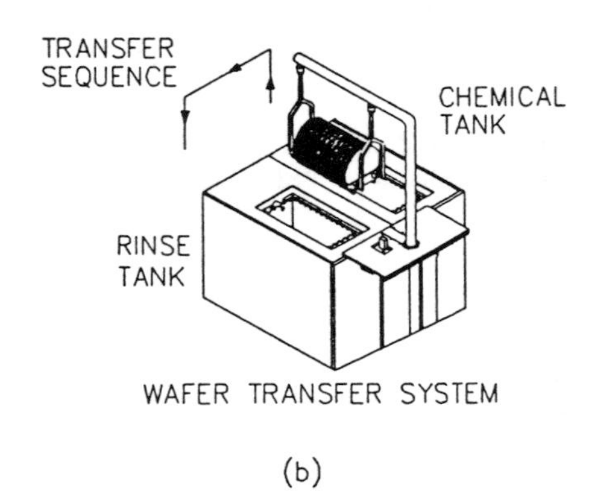

Figure 1 The wafer gap cell replaces two wafers in the reduced cassette (a). For the rinsing experiments a full cassette of wafers with the wafer gap conductivity cell are transferred from the chemical tank to the rinse tank (b).

EXPERIMENTAL

The apparatus used for these experiments consists of a wafer gap conductivity cell, a reduced cassette for fifty 200 mm wafers spaced at 6.35 mm wafer pitch, a tank to recirculate process chemicals to be rinsed, a rinse tank flow meter, and a rinse tank. The flow field in the rinse tank was varied by altering the geometry of the fluid inlets, configurations A, B, C, and D. Wafers were manually transferred between tanks for rinsing experiments with NaCl. These experiments were run on components used in the Santa Clara Plastics model SPS 9400 equipment. Rinsing experiments with sulfuric acid, HF, SC-1 and SC-2 process chemistries were run at room temperature on an SPS 9400 with automated transfers between tanks. This equipment is installed in a class 1 mini-environment at the Santa Clara Plastics Applications Laboratory.

Electrodes for the wafer gap conductivity cell were fabricated from two 200 mm silicon wafers coated with 2,000 nm chromium and 20,000 nm gold deposited by metal evaporation. Spacers fabricated from PTFE were used to keep the electrode surface of the conductivity cell parallel and fix the gap at 5.7 mm, the same pitch as wafers in the reduced cassette. The cell constant for this electrode configuration was calculated as 0.0018 cm^{-1} and determined experimentally as 0.0017 cm^{-1}. For each rinse test, the cell replaced 2 wafers of the 50 needed for a fully loaded reduced cassette. Although, the conductivity cell can be interchanged with any two wafers in the cassette, the cell was positioned in slots 25 and 26 for most of the tests. A 2 channel AC conductivity meter was used to monitor the conductivity as a function of time. Channel one was used for the wafer gap conductivity cell and channel two was used for a wall mounted conductivity probe located at the tank weir.

Shortly after the conductivity cell was fabricated and tested, it was evident that this technique would be ideal for experimentally verifying new rinse tank designs and investigating the rinsing of RCA chemistries. A typical rinse experiment is described as follows: 1) Soak a reduced cassette filled with dummy wafers and the conductivity cell in the test chemistry (NaCl, H_2SO_4, HF, SC-1 or SC-2); 2)Transfer the wafers and cassette to the rinse tank (configuration A, B, C, or D); 3) Measure the conductivity as a function of time; 4) Repeat steps 1 thru 3 for each test parameter (chemistry, configuration, and flow). The test matrix is shown in Table I.

Experiment	System Set Up	Parameter	Setting
1	Fixed Rinse Tank Design	NaCl[ppm]	100, 500, 1000
		Flow	Low, High
2	Fixed NaCl [1000 ppm]	Rinse Tank	A, B, C, D
		Flow	Low, High
3	Fixed Rinse Tank	Chemistry	H_2SO_4, HF, SC-1,SC-2
		Flow	High

Table I Experimental Test Matrix

BACKGROUND

In this section, we define the rinse set up and how it is modeled, evaluate what rinsing parameters are needed to define a rinse efficiency metric.

The primary mass transport mechanisms that describe rinsing chemical residues from an array of wafers are: 1) The out diffusion from the wafer surface into the wafer gap; and 2) The convective flow from the wafer gap out of the tank. The focus of this work was to investigate convective flow through the wafer gap. Analyzing the configuration and flow field of a rinse tank with wafers, the governing convection-diffusion equation in the wafer gap is:

$$\frac{\partial c}{\partial t}+\upsilon_x\frac{\partial c}{\partial x}+\upsilon_z\frac{\partial c}{\partial z}=D\left(\frac{\partial^2 c}{\partial x^2}+\frac{\partial^2 c}{\partial z^2}\right) \tag{1}$$

Little can be done to improve the diffusion of the solute into the solvent. For RCA chemistries the diffusion coefficients (D) are very close when compared, see Cussler[7]. Thus, neglecting diffusion in both directions and convection in the transverse direction (x), the convection-diffusion equation reduces to:

$$\frac{\partial c}{\partial t}+\upsilon_z\frac{\partial c}{\partial z}=0 \tag{2}$$

The solution of this equation has the form

$$c(t)=c_o e^{-t/\tau} \tag{3}$$

where c(t) is the concentration of ions in the wafer gap as a function of time, t; c_o is the concentration of ions in the cell volume at t = 0; and τ is a characteristic time constant for rinsing.

Additional rinsing parameters can be computed by analyzing the rinse curves. These are the average velocity flow through the wafer gap, dimensional groups (k, the mass transfer coefficient) and non-dimensional groups (Reynolds, Peclet, and Sherwood number). To compute these numbers, first compute the velocity in the wafer gap. The flow through the gap is computed by integrating the time rate of change of concentration and multiplied by the cell volume, the result is the mass average velocity through the cell. Since the fluid system is assumed to be at constant density, this is proportional to the volume average velocity through the gap. The average wafer gap velocity is then calculated.

$$v=\frac{1}{t_f}\sum_{i=0}^{t_f}\frac{\Delta c_i}{c_i}\frac{V_{cell}}{ld} \quad (4)$$

Combining the velocity with the other dimensional quantities, a set of non-dimensional parameters can then be determined, see Table III.

RESULTS

The results of experiment 1 are shown in Figure 2 and Table II. There is a good correlation between the concentration measured by the wafer gap conductivity cell and the results predicted by equation 3. It is possible to calculate the carry over thickness, from C_o and the C_{co} (the carry over concentration of NaCl). Assuming that: $C_o*V_{cell} = C_{co}*V_{carry\ over}$ and $V_{carry\ over}$; and the carry over thickness is independent of carry over concentration. From this analysis the carry over thickness is computed as 19 μm. The these values agree with those predicted by Rosato et. al. [1] and have been confirmed with tests run to determine the bulk carry over volume for dilute chemistries per cassette transfer for a reduced cassette.

C_{co} [ppm]	Co [ppm]	τ [sec]
100	0.0447	39.5
500	0.8470	36.0
1000	3.0246	34.7

Table II Constants determined by fitting the data into equation (3) and solving for C_o and τ.

The rinsing behavior of tanks A, B, C, and D were evaluated in the second experiment, see Figure 3. Rinse parameters C_o, τ, the wafer gap velocity, Pe, Re, and Sh numbers were computed and they are shown in Table IV. Inspection of the dimensionless groups computed for this data reveals the convective flow through the wafer gap dominates mass transfer correlation compared to diffusion. Note: τ is reduced by approximately 1/2 when the flow is doubled (high flow = 2 X low flow). From Figure 3 it is evident that tank D does not rinse down as fast as the other tank configurations. Using the experimentally determined τ for each configuration, it is concluded that rinse efficiency of configurations A and C are comparable. On a percentage basis, the τ for tanks A and C are 47 % better than tank D. When the wafer gap velocity for tanks A and C are compared to D there is a 29% difference between the tanks. Additional experiments are planned to investigate the correlation between dimensional groups and the rinse system.

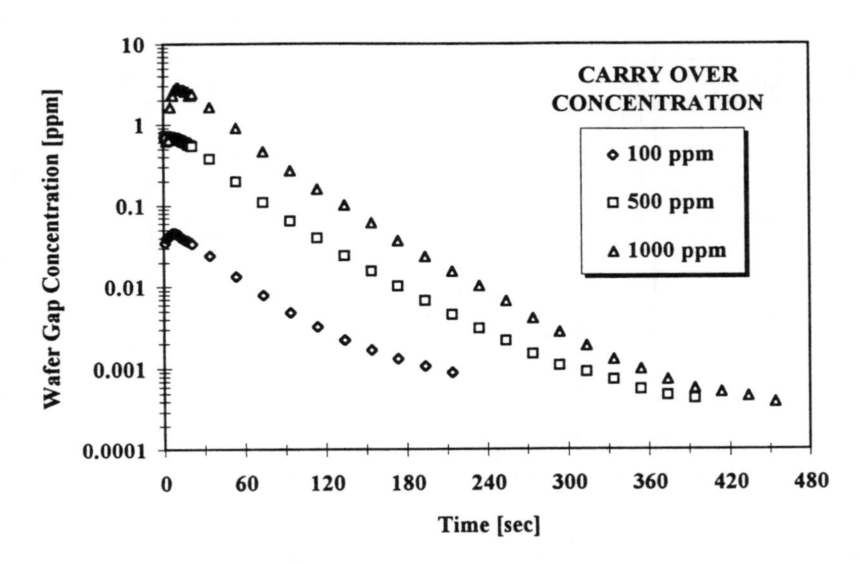

Figure 2 Wafer gap concentration for 100, 500, and 1000 ppm NaCl rinsing experiments.

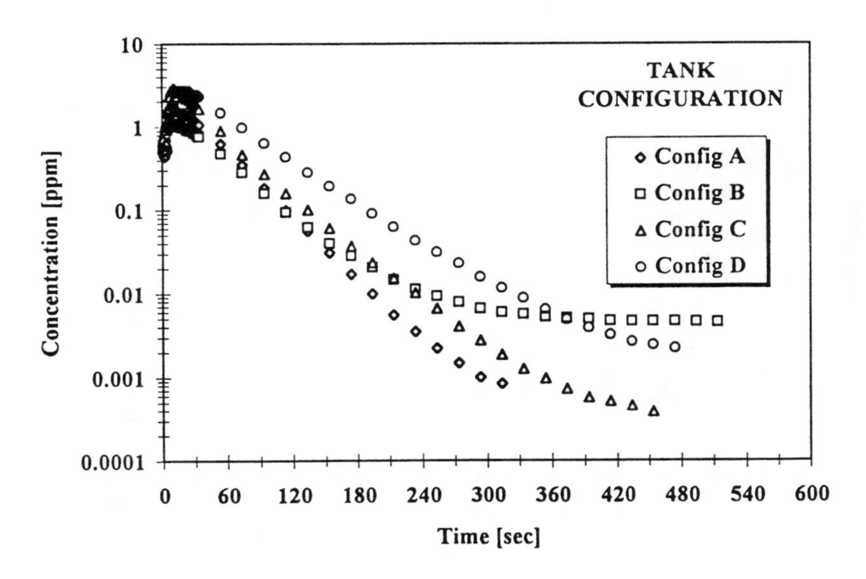

Figure 3 Rinse curves for tank configurations A, B, C, and D

	Low Flow				High Flow			
Tank Configuration	A	B	C	D	A	B	C	D
C_o, ppm	2.62	3.95	3.85	3.46	2.78	1.75	4.16	4.28
τ, sec	63.5	59.5	70.7	103.5	34.8	39.5	34.7	50.1
wafer gap velocity, cm/s	0.20	0.22	0.20	0.09	0.39	0.33	0.39	0.3
Peclet Number, $P_e \; \frac{vl}{D}$	6978	7880	6981	3169	13993	11686	13785	10820
Reynolds Number, $R_e \; \frac{vl}{\nu}$	14	15	14	6	27	23	27	21
Sherwood Number, $S_h \; \frac{kl}{D}$	2211	2436	2211	1176	3857	3339	3811	3140
k, cm/s	0.061	0.068	0.061	0.032	0.10	0.09	0.11	0.09

Table III Rinsing parameters computed for tank configurations A, B, C, and D. These parameters were computed from equations (3) and (4). For Pe, Re, and Sh: l = wafer gap [cm]; D = Diffusion constant [cm^2/s]; ν = kinematic viscosity[cm^2/s]; v = velocity [cm/s]; k = mass transfer coefficient [cm/s].

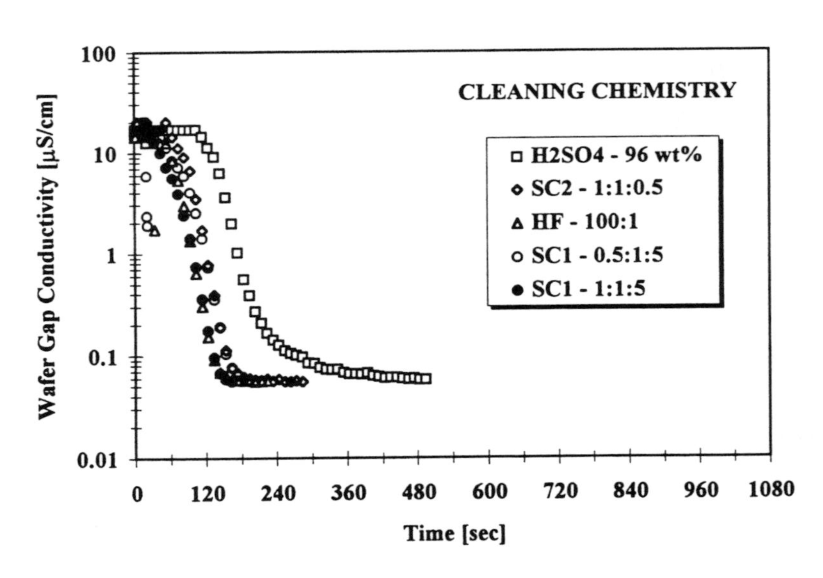

Figure 4 Rinse curves for RCA chemistries, Sulfuric acid, HF, SC-1 and SC-2

Rinse curves for wafers soaked in RCA chemistries are shown in Figure 4. Each conductivity curve is characterized by a plateau region (I), a region (II) that can be modeled by equation (3) anda transition region (III) going to zero concentration. The duration of the plateau region (I) ranges is about 80 seconds for HF, SC1 and SC2 chemistries and 260 seconds for sulfuric acid. Additional experiments need to be run to determine the significance of the plateau region (I). It is speculated that the time required to transition from the plateau region (I) to region (II) is proportional to the concentration of the carry over chemistry and the thickness of the carry over layer. From this data it is apparent that the duration of the plateau region for sulfuric acid is a significant portion of the time required to rinse. Using the wafer gap conductivity cell we find the plateau duration is reduced by 100 seconds by quick dump rinsing[8]. Rinse times can be further reduced by minimizing the quantity of sulfuric acid dragged into the tank by careful design of the robot transfer between tanks. The transition from region (I) to (II) and (II) to (III) depends on the interaction between the process chemistry and the wafer surface. For (I) and (II) the transition is not as sharp for weak bases (SC-1) compared to weak acids (HF) or strong acids (HCl or H_2SO_4). For transitions between (II) and (III), chemistries with strong acids (H_2SO_4) are not as sharp as weak acids or bases. Further experiments are needed to understand the role that chemical bonding has on the rinse process.

CONCLUSIONS

Few instruments can quantify the performance of rinse tank designs or rinsing processes. Based on the results of this study, it is concluded that the wafer gap conductivity cell is a good tool for evaluating new tank designs, processes, and determine rinsing performance parameters. A wafer gap cell is useful because mass transport in the wafer gap can be modeled from first principles, the conductivity cell replaces wafers in a cassette used for batch processing, the output from the conductivity cell can be converted to ionic concentrations in the cell volume, and the conductivity cell can be used to study rinsing of model solutions like NaCl (an ionic electrolyte) or and chemistries in the RCA wafer cleaning process.

REFERENCES

1. J.J. Rosato, R.N. Walters, R.M. Hall, P.G. Lindquist, R.G. Spearow, and C.R. Helms in Proceedings of the 3rd International Symposium on Cleaning Technology in Semiconductor Device Manufacturing, edited by J. Ruzyllo and R. Novack, (Electrochemical Society 184th Meeting, vol 94-7), 1993.
2. R.M. Hall, presented at the Santa Clara Plastics 2nd Symposium on Semiconductor Wafer Cleaning, Boise, ID, 1994.
3. R.M. Hall, J.J. Rosato, P.G. Lindquist, T. Jarvis, J.D. Kelly, and R.N. Walters, presented at the Balaz Symposium on Ultra Pure Water and Chemical Conference, Santa Clara, CA, 1995.
4. S.N. Kempka, J.R. Torcznski, A.S. Geller, J.J. Rosato, R.N. Waters, and S.S. Sibbet, presented at Micro 94, San Jose, CA, pp 225-234, 1994.
5. A. Tonti in Proceedings of the 2nd International Symposium on Cleaning Technology in Semiconductor Device Manufacturing, edited by J. Ruzyllo and R. Novack, (Electrochemical Society Meeting, vol 92-12), pp. 41-47, 1993.
6. R.D. Olen, R.D. Thornton, presented at the 2nd annual Semiconductor Pure Water Conference, San Jose, CA, 1983; Thornton Associates marketing information, Waltham, MA, 1984.
7. E.L. Cussler, Diffusion Mass Transfer in Fluid Systems, (Cambridge University Press, Cambridge, U.K., 1984).
8. P.G. Lindquist, 1995 (unpublished).

Electrochemical Aspects of Etching and Passivation of Silicon in Alkaline Solutions

Joong S. Jeon and Srini Raghavan
Dept. of Materials Science and Engineering
University of Arizona

ABSTRACT

Electrochemical polarization experiments were performed on Si wafers in ammoniacal solutions maintained at a pH in the range of 9.5 to 11.5. Anodic polarization of silicon yielded curves which are typical for materials that undergo passivation. The values of open circuit potential and passivation potential for p-type Si wafers were more anodic than for the n-type Si wafers. Corrosion current density of p-type Si wafers of low resistivity was lower than that of wafers of high resistivity. Corrosion current densities correlated well with surface roughness induced in alkaline solutions. Addition of surfactant or H_2O_2 to alkaline solutions reduced critical current density for passivation and corrosion current density.

I. Introduction

The electrochemical etching of silicon in alkaline solutions is an important method that is used in the fabrication of electronic microstructural devices such as microsensors, tranducers and actuators. For the electrochemical etching of silicon, hydrofluoric acid [1] and alkaline solutions such as aqueous potassium hydroxide [2], aqueous sodium hydroxide [3], ethylene diamine pyrocatechol [4] and aqueous hydrazine [5] have been used. Substrate characteristics that influence the etch rate and anisotropy have been widely studied. It has been established that highly doped silicon ($> 10^{20}cm^{-3}$) acts as an etch-stop and silicon (111) plane etches orders of magnitude slower than the (100) or (110) crystal plane.

Alkaline solutions based on NH_4OH and QAH (quaternary ammonium hydroxides) such as choline and TMAH (trimethyl-2-hydroxyethyl ammonium hydroxide) are used in wafer cleaning. For example, classical SC1 solutions containing NH_4OH, H_2O_2 and H_2O at a ratio of 1:1:5 are used to reduce particulate contaminants on wafers. QAH/H_2O_2 or QAH/surfactant solutions have been considered as suitable alternatives to SC1 solutions. Even though it well known that H_2O_2 and surfactant reduce surface roughness induced in alkaline solutions, fundamental information on their function in reducing etching rate is not available.

In this research, the dissolution (etching) of silicon in ammoniacal solutions and the effect of H_2O_2 or surfactant on dissolution have been investigated using an electrochemical polarization technique. Specifically, the corrosion rate of silicon has been measured by carrying out Tafel polarization experiments.

II. Experimental Materials and Methods

Materials: For the electrochemical experiments, as shown in Table 1, n(100) and p(100) silicon wafers with resistivities in the range of 0.01 to 60 Ω-cm were used. Prior to their use, these samples were cleaned in piranha followed by HF and then rinsed in DI water. Octylphenol polyethyleneoxide (OPEO) surfactant represented by the chemical structure, C_8H_{17}-C_6H_4-

Mat. Res. Soc. Symp. Proc. Vol. 386 © 1995 Materials Research Society

$(OCH_2CH_2)_{9.5}$-OH, was used to investigate the effect of surfactant on the electrochemical behavior of silicon.

Experimental Methods: For the electrochemical investigations of silicon in alkaline solutions, the pH value of solutions was adjusted to 9.5, 10.5 or 11.5 with NH_4OH. Potentiodynamic polarization experiments were performed using an EG & G Potentiostat/Galvanostat (Model 273A) at a scan rate of 0.5 mV/sec. Silicon samples (1 cm^2) were exposed to solutions using a flat cell (EG & G, K0235), and the potential of silicon was measured with respect to a standard calomel electrode (SCE). A flat sheet of Pt/Nb was used as a counter electrode. Digital Nanoscope III was used to measure the root-mean-square surface roughness (R_{rms}) of silicon samples conditioned in alkaline solutions.

Table 1. Type and resitivity of silicon wafers used for experiments.

Type	Orientation	Resitivity (Ω-cm)	Symbol
p	(100)	0.01-0.02	TL
		0.15-0.30	SL
		5-10	TM
		30-60	SH
n		0.02-0.04	NL
		2-5	NH

III. Results and Disscusion

The anodic polarization curves for p- and n-type silicon wafers of different resistivities are plotted in Fig. 1. These polarization curves were obtained in a solution of pH=10.5 adjusted with NH_4OH. Similar to the results reported in the literatures for KOH solutions [6], anodic polarization curves for the p-type exhibited a much sharper passivation behavior than n-type silicon. The critical current density for passivation (i_{crit}) for p-type seemed to be dependent on the doping level. It may be seen from Fig. 2 that the open circuit potentials for p-type wafer became more negative with a decrease in doping level (higher surface resistivity), and the critical current density for passivation decreased with an increase in doping level. In contrast, open circuit potential (OCP) as well as critical current density for n-type did not depend much on doping level. The anodic and cathodic reactions that establish the OCP may be represented as follows [5]:

(at anode) $Si + 4OH^- \rightarrow Si(OH)_4 + 4e^-$

(at cathode) $4H_2O + 4e^- \rightarrow 4OH^- + 2H_2$

Most of the literature reports on silicon etching in alkaline solutions have focused on large values (i.e., 2 V) of anodic and cathodic bias. The main purpose of this work was to evaluate the corrosion current density of silicon samples of different resistivities through the construction of Tafel plots. Hence, anodic and cathodic polarization were typically limited to about 0.25V on either side of OCP. The Tafel plots for p-type samples of different resistivities are shown in Fig. 2. Because of restricted Tafel region in the anodic polarization range, cathodic polarization curves were used to calculate the corrosion current density (i_{corr}). This analysis was done by extrapolating cathodic Tafel region to OCP. Ideally, at least one decade of linearity in the E vs. log *i* plot is desirable in the Tafel region for the calculation of corrosion rate. Unfortunately, the Tafel region in the cathodic polarization curves was limited to half a decade in most cases. Using this procedure, it was found that corrosion current density of heavily doped Si sample (0.01-0.02 Ω-cm) was roughly 1.6×10^{-6} A/cm^2. Lightly doped samples (above 5 Ω-cm) were characterized by i_{corr} of approximately 3.1×10^{-6} A/cm^2.

AFM micrographs of two silicon samples of different resistivities conditioned in an ammoniacal solution at a pH of 10.5 are shown in Fig. 3. Roughness values (R_{rms}) of 2.1, 5.7 and 6.2 nm were measured for TL, TM and SH samples. A comparison of corrosion current density with surface roughness value clearly indicates that a high corrosion current density resulted in a higher surface roughness.

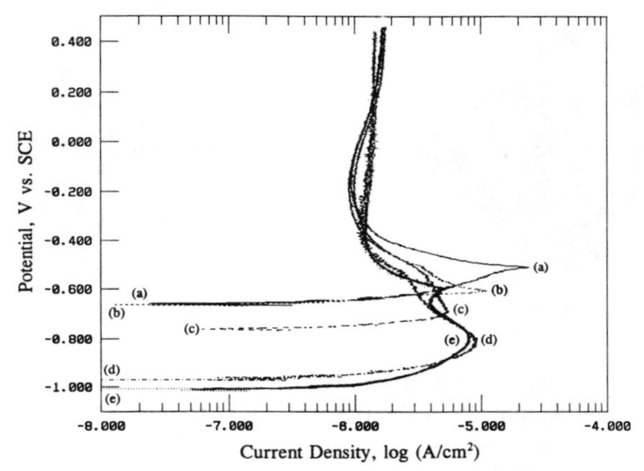

Fig. 1, Anodic potentiodynamic polarization curves for silicon wafers in an ammonical solution at a pH of 10.5. [(a) TM, (b) SL, (c) TL, (d) NL and (e) NH]

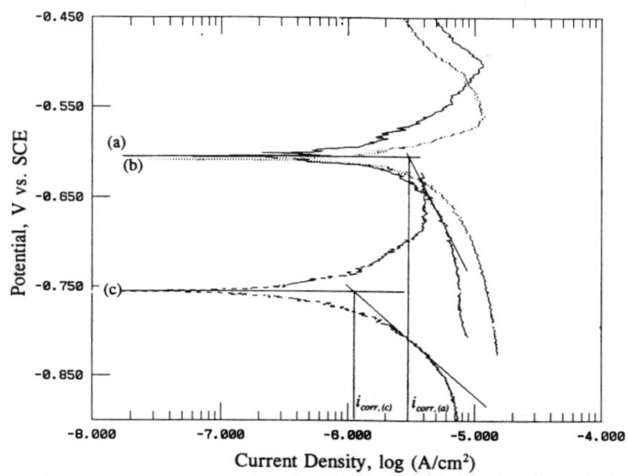

Fig. 2, Potentiodynamic polarization curves for silicon samples [(a) SH, (b) TM and (c) TL] of different resistivities in ammoniacal solutions at a pH of 10.5.

Fig. 4 shows the effect of pH on anodic and cathodic polarization curves for TM silicon samples. The values of i_{crit} increased with an increase in pH. The OCP for silicon became more negative with an increase in pH. Since hydrogen peroxide is an ingredient in most NH_4OH

based cleaning solutions used for Si wafer cleaning, the effect of H_2O_2 addition to NH_4OH based solutions was also investigated using an electrochemical technique. For a solution containing 5 part of H_2O and 1 part of NH_4OH, i_{corr} was calculated to be approximately 5.4×10^{-6} A/cm^2 from the data in Fig. 5. Upon addition of H_2O_2 to this solution at a ratio of a standard SC1 solution, marked changes were observed in anodic and cathodic polarization behaviors, and the value of i_{corr} was reduced to 0.5×10^{-6} A/cm^2. The addition of H_2O_2 to NH_4OH-H_2O solution allowed the passivation of Si to occur at lower current densities and at lower ovetrpotential, and the cathodic polarization curve showed a limiting current density typical for O_2 reduction.

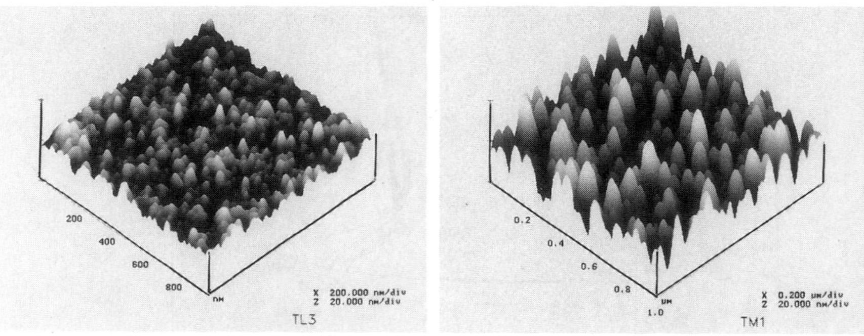

Fig. 3, AFM micrographs for silicon samples of different resistivities conditioned in ammoniacal solutions (pH=10.5) for 10 min [(left) TL and (right) TM].

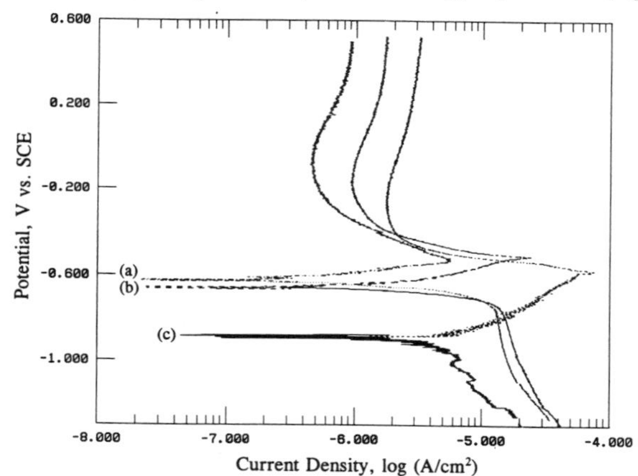

Fig. 4, Effect of pH value of ammoniacal solution on potentiodynamic polarization curves of silicon samples (TM). [pH: (a) 9.5, (b) 10.5 and (c) 11.5]

In order to characterize the role of surfactant in alkaline solutions, the anodic and cathodic polarization curves were obtained in NH_4OH solutions at a pH of 10.5 in the presence of different levels of a nonionic surfactant (OPEO). The results are shown in Fig. 6. It may be seen that the addition of surfactant did not affect the slope of anodic or cathodic polarization curves. The Tafel regions in the cathodic polarization curves were slightly shifted due to a shift in OCP. The most significant feature observed in this figure is the reduction in the values of

i_{crit} as the surfactant level increased. The calculated values of i_{corr} decreased from 3.1×10^{-6} A/cm^2 in the absence of surfactant to 0.3×10^{-6} A/cm^2 in the surfactant (200 ppm) containing solution. The AFM micrographs given in Fig. 7 show a dramatic reduction in surface roughness on the addition of surfactant. The reduction in surface roughness is due to reduction in etch rate of Si in the presence of added surfactant [7].

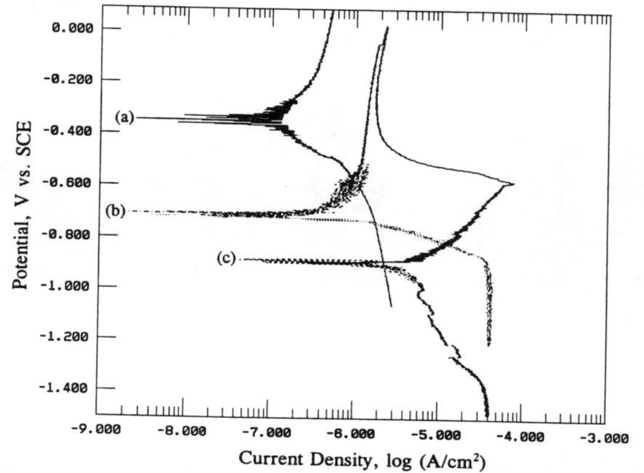

Fig. 5, Potentiodynamic polarization curves for silicon samples (TM) in (a) 5:1 H_2O:H_2O_2 (pH$\approx$4.2), (b) 5:1:1 H_2O:H_2O_2:NH_4OH (pH$\approx$11.5) and (c) 5:1 H_2O:NH_4OH (pH$\approx$10.3) solutions.

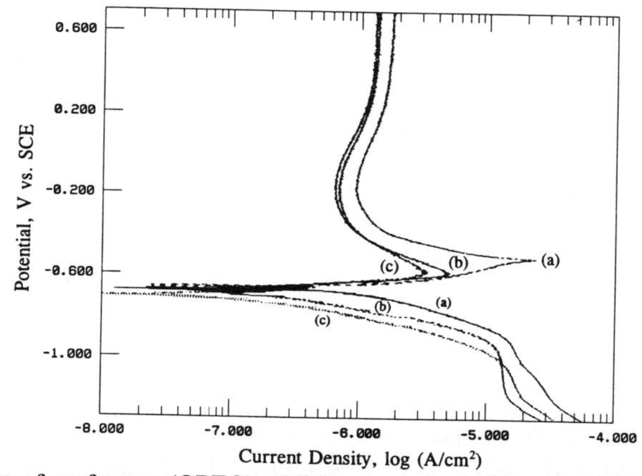

Fig. 6, Effect of surfactant (OPEO) addition on potentiodynamic polarization curves of silicon samples in ammoniacal solutions (pH=10.5). [surfactant concentration: (a) 0 ppm, (b) 10 ppm and (c) 200 ppm]

At this stage, an attempt will be made to explain the role of surfactant in reducing the etch rate of silicon. The anodic reactions responsible for etching of silicon may be proposed as follows [8]:

$$\text{-Si-}\dot{\text{Si}}(OH)_2 + H_2O \rightarrow \text{-}\dot{\text{Si}}(OH) + \text{H-}\dot{\text{Si}}(OH)_2 \quad (1)$$

$$-Si(OH) + H\text{-}Si(OH)_2 + OH^- \rightarrow -Si(OH) + Si(OH)_3 + e^- + 1/2\ H_2 \quad (2)$$

In reaction (1), $-Si\text{-}Si(OH)_2$ represents the hydroxide formed in the previous etching step. Initially water attacks the Si-Si bond producing Si(OH) and $HSi(OH)_2$ groups. The SiH bond can be further attacked by OH^- resulting in $Si(OH)_3$, electron and 1/2 H_2 as shown in reaction (2). In other words, the SiH group is active and continuously attacked by OH^- group electrochemically. The improved passivation due to surfactant addition may be due to the preferential adsorption of surfactant on the hydrophobic Si-H sites.

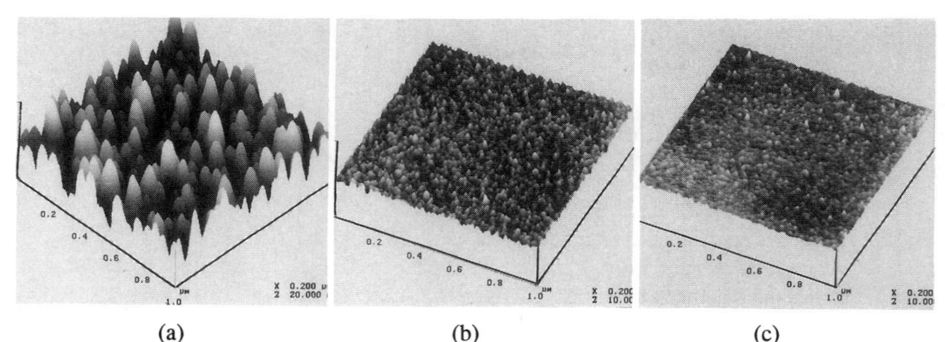

(a) (b) (c)

Fig. 7, AFM micrographs showing the effect of surfactant addition on surface roughness induced in ammoniacal solutions (pH=10.5 and immersion time: 10 min). [surfactant concentration: (a) 0 ppm, (b) 10 ppm and (c) 200 ppm]

IV. Conclusions

Silicon passivated in ammoniacal solutions at high anodic overvoltages (0.3V from OCP). Addition of H_2O_2 or a nonionic surfactant to ammoniacal solutions reduced the critical current density for passivation. Surface roughness induced in alkaline solutions was dependent on the surface resistivity of silicon wafers, and surfactant addition reduced surface roughness.

Acknowledgment: The authors wish to acknowledge the Center of Microcontamination Control at the University of Arizona for the financial support to carry out this work.

REFERENCES

[1] X. G. Zhang, S. D. Collins, and R. L. Smith, J. Electrochem. Soc., **136**, 1561 (1989).

[2] R. L. Smith, B. Kloeck, and S. D. Collins, J. Electrochem. Soc., **135**, 2001 (1988).

[3] P. Allongue, V. Costa-Kieling, and H. Gerischer, J. Electrochem. Soc., **140**, 1009 (1993).

[4] H. Siedel, L. Csepregi, A. Heuberger, and H. Baumgartel, J. Electrochem. Soc., **137**, 3612 (1990).

[5] K. B. Sundaram and H. Chang, J. Electrochem. Soc., **140**, 1592 (1993).

[6] L. Chen, M. Chen, C. Lien, and C. Wan, J. Electrochem. Soc., **142**, 170 (1995).

[7] J. S. Jeon, S. Raghavan, and R. P. Sperline, J. Electrochem. Soc., **142**, 621 (1995).

[8] E. D. Palik, O. J. Glembocki, and I. Heard, Jr., J. Electrochem. Soc., **134**, 404 (1987).

REMOVAL OF METALLIC CONTAMINANTS AND NATIVE OXIDE FROM SILICON WAFER SURFACE BY PURE WATER CONTAINING A LITTLE DISSOLVED OXYGEN

Yuka HAYAMI, Miki T. SUZUKI, Yoshiko OKUI, Hiroki OGAWA, and Shuzo FUJIMURA

Process Development Division C850, FUJITSU LIMITED
1015 Kamikodanaka, Nakahara-ku, Kawasaki, 211, JAPAN

ABSTRACT

Cleaning effects of pure water containing dissolved oxygen of very low concentration (LDO water) to metallic contaminants on silicon wafer surface were confirmed. To maintain the concentration of the dissolved oxygen in water, experiments were performed in a glove box in which ambience was controlled so as to satisfy Henry's law between the water and the ambient gas. In the experiment using intensionally contaminated wafers, residual metal contaminants except copper on Si-surface decreased from 10^{14} atoms/cm^2 to 10^{11} atoms/cm^2 after the 1ppb hot LDO treatment at boiling point. This effect depended on the concentration of dissolve oxygen, treatment temperature, and rinsing time. Contact angle of the wafer surface increased gradually from about 10 [deg] with decrease in the residual metals and jumped up to about 90 [deg]. when the amount of residual metals reached to minimum. Then absorption peak of Si-O bonds in FT-IR-RAS spectra also disappeared. These results therefore show that hot LDO water removed metal contaminants from the wafer surface with etching of the native oxide.

INTRODUCTION

Cleaning processes have been indispensable to improve manufacturing yield, performance, and reliability of ULSI devices. Requirements of cleaning processes are removal of particles, organic and metallic contaminants, and control of surface condition including native oxide removal [1)]. The RCA standard clean [2)], which was determined phenomenologically, has been used conventionally but may not satisfy requirements for future devices. Thus we need to restudy characteristics of cleaning techniques and develop a new cleaning process based on scientific approach with physical and chemical analysis.

Although DI water is the most fundamental agent used for wafer cleaning processes, it has not been investigated in detail in the past. This was because DI-water believed not to influence the chemical condition and morphology of Si-wafer surface. In these days, however, several studies on the DI-water interaction with Si-wafer surface have appeared [3-10)]. In particular, reports concerning to DI pure water containing a little dissolved oxygen (LDO water) were very attractive [9,10)]. Thus we tried to apply the LDO water to cleaning of silicon wafer surface and characterize the process.

EXPERIMENTAL

To clarify the removal effects of metallic contaminants and chemical oxides from the Si wafer surface using LDO water, we measured the adsorption of metals on the Si-wafer surface by AAS (atomic absorption spectrometry) and ICP-MS (Inductively Coupled Plasma Mass Spectrometry) [11)] and surveyed the wafer surface by FT-IR RAS (Fourier Transform Infrared Reflection absorption Spectroscopy) and the contact angle method. ICP-MS we used was SPQ6500 manufactured by

Mat. Res. Soc. Symp. Proc. Vol. 386 © 1995 Materials Research Society

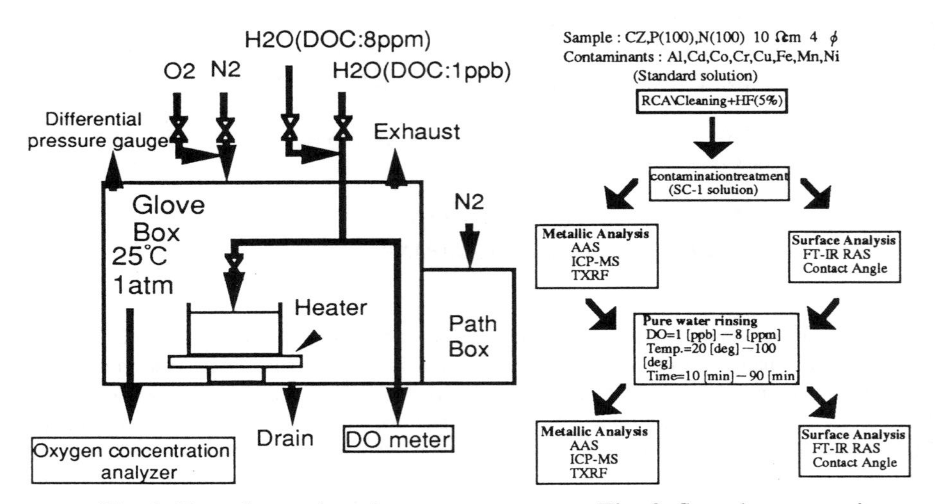

Fig. 1 Experimental environment Fig. 2 Sample preparation

SEIKO Instruments Inc. and AAS was Z-8000 manufactured by HITACHI. FT-IR spectrometer used for this study was JIR-6500 manufactured by JEOL. Moreover, to confirm the condition of metals adsorbed, we observed the depth profile of metals on the Si-wafer surface using X-ray photoelectron spectrometry (XPS) and Total Reflectance X-ray Fluorescence (TRXRF).

All experiments in this study were performed in a close glove box shown in Fig. 1. The LDO water containing a 1ppb dissolve oxygen was supplied by the MHF degassing system developed by Mitsubishi Rayon Engineering Co.,LTD and Nomura Micro Science Co.,LTD. To change the oxygen concentration of the LDO water, we mixed the 1ppb LDO water with conventional pure water containing 8ppm dissolved oxygen. We measured the concentration of dissolved oxygen in water with a DO-meter, model 3500, manufactured by Orbisphere Laboratory Co.,LTD. According to Henry's Law, the dissolved oxygen in pure water is proportional to the partial gas pressure of oxygen in the glove box. Therefore we controlled the oxygen concentration in the glove box ambient to keep the dissolve oxygen concentration in the LDO at a constant value. The oxygen concentration in the glove box environment was monitored with oxygen concentration analyzer LC-700L manufactured by Torey Engineering Co.,LTD and controlled by changing flow rate of nitrogen and oxygen gases flowing into the glove box.

Sample wafers used for this study were 4 inch P and N-type Czochralski(CZ) Si(100) with a resistivity of 10 Ω -cm. Figure 2 shows the experimental procedure. At first, the sample wafers were treated with 5% HF to remove native oxides after the conventional RCA cleaning. Next, we contaminated the sample wafers intentionally by immersing them into SC-1 solution added the standard solutions for AAS containing metals [12]. Metallic contaminants added intentionally were Al, Cd, Co, Cr, Cu, Fe, Mg, Mn, and Ni. After the intentional contamination, the sample wafers were immersed into the hot LDO water. Then we varied rinsing time, treatment temperature and dissolved oxygen concentration.

The amount of metals adsorbed on the wafer surface was estimated by etching the sample surface with a HF/HNO_3 mixture and measuring the metals in the mixture with AAS and ICP-MS. We compared the amounts of metallic contamination and the state of the Si-surface with before and after

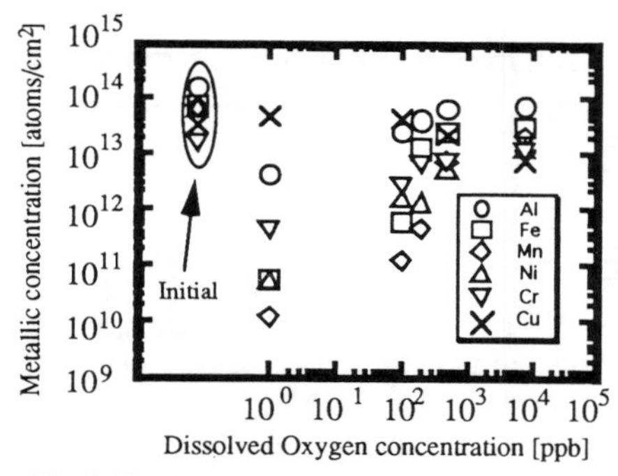

Fig. 3 The dependence of metallic contamination removal on dissolved oxygen in pure water

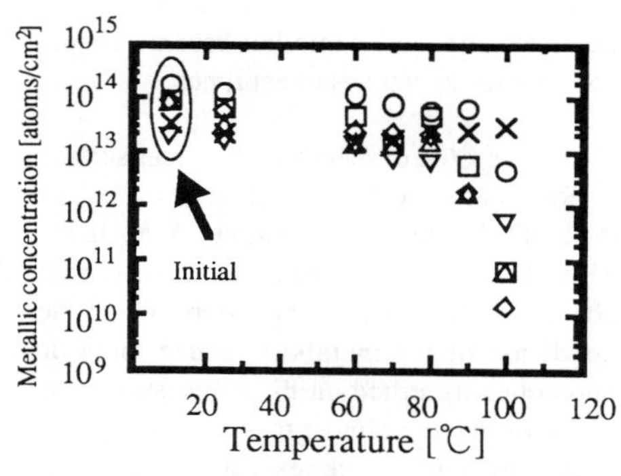

Fig.4 The dependence of metallic contamination removal on rinsing temperature

LDO pure water treatment.

RESULTS AND DISCUSSION

1. The effects of metallic contamination removal

Figure 3 shows the dependence of the amounts of residual metallic contamination on the dissolved oxygen concentration in the LDO water after the hot LDO water treatment. The dissolved oxygen concentration is measured at room temperature. This is because measuring the dissolved oxygen concentration in hot water is impossible. As shown in Figure 3, the amounts of the residual metallic contaminants, except Cu, decreased from about 10^{14} atoms/cm^2 to about 10^{11} atoms/cm^2 after the 1ppb hot LDO water treatment. In the hot water treatment using conventional pure water containing dissolved oxygen of 8 ppm, however, the amounts of residual metallic contaminants decreased a little. Therefore the cleaning effect of the hot LDO water depends the dissolved oxygen concentration obviously.

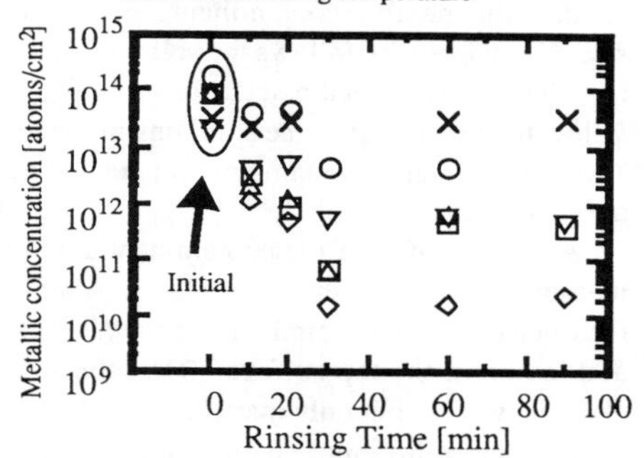

Fig. 5 The dependence of metallic contamination removal on treatment time

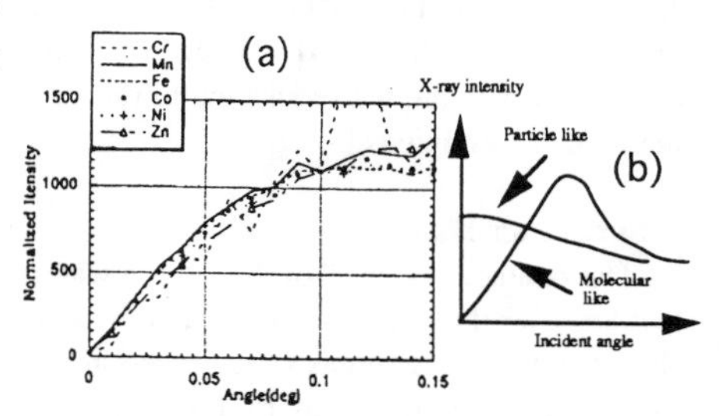

Fig. 6 Depth profile of metals - TXRF -

Figure 4 shows the dependence of the amounts of residual metals on treatment temperature for a 30 min cleaning by the 1 ppb hot LDO water. From fig. 4 Cu is not reduced at all, Cr by a factor of 30, Al by 50, Ni by 10^3, Fe by 10^3 and Mn by almost 10^4. As demonstrated the figure, the amounts of metals remaining on the wafer surface decreased dramatically between 90℃ and boiling-point. Thus, rinsing at the boiling-point is most effective for metal removal.

In general, the cleaning efficiency for a wet process strongly depends on the treatment temperature. Thus examining the temperature dependence of the LDO cleaning is a meaningful approach. Figure 5 shows the results of the experiment. The amounts of metals remaining on the silicon wafer surface continuously decreased with increasing treatment time to 30 min but became almost constant for longer treatment times.

2. Depth profile of metals on Si-wafer surface

Results for the TRXRF observation before the LDO cleaning is shown in Fig. 6 (a). In the TRXRF measurement, the X-ray intensity with change in the incident angle depends on the condition of the metals[13]. If the metallic contaminants exited on the wafer surface as particles, having almost the same size as the chemical oxide thickness, the X-ray intensity follows the dashed line in Fig. 6 (b). On the other hand, if the metallic contaminants are molecular size, X-ray intensity follows the solid line in Fig. 6 (b). The experimental results shown in Fig. 6 (a) follow the initial part of the solid line in Fig. 6 (b). Therefore, metallic contaminant on the Si-surface are molecules.

A depth profile of the contaminant aluminum measured by XPS is shown in Fig. 7. Aluminum distributed almost uniformly in the chemical oxide, and is not on the top surface. Thus the surface cleaning washes out only materials on the surface. It is probably difficult to remove metals that are in native oxide form. Thus above results suggest that the hot LDO water treatment changed the chemical condition and morphology of the wafer surface to remove metallic contaminants.

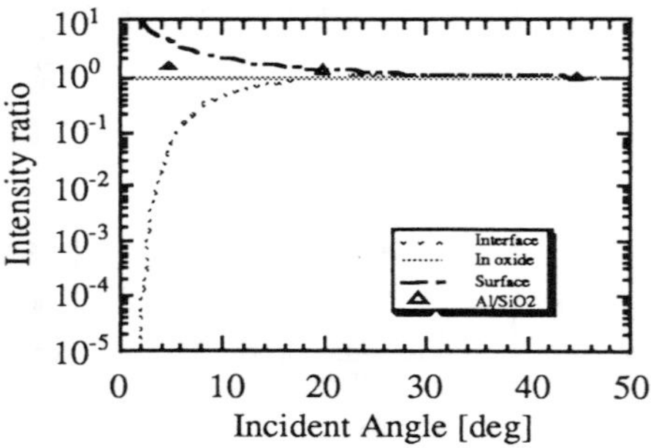

Fig. 7 Depth profile of Al in Si/SiO2- XPS -

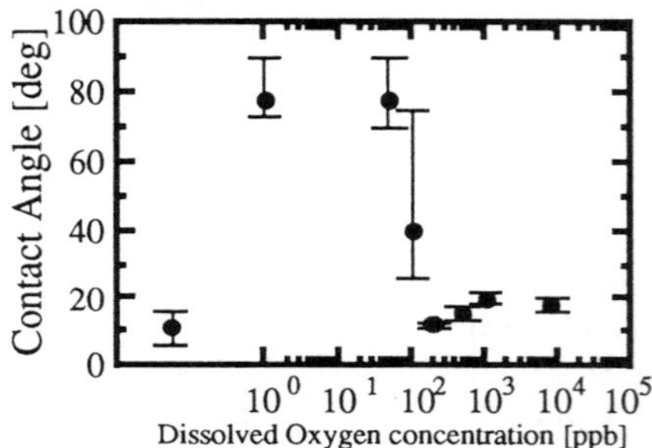

Fig. 8 The dependence of contact-angle on dissolved oxygen in pure water

3. Surface analysis

The contact angles [14] of the silicon wafer surface after the hot LDO water treatment is shown in Fig. 8. Samples were treated in the same way as the experiment for Fig. 3. As the concentration of the dissolved oxygen lowered, the value of contact-angle started to increase from 100 ppb and reached to about 90 degrees, this is a hydrophobic surface. The contact-angle of the silicon surface reflects the extent of oxide formation of the outermost surface. Thus this result indicates that the native oxide has etched off, or the top layer of the native oxide has turned to silicon.

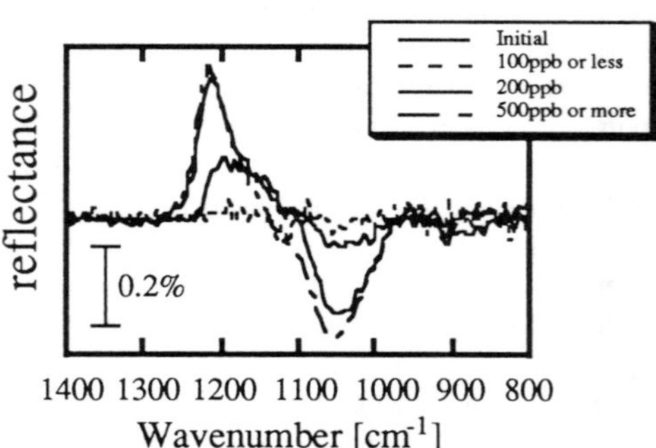

Fig. 9 IR absrption spectra - DOC dependence -

Figure 9 shows the change in the IR-RAS spectra after the hot LDO water treatment, using several LDO waters containing different amount of dissolved oxygen. The peak of about 1200 cm^{-1} and

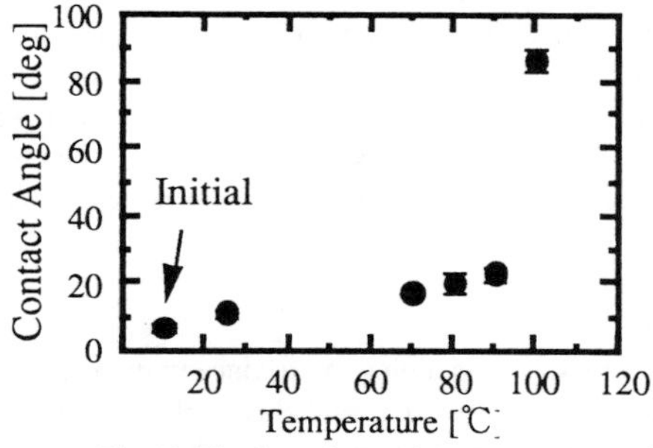

Fig. 10 The dependence of contact-angle on rinsing temperature

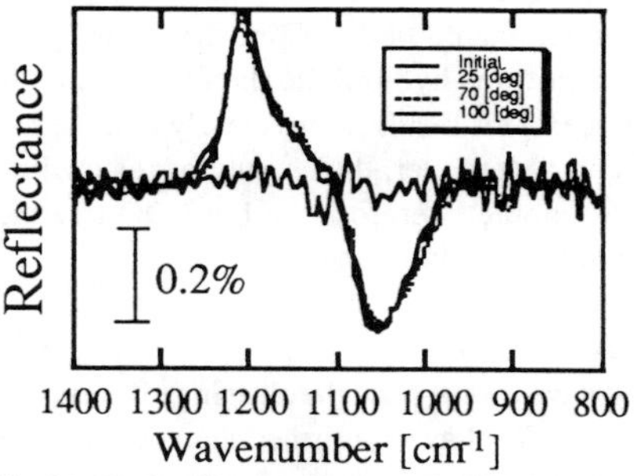

Fig. 11 IR absrption spectra - temperature dependence -

1050 cm^{-1} were assigned as the absorption of the longitudinal optical phonon (LO phonon) and the transverse optical phonon (TO phonon) of stretching vibration mode of Si-O bond. The peak area of the TO phonon is proportional to the amounts of Si-O bonds. In Fig. 9, features of the Si-O have remained after the treatments of the LDO water containing more dissolved oxygen than 500 ppb. However, these signals of Si-O bonds have disappeared after the treatments using the LDO water containing less dissolved oxygen than 100 ppb. Therefore the results shown in Fig. 8 and 9 indicate that the chemical oxide did not exist on the wafer surface when the contact angle reached to about 90 degrees.

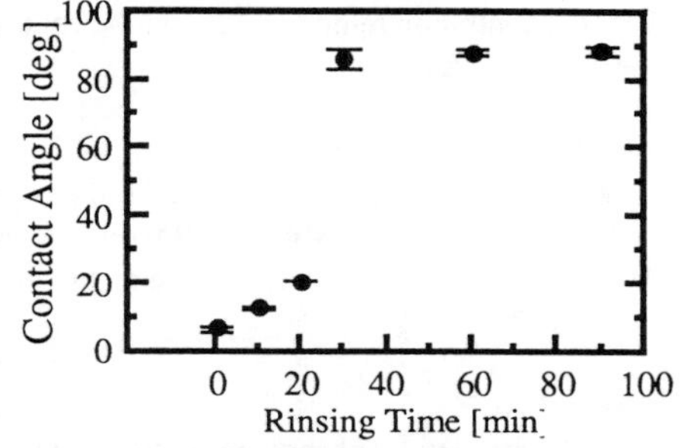

Fig. 12 The dependence of contact-angle on treatment time

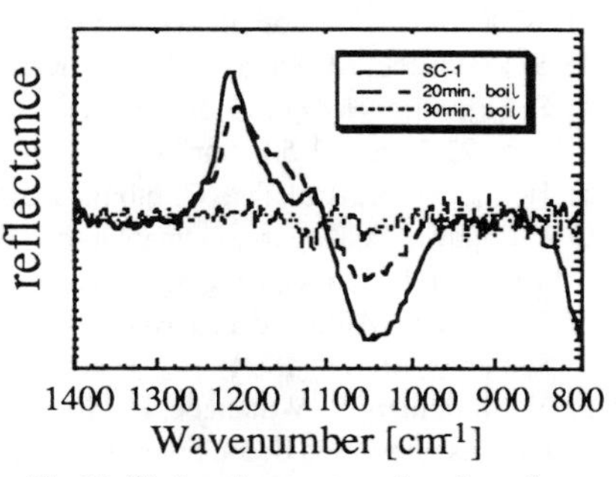

Fig. 13 IR absrption spectra - time dependence -

Figure 10 shows the contact-angle change in the wafer surface with changing in treatment temperature. In this experiment, the concentration of dissolved oxygen in the LDO water was 1ppb and immersion time was 30 min. The value of contact-angle increase dramatically between 90℃ and boiling-point i.e. the outer surface of the samples turned from hydrophilic to hydrophobic between 90 ℃ and boiling-point. Under the same condition, the IR-RAS spectra changed as shown in Fig. 11. Since the TO phonon area did not decrease after treatment under 90℃, the initial chemical oxide remained with no change. However we could detect no feature of the TO phonon after the LDO water treatment at the boiling-point. These results support the interpretation concerning to the results of contact-angle.

Figure 12 shows the contact-angle change in the wafer surface with increasing immersion time. In this experiment, the 1 ppb LDO water was used at the boiling-point. The surface condition of the silicon wafers turned from hydrophilic to hydrophobic after LDO water treatment longer than 30 min. Figure 13 shows the IR-absorption spectra change with increasing immersion time. The peaks of the LO and TO phonon monotonously decreased with treatment time and disappeared after longer treatment than 30 min. Comparing Fig. 12 with Fig. 13, the sample surface with less than 20 min treatment was hydrophilic though the amount of Si-O bonds were half or less of that of the initial surface. This result shows clearly that the hot LDO water etched the native oxide but changed it to silicon. Moreover, the amount of residual metals was minimized when the wafer surface became hydrophobic, therefore the hot LDO treatment etched the chemical oxide and removed metallic contaminants simultaneously.

CONCLUSION

We have investigated the cleaning efficiency of pure water containing low a concentration of dissolved oxygen. We have confirmed that metallic contaminants, except Cu, and chemical oxide on a silicon wafer surface are removed by the hot water rinsing at the boiling-point using the LDO water. The dissolved oxygen concentration of the LDO water used in our experiments was below 100 ppb. This concentration removed the metallic contaminants, except Cu, by etching the chemical oxide on the silicon wafer.

ACKNOWLEDGMENTS

The authors would like to thank Dr. Horiike of Toyo university for providing experimental equipment and support, S. Ishii of Mitsubisi Rayon Engineering Co.,Ltd for technical support and supplying the MHF degassing system, and M. Abe of Nomura Micro Science Co.,Ltd for technical advice on LDO pure-water.

REFERENCE

1) T. Ohmi : 8th Workshop on ULSI Ultra Clean Technology, Advanced Wet Chemical Processing 1,Tokyo, JAPAN, 1990 (Ultra Clean Society, 1990), pp. 5-15
2) W. Kern, D. A. Puotien : RCA Review 31, pp187-205, June (1970)
3) S. Watanabe, M. Shigeno, N. Nakayama and T. Ito : Jpn. Appl. Phys. 30, pp. 3575 (1991)
4) S. Fujimura, H. Ogawa, K. Ishikawa, C. Inomata and H. Mori : Technicalreport of IEICE, SDM 93-7, pp. 43-50, April (1993)
5) H. Ogawa, K. Ishikawa, C. Inomata and S. Fujimura : Submitted to J. Appl. Phys.
6) T. Isagawa, M. Kogure, T. Imaoka and T. Ohmi : 15th Symp. on Ultra Clean Technology, Advanced Wet Chemical Processing 3, Tokyo, JAPAN, 1992 (Ultra Clean Society, 1992) pp.199-215
7) T. Koito, H. Aoki and T. Toyoda : Ext. Abstr., The Japan Society of Applied Physics and Related Societies, March (1994)
8) Y. Shiramizu, K. Watanabe, H. Aoki and H. Kitajima : Ext. Abst., The Japan Society of Applied Physics and Related Societies, March (1994)
9) S. Watanabe, Y. Sugita : Appl. Phys. Lett. 66(14), pp. 1797-1799, April (1995)
10) H. Ogawa, K. Ishikawa, M. T. Suzuki, Y. Hayami and S.Fujimura : Jpn. Appl.Phys. 34, pp. 732-736 (1995)
11) H. Ogawa, C. Inomata, K. Ishikawa, S. Fujimura and H. Mori : Proc. of the First International symp. on control of semiconductor interface, Karuizawa, JAPAN, 1993, (ELSEVIER, 1994), pp. 383-388
12) Y. Mori, K. Uemura, K. Shimono : ECS proceedings PV 94-10, pp258-269, October (1994)
13) W. Berneike, et al. : Fresenius Z Anal. Chem. 333, 524 (1989)
14) T. Ohmi : Institute of Environmental Proceeding (1993)

INVESTIGATION OF PRE-TUNGSTEN SILICIDE DEPOSITION WET CHEMICAL PROCESSING

A. PHILIPOSSIAN, M. MOINPOUR, R. WILKINSON AND V.H.C. WATT
Intel Corporation, Santa Clara, CA 95052 USA

ABSTRACT

Removing the native oxide from the poly-Si surface prior to WSi_x deposition is essential for achieving high quality silicides as well as sufficient film adhesion, particularly after high temperature anneal or oxidation. Contact angle studies have been used to determine initial and time-dependent surface characteristics of several types of silicon surfaces following immersions in HF-based etchants for varying amounts of time. The morphological characteristics of the surfaces before and after exposure to etchants, as well as the relative etch rates and wetting capabilities of the etchants have been used to explain the following results: With respect to initial contact angle studies, the implanted & annealed polycrystalline silicon surface has the lowest contact angle followed by polycrystalline and monocrystalline surfaces. Longer immersion times yield lower initial contact angles. The 0.1% lightly-buffered HF solution results in the highest contact angle followed by the 1% buffered HF solution with surfactant, and the 1% HF solution. With respect to contact angle changes during ambient air exposure time, the as-deposited polycrystalline silicon surface is most stable followed by monocrystalline, and implanted & annealed polycrystalline silicon surfaces. Longer immersion times improve surface stability while the 0.1% lightly-buffered HF solution results in the most stable surface followed by the 1% buffered HF solution with surfactant, and the 1% HF solution.

INTRODUCTION AND MOTIVATION

Ultra-thin refractory metal silicide films, such as WSi_x, deposited over in-situ doped or implanted poly-Si, offer wide applications in VLSI and ULSI technologies. CVD WSi_x films have been widely employed for ultra thin gate, bit-line and word-line interconnects due to their lower resistance [1]. It is well established that the integrity of the interface between tungsten silicide and the underlying surface affects the stability, as well as the electrical performance of the film. The cleanliness and chemical stability of this interface becomes even more critical when one uses dichlorosilane (DCS) based chemistry for WSi_x deposition [2]. An effective pre-tungsten silicide deposition wet surface preparation process needs to fulfill the following two requirements: (1) the process should render the to-be-deposited surface chemically passivated by effectively removing contaminants (i.e. native and chemical oxides, particles, and other foreign matter), and (2) in open manufacturing environments, the process should ensure that the surface is stable over time during ambient air exposure prior to tungsten silicide deposition.

In this study, the contact angle measurement technique has been used to determine initial and time-dependent surface characteristics of several types of silicon surfaces following immersions in HF-based etchants for varying amounts of time. The technique has been proven to be successful in monitoring initial surface passivation and changes in silicon surface energies [3-5].

Mat. Res. Soc. Symp. Proc. Vol. 386 © 1995 Materials Research Society

EXPERIMENTAL

A 3-factor, 3-level D-Optimal experimental design was adopted to study the main effects. The three different wafer surfaces studied included:

- P<100> monocrystalline silicon (Surface A)
- LPCVD polycrystalline silicon (Surface B)
- Phosphorous implanted & annealed LPCVD polycrystalline (Surface C)

The polycrystalline silicon film was deposited using a commercially available hot-wall vertical LPCVD reactor. Implantation energy and dose were 30 KeV and 5E15/cm^2, respectively. Annealing was for 30 minutes in a commercially available hot-wall vertical thermal reactor using nitrogen gas. The status of each surface prior to HF immersion was as follows: Surface A was covered by native oxide having an apparent thickness of 10 Angstroms. As for Surfaces B and C, given the high processing temperatures and subsequent removal of hot wafers from the reactor chamber into ambient air, the wafers were covered with approximately 20 Angstroms of 'densified' native oxide. The annealing temperature was 850 °C.

The three different etchants used in this study were:

- 1.0% HF solution (Etchant 1)
- 1.0% buffered HF solution with surfactant (Etchant 2)
- 0.1% lightly-buffered HF solution (Etchant 3)

A 2-minute etch using Etchant 1 was arbitrarily considered as the reference process. To explore the effect of etching time, 1-minute and 3-minute immersion times were also considered. Since Etchant 2 and Etchant 3 each resulted in thermal oxide etch rates which were different than Etchant 1, immersion times were normalized (i.e. Low, Medium and High) based on a constant amount of thermal oxide removed from the silicon surface at room temperature (Table I):

Table I: Immersion times (in minutes) normalized to the thermal oxide etch rate of each etchant

	Etchant 1	Etchant 2	Etchant 3
Low	1	2.22	0.38
Medium	2	4.44	0.76
High	3	6.66	1.14

For each etching experiment, two 200 mm wafers were used. Following etching, the wafers were rinsed in ultra pure water for 10 minutes, dried with nitrogen and measured within 1 minute of ambient air exposure. The wafers were stored in a closed polypropylene box and measured again for contact angle after 30, 60, 90, 120, 150, 240, 300 and 1440 minutes of ambient air exposure time. During wafer storage, the box was not purged with any gases. A conventional goniometer was used for contact angle measurements. A micro-syringe was used to consistently deliver 0.0025 ml of the water to the substrate. Photographs of the droplets were taken at a droplet age of one minute and the respective angles were determined [3].

RESULTS AND DISCUSSION

Typical plots of contact angle vs. ambient air exposure time are shown in Fig. 1 for various processing conditions (see Table II).

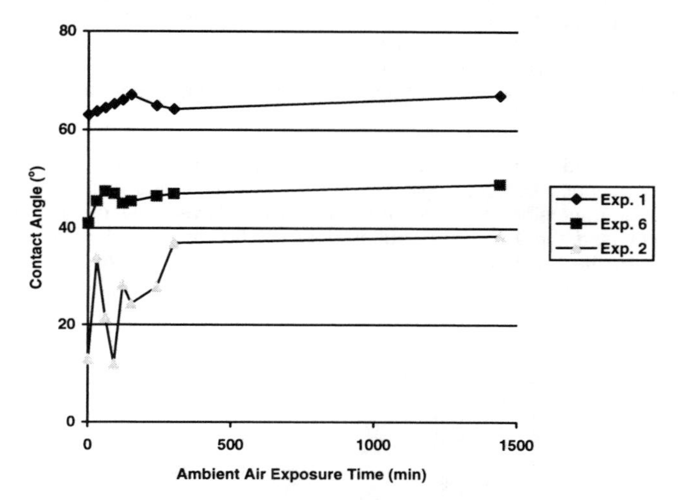

Figure 1: Contact angle as a function of ambient air exposure time for various processing conditions.

Generally speaking, during the first 120 minutes of exposure to ambient air, the surface undergoes rapid changes. After 120 minutes, the surface seems to remain essentially constant over time. Table II summarizes the completed experimental matrix. The responses investigated in this study are the initial contact angle, and the *average* rate of change of contact angle over time during the first 120 minutes of ambient air exposure:

Table II: Experimental worksheet and results

Experiment	Etchant	Immersion Time	Surface	Initial Contact Angle (°)	Change in Contact Angle (°/hr)
1	3	Low	A	63	1.4
2	1	Medium	A	13	7.1
3	1	High	C	12	1.8
4	3	High	B	31	0.7
5	2	High	A	43	1.5
6	3	Medium	C	41	1.5
7	2	Medium	B	37	0.6
8	1	Low	B	46	2.0
9	2	Low	C	21	7.0

Initial Contact Angle Studies

With an R-Square of 0.72, the linear model indicates that surface type, etchant type and immersion time all affect initial contact angle. It should be noted that since no chemical or physical analysis of as-treated surfaces were performed in this study, the passivation and roughness components could not be de-coupled by initial contact angle information alone. In the next section where contact angle changes over time are discussed, the energy required to alter surface morphology is significantly higher than that available at ambient conditions, therefore any changes in contact angle can be attributed to changes in surface passivation.

According to the model, implanted & annealed polycrystalline silicon surface has the lowest air/water/surface contact angle followed by polycrystalline and monocrystalline surfaces. Longer immersion times are shown to result in lower initial contact angles. As for the etchants used, the 0.1% lightly-buffered HF solution results in the highest contact angle followed by the 1% buffered HF solution with surfactant, and the 1% HF solution. Figs. 2 to 4 summarize these trends quantitatively:

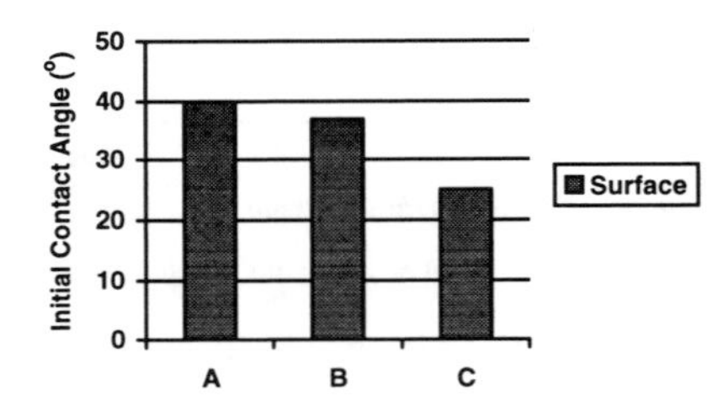

Figure 2: Initial contact angle associated with various surface types (values of initial contact angle are derived from statistical model based on main effects only).

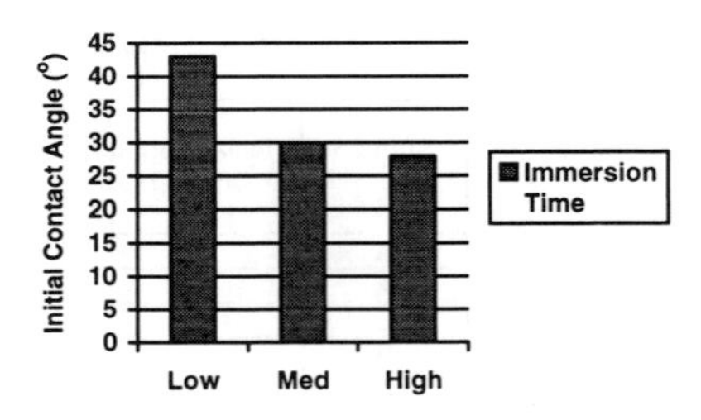

Figure 3: Initial contact angle associated with various immersion times (values of initial contact angle are derived from statistical model based on main effects only).

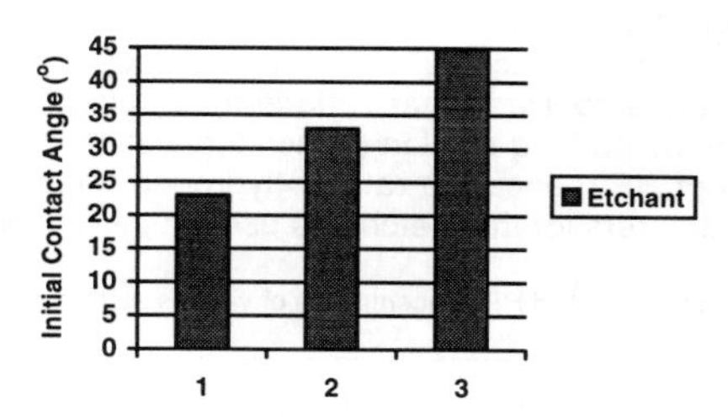

Figure 4: Initial contact angle associated with various etchants (values of initial contact angle are derived from statistical model based on main effects only).

These trends can be explained by taking into account the morphological characteristics of the surfaces before and after exposure to etchants, as well as the relative etch rates and wetting capabilities of the etchants.

Several orientations are formed in polycrystalline silicon which can grow at different rates and contribute to surface micro-roughness [6]. Defects (i.e. twins, voids and grain boundaries) are also present and have higher free energies compared to the crystalline regions of the silicon [6]. When exposed to etchants, such surface heterogeneities tend to etch preferentially leading to pitting and greater surface roughening. Moreover, following implantation and anneal, phosphorus segregates at grain boundaries [7] and reduces the activation energy for grain growth [8]. Similar trends gave been observed in this study (Table III). The combination of lower energy grain boundaries with enhanced surface roughness is consistent with the lower initial contact angle (i.e. greater wetting) seen in Fig. 2.

Table III: Average grain size (in Angstroms measured by Atomic Force Microscopy) of various silicon surfaces

Surface	Average Grain Size (Å)
Monocrystalline Silicon	N/A (atomically smooth)
As-Deposited Polycrystalline Silicon	645
Implanted & Annealed Polycrystalline Silicon	1060

With regards to immersion time, previous reports have indicated that longer immersion times improve surface passivation by promoting hydrogen terminated bonds at the surface [9]. While we believe this to be true (see discussion below on contact angle change vs. ambient air exposure time), as a surface is etched for long periods of time, the preferential etching due to higher energy sites will become more pronounced, and the depths of the pits created will become greater, thus enhancing surface wetting and resulting in lower contact angle. In contact angle studies, the effect of the surface energetics due to roughening may outweigh the effect due to hydrogen termination.

With regards to etchants, it has been reported that etching with surfactonated buffered HF solutions lead to smoother surfaces than unsurfactonated solutions [4, 5]. This phenomenon has been explained by the reaction of silicon with HF which generates hydrogen bubbles that adsorb onto the silicon surface, and lead to non-uniform etching through micro-masking. Use of surfactonated solutions result in smoother surfaces due to improved wetting of the surface and effective dislodging of adsorbed hydrogen bubbles. One can therefore conclude that surface morphology as a function of etchant type should depend on the surface tension (hence the wetting capability) of the etchant, and the concentration of HF (hence the rate of hydrogen bubble generation). Table IV summarizes these parameters for the 3 etchants used in this study:

Table IV: Surface tension and HF concentration of various etchants used in this study

Etchant	Surface Tension (dyne/cm)	HF Concentration (%)
1	77.1	1.0
2	39.0	1.0
3	77.8	0.1

The data in Table IV indicate that Etchant 1 should indeed result in a lower contact angle when compared to Etchant 2 due to its higher surface tension and therefore poorer wettability. In the case of Etchant 3 (lightly-buffered HF solution), even though its surface tension is highest, the low HF content reduces the amount of hydrogen bubble formation and thus surface roughness. These are consistent with the trends reported in Fig. 4.

Contact Angle vs. Time Studies

With an R-Square of 0.75, the linear model indicates that surface type, etchant type and immersion time all affect contact angle change over time. As noted previously, the energy required to alter surface morphology is significantly higher than that available at ambient conditions, therefore any changes in contact angle can be attributed to changes in surface passivation.

According to the model, as-deposited polycrystalline silicon is the most stable surface during ambient air exposure followed by monocrystalline silicon and implanted & annealed polycrystalline silicon. Longer immersion times tend to improve surface stability. As for the etchants used, the 0.1% lightly-buffered HF solution results in the most stable surface followed by the 1% buffered HF solution with surfactant, and the 1% HF solution. Figs. 5 to 7 summarize these trends quantitatively:

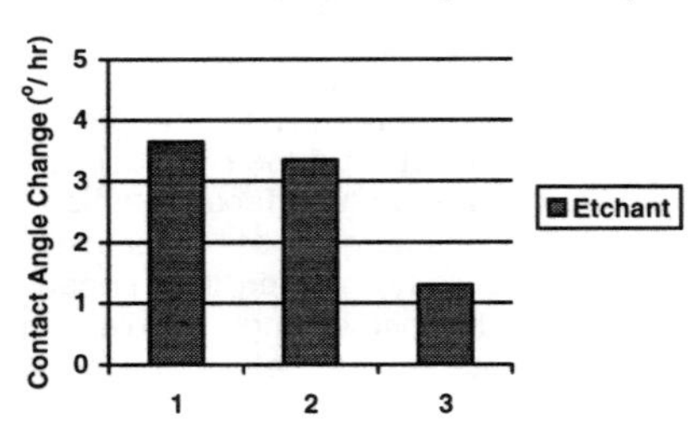

Figure 5: Contact angle change over time associated with various etchants (values of contact angle change over time are derived from statistical model based on main effects only).

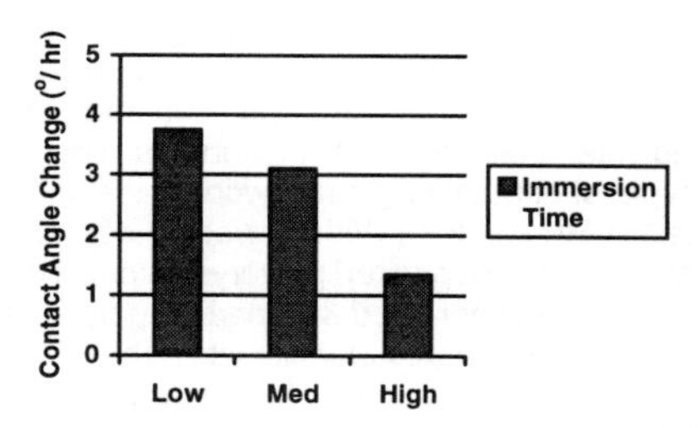

Figure 6: Contact angle change over time associated with various immersion times (values of contact angle change over time are derived from statistical model based on main effects only).

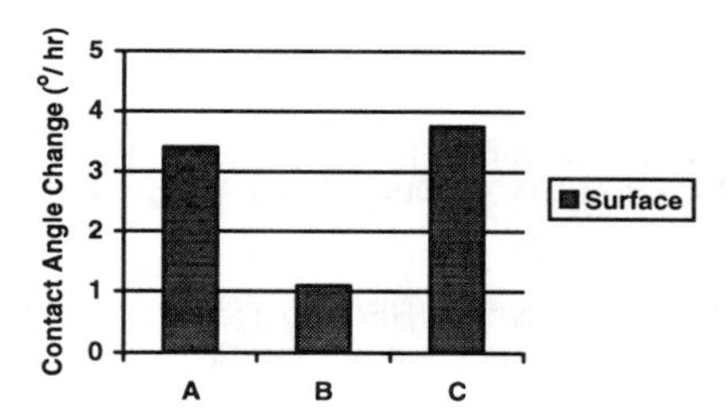

Figure 7: Contact angle change over time associated with surfaces etchants (values of contact angle change over time are derived from statistical model based on main effects only).

With regards to etchants, the trend shown in Fig. 7 is the reverse of the one reported in Fig. 4. This suggests that using a 0.1% lightly-buffered HF solution not only results in the smoothest surface (Fig. 4 and discussions thereafter), but may also result in a highly passivated surface. The latter is consistent with the trend observed in Fig. 7 since an initially well passivated surface (i.e. a high initial contact angle), would be less sensitive to changes from exposure to ambient air.

The improved surface stability associated with high immersion times can be explained by the increase in hydrogen passivation as a function of etching [9]. In this instance, there is no potential competition between surface morphology and passivation since surface structure is not expected to be changing at ambient conditions.

With regards to effect of various surfaces on contact angle change over time, the results cannot be explained at this time and experiments are underway to not only examine each surface independently, but to consider 2-factor interactions as well.

CONCLUSIONS

Contact angle studies have been used to determine initial and time-dependent surface characteristics of several types of silicon surfaces following immersions in HF-based etchants for varying amounts of time. The morphological characteristics of the surfaces before and after exposure to etchants, as well as the relative etch rates and wetting capabilities of the etchants have been used to explain the following results: With respect to initial contact angle studies, the implanted & annealed polycrystalline silicon surface has the lowest contact angle followed by polycrystalline and monocrystalline surfaces. Longer immersion times yield lower initial contact angles. The 0.1% lightly-buffered HF solution results in the highest contact angle followed by the 1% buffered HF solution with surfactant, and the 1% HF solution. With respect to contact angle changes during ambient air exposure time, the as-deposited polycrystalline silicon surface is most stable followed by monocrystalline, and implanted & annealed polycrystalline silicon surfaces. Longer immersion times improve surface stability while the 0.1% lightly-buffered HF solution results in the most stable surface followed by the 1% buffered HF solution with surfactant, and the 1% HF solution.

REFERENCES

1. Chu, C., Moinpour, M., Cham, J., Lu, W., Hwang, D., Watt, V.H.C., Sadjadi, R., Gaynor, W., and Moghadam, F., 'DCS Based CVD Tungsten Silicide Technology for Sub-Micron Flash Memory Applications', *IEEE/VMIC Conf. Proc.,* (1994).

2. Moinpour, M., Moghadam, F., Cham, J., and Lu, W., 'Composition and Structure of As-deposited and Oxidized DCS-Based CVD WSi_x Films', *Mat. Res. Soc. Symp. Proc.,* Vol. 337, (1994).

3. Philipossian, A., 'The Activity of HF/H2O Treated Silicon Surfaces in Ambient Air Before and After Gate Oxidation,' *J. Electrochem. Soc.,* Vol. 139, No. 10 (1992).

4. Kikuyama, H., N. Miki, J. Takano, and T. Ohmi, 'Developing Property Controlled, High-Purity Buffered Hydrogen Fluorides for ULSI Processing,' *Medical Device and Diagnostic Industry*, April (1989).

5. Kikuyama, H., N. Miki, K. Saka, J. Takano,I. Kawanbe, M. Miyashita, and T. Ohmi, 'Surface Active Buffered Hydrogen Fluoride Having Excellent Wetability for ULSI Processing,' *IEEE Transactions on Semiconductor Manufacturing*, Vol. 3, No. 3, August (1990).

6. Cerva, H., and H. Oppoizer, 'Microstructure and Interfaces of Polysilicon in Integrated Circuits,' *Springer Proceedings in Physics*, Vol. 35 (1989).

7. Duffy, M., J. McGinn, J. Shaw, R. Smith, R., R. Soltis, and G. Harbeke, 'LPCVD Polycrystalline Silicon: Growth and Physical Properties of Diffusion-Doped, Ion-Implanted and Undoped Films,' *RCA Review*, Vol. 44, June (1983).

8. Mei, L., M. Rivier, Y. Kwark, and R. Dutton, 'Grain-Growth Mechanisms in Polysilicon,' *J. Electrochem. Soc.,* Vol. 129, No. 8 (1982).

9. Verhaverbeke, S., 'Dielectric Breakdown of Thermally Grown Oxide Layers,' Ph.D. Dissertation, Catholic University of Leuven, Belgium (1993).

Part II

Chemical Mechanical Polishing and Post-CMP Cleaning

Characterization of Defects Produced in TEOS Thin Films due to Chemical-Mechanical Polishing (CMP)

F.B. Kaufman, S.A. Cohen and M. A. Jaso
IBM T. J. Watson Research Center and Semiconductor Research and Development Center, Yorktown Heights, N.Y. 10598

Abstract

Time zero breakdown electrical measurements, SIMS, Bias Temperature Stress (BTS), and Triangular Voltage Sweep (TVS) techniques performed on MOS capacitors have been used to characterize two types of defects produced in TEOS thin films subjected to CMP process planarization. Microcracking of the insulator, resulting in degradation in breakdown characteristics, and uptake of K^+ ions (from the polishing slurry) are found to occur. These defects are thought to be caused by the chemical and mechanical stress to which the oxide surface is subjected during the CMP process. The effects of polish pad type, non K^+ ion containing slurry and post-CMP wet etch steps were all found to influence the extent of damage observed in polished films.

Introduction

Chemical-mechanical polishing (CMP) of insulating films, such as SiO_2, has been shown to be a useful method of planarization at interlevel dielectric (1,2,3) and shallow trench isolation (4) process levels. Integrated with other processes, the technique has been applied to the fabrication of CMOS (5), bipolar (6) and Josephson junction (7) devices. While major emphasis has been placed on enhancing the ultimate manufacturability (8) of this technique in terms of removal rate reproducibility, planarization efficiency and high within wafer material removal uniformity, a detailed understanding of the predominant mode(s) of generating surface or subsurface defects is not currently available. The latter is of interest for several reasons. The CMP process subjects the insulator to high mechanical (applied pressure, abrasion) and chemical (high pH) stresses. In addition, earlier work has shown significant degradation in time-zero breakdown (9) and I-V (10) electrical characteristics of thin ($<$ 130 nm) PECVD tetraethylorthosilicate (TEOS) or thermal oxide films following CMP processing. Moreover, as insulator films scale to smaller thicknesses, and as polished surfaces are subjected to greater chemical stress (ie. in metal CMP), the CMP-induced generation of (surface/subsurface) defects is expected to become more critical to the ultimate performance and reliability of the fabricated device.

Experimental

Chemical-mechanical polishing on 125 mm wafers was done using a Westech model 372 polisher, using industry-standard process conditions (3), with a nonoptimized pad conditioning process used between wafers. Details of the time-zero breakdown and TVS measurements on Al dot/SiO_2/ Si/Al MOS structures can be found in the relevant literature (9,11). A timed etch of the insulator surface was typically done in 50:1 DHF to remove from 1 up to 70 nm of material.

Effect(s) of CMP on electrical properties

In Figure 1 is shown the change in MOS Al dot time zero breakdown field for a 500 nm thick blanket TEOS film after CMP polishing with a stiff, glass-bead embedded polymer polishing pad where approximately 50 nm of material has been removed during CMP. Whereas the deposited film has excellent breakdown characteristics, the polished film is found to have approximately 20% of the devices shorted, with the remaining devices showing significantly reduced breakdown strength. Comparable electrical degradation has been observed for other types of insulating materials (thermal oxide, PSG, LPCVD TEOS) for films in the 100-700 nm thickness regime polished under similar

Mat. Res. Soc. Symp. Proc. Vol. 386 © 1995 Materials Research Society

conditions. Calculated (12) defect densities leading to low field failures were typically at least 20 times those in unpolished films as a result of time-zero breakdown measured on 64 mil diameter dots. Similar results were observed on patterned wafers where electrical measurements were performed on dots placed over partially planarized 400 nm high steps (13).

CMP process effects: Polish pad dependence.

Comparison of breakdown characteristics for 4 different types of polishing pads (Figure 2) shows interesting differences among pad types. Compared to the as-deposited, unpolished, TEOS film we observe the highest degree of degradation in breakdown behavior in the glass-bead polymer pad. The remaining pads showed successively smaller amounts of damage going from the polymer pad to the stacked polymer pad to the felt pad. This data suggests that harder, less compressible polishing pads cause greater electrically detectable damage in the polished insulator. Significantly, the greatest damage to the oxide, both in terms of average breakdown strength and percentage of low field failures was caused by the glass-bead embedded polymer polishing pad. This latter result may also be relevant to the defect consequences of using poorly conditioned polishing pads, where a second phase of polish debris and slurry suspended in the polish pad causes glazing of the surface.

Mechanism of CMP-induced defect generation

Defects in silica glass are typically discussed in terms of stress-induced diffusion of water into the glass (14) with the inititation of cracks in the glass under subcritical loads (15). Transmission electron microscopy (16) of chemical-mechanical polished PECVD SiO_2 shows chemical/structural modification of the oxide extending up to 200 nm below the surface of the polished oxide. In addition, abrasion-damaged oxide glass surfaces (17) can exhibit growth of a network of cracks when immersed in alkali-ion containing basic solutions. Previous workers have suggested that the precise details of an optical polish process can be a major determinant in the generation of subsurface damage fissures in fused silica (18).

In order to gain further information regarding these electrically detectable defects, we have used a 30 second BHF treatment (removes 10 nm of the polished oxide) following CMP to attempt to modify and highlight the damaged layer(s). The resulting changes in the breakdown behavior for the felt, polymer and glass bead-polymer polishing pads are shown in Figures 3,4 and 5, respectively. We observed, in all cases, an increase in the electrically detectable damage, both in terms of extent of low field failures and a reduction in average breakdown fields. In addition, the extent of additional damage (post etch) detected followed the trend originally determined, see above. Unpolished films etched under identical conditions show no significant changes in breakdown. These data are consistent with the presence of CMP process-induced microcracks which become deeper after the short wet etching step. Based on the thickness dependence of numbers of shorts observed for polished and polish-etched films, we estimate that the depth of the microcracks is in the 50-250 nm range. The data also suggests that the CMP-damaged oxide etches faster than nondamaged oxide.

Reaction of water molecules with silica glass is known (19) to lead to scission of Si-O bonds forming silanol species, SiOH. The collapse of the measured contact angle (surface water droplet), see Figure 6, (20) after as little as 15 seconds of CMP, and an absorption increase at 3600 cm-1 in the ATR-FTIR spectrum (21) of polished TEOS are all consistent with the presence of silanols. Such species are capable of binding alkali (X) ions, such as sodium or potassium, via formation of SiO^-X^+ salts. SIMS depth profiles of polished TEOS films (Figure 7) indicate that K^+ is incorporated into these films at concentration levels up to 10^{12} atoms/cm^2, up to 2 orders of magnitude higher than we have typically found in unpolished films. The silica containing polishing slurry used in these CMP experiments contained K^+ ion at approximately a 2000 ppm level.

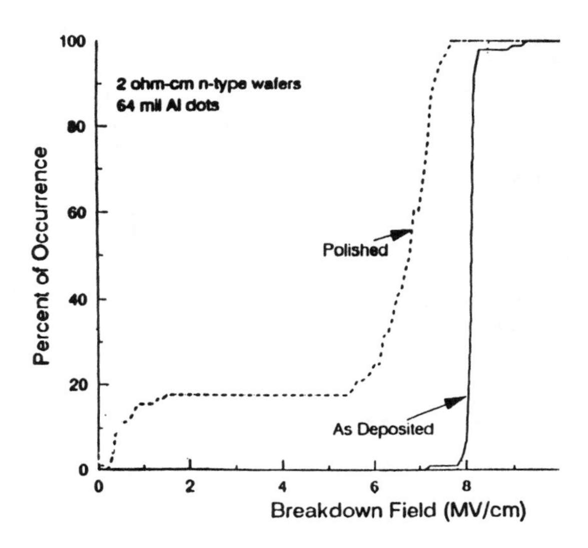

Figure 1. MOS Al dot cumulative breakdown fields for 500 nm TEOS films showing changes from initial films as a function of CMP with a glass bead/polymer pad. For CMP conditions, see text.

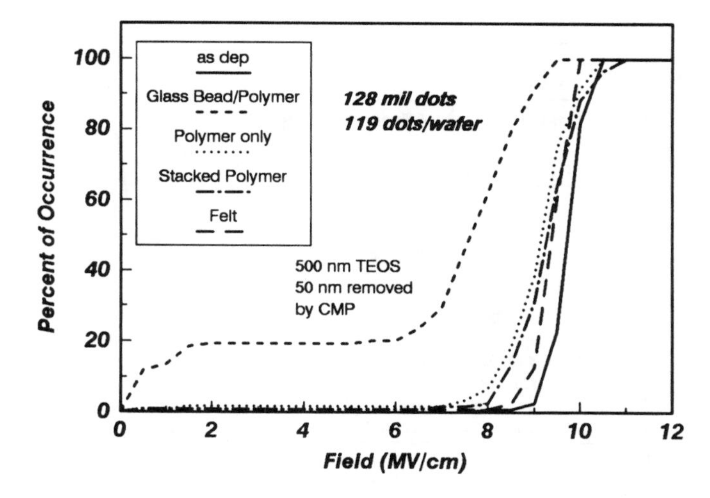

Figure 2. MOS Al dot cumulative breakdown fields for 500 nm TEOS films showing changes from initial films after CMP, with 3 different kinds of polish pads

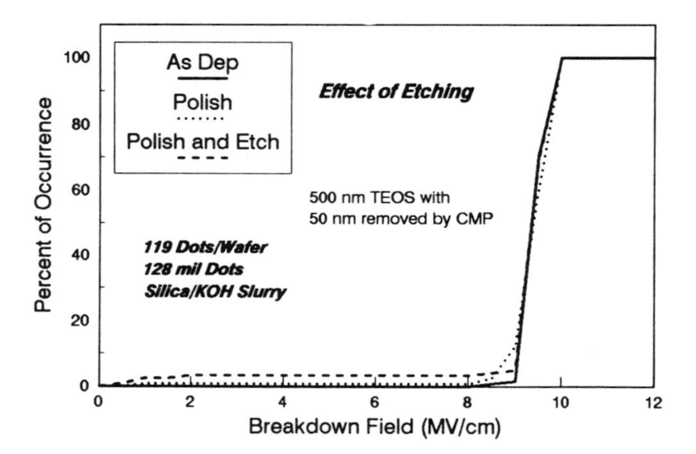

Figure 3. MOS Al dot cumulative breakdown fields for 500 nm TEOS films showing small changes in the initial films after CMP and CMP-etch with a felt pad.

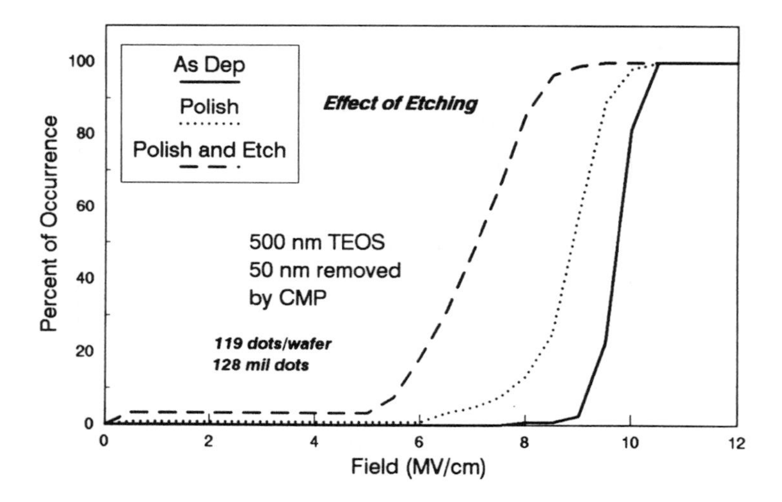

Figure 4. MOS Al dot cumulative breakdown fields for 500 nm TEOS films showing significant changes in the films after CMP and CMP-etch with a polymer pad.

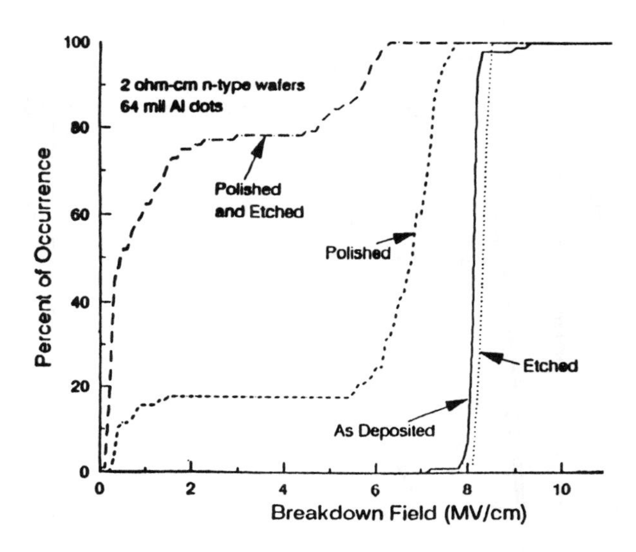

Figure 5. Same as figure 1 with the addition of cumulative breakdown data after post CMP etching.

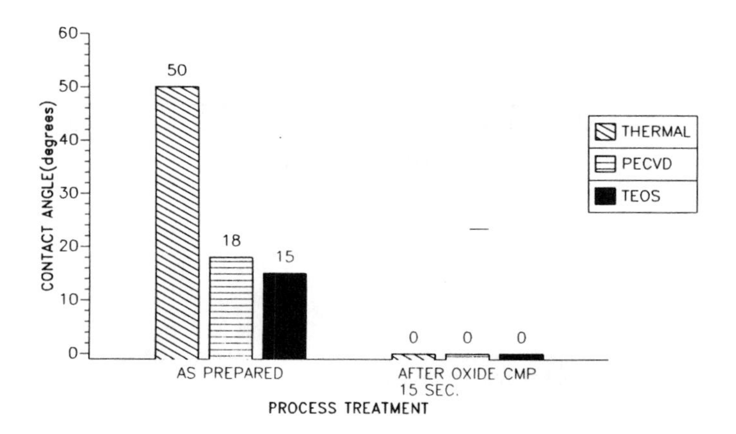

Figure 6. Contact angles of thermal, PECVD, and TEOS oxides before and after CMP indicating changes in wetting caused by a 15 second CMP step.

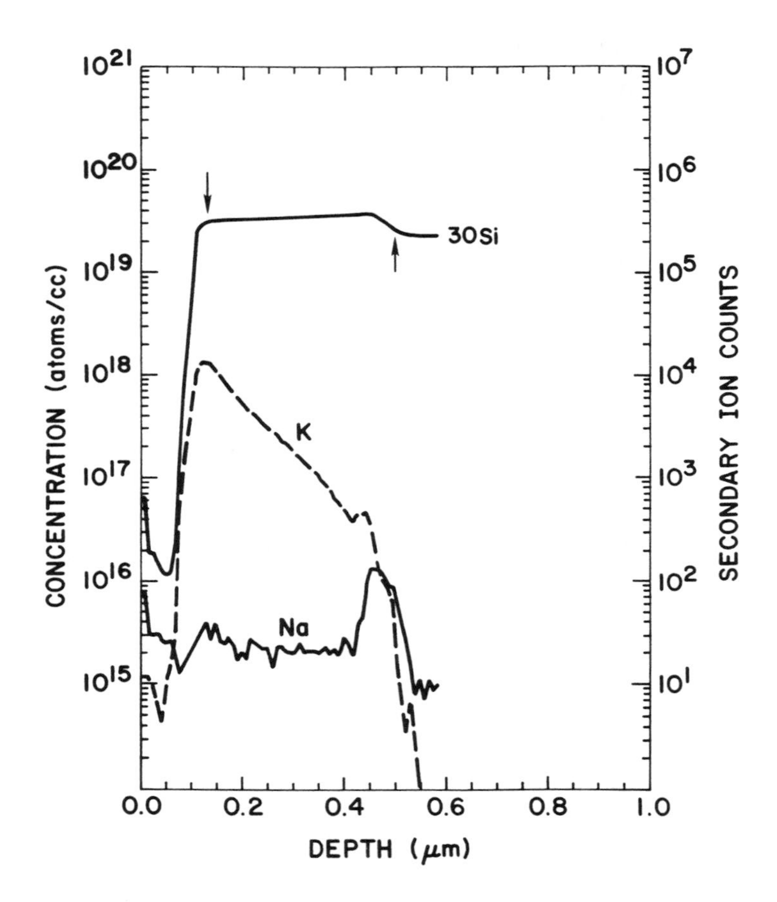

Figure 7. SIMS depth profiles of polished 470 nm thick TEOS Al dot MOS structures. Arrows indicate Al-SiO_2 (down) and SiO_2Si (up interfaces).

SIMS cannot be used to unambiguously define either the total amount of potassium ion introduced during the CMP or provide a truly realistic depth profile. This is because the SIMS experiment neglects both the surface and near-surface (approx 100 A) ionic species present, and, the O^{2+} ion probe beam causes positive ions such as Na^+ and K^+ to move into the insulator during the experiment due to electrostatic repulsion. However, we have found that SIMS is very useful in delineating relative changes in alkali ion content of polished oxide films. To get additional information about the positive ion defect in these polished films, we examined the effect of an external field (1.5 MV/cm) applied to selected MOS dots at a temperature of 200 C for 5 minutes. In Figure 8 are shown the SIMS profiles of the initial, polished, oxide film (Fig. 8A) and the SIMS profile after application (Fig. 8b) of the bias (top MOS dot positively charged), T, stress. We observed that under these conditions both Na^+ and K^+ ions are driven to the bottom electrode (right hand side of figure) of the MOS structure and, as in an earlier study involving alkali contamination of SiO_2 during gate metallization (22), the Na^+ ion is observed to be considerably more mobile than K^+ ion in these CMP processed samples.

Electrical measurements of mobile alkali ion content in polished TEOS films

SIMS analysis of polished oxide is not an ideal measurement technique due to cost and the technical considerations mentioned above. We have found, however, that appropriate electrical measurements performed on MOS dot structures provide a simpler, and, potentially automated means of measuring mobile ion contaminants introduced during the chemical-mechanical polishing process. In this manner the effects of contaminants introduced, unintentionally, through use of new polishing pads. or polishing slurries, or clean procedures, can be readily assessed. The selection of either the Bias Temperature Stress (BTS) or the Triangular Voltage Sweep (TVS) technique to measure the Na^+ and K^+ content of polished TEOS was determined by the after-polish film thickness. For films whose thickness is 300 nm or less, TVS was used to detect both types of ions. For films greater than 300 nm thick, BTS was used for K^+ measurement, and when K^+ is low or absent, TVS was used to achieve high resolution measurement of Na^+ content.

The BTS technique consists of a 1 MHz C-V before and after the appplication of a +2 MV/cm, 5 minute, 300°C stress. Mobile charge is calculated from the difference in flatband voltages before and after the stress. The TVS technique consists of a voltage ramp at 300°C where the ion flow is superimposed on the displacement current of the capacitor under test. Although an easy technique, BTS can have poor resolution when applied to thin TEOS films ($<$ 300 nm thick) or low mobile charge concentrations because of a stretch-out in the C-V curves after stress. This stretch-out is due to increases in the interface state density created by the stress-induced instability of the TEOS/Si interface. This adds a component to the flatband shift which is unrelated to the mobile charge. For films 300 nm or less in thickness, however, TVS can easily detect Na^+ and K^+ (Figure 9) which appear as separate peaks due to the mobility differences previously observed in the SIMS experiments. The figure shows individual, well resolved, Na^+ and K^+ peaks with a reduction in total mobile charge concentration, by a factor of almost three orders of magnitude, when the polished sample was subjected to a BHF etch step prior to metallization.

For thick ($>$ 300 nm) films, it is necessary to use the BTS technique (see Figure 10) to measure K^+ introduced by the slurry. For these films the flatband shift is large (several volts) compared to the small shift (100-200 mV) created by the C-V stretch-out and the mobility of K^+ is too low to provide an appreciable current during TVS. Using the appropriate electrical techniques, it was possible to measure the total amount of mobile charge (Figure 11) introduced by two kinds of KOH containing slurries and, an Ammonium hydroxide containing slurry. In all three cases, we observed a reduction in ionic content when a post-polish etch step was included in the processing sequence.

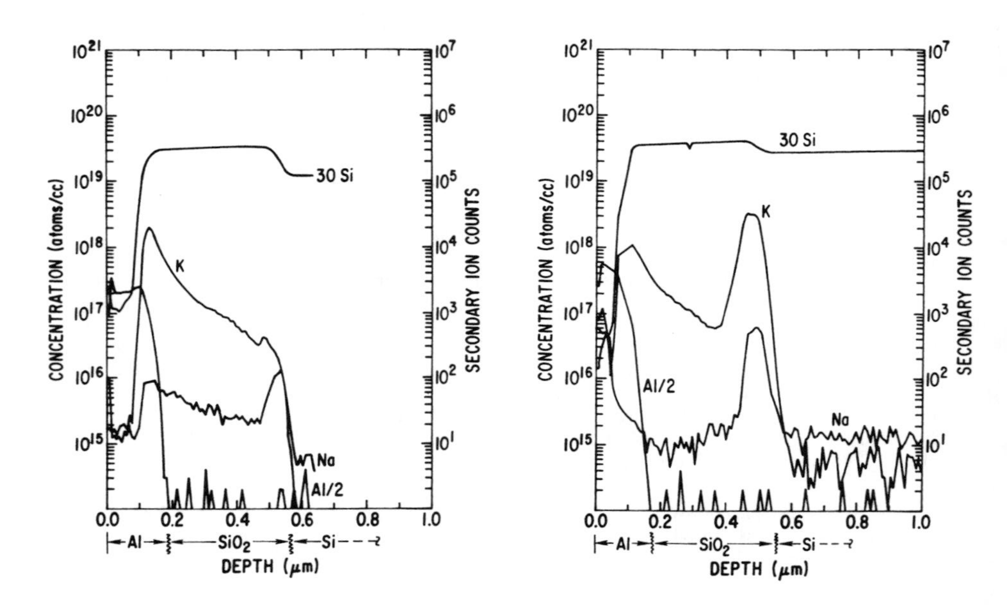

Figure 8. SIMS depth profiles of TEOS Al dot MOS structures. Left, (a). SIMS after CMP. Right,(b). SIMS after CMP on same wafer as (a) following the application of a bias/stress treatment applied to individual dot, see text for details.

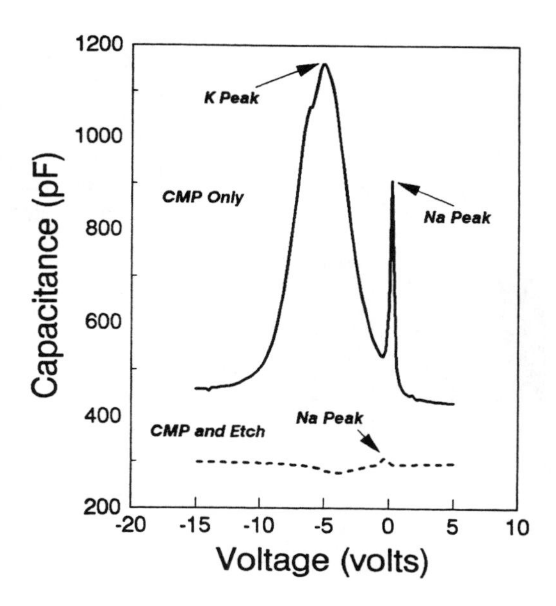

Figure 9. TVS plot of polished TEOS with a final thickness of 200 nm. The polished sample shows a total mobile charge concentratiiion of 1x10¹2./cm^2 with the K^+ contributing 90% of that charge. The polished and etched sample has a Na^+ peak which only corresponds to a mobile charge concentration of $5x10^9$ Mobile charge is proportional to the area under the curve.

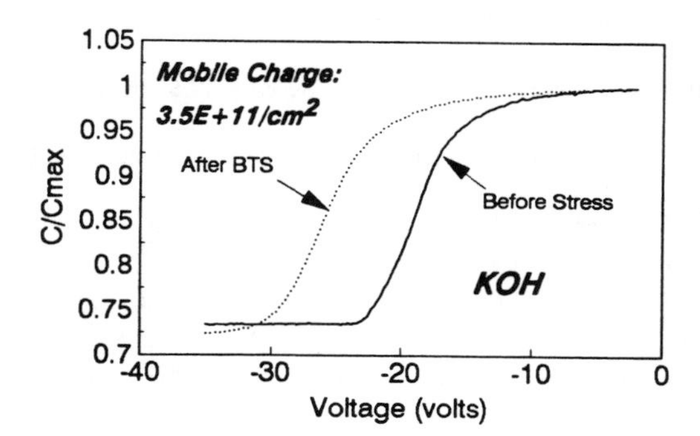

Figure 10. Bias-temperature-stress (BTS) data showing total mobile charge in 500 nm CMP TEOS before and after a 1.5 MV/cm, 200°C, 5 minute stress.

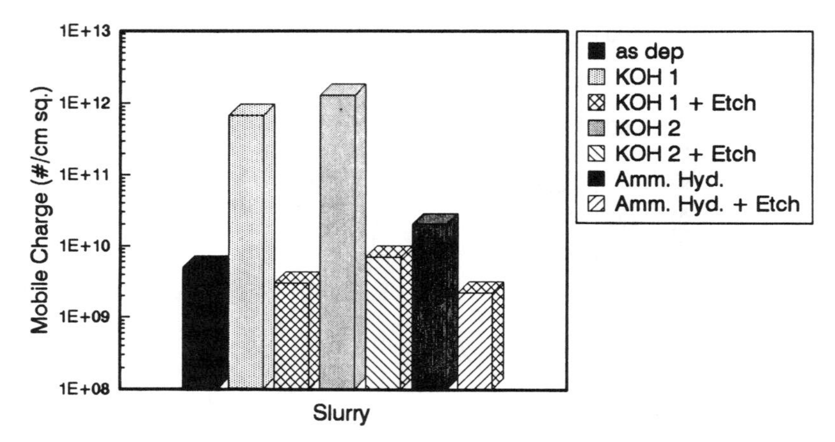

Figure 11. Summary of mobile charge distribution after CMP showing differences between various slurry and etching combinations. In the etch cleanup step, 15 nm of oxide was removed.

Conclusions

CMP processing of insulators such as TEOS has been shown to lead to electrical degradation via formation of microcracks (low field shorting) and incorporation of K^+ ions. Softer polishing pads and purer polishing slurries significantly reduce the incidence of these defects. Electrical measurements on polished oxide films have been shown to be a useful means of characterizing these polish-induced defects.

References

1. W.J. Patrick, W.L. Guthrie, C.L. Standley and P.M. Schiable,J. Electrochem. Soc. 138 1778 (1991)
2. P. Renteln, M.E. Thomas and J.M. Pierce, Proc. Seventh VLSI Interconnection Conference, p.57 (1990)
3. S.Sivaram, H. Bath, R. Leggett, A. Maury, K. Monnig and R. Tolles, Solid State Tech.,p. 87 (1992)
4. B. Davari, C.W. Koburger, R. Schultz et al., IEDM Tech. Digest p.61 (1989)
5. H. Landis, P. Burke, W. Cote et al., Thin Solid Films, 220 1 (1992)
6. W.L. Guthrie, W.J. Patrick, E. Levine et al., IBM J.of Res. Devel. 36 845 (1992)
7. M. Ketchen, W.J. Gallagher, D. Pearson et al., Appl. Phys. Letts., 59 2609 (1991)
8. SRC Topical Research Conference on Chem-Mechanical Polishing for Planarization, M.R. Witty, Editor, September 1992
9. S.A. Cohen, M. A. Jaso and A.A. Bright. J. Electrochem. Soc., 139 3572 (1992)
10. S. Murarka, S-H Ko, M. Tomazawa, P-J Ding and W. A. Lanford, presented at 1994 Spring MRS Meeting.

11. E.H. Nicollian and J.R. Brews, MOS Physics and Technology. New York: Wiley (1982)
12. Defect Density = -ln(Yield)/Capacitor Area, Yield= Percent good devices; good means breakdown fields >/= 6 MV/cm
13. S.A. Cohen and F.B. Kaufman, to be published.
14. K. Hirao and M. Tomazawa, J. Am. Ceram. Soc., 70, 377 (1987)
15. M. Tomazawa, W-T Han and W.A. Lanford, ibid., 74 2573 (1991).
16. J. A. Trogolo and K. Rajan, J. of Materials Science, 29 4554 (1994)
17. M.R. Bache and D.G. Holloway, Glass Tech. 31 126 (1990)
18. A.A. Tesar, N.J. Brown, J.R. Taylor and C.J. Stolz, Proc of SPIE 1441 154 (1991)
19. R.K. Iler, The Chemistry of Silica, John Wiley and Sons, 1979.
20. R. R. Thomas, F.B. Kaufman, T. Kirleis and R. Belsky, to be published
21. N.R. Klymko and F.B. Kaufman, unpublished.
22. G.F. Derbenwick, J. Appl. Phys. 48 1127 (1977).

IMPACT OF CHEMOMECHANICAL POLISHING ON THE CHEMICAL COMPOSITION AND MORPHOLOGY OF THE SILICON SURFACE

Hermann FUSSTETTER, Anton SCHNEGG, Dieter GRÄF, Helmut KIRSCHNER, Michael BROHL and Peter WAGNER
Wacker-Chemitronic GmbH, Central R&D, P.O. Box 1140, D-84479 Burghausen, Germany.

ABSTRACT

The polishing technology used for manufacturing ultraflat and smooth Si surfaces on a large scale is the chemomechanical polishing (CMP) technique. This technique combines the chemical corrosive removal of silicon atoms and the mechanical transport of the agents. The removal rates strongly depend on the interaction of mechanical parameters and the chemistry involved in the polishing process like the pH of the alkaline polishing slurry used. Removal of Si during CMP is explained by a nucleophilic attack of OH^- to silicon atoms catalyzing the corrosive reaction of H_2O resulting in cleavage of silicon backbonds. Characterization of the surface chemistry of the silicon wafer after polishing by X-Ray Photoelectron Spectroscopy and High-Resolution Electron Energy Loss Spectroscopy reveals an oxide free, predominantly hydride covered silicon surface displaying hydrophobic properties. Morphological features like microroughness as well as localized surface irregularities on the silicon surface, also referred to as Light Point Defects, depend on different strongly interacting process parameters. Microroughness is reduced by CMP by several orders of magnitude as characterized by lightscattering techniques and Atomic Force Microscopy.

INTRODUCTION

The semiconductor industry requires silicon wafers with extremely tight specifications with respect to flatness and surface uniformity. Thickness variations in the sub-μm range as well as roughness RMS (root mean square) values in the sub-Angstrom regime are a prerequisite for efficient manufacturing of advanced integrated circuits. Polishing as the last step of the wafer shaping process finally determines the geometry and surface morphology of the silicon wafers.

The polishing technology used for manufacturing ultraflat and smooth Si surfaces on a large scale has been chemomechanical polishing (CMP) since more than 25 years. This technique combines the mechanical action of a rotating polymeric polishing pad with the chemical activity of an alkaline polishing slurry containing silica particles [1-4]. A variety of different polishing machines offers single side polishing with and without wax mounting as well as double side polishing in a free float polishing process [5]. Removal rates of hundreds of monolayers of silicon per second can be obtained depending on the pH of the polishing slurry and the applied process conditions.

In the present paper the impact of different chemical and mechanical parameters of the polishing process and the influence of intrinsic properties of the silicon wafer like orientation and doping on the stock removal rate of Si are discussed as well as the mechanism involved. Surface chemistry after polishing is characterized by ultra-high vacuum (UHV) techniques like

Mat. Res. Soc. Symp. Proc. Vol. 386 © 1995 Materials Research Society

X-Ray Photoelectron Spectroscopy (XPS) and High-Resolution Electron Energy Loss Spectroscopy (HREELS). Aspects of surface morphology including microroughness as well as localized surface irregularities, also referred to as Light Point Defects (LPDs), which can consist of particles, pits, locally increased microroughness or micro scratches, are identified by lightscattering equipment and characterized by high resolution techniques like Atomic Force Microscopy (AFM).

INFLUENCE OF POLISHING PARAMETERS ON SILICON REMOVAL RATES

The combination of chemical and mechanical action during polishing is the key for obtaining superior surface quality. With the absence of the mechanical component silicon removal rates are lower and the chemical attack results in anisotropic etching effects due to different etching speed for the various crystal orientations generating characteristic etch patterns [6].

Major impact on the silicon removal rate is observed by variation of the pH value of the polishing slurry (Fig. 1). With increasing pH the stock removal increases until a pH of around 12.5 is reached, indicating the importance of the OH^- ion for the mechanism of silicon removal. Economical polishing therefore uses an alkaline polishing slurry to obtain high silicon removal rates.

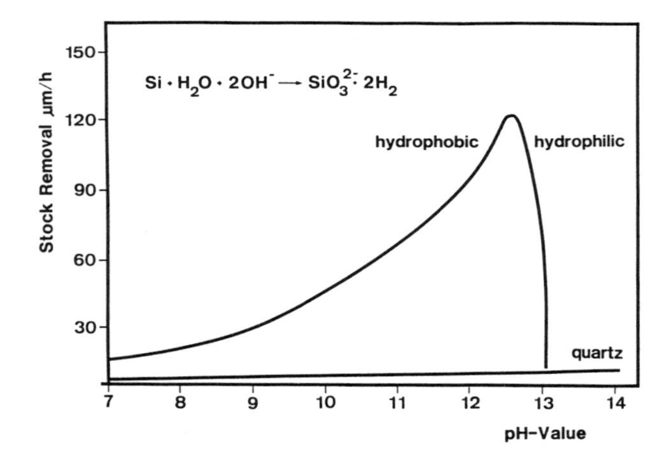

Fig.1

The stock removal rate for silicon increases with the pH of the polishing slurry and shows a maximum value for a pH of 12.5. At higher pH SiO_2 is formed which results in a strong decline of the removal rates approaching the equivalent values for polishing of quartz.

At a pH higher than 12.5 a strong decline in the silicon removal rate is observed reaching the much lower values obtained for polishing of quartz as also indicated in the figure. In this transition range the silicon wafer is oxidized according to the high pH (Nernst equation) and the surface condition changes from hydrophobic to hydrophilic which implies a change of the reaction mechanism to resolving the SiO_2 like in quartz polishing.

Contact angle measurements show a similar dependence on the pH value as observed for the silicon removal rates. A maximum of the contact angle which is equivalent to the highest degree of hydrophobicity is observed for a pH around 12 [7].

Besides the pH value another important parameter strongly influencing the silicon removal rate is the temperature of the polishing slurry. Karaki et al. [2] have reported a logarithmic increase of the silicon removal rate with temperature and have determined characteristic activation energies for the polishing reaction.

Mechanical parameters like the polishing pressure also influence the activation energy. A decrease of the activation energy with increasing polishing pressure is observed resulting in an increase of silicon removal rates. Polishing pressure as well as the flow rate of the polishing slurry and rotation of the polishing pad determine the fluidynamics of the polishing process. The slurry flow controls the supply and the exchange of chemical compounds at the silicon surface and the removal of the reaction products. In parallel the mechanical action causes frictional forces which in turn lead to an increase of the silicon surface temperature, determined as a dominant factor for silicon removal rates.

The polishing slurry with defined pH and temperature for achieving proper polishing conditions contains silica particles which are a prerequisite for the removal mechanism. Increasing the concentration of silica particles was found to increase silicon removal and could be ascribed to a reduction of the activation energy of the polishing process [8]. Another important parameter influencing the performance of the polishing process is given by the polishing pad itself. Different pad dressings as well as aging effects influence structural features of the pad and result in a variation of removal rates [9].

WAFER ORIENTATION AND DOPING

The silicon removal rate is also influenced by intrinsic properties of the wafers. A strong anisotropy with respect to wafer orientation and doping concentration is observed [6]. Si(111) surfaces are etched slower by one or two orders of magnitude as compared to Si(100) generating characteristic etch patterns.

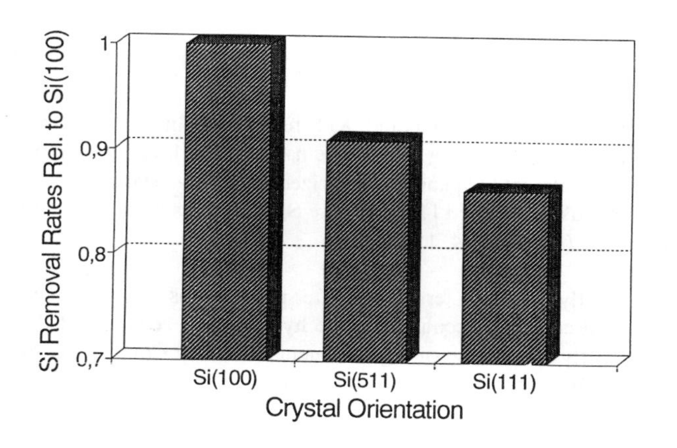

Fig.2

Silicon removal rates for CMP for different surface orientation decreasing in the order Si(100), Si(511) and Si(111). Values are normalized to the removal rate of the Si(100) surface.

Similarly, high boron concentration exceeding $1*10^{19}/cm^3$ is found to result in a strong reduction of etch rates [10].

The selectivity of the chemical reaction is strongly reduced in CMP due to lowering of the effective activation energy by adjusting the appropriate processing conditions. Consequently, the difference in silicon removal rates during polishing is much less pronounced. The silicon removal rate for Si(111) surfaces is observed to be only about 10% less than for Si(100), Si(511) surface orientation was found to be in between (Fig. 2).

Similarly the influence of boron doping concentration is much less pronounced as compared to etching (Fig. 3). Highly boron doped wafers in the range of $10^{19}/cm^3$,

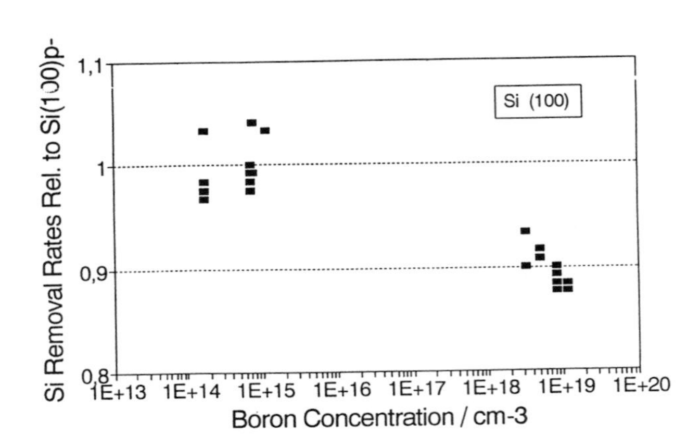

Fig.3

Silicon removal rates for different boron doping concentrations of Si (100) wafers. Removal rates are lower for highly doped p+ wafers typically used as substrates for epitaxial application in comparison to low boron doped p-. Values are normalized to p- wafers.

typically used as substrates for epitaxially grown silicon layers, show removal rates slightly lower as compared to boron doped wafers in the range $10^{14}/cm^3$-$10^{16}/cm^3$ used for different device applications. No reduction of the removal rates was observed for highly doped n-type silicon.

SURFACE CHEMISTRY

The silicon wafer surface after polishing is hydrophobic. XPS reveals an almost oxide free surface after polishing and subsequent water rinse (Fig. 4b,c). The native oxide layer present on the wafer surface after a typical RCA treatment [11] and characterized by a SiO_2 peak shifted to higher binding energies in XPS (Fig. 4d) is removed [12, 13]. The Si 2p spectra after polishing resemble those of an HF treated surface (Fig. 4a) [14, 15].

The silicon surface is predominantly hydrogen terminated after polishing as was examined with HREELS (Fig. 5). The hydrogen coverage accounts for the hydrophobic behavior of the wafer surface as was studied extensively in the case of HF treatments [16, 17]. Comparison of HREELS spectra of Si(111) and Si(100) wafers display some differences.

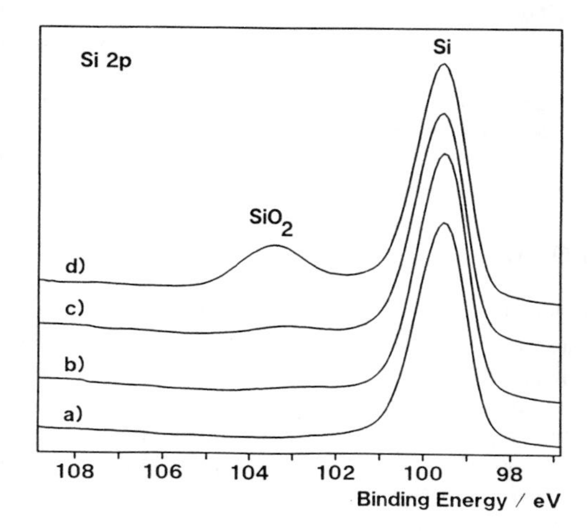

Fig.4

XPS spectra of the Si 2p line after polishing (b: Si(111), c: Si(100)) in comparison to HF dip (a) and RCA clean (d) of a Si (100) surface. The RCA clean generates a native oxide layer on top of the silicon wafer with an oxide thickness of about 7Å, indicted by the shifted SiO_2 component. After polishing and a short subsequent water rinse almost no oxide is found, similar to the silicon surface after HF treatment.

On Si(111) a rather sharp Si-H stretching vibration at around 2070 cm^{-1} is observed indicating predominantly monohydride surface termination [18, 19]. On Si(100) the corresponding peak is much wider and shifted to higher wave numbers consisting of contributions of different valent $Si\text{-}H_x$ portions as well as of oxygen backbonded Si-H [20]. Infrared studies confirm these observations [7].

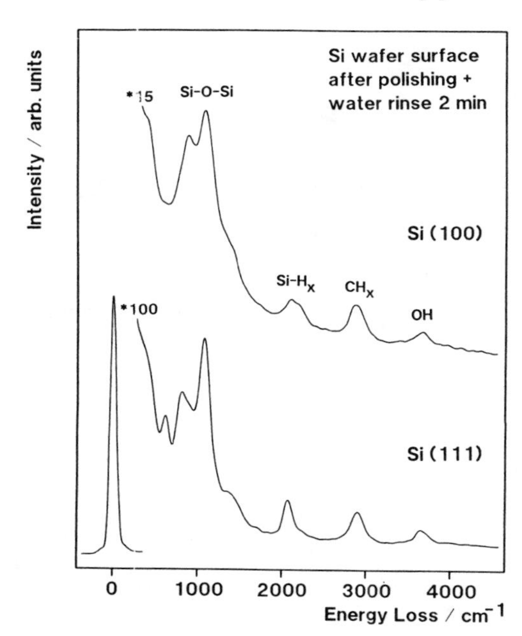

Fig.5

HREELS spectra of Si(100) and Si(111) surfaces after polishing followed by a short water rinse. Both wafer surfaces show pronounced Si-H stretching vibrations (at around 2100 cm^{-1}). The peak of the Si-H stretching vibration on Si(111) occurs at a slightly lower frequency and is more narrow, indicating predominantly monohydride termination. On Si(100) a much broader structure is observed due to various silicon hydride components and oxygen backbonds. The hydrogen termination is responsible for the hydrophobic behavior of these surfaces.

After polishing wafers have to be rinsed to remove remaining slurry from the wafer surface. This water rinse does not influence surface chemistry significantly as was also checked by prolonged water rinsing. Previous investigations of HF treated silicon wafers after various storage times in water have shown only a very slow exchange reaction of surface Si-H by Si-OH on a time frame of hours which could be traced as a preliminary stage for building up an oxide layer [16].

MECHANISM OF SILICON REMOVAL

It has been demonstrated, that the pH value of the polishing slurry has a very strong influence on the silicon removal rate (see Fig. 1). It was further observed, that the OH^- concentration decreases during CMP if the pH of the slurry is not stabilized by buffering. This implies a dominant role of the OH^- ion in the silicon removal mechanism [21].

Another observation during polishing regards the surface condition. At a pH below 12.5 the surface is hydrophobic, chemical analysis of the surface reveals predominantly hydrogen coverage (see Figs. 4 and 5). For higher pH the surface becomes hydrophilic, an oxide is formed and the silicon removal rate drops abruptly to that of polishing of quartz.

We therefore assume a silicon surface in contact with the polishing slurry terminated by some species X, which can be H, OH or OSi as a starting condition for discussing the silicon removal mechanism. These substituents presenting a higher electronegativity as compared to Si induce a polarization of Si-Si backbonds. In case of hydrogen terminated silicon surface atoms the reaction continues to form Si-OH with a reaction rate depending on the pH of the polishing slurry.

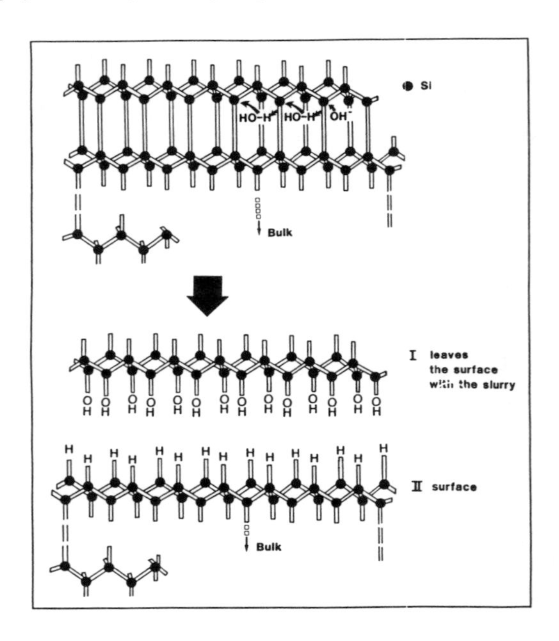

Fig.6 The polishing mechanism suggested involves the corrosive attack of H_2O, catalyzed by a nucleophilic reaction of OH^-. The cleavage of Si-Si backbonds results in the removal of silicon in a lower oxidation state and a remaining hydrophobic, predominantly hydrogen terminated silicon surface.

The silicon removal is initiated by a nucleophilic attack of OH^- ions at a positively polarized Si atom, breaking a Si-Si bond and forming Si-OH (Fig. 6) [21]. The remaining Si atom reacts with a proton supplied by the ambient water forming a Si-H bond and leaving another OH^- available for a further attack on positively polarized Si atoms. With this mechanism the cleavage of Si-Si bonds by water continues producing as reaction products Si-OH containing species leaving the surface and a remaining hydrophobic Si surface terminated by hydrogen.

It should be emphasized, that silicon is not removed from the wafer surface by complete oxidation to SiO_3^{2-}. A brown deposit can be found on the polishing pad by strongly reducing the slurry flow rate which consists of silicon in a lower oxidation state and which can be removed by slightly oxidizing agents [22]. This indicates that the complete oxidation of the silicon species removed from the surface by polishing does not occur during the silicon removal step but rather in subsequent oxidation steps in the polishing slurry or on the polishing pad. The overall reaction of the oxidation of silicon to SiO_3^{2-} accounts for the observed consumption of OH^-.

Even the occurence of SiH_4 during the reaction of water with the silicon surface could be detected [23] demonstrating the corrosive attack of water cleaving Si-Si bonds. This is a further indication that complete oxidation of silicon is not required in order to remove silicon atoms from a wafer surface.

As mentioned previously, polishing of silicon requires the presence of silica particles to obtain silicon removal rates as given e.g. in Fig. 1. The silica particles in the polishing slurry carry a high surface charge which is transferred to the silicon surface.

The surface charge on silica particles with a diameter of 32 nm was investigated for different pH values of the polishing slurry (Fig. 7). The number of OH^- ions is also shown for water at the same pH with a volume equivalent to the size of the silica particles. For a given pH, the OH^- surface concentration on the silica particles is higher by several orders of magnitude as compared to the equivalent volume of the alkaline solution. This high surface concentration of OH^- ions facilitates the attack on the negatively charged silicon surface atoms and the cleavage of Si-Si bonds.

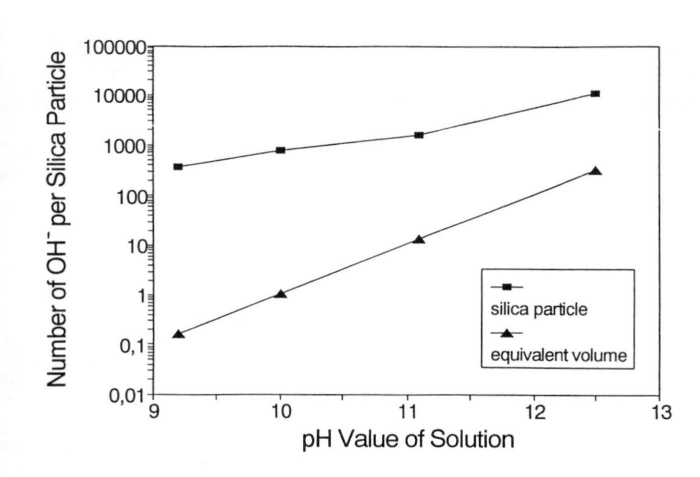

Fig.7

Surface charge on silica particles with a diameter of 32 nm for different pH values of the polishing solution as well as the number of OH^- ions for a volume of water equivalent to the size of the silica particles.

The differences in removal rates for various crystal orientations (see Fig. 2) as well as the dependence on the doping concentration (see Fig. 3) are due to a change in chemical reactivity.

For Si(111) with three backbonds and only one surface bond to another species like H or OH, polarization of the backbonds is reduced as compared to Si(100). Therefore, the initial reaction of the nucleophilic attack of OH^- to surface Si (111) is retarded. The corresponding activation energies observed for KOH etching (20-40%) are reported to be about 0.6 eV for Si(100) [6, 8] and about 0.7 eV for Si(111) [6] accounting for the reduced removal rates for Si(111).

In a band model description the oxidation of a Si atom corresponds to the injection of an electron into the conduction band as was discussed in the case of caustic etching [6]. Due to the downward bending of the bands at the silicon/slurry interface electrons injected remain localized on the Si surface being available for a transfer to water molecules to form Si-H and hydroxide ions available for further reaction.

For high boron concentration in the range of $10^{19}/cm^3$ and beyond the negatively charged surface layer becomes very narrow. The lifetime of electrons injected into the conduction band during silicon oxidation is reduced due to the high hole concentration increasing the recombination probability. Consequently, these electrons are not available for a further reaction with water molecules resulting in a retardation of the removal mechanism [10].

For polishing, the anisotropic behavior concerning crystal orientation and doping concentration is much less pronounced as compared to etching. The mechanical component in CMP strongly reduces the selectivity of the chemical reaction. This results in a lowering of the activation energy and increased silicon removal during CMP as compared to pure chemical etching. A reduction of the activation energy to about 0.2 eV for Si(100) is reported by appropriate increase of the polishing pressure [8].

SURFACE MORPHOLOGY

Polishing leads to a strong reduction of the roughness of the silicon wafer. The controlled flow of the polishing slurry between the polishing pad and the silicon surface results in locally enhanced removal of surface elevations and reduced attack at areas with surface depressions. This selectivity with respect to surface morphology enables a continuous reduction of the overall surface roughness.

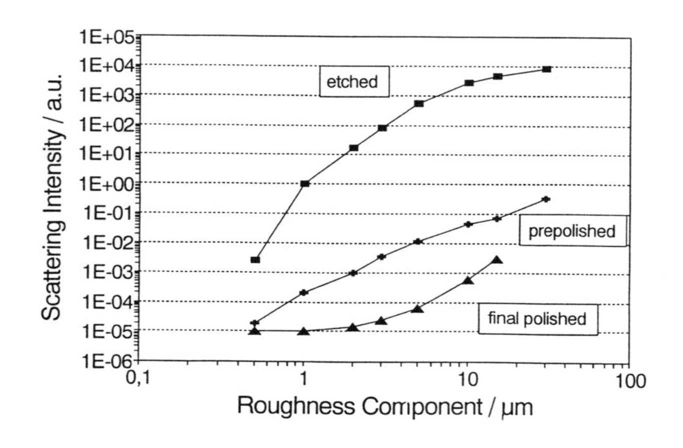

Fig.8

Reduction of surface roughness by polishing for different roughness components extracted from ARLS. Lightscattering intensity drops by several orders of magnitude when going from an etched wafer to a prepolished and final polished wafer surface, respectively.

Angle resolved lightscattering spectroscopy (ARLS) is a sensitive tool providing information on the smoothness of the wafer surface with respect to different roughness components [24]. The relative lightscattering intensity drops by several orders of magnitude when starting with an etched wafer and going to a prepolished and final polished wafer surface (Fig. 8).

After final polishing distinct scattering peaks can be observed on wafers having a slight misorientation demonstrating the existence of atomic steps on the wafer surface with a long range regularity [25]. This also indicates the high surface quality obtained after polishing as well as the sensitivity of the lightscattering techniques.

AFM measurements illustrate the reduction of the overall roughness components (Fig. 9). The corresponding RMS values obtained for the different wafer scans shown are about 1500Å for the etched wafer, 3Å for the prepolished and 0.6Å for the final polished wafer surface, respectively. Note the different lateral and vertical scales used in order to visualize the characteristic texture of the different wafer surfaces. The ratio of the lightscattering intensity for a given roughness component correlates with the square root of the corresponding RMS value.

Surface morphology is characterized by an overall distribution of microroughness for different roughness correlation lengths as well as by localized surface irregularities on the silicon wafer surface. Such irregularities, called LPD in a generic way, can be localized by commercially available scanning lightscattering techniques [26, 27] and may consist of particles, pits or crystal originated particles (COPs) [28, 29].

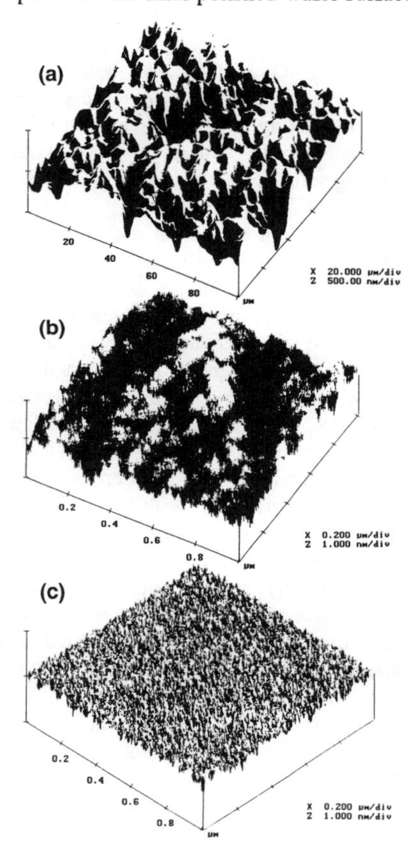

Fig.9 Comparison of AFM maps of wafer surfaces of different roughness after etching (a), prepolishing (b) and final polishing (c). Note the different scales for the lateral and vertical scan width. The corresponding RMS values for the scans showed are 1500Å for the etched wafer, 3Å for the prepolished and 0.6Å for the final polished wafer surface.

Characterization of COPs after polishing by AFM reveals shallow surface depressions with a rather smooth curvature (Fig. 10a). The origin of these defects was traced back by variation of crystal growth conditions resulting in a change of density and defect size distribution [30, 31]. The appearance and shape of the COP after polishing demonstrates the two components involved in CMP: The chemical etching of crystal defects in combination with the surface smoothening obtained by the fluidynamics of the polishing action.

A commonly applied procedure to delineate COPs is to treat wafers by immersion in hot SC1 (Fig. 10b; $NH_4OH:H_2O_2:H_2O=1:1:5$ solution at 85°C applied for 4h). The sharply edged pits originating from this treatment reflect the anisotropic action of the alkaline etching solution.

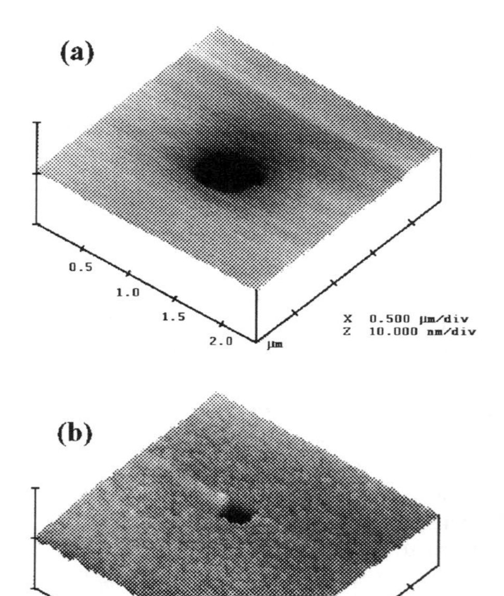

Fig.10 AFM images of a LPD after polishing (a) and a COP after SC1 treatment (b). Polishing results in a shallow surface depression with very smooth curvature whereas etching with the alkaline SC1 solution generates a rather sharply bordered pit.

Inspection of wafers with respect to LPD is typically performed by scanning lightscattering techniques which allow a full wafer mapping of a 200 mm wafer in less than 30 seconds. The measured lightscattering intensity is converted into an equivalent 'particle size' as calibrated by latex spheres (units of LSE, latex sphere equivalents).

COPs after SC1 treatment are sized rather accurately with respect to geometrical dimensions. LPDs after polishing, however, are underestimated systematically because of their shallow appearance and smooth curvature, strongly deviating from a sphere. In the latter case sizing also depends on the equipment used due to different solid angles of the detection unit which also complicates correlations between different instruments [30, 32].

The impact of LPDs occurring as surface defects originating from crystal defects was studied with respect to electrical properties. No correlation was found between yield and LPD within a wide range of LPD densities, whereas a clear dependence could be found for different substrate materials with various amounts of crystal defects [33]. As a consequence, the mere appearance of smooth surface depressions originating from crystal defects is not influencing the electrical performance of silicon wafers.

SUMMARY

The interaction of a variety of mechanical and chemical parameters determines the silicon removal rates and the quality of the wafer surface obtained after chemomechanical polishing. Strong influence is observed by the pH of the polishing slurry confirming the important role of the OH^- in the removal mechanism. The oxidation of silicon atoms by OH^- accompanied by polarization of silicon backbonds and the reduction of water molecules breaking Si-Si backbonds results in the removal of silicon in a lower oxidation state and the formation of a hydrogen terminated hydrophobic silicon surface. CMP yields a reduction of microroughness by several orders of magnitude. The appearance of localized surface irregularities on the silicon surface displays the etching and smoothening effect of the polishing action.

ACKNOWLEDGMENTS

This work has been supported by the Federal Department of Research and Technology of Germany under contract number M 2793 F. The authors alone are responsible for the contents. The contributions and discussions with a number of people involved in different fields of polishing and characterization is highly appreciated, specially to mention T. Altmann, T. Buschhardt, E. Feuchtinger, R. Friedl, H. Hennhöfer, H. Krämer, H. Piontek and R. Rurländer.

REFERENCES

1. E. Mendel, Solid State Technol. **10** (1967) 27.
2. T. Karaki, S. Miyake, J. Watanabe, Bull. Jpn. Soc. of Prec. Engg. **12** (1978) 207.
3. A. Schnegg, M. Grundner, H. Jacob, in Semiconductor Silicon 1986, H.R. Huff, T. Abe, B. Kolbesen PV 86-4, The Electrochemical Society Softbound Proceedings Series, Pennington, NJ (1986) 198.
4. L. Feng-wei, C. Guo-chen, W. Guang-yu, Semiconductor Silicon (1986) Proc. 5th Int. Symp. 183.
5. R. Iscoff, Semicond. Int. (1993) 72.
6. H. Seidel, L. Csepregi, A. Heuberger, H. Baumgärtel, J. Electrochem. Soc. **137** (1990) 3612.
7. G.J. Pietsch, G.S. Higashi, Y.J. Chabal, Appl. Phys. Lett. **64** (1994) 3115.
8. T. Karaki, S. Miyake, J. Watanabe, Bull. Jpn. Soc. of Prec. Engg. **15** (1981) 14.
9. E. Mendel, P. Kaplan, A.V. Patsis, IBM Techn. Rep. 22.2341 (1980).
10. H. Seidel, L. Csepregi, A. Heuberger, H. Baumgärtel, J. Electrochem. Soc. **137** (1990) 3626.
11. W. Kern and D.A. Puotinen, RCA Rev. **31**, 187 (1970).
12. M. Grundner and H. Jacob, Appl. Phys. **A39**, 73 (1986).
13. M. Grundner, D. Gräf, P.O. Hahn and A. Schnegg, Solid State Technology, 69 (1991).
14. D. Gräf, M. Grundner, R. Schulz, and L. Mühlhoff, J. Appl. Phys. **68**, 5155 (1990).
15. V. A. Burrows, Y. J. Chabal, G. S. Higashi, K. Raghavachari, and S. B. Christman, Appl. Phys. Lett. **53**, 998 (1988).
16. D. Gräf, M. Grundner, and R. Schulz, J. Vac. Sci. Technol. **A7**, 808 (1989).
17. G.S. Higashi, Y.J. Chabal, G.W. Trucks, and K. Raghavachari, Appl. Phys. Lett. **56**, 656 (1990).
18. H. Ibach, D. Bruchmann, H. Wagner, Appl. Phys. **A29**, 113 (1982).
19. H. Froitzheim, U. Köhler, and H. Lammering, Surf. Sci. **149**, 537 (1985).

20. J. A. Schaefer, F. Stucki, D.J. Frankel, W. Göpel, and G.J. Lapeyre, J. Vac. Sci. Technol. **B2**, 359 (1984).
21. H. Prigge, P. Gerlach, P.O. Hahn, A. Schnegg, and H. Jacob, J. Electrochem. Soc. **138** (1991) 1385.
22. A. Schnegg, I. Lampert, and H. Jacob, Electrochem. Soc. Ext. Abstr. 85-1, No. 271, Toronto (1985).
23. I. Lampert, H. Fußstetter, and H. Jacob, J. Electr. Soc. **133**, 1472 (1986).
24. P.O. Hahn, M. Grundner, A. Schnegg, H. Jacob, in The Physics and Chemistry of SiO_2 and the Si-SiO_2 Interface, eds. C.R.Helms and B.E. Deal, Plenum Publishing Corp. (1988) 401.
25. P.O. Hahn and M.Kerstan, SPIE Proc. Vol.1009, 172 (1988).
26. E.F. Steigmeier and H. Auderset, Appl. Phys. **A50** (1990) 531.
27. T. Abe, E.F. Steigmeier, W. Hagleitner, and A.J. Pidduck, Jpn. J. Appl. Phys. **31** (1992) 721.
28. J. Ryuta, E. Morita, T. Tanaka, and Y. Shimanuki, Jpn. J. Appl. Phys. **29** L1947 (1990).
29. J. Ryuta, E. Morita, T. Tanaka, and Y. Shimanuki, Jpn. J. Appl. Phys. **31** L293 (1992).
30. M. Brohl, D. Gräf, P. Wagner, U. Lambert, H. A. Gerber, H. Piontek; ECS Fall Meeting Volume 94-2, Miami Beach, Oct. 9-14, 1994, p. 619.
31. H. Klingshirn, and P. Gerlach, UCS **3**, 407 (1991).
32. P. Wagner, M. Brohl, D. Gräf, U. Lambert; Mat. Res. Soc. Symp. Proc. (1995), to be published.
33. D. Gräf, M. Brohl, S. Bauer-Mayer, A. Ehlert, P. Wagner, and A. Schnegg; Mat. Res. Soc. Symp. Proc. Vol. 315, 23 (1993).

POST-CMP CLEANING OF W AND SiO_2: A MODEL STUDY

Igor J. Malik, Jackie Zhang, Alan J. Jensen, Jeffrey J. Farber, Wilbur C. Krusell
OnTrak Systems, Inc., 1753 S. Main St., Milpitas, CA 95035

Srini Raghavan, Chilkunda Rajhunath
University of Arizona, Dept. of Materials Science and Engineering, Tucson, AZ 85721

ABSTRACT

Chemical-Mechanical Planarization (CMP) of SiO_2 is performed using alkaline silica slurries while CMP of tungsten (W) utilizes acidic slurries with alumina as the abrasive. Proposed mechanisms for the two CMP processes, with more emphasis on SiO_2-CMP, have been discussed in literature. However, much less is known about the removal mechanism of residual slurry particles from the planarized surfaces - a crucial step for subsequent device processing. We discuss the chemical and physical basis of post-CMP cleaning by double-side scrubbing using polyvinyl alcohol (PVA) brushes and show how the interactions between the wafer surface, slurry, and the brush material affect the overall cleaning efficiency. Using the zeta potential concept the common features for cleaning surfaces after SiO_2-CMP and W-CMP are established and the differences between these two systems are highlighted. We present surface particle levels for two model systems as a function of cleaning chemistries and discuss their influence on post-CMP surface metal levels.

INTRODUCTION

Aqueous cleaning is one of the most frequently used steps in semiconductor manufacturing. CMP - an increasingly important technology in multilevel metallization [1,2] - is a typical semiconductor manufacturing process step that requires subsequent cleaning. This post-CMP cleaning step is the topic of this paper.

An established view in wet Si cleaning [3] is that alkaline baths (e.g., SC-1) have high particle removal efficiency but low surface metal removal efficiency while acidic baths (e.g., SC-2) remove surface metals but often re-deposit particles. The experiments described in this work reflect this view: We compare the cleaning efficiencies (with emphasis on particles) for cleaning processes characterized by different pH values. The alkaline cleaning solutions used in this work contained ammonia, the acidic ones contained citric acid - an effective chelating agent for surface metal removal [4]. The cleaning was performed by double-side scrubbing, a well established method for post-CMP cleaning [5-9].

Two CMP processes of high importance to the semiconductor industry at this time are SiO_2 and W planarization [2]. SiO_2-CMP is performed using slurries with fumed silica of colloidal size under alkaline conditions and the final surface to be cleaned is a planarized SiO_2 surface. On the other hand, acidic colloidal alumina slurries are used in W-CMP and the final surface for cleaning is predominantly planarized SiO_2 with small W-areas (plugs). In our model study we approximate both of these surfaces with native-oxide-covered Si wafers. The post-CMP contamination is simulated by a well-controlled slurry-dip procedure [10] designed to closely mimic the environment the

Mat. Res. Soc. Symp. Proc. Vol. 386 © 1995 Materials Research Society

wafers encounter immediately after termination of a CMP step using a particular slurry. The chemical conditions on the wafers during the double-side scrubbing were varied by using dilute solutions of ammonia and citric acid that cover a pH range from 3.2 to 10.4. Native oxide-covered Si wafers are chemically equivalent to the post-CMP SiO_2 surfaces while having lower haze allowing more reliable surface particle counting by methods based on light scattering.

The trends in the efficiency of particle removal can be explained by electrostatic interactions between the surface, brush, and the slurry. We use the zeta potentials of the materials of inteterest to explain these trends. This approach is similar to the treatment of charged particles in process fluids [11] but with the addition of the brush material to obtain a more complete picture.

EXPERIMENTAL

The experiments were performed in a Class 10 cleanroom. 150 mm diameter epitaxial Si(100) wafers covered with native oxide were used in the experiments. Particles were counted on 50 as-received wafers before the start of the experiments; the average count was 278 at 0.15 μm with large differences between wafers (min.=12, max=3744). The wafers were then contaminated in a controlled way using the slurry-dip procedure [10]: a 1 minute immersion in a 1:14 = slurry:DIW (deionized water) mixture by volume followed by a 1 minute DIW rinse and a wet transfer to the input of a double-side scrubber. The slurries used were CAB-O-SPERSE SC-1 (Cabot/Rippey, average particle size = 30 nm, pH=10.1 after 1:14 dilution) - an alkaline silica slurry used for SiO_2-CMP, and MSW 1000 (Rodel, average particle size=230 nm, pH=4.7 after 1:14 dilution) - an acidic alumina slurry used in W-CMP.

The wafer cleaning was performed in an OnTrak Systems double-side scrubber DSS-200. Figure 1 illustrates its layout and outlines the wafer cleaning process in DSS-200: a cassette of wafers is placed in a wet input station (DIW sprays), from here wafers are sent individually through two brush boxes and a spin/rinse/dry module

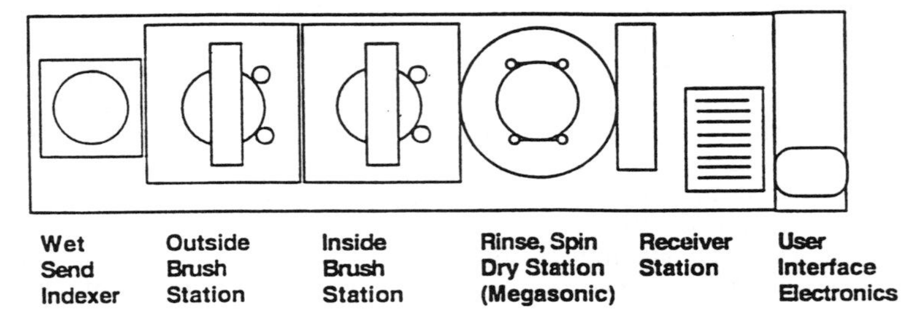

Figure 1. DSS-200 System Configuration

into an output cassette. Inside the brush boxes the wafers are scrubbed with polyvinyl alcohol (PVA) brushes (Rippey/Kanebo). DIW is supplied through the brush cores and penetrates the porous structure of PVA. The surface of the brushes has round-shaped nodules with open spaces between them to allow a sufficient DIW flow across the wafer surface. Chemicals can be delivered to the brushes from teflon drip manifolds; the chemicals are delivered at a rate of approximately 1 ml/s. The cleaning cycle inside each brush box is set to 40 s. Three cleaning recipes were used in this study:

1. DIW only,
2. 2% ammonia solution drip, and
3. 0.5% citric acid solution drip.

The 2% ammonia solution was prepared from 29% gigabit grade ammonium hydroxide (Ashland Chemicals). [The concentrations are by weight, calculated as NH_3.] The 0.5% citric acid solution (by volume) was prepared by measuring the volume of Enzyme grade anhydrous citric acid (Fisher Scientific) in a volumetric cylinder. pH of the liquid layer on the wafer surface was measured by an Accumet 1001 pH meter (Fisher Scientific). The chemicals (recipes 2 and 3) were dripped on the brushes for 40 s in the 1st brush box and 30 s in the 2nd brush box. The chemicals were delivered to the PVA brushes through a point-of-use 0.1 µm hydrophilic nylon filter (N66, Pall).

The surface particles were counted by Tencor 6420; this is an unpatterned wafer inspection station with an angle of incidence of a 488 nm laser beam 70° from the surface normal [12]. All particle counts reported in this paper were obtained from medium scan rate maps with p-p (incident beam - detector) polarization setting. The threshold was set at 0.15 µm based on a polystyrene latex (PSL) spheres calibration curve for bare Si.

The electrokinetic (zeta) potential of PVA, SiO_2, Al_2O_3, and W particles was measured using an electrophoretic method. A Penn Kem 501 Laser Zee Meter was used to make the electrophoretic measurements in the pH range of 2.0 to 10.5. Solution pH adjustments were done using HCl and KOH, or citric acid and NH_4OH.

RESULTS AND DISCUSSION

Fig. 2 shows surface particle data from a group of wafers cleaned after the CAB-O-SPERSE SC-1 slurry dip; wafers 1-13 were cleaned in DSS-200 using DIW only while citric acid drip was used for wafers 14-23. There is a clear difference between the two groups of wafers: the average count for the DIW-only wafers is 13 while the average count rises to 35 for the wafers cleaned with citric acid. It is interesting to note that we did not observe such a clear difference in particle counts on a limited number of post-CMP SiO_2 wafers (see Table II); studies using larger sample sizes are planned.

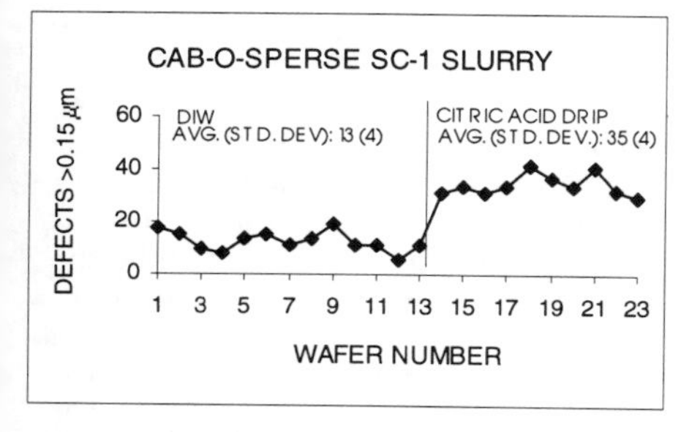

Fig. 2. Post-cleaning defects for SC-1 slurry. DIW Only (wafers 1-13), citric acid drip (wafers 14-23).

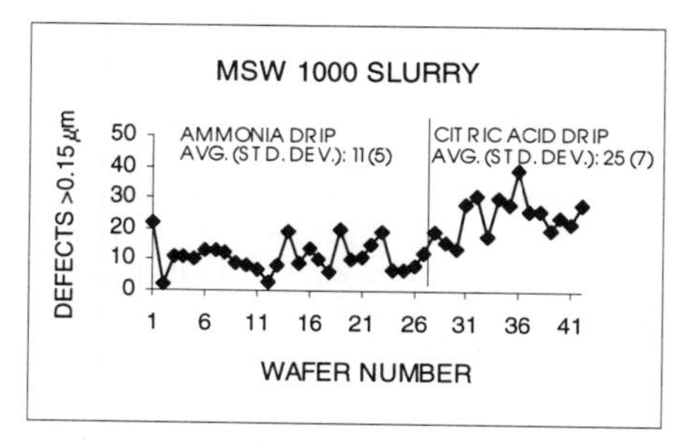

Fig. 3. Post-cleaning defects for MSW 1000 slurry: Ammonia drip (wafers 1-27), citric acid drip (wafers 28-42).

Fig. 3 shows particle data for two cassettes of wafers cleaned after the MSW 1000 slurry dip; wafers 1-27 were cleaned in DSS-200 using ammonia drip recipe. After the first 27 wafers the brushes were flooded with citric acid solution to wash out the remaining ammonia solution. Then the rest of the wafers was cleaned with citric acid drip recipe. The average particle count is 11 for the first group of wafers, and 25 for the second group.

The trends illustrated in Figs. 2 and 3 were consistent throughout the whole set of experiments. The results from these experiments are summarized in Table I: wafers as received are on the first line; the two groups of wafers that were slurry-dipped but not DSS-cleaned were sent from the double-side scrubber input (without the brushes touching the wafers) directly into the spin/rinse/dry station; this is an equivalent of a rinse-dry cleaning process - clearly insufficient for post-CMP cleaning.

TABLE I

Average particle counts (>0.15 μm, Tencor 6420) as a function of treatment

slurry dip	DSS-clean	# of wafers	approx. pH	rinse +dry	#of particles (xbar±σ)	
none	none	50	-	no	278	±597
Rippey SC-1	none	10	-	yes	879	±540
Rippey SC-1	citric	173	3.2	yes	44	±17
Rippey SC-1	DIW	88	6.6	yes	9	±4
Rippey SC-1	ammonia	99	10.4	yes	7	±4
Rodel MSW 1000	none	3	-	yes	3805	±1195
Rodel MSW 1000	citric	15	3.2	yes	25	±7
Rodel MSW 1000	ammonia	27	10.4	yes	11	±5

Figs. 2 and 3 and Table I show clear differences in the final surface particles count as a function of the cleaning process. The alkaline cleaning conditions result in lower particle counts. Although the average particle count and the std. dev. (σ) between DIW-only and ammonia-drip process are very close in this set of data, historically, based on a large amount of data, DIW-only process usually has shown lower stability (higher σ).

We interpret the results in terms of differences in zeta potentials of the slurry particles, brush and wafer surface. Fig. 4 shows the zeta potential - pH profiles for SiO_2 (SiO_2-CMP slurry particles, wafer surface), Al_2O_3 (W-CMP slurry particles), PVA (brush material), and W (abraded/chipped particles during W-CMP). For electrostatic interactions to be of the greatest benefit in the cleaning action, the zeta potential values of all three components of the system (wafer, contaminants, brush) should be of the same sign and large absolute value. In this way, repulsion will be the dominating interaction between the contaminant and both the wafer and the brush and the chances for the particle dislodged from the surface by the brush-cleaning action to be removed from the wafer surface by a stream of liquid without reattachment will be maximized. It is important to maintain the repulsive nature of the interaction not only between the particle and the surface but also between the particle and the brush. If this is not the case particles would load the brush and after an initial period of good particle removal from the wafer (and at the same time a period of brush loading with particles) the brush would start serving as a particle source. While the data presented in this study are consistent with the zeta potential considerations, capillary interactions [13] may also play a role in the overall cleaning efficiency.

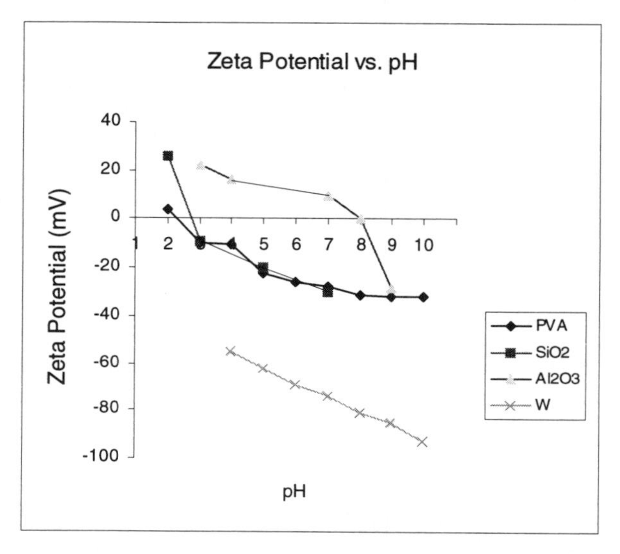

Figure 4. Zeta Potential as function of pH for PVA, SiO_2, Al_2O_3, and W.

It is clear from Fig. 4 that the repulsive interactions will be maximized at high pH. Therefore, for the lowest particle counts the cleaning should be performed under alkaline conditions as shown in Figs. 2 and 3. However, there are additional considerations to particle performance when evaluating a wafer cleaning process. If surface metals need to be brought to low levels acidic cleaning is preferable. This point is illustrated in Table II which shows surface metal data for polished thermal oxide (using Rodel ILD 1300 slurry) after two diferent DSS-cleaning processes [14].

TABLE II.

Post-CMP Thermal Oxide Surface Metals in 10^{10} atoms/cm^2 by TXRF

	K	Ca	Fe	Ni	Cu	Zn	Defects>0.3μm Tencor 7600
as oxidized, no clean	<20	20	4	<2	<2	<3	-
CMP, ammonia drip DSS clean	<20	90	10	<2	20	200	5
CMP, citric acid drip DSS clean	<20	<10	2	<2	<2	<3	3

The benefits of citric acid for surface metals reduction are obvious. We conclude by stating that a cleaning process can be tailored to the particular needs of a process engineer - e.g., if surface metal contamination is considered highly important it may be desirable to accept slightly higher particle counts while reducing the surface metals substantially.

SUMMARY

Double-side scrubbing using PVA brushes is an effective post-CMP cleaning process. We presented a model study for cleaning wafer surfaces after SiO_2-CMP and W-CMP. For slurry particle removal a suitable choice of cleaning chemistry is important. Chemistry controls the electrokinetic interactions in the surface-brush-particle system and the particle detachment mechanism from both the surface and the brush. Although the experimental results are consistent with the presented model more experimental work together with a quantitative analysis is required for a satisfactory understanding of post-CMP cleaning.

ACKNOWLEDGMENTS

The authors would like to acknowledge M. Ravkin, F. Mohr, J. M. de Larios (OnTrak Systems), R. Shah, S. H. Peterman, and J. Stavins (Sematech) for insightful discussions, and J. Metz (Charles Evans & Associates) for performing the TXRF measurements.

REFERENCES

1. Handbook of Multilevel Metallization for Integrated Circuits: Materials, Technology, and Applications, S. R. Wilson, C. J. Trace, and J. L. Freeman, Jr., Eds., Noyes Publications (1993).
2. W. C. O'Mara, Semiconductor International, 140 (July 1994).
3. W. Kern, J. Electrochem. Soc. **137**, 1887 (1990).
4. H. Akiya, S. Kuwano, T. Matsumoto, H. Muraoka, M. Itsumi, and N. Yabumoto, J. Electrochem. Soc. **141**, L139 (1994).
5. S. R. Roy, I. Ali, G. Shinn, N. Furusawa, R. Shah, S. Peterman, K. Witt, S. Eastman, and P. Kumar, J. Electrochem. Soc. **142**, 216 (1995).
6. W. C. Krusell and J. M. de Larios, SEMICON Korea 95 Proceedings, p.39, SEMI (1995).
7. D. L. Hetherington, P. J. Resnick, R. P. Timon, B. L. Draper, M. Ravkin, J. M. de Larios, W. C. Krusell, and A. F. Madhani, Proceedings: First International Dielectrics for VLSI/ULSI Multilevel Interconnection Conference (DUMIC), p.156, DUMIC Catalog Number 95 ISMIC-101D, Library of Congress No. 89-644090 (1995).
8. D. I. Golland, P. D. Albrecht, W. C. Krusell, and F. A. Puerto, Semiconductor International, Sept. 1987.
9. W. C. Krusell, SEMICON WEST 1994 Proceedings, Planarization Technology: Chemical Mechanical Polishing (CMP), p.108, SEMI (1994).
10. OnTrak Systems, Inc., DSS-200 Process Acceptance Procedure.
11. I. Ali, S. Raghavan, and S. H. Risbud, Microcont., 92 (April 1990).
12. R. S. Howland, Semiconductor International, 164 (July 1994).
13. M. B. Ranade, Aerosol Science and Technology **7**, 161 (1987).
14. I. J. Malik, A. J. Jensen, D. J. Hymes, and W. C. Krusell, unpublished data.

CLEANING OF SiO_2: DIFFERENCES BETWEEN THERMAL AND DEPOSITED OXIDES

M. Ravkin, J.J. Farber, I.J. Malik, J. Zhang, A.J. Jensen, J.M. de Larios, W.C. Krusell,
OnTrak Systems, Inc., Milpitas, CA 95035

Abstract

The effects described in this paper are specific to cleaning with mechanical brush scrubbing. Oxides, both thermally grown and deposited are common cleaning applications for mechanical brush scrubbing. Thermally grown oxides present higher final defect counts after scrubbing with deionized water compared to deposited oxides. In this paper we present our results on the cleaning of unpolished and polished oxide surfaces and show differences in these results to be dependent on the chemistry used for cleaning and the degree of hydrophilicity of the wafer surface.

Introduction

Due to the increased presence of mechanical brush scrubbers in the fab brought about by the wide acceptance of CMP, the surfaces of many layers commonly employed in the IC manufacturing process are also being cleaned with the scrubber. Scrubbers equipped with Polyvinyl Alcohol (PVA) brushes that come in direct contact with the wafer surface have shown excellent particle removal performance [1,3]. In particular, the OnTrak double-sided scrubber has demonstrated particle removal ability on as-deposited and polished oxide surfaces while being neutral to metallic contamination [1,2]. As-grown thermal oxide is another surface that often requires scrubbing to reduce particle contamination prior to the next step in the device process. In many cases, scrubbing of the thermal oxide with deionized water (DIW) has yielded increased particle counts after scrubbing. On the other hand, scrubbing with the addition of a small quantity of dilute ammonium hydroxide has shown good results effectively removing particles from the wafer surface (Figure 1).

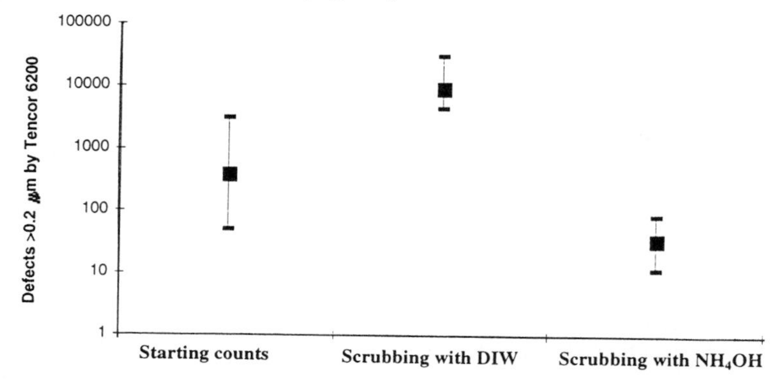

Figure 1 Average post DSS defect counts from different evaluations on thermally oxidized wafers. The effect of 2% ammonium hydroxide dispensed on the brush vs. the DIW cleaning.

Mat. Res. Soc. Symp. Proc. Vol. 386 © 1995 Materials Research Society

This behavior is specific to as-grown thermal oxide processed on a scrubber and not being subjected to any chemical pre-cleaning. The response of TEOS wafers to mechanical brush scrubbing is different from that of thermal oxide wafers. Typical results for scrubbing TEOS or PETEOS wafers with DIW or dilute ammonium hydroxide are shown in Figure 2.

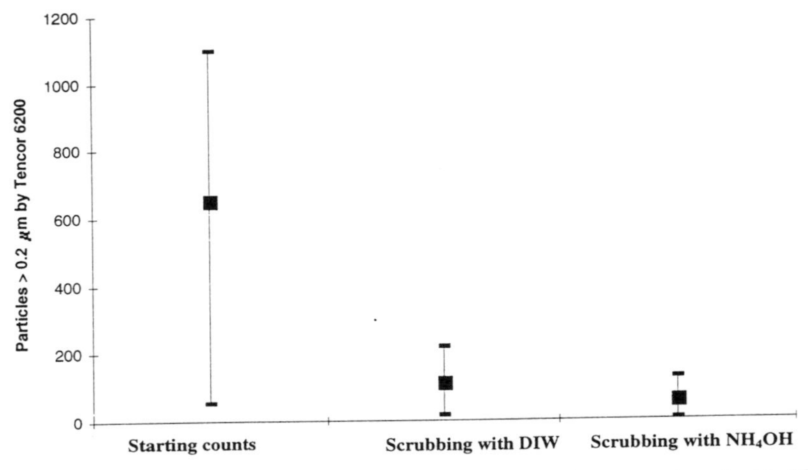

Figure 2 Average post DSS defect counts from different evaluations on as-deposited TEOS wafers scrubbed with DIW and 2% ammonium hydroxide.

The number of defects (light point defects- LPD) detected by the Surfscan instrument after DIW scrubbing thermally oxidized wafers tends to increase. A similar trend has been noticed while scrubbing hydrophobic silicon wafers. In this work we attempt to quantify particle levels on wafers processed through the scrubber and relate results to the wafer surface state as revealed by contact angle measurements.

Experimental

All experiments were performed in either a class 10 or class 1 clean room. All defect counts were obtained on a Tencor Surfscan 6420 or 6200. The calibration of the Tencor instruments was performed for every oxide thickness group by depositing PSL spheres on a wafer with the appropriate oxide thickness. The scrubber used in this study was a DSS 200 double-sided scrubber from OnTrak Systems. The scrubber is equipped with cylindrical PVA brushes manufactured by Rippey/Kanebo that come in direct contact with the wafer surface. Deionized water is supplied through the brush core and flows through the brush onto the wafer surface. Dilute ammonium hydroxide or other chemicals may be supplied from a bulk delivery system or from a pressurized canister with a Teflon lining manufactured by Fisher. Chemicals are dispensed on the brush either through drip or spray manifolds mounted in the brush station. The complete description of this scrubber may be found elsewhere [1, 3, 5]. Pre-scrub cleans with SC-1 and HF were performed in a Verteq wet bench. Concentration of the SC-1 solution in these experiments was 10:2:1, DIW: H_2O_2: NH_4OH respectively. The HF concentration was 0.5% for all tests. Oxide polishing was performed on either a Cybec 3900 or Westech 472 polisher using fumed silica slurry. The polishing pad used in these experiments was a Rodel IC1000/Suba IV stack.

Results and Discussion

We evaluated the effect of the hydrophobic versus hydrophilic silicon surface state on final (post DSS) defect level. Wafers after SC-1 wet cleaning were dipped in diluted (0.5%) hydrofluoric acid (HF) to remove chemical oxide from one half of the wafer. Wafers were rinsed in DIW and processed on the scrubber. All wafers yielded similar results. The characteristic defect map from the Tencor 6420 surfscan is shown in Figure 3. It is apparent from the defect distribution that the hydrophilic part of the wafer yielded low defect counts while the hydrophobic part showed high LPD count. Results were verified on the optical microscope and particles of various sizes were detected to confirm LPD counts.

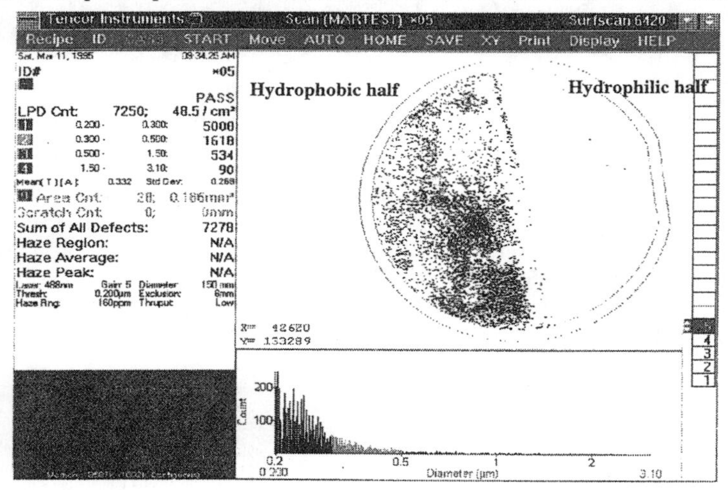

Figure 3 The defect map of the partially hydrophobic (H-terminated) silicon wafer processed on the scrubber.

A group of thermally oxidized wafers was sequentially scrubbed with DIW and ammonium hydroxide and contact angle measurements of a water droplet on the wafer surfaces were performed after each process step. First, incoming wafers (as-grown) were processed on a scrubber with the brushes "up"- no contact between a brush and a wafer. No significant difference in LPD counts was detected. Defects were counted using a Surfscan instrument. Both defect and contact angle data is presented in Figure 4. The results demonstrate dependence of post DSS defect level on the state of the wafer surface. The surface of as-grown thermal oxide is generally hydrophobic with typical contact angles ranging from 30 to 50 degrees depending on the oxidation conditions and the time and humidity of the ambient. Ammonium hydroxide and other strong bases are known to facilitate hydration of the oxide surface [4].

To further verify the effect of surface hydration separate groups of wafers were polished to remove approximately 2000 Å of thermal oxide with the Cabot SS-12 slurry. The polished surfaces yielded contact angles close to zero and the number of LPDs detected were consequently quite low. The low contact angle could again be explained by the hydration of the oxide in the presence of KOH stabilized slurry. Similar results were obtained when ILD 1300 from Rodel - NH_4OH stabilized slurry was used for polishing. Results are summarized in Figure 5 and are typical for oxide wafers polished in one step on the IC1000/Suba IV stack [2, 5]. In this case defect counts were obtained on the Tencor Surfscan 6200.

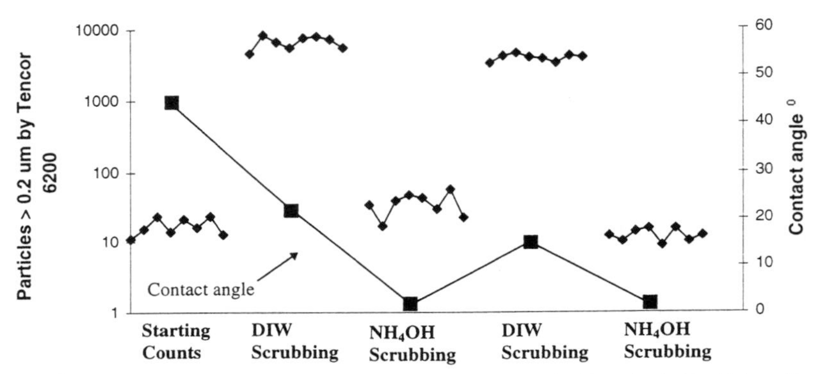

Figure 4 Sequential cleaning of as-grown oxide with DIW and ammonium hydroxide on the double-sided scrubber DSS 200. The change in the Light Point Defect density correlates with the change in the contact angle.

Etching of thermally oxidized wafers in HF solution followed by DIW rinse also hydrolyzes the oxide surface. The oxide surface hydrolyzes during wafer rinse by replacing fluorine with OH groups [6, 7] (Figure 5). Therefore results clearly indicate the dependence of light point defects (LPD) found on the wafer on the hydrophilic or hydrophobic state of the surface.

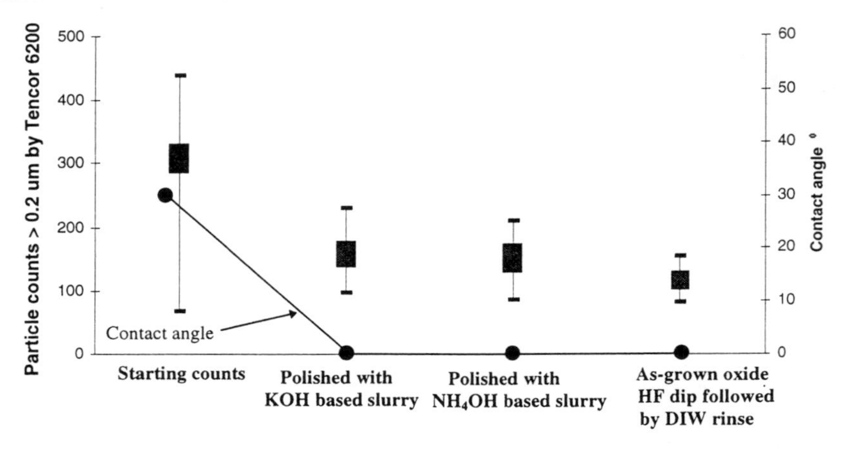

Figure 5 LPD counts and contact angle on the polished and HF rinsed thermal oxide wafers scrubbed with DIW.

PVA brushes always contain particles within their pore network and effectively exchange particles with the wafer surface when brought into intimate contact during cleaning cycle. The attachment of particles for hydrophobic surfaces may occur through the electrostatic attraction of opposite charged surface and particles and through the multiple solid-liquid interfaces present on the wafer surface during scrubbing [7, 3]. Hydrophilic surfaces are covered with a film of water and therefore are less susceptible for particle deposition. The reason for the surface to be hydrophobic may be different such as hydrogen termination of silicon or siloxane structure of thermal oxide [4] but effects are similar for particle deposition. Figure 6 shows the Tencor Surfscan 6420 LPD defect map for the case when a PVA brush was brought into contact with a thermally oxidized wafer with measured contact angle of 30^0. In this case neither the brush nor the wafer were rotating. The marks from the brush nodules are clearly seen on the wafer surface. Similar results were obtained for the hydrophobic silicon wafer. On the other hand wafer with the polished thermal oxide surface (contact angle ~0^0) processed in the same way show no difference in LPD counts.

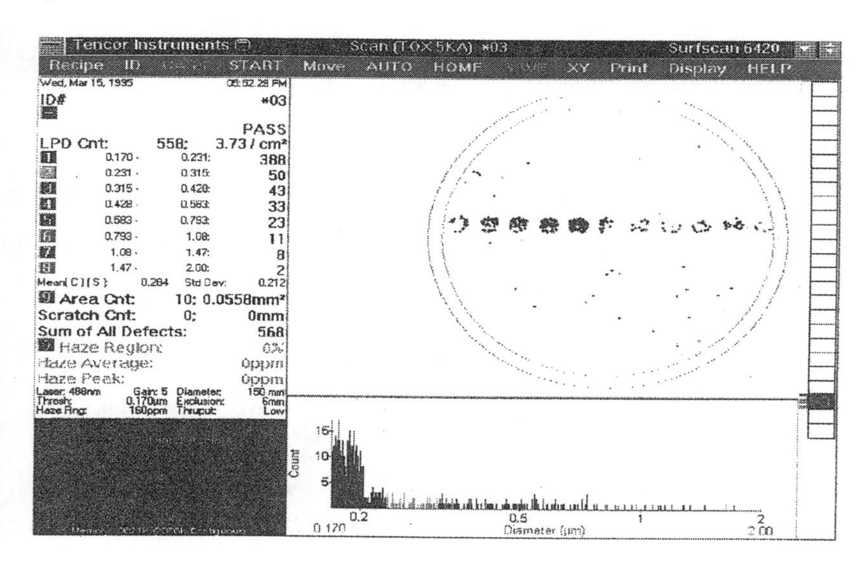

Figure 6 Particles deposited on the as-grown oxide (hydrophobic) wafer surface from the PVA brush. Both the brush and the wafer were not rotating and the brush was brought in contact with the wafer to the nominal compression setting.

Conclusion

Hydrophilic surfaces are readily cleaned by the Polyvinyl Alcohol (PVA) brushes via removal of the particle from the wafer surface, trapping it within the brush and releasing into the water layer on the wafer surface. Hydrophobic wafers attract particles from the brush where the deposition of particles may occur directly from the brush or through the multiple solid-liquid interfaces always present on the hydrophobic surfaces during scrubbing. Differences in mechanical brush cleaning of thermal and deposited oxides could be explained by the different degrees of hydrophilicity of these oxides. Thermally grown oxide presents siloxane surface composed of silicon-oxygen bonds and shows properties of the hydrophobic surface. This does not rule out the possibility of organic contamination on the oxide surface also resulting in the phobic surface. Deposited oxides contain large numbers of silanol (OH) groups and are therefore hydrophilic.

Mechanical brush scrubbing is a proven technique that shows good results when cleaning hydrophilic surfaces. For a hydrophobic surface the use of appropriate chemistry (APM, SPM, O_3) is required to change the surface state prior to scrubbing.

Chemical-mechanical polishing (CMP) of oxide layers using basic slurries effectively hydrolyzes oxide surface consequently making it suitable for mechanical brush scrubbing. The use of dilute ammonium hydroxide in the scrubber also helps to hydrolyze oxide and achieve good cleaning results.

Acknowledgments

We would like to acknowledge Dale Hetherington from Sandia National Laboratories for his help with measurements and participation in discussions.

References

1. W.C. Krusell, J.M. de Larios, The Resurgence of Mechanical Brush Scrubbing, Semicon Korea, Presentation.
2. D.L. Hetherington, P.J. Resnick, R.P. Timon, B.L. Draper, M. Ravkin, J.M. de Larios, W.C. Krusell, A.F. Madhani, (February, 1995 VUMIC).
3. I.J. Malik, J. Zhang, A.J. Jensen, J.J. Farber, W.C. Krusell, O2. 4, MRS, (1995 Spring).
4. R.K. Iler, The Chemistry of Silica, pp. 643-700, (John Wiley & Sons, NY, 1979).
5. S.R. Roy, I. Ali, G. Shinn, N. Furusawa, R. Shahn, S. Peterman, K. Witt, S. Eastman, J. Electrochem. Soc., Vol. 142, No. 1, (January 1995).
6. L. Mouche, F. Tardif and J. Derrien, J. Electrochem. Soc., Vol. 141, No. 6, 1684, (June 1994)
7. S. Verhaverbeke, R. Messoussi and T. Ohmi, UCPSS, P35, (1994).

Post CMP Cleaning Using Ice Scrubber Cleaning.

N.Takenaka, Y.Satoh, A.Ishihama and K.Sakiyama
VLSI Development Laboratories, SHARP Corporation
2613-1, Ichinomoto-cho, Tenri, Nara, 632 , JAPAN

ABSTRACT

The surface cleaning technology with the use of ice scrubber cleaning has been developed to remove the particles after Chemical Mechanical Polishing (CMP) process. The ice particles with a high speed nearly equal to the sound velocity bombarded the Si wafer surface, as a result, the residue from the slurry solution was reduced from ~5/cm^2 to ~0.05/cm^2 and the metal impurities are completely eliminated below the defect limitation for ICP mass spectroscopy. The charge build-up damage due to the high speed particles is not introduced into the the MOS capacitors. This technology is quite effective, compared with the conventional brush scrubber method and is applicable for the cleaning process below the quarter micron devices.

1. INTRODUCTION:

When the scale down of the devices below quarter micron, the surface planarization technology becomes very important. Chemical mechanical polishing (CMP) has been studied[1-4]. Recently it has attracted much attention to the application for the planarization of the inter-metal dielectric layers[5-8], damascene process[9,10], polysilicon etch back process[11] and fabricating largearea ultrathin silicon-on-insulator (SOI) layers[12]. However, the cleaning of particles from the polished wafers is key issue and must be solved in order to be widely applicable for silicon process. Among after cleaning technology of CMP, brush scrubber cleaning is conventionally used for the cleaning method. However, the efficiency to remove particles are largely degraded, because the brush is consumed by using repeatedly, so it is difficult to keep the stabile cleaning efficiency. On the other hand, the ice scrubber technology[13] supplied the ice particles stably, so, it is effective to keep the constant cleaning efficiency. And the particles due to the slurry solution eliminated completely, even though, the solvent contained fumed silica are evaporated and the solid-state particles are strongly adsorbed on the Si wafer. This paper describes the effectiveness of this ice scrubber cleaning.

2. EXPERIMENTAL:

Figure 1.shows the ice scrubber cleaning equipment. It consists of five units, a wafer loading unit, an ice generating unit, a cleaning unit, a chemical cleaning with drying unit and a wafer unloading unit. In the loading unit, the wafer carrier is set in DI water bath to keep the polished wafer surface wet. Figure 2 give the detail of the cleaning unit and the ice generating unit. In the ice generating unit, small ice particles are produced. The hoppers are kept at low temperature (from -110 degree to -140 degree) by liquid N_2. DI water spray through a spiral spray nozzle into the freezing hopper. At the same time, liquid N_2 is also supplied into it. And ice particles which have between 10 micron and 300 micron diameters are produced.

Mat. Res. Soc. Symp. Proc. Vol. 386

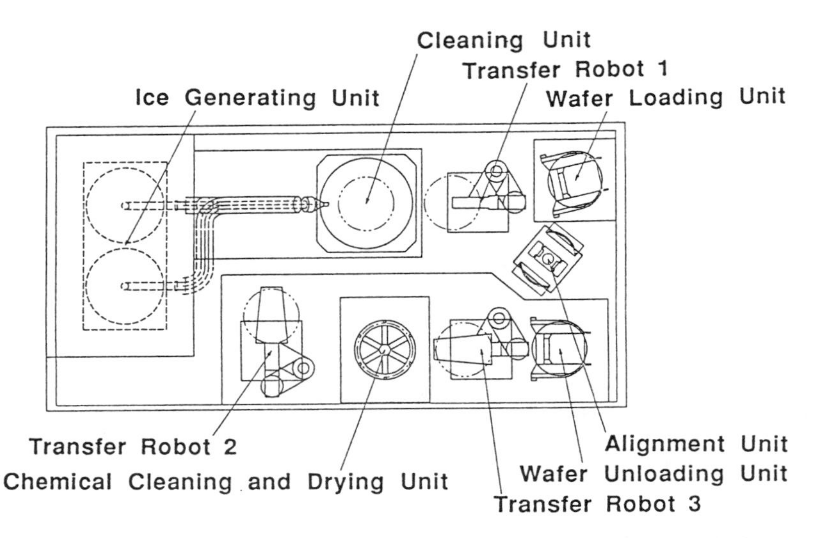

Figure 1. Schematic diagram of the Ice Scrubber Cleaning equipment.

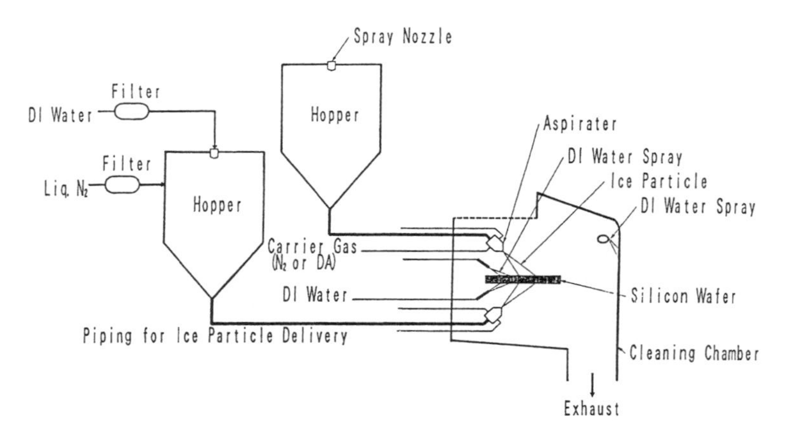

Figure 2. The detailed construction of the cleaning unit.

Ice particles produced in the freezing hopper are jetted from the two jet nozzles on the high speed dry air flow to wafer surface and backside. The speed is nearly equal to the velocity of sound. And the high speed ice particles bombard the particles on the wafer. They knock particles out of the wafer surface. And to prevent readsorption, DI water are spraied to the wafer and the wall. Particles removed from the wafer are carried out by the DI water flow. And the wafer moved horizontally with rotation in order that ice particles cover the whole wafer surface spirally. Next in the chemical cleaning unit, a wafer is cleaned with 1% HF solution, rinsed with DI water and finally dried by high speed rotation.

The dry surface wafers were prepared to study the effect of removing slurry particles. Thermal oxide wafers are dipped into the slurry which contains fumed silica in KOH solution. And next the wafers were rinsed in running DI water during 4 hours. Finally they were dried by keeping in the air and heated on a hot plate. These wafers were used as standard samples to study the properties of ice scrubber cleaning. The reason for making standard samples was described below. If the wafer surface is kept wet condition, it is easy to remove the slurry particles. So the standard samples which have dry surfaces are used to clearly evaluate the dependence of the cleaning method (ice scrubber or brush scrubber), and the cleaning conditions. In this study, chemical cleaning has not done to study the effect of only ice scrubber cleaning.

The slurry particles that remain on the wafer were measured over 0.3 micron diameter by using laser particle counter : WIS-850 (HITACHI).

3. RESULTS AND DISCUSSIONS

3-1. Removal efficiency

Figure 3 shows the stability of cleaning efficiency by ice scrubber cleaning with the standard samples. Even in continuous 1000 wafers run (during 7 days), no increase of particles were observed. The overall counts were very low. And fundamentally ice scrubber cleaning can easily keep this cleaning efficiency level forever, because it has no consumable.

Figure 4 shows the dependence of remaining particle counts on the speed of jetting ice particles. As the ice jets acceleration incleased, the remaining particle counts decreased. This means that the efficiency of removing the slurry particles is depended on the speed of the.ice particles.

Figure 5 shows the particle counts dependence of cleaning time. Even though cleaning time become longer, the particle counts are almost same level. If the ice particles does not have enough efficiency to remove the slurry particles, it is not effective to put off the cleaning time.

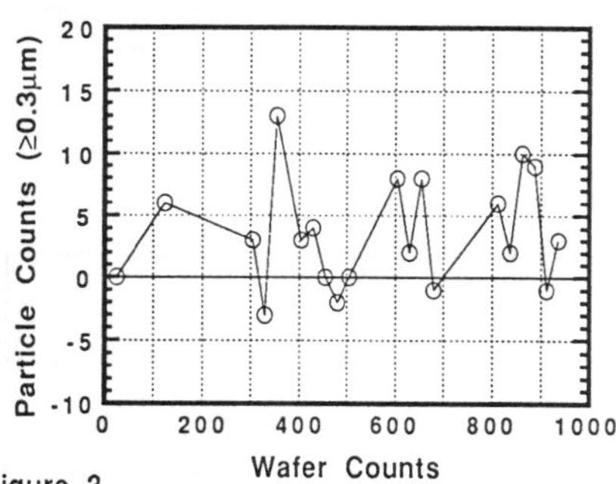

Figure 3.
Wafer counts dependence on particle counts. (after cleaning-befor slurry dip)

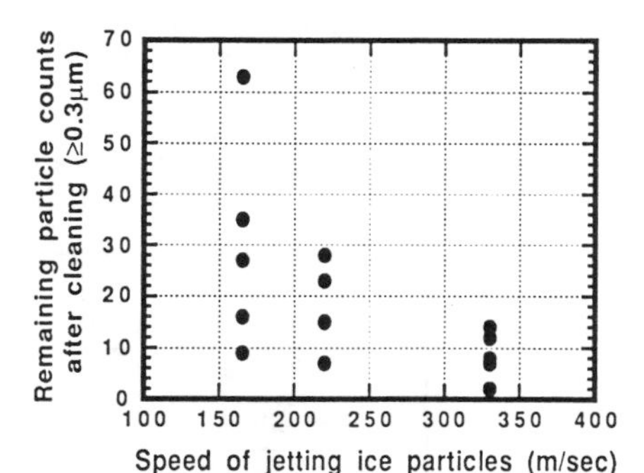

Figure 4
Dependence of particle counts on the speed of jetting ice particles.

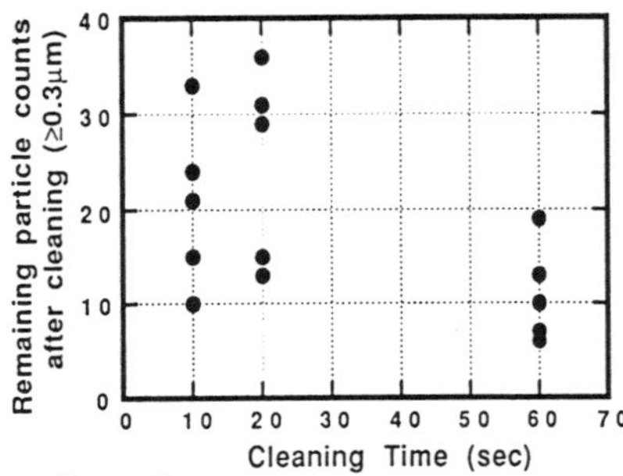

Figure 5.
Dependence of particle counts on the cleaning time

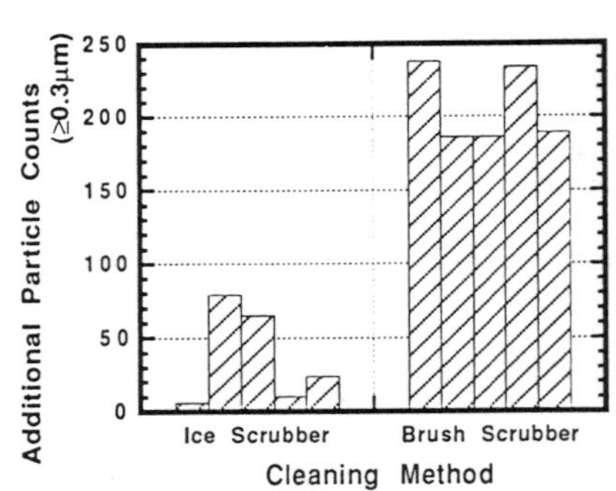

Figure 6.
Dependence of particle counts on cleaning method using dry surface wafers.

Figure 6 shows the cleaning results comparing with the brush scrubber. The advantage of ice scrubber is clear.

These results suggested the matter as shown below. The adhesion force of slurry particles to a wafer surface has a distribution. If a particle has stronger adhesion force than the force of ice bombardment, it remains on the surface. And it is difficult to reduce their adhesion by lower removal efficiency, so the extension of cleaning time is not so effective. But it is possible to remove the particles which have stronger adhesion force with the ice that has faster accelation. Because the ice particles accelerated faster have stronger bonbardment force.

In addition, according to the result of Figure 6, ice scrubber has higher removal efficiency than brush scrubber.

3-2. Contamination

Table 1 shows the removal efficiency of metal contaminations measured by ICP mass analysis. After the ice scrubber treatment, the contaminations on the polished wafer were reduced to the same level as the reference wafer. This result means the ice scrubber cleaning is also effective method to remove metal contaminations.

Table 1. Contaminations measured on the wafer before and after cleaning by using ICP mass analysis.

Item	Element (*10^{10} atoms/cm^2)					
	K	Na	Ca	Fe	Ni	Cu
Reference	4.7	5.9	<0.2	<1.0	<0.5	<0.05
after CMP	10.4	113	3	12	<0.5	5.7
Ice Scrubber	2.5	0.5	<0.2	<1.0	<0.5	<0.05
Brush Scrubber	4.1	20.5	<0.2	2	<0.5	<0.05

3-3. Damage

Figure 7 shows the yield of the MOS capacitor connected to various antennas. The test samples have patterned poly-Si electrode on the gate oxide which thickness is 95Å. And the antenna area ratios of poly-Si area to gate oxide area are from 10^3 to 10^6. After fabricating the samples, the ice scrubber were performed. The property of MOS capacitor is not degraded, even though the test pattern with the antenna area ratio of 10^6 is used. It is believed that by mixing CO_2 gas with ice particles flow, ice scrubber is free from charge up damage.

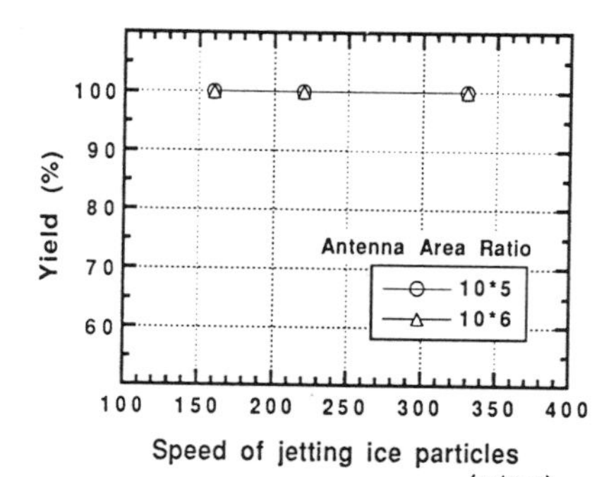

Figure 7.
Dependence of yield on the speed of jetting ice particles.

4. CONCLUSIONS:

As a new cleaning method for polished wafer using slurry, we have been investigated the characterization of ice scrubber cleaning. It jets ice particles onto a polished wafer surface and backside. Applying this method to CMP wafers is confirmed that this is very effective for not only removing particles but also contaminations. Yet it does not cause the charge build-up damage.

ACKNOWLEDGMENTS:

The authors would like to thank S.Takeda, M.Tada, T.Hiroi and Y.Hirakata in Taiyo Sanso Co. for their support in experiments and useful suggestions. We are grateful to the staffs of VLSI Development Laboratories for their help.

REFERENCES:

[1] J.Warnock, J.Electrochem.Soc., 138,8,(1991), p.2398
[2] Lee.M.cook, Journal of Non-Crystallinc Solids, 120,(1990), p.152
[3] C.Fruitman, M.Desai, VMIC Conference, 103, (1994), p.218
[4] W.J.Patrick, et.al., J.Electrochem. Soc., 138, 6, (1991), p.1778
[5] S.Poon, A.R.Sitarman, et.al., VMIC Conference, (1993), p.59
[6] Y.Hayashi, S.Takahashi, et.al., Jpn.J.Appl.Phys. 32, (1993), p.1060
[7] W.Y-C.Lai, J.F.Miner, et.al., VMIC Conference, 103, (1994), p.179
[8] Y.Hayashi, S.Takahashi, Jpn. J. Appl. Phys., 32, (1993), p.1060
[9] S.Lakshminarayanan, et.al., VMIC Conference, 103, (1994), p.49
[10] R.V.Joshi, IEEE Electron Device Letters, 14, 3, (1993), p.129
[11] C.L.Keast, et.al., VMIC Conference,103, (1994), p.204
[12] F.Sugimoto, et.al., Jpn.J.Appl.Phys., 34, (1995), p.30
[13] T.Ohmori, H.Komiya, Ultra Clean Technology, 2, (1990), p.213

EFFECT OF SC-1 PROCESS PARAMETERS ON PARTICLE REMOVAL AND SURFACE METALLIC CONTAMINATION

R. Mark Hall, John J. Rosato, Taura Jarvis, Thad Parry, and Paul G. Lindquist, Santa Clara Plastics, 400 Benjamin lane, Boise, ID 83704

ABSTRACT

The effect of bath temperature, megasonic power, and $NH_4OH:H_2O_2$ ratio are studied for particle removal efficiency, surface roughness, and surface Fe concentration in SC-1 cleaning solutions. Experimental results are presented which show removal efficiencies better than 97% on bare silicon wafers for optimized process conditions. These results are related to the etch rate of thermal oxides and a model is developed for reducing surface roughness and minimizing Fe contamination levels while maximizing particle removal efficiency.

INTRODUCTION

As device geometries approach 0.25 μm dimensions, the low levels of contamination required to achieve high yielding wafers continue to stretch cleaning technologies. The SC-1 cleaning sequence has been used widely in pre-diffusion cleans as well as many etch and stripping applications for efficient particle removal. Cleaning efficiencies better than 97% have been reported for bare silicon wafers in SC-1 solutions[1-2]. It has also been shown that SC-1 cleaning efficiency is strongly dependant on the surface properties of the wafer as well as bath conditions and chemistry[1-3]. The surface attributes of a wafer depend on the particle, metallic, and surface roughness levels which remain after a cleaning sequence. These attributes can often have a dramatic effect on gate oxide integrity and device yield/reliability[3-4]. In this paper we discuss the process parameters relating to particle removal, surface roughness, and surface metallic contamination for silicon wafers cleaned in an SC-1 cleaning sequence.

EXPERIMENTAL

In order to better understand the effect of different SC-1 cleaning conditions, a design of experiments (DOE) approach was used to evaluate the effect of temperature, megasonic power, and chemical ratio in the SC-1 bath. Performance responses in this experiment included etch rate of thermal oxides, surface roughness, particle removal efficiency, and surface Fe concentration. Table 1 contains a summary of the experimental factors and responses in the DOE. Each run in the DOE was processed for 10 min. in an SC-1 cleaning solution (ppb grade) followed by a rinse in UHP DI water and an IPA Vapor Jet Dry. This sequence was carried out with a Santa Clara Plastics 9400 surface preparation system enclosed in a class-1 minienvironment. The megasonic transducer in the SC-1 tank had a curved lens for radial power distribution and was set a frequency of 860 KHz. No moving parts were required for effective distribution of megasonic energy.

Mat. Res. Soc. Symp. Proc. Vol. 386

Table 1: *Experimental Factors and Responses in Designed Experiment*

Factor	Units	Range	Response	Units
Temperature	'C	30 to 70	Removal Eff	%
Megasonic Power	Watts	0 to 500	SiO2 Etch Rate	A/min
NH4OH:H2O2 Ratio		0.01 to 1.0	RMS Roughness	A
			Fe Surface Conc.	Atoms/cm-2

Particle performance was measured by placing 3 previously contaminated 200 mm wafers (P-type) in a quartz reduced cassette. Initial particle counts ranged from 300 to 1500 particles/wafer, and the contamination was composed of liquid and airborne particles. The remainder of the slots in the 50 wafer cassette were occupied with filler wafers. A reduced profile quartz carrier was used to help eliminate megasonic shadowing and improve fluid dynamics in the process tank. Particles on the wafer surface were detected with a Tencor 6200 surfscan for diameters greater than or equal to 0.16μm, and an edge exclusion of 3mm was used. Wafers were scanned directly before and after the cleaning sequence, and the removal efficiency was calculated as follows:

$$\mathrm{Eff}(\%) = [(P_i - P_f)/P_i] \cdot 100 \tag{1}$$

where P_i and P_f represent the initial and final particle count on the wafer surface respectively. The IPA dryer used in the cleaning sequence had consistently shown a zero net adder performance and, therefore, was not a factor in removal efficiency calculations. Surface μ-roughness was measured by atomic force microscopy with etched silicon tips to provide high resolution imaging. Roughness analysis were performed on 1μm x 1μm size areas, and the rms roughness was given by:

$$R_q = \sqrt{(\Sigma(Z_i - Z_{avg})^2/N)} \tag{2}$$

where N is the number of samples in the image, Z_{avg} is the average roughness value in the image, and Z_i is the current value of roughness. Etch rate was measured with thermal oxide monitor wafers using ellipsometry to determine the amount of oxide removed during a cleaning sequence. Finally, Fe contamination was measured using a surface photovoltage technique. After the cleaning sequence, surface metallic impurities were driven into the bulk of the wafer with a rapid thermal anneal at 1100 °C for 4 min. followed by a low temperature anneal at 80 °C to promote Fe-B pairs. The bulk Fe concentration was given by[5]:

$$\mathrm{Fe}(\mathrm{cm}^{-3}) = 1.05 \times 10^{16} (L_1^{-2} - L_o^{-2}) \tag{3}$$

where L_1 and L_o are the final and initial diffusion lengths respectively.

RESULTS AND DISCUSSION

1. Particle Removal Efficiency

There are several factors which influence particle adhesion and removal in the SC-1 chemistry, however, it is generally accepted that the removal force must be on the same order of magnitude as the adhesion force. Adhesion forces (capillary, van der Waals) are typically proportional to the particle diameter, whereas the removal forces (drag, gravity) usually scale as multiples of the particle diameter[6]. Consequently, it becomes increasingly difficult to remove particles from a wafer as the particle diameter decreases. This is illustrated in Figure 1 for wafers cleaned in an SC-1 cleaning sequence with and without megasonic energy. It is interesting to note that wafers cleaned without megasonic energy showed a rapid decline in removal efficiency for particles less than 0.2 μm in size, while wafers cleaned with megasonic energy showed improved particle removal efficiencies for all particle diameters measured.

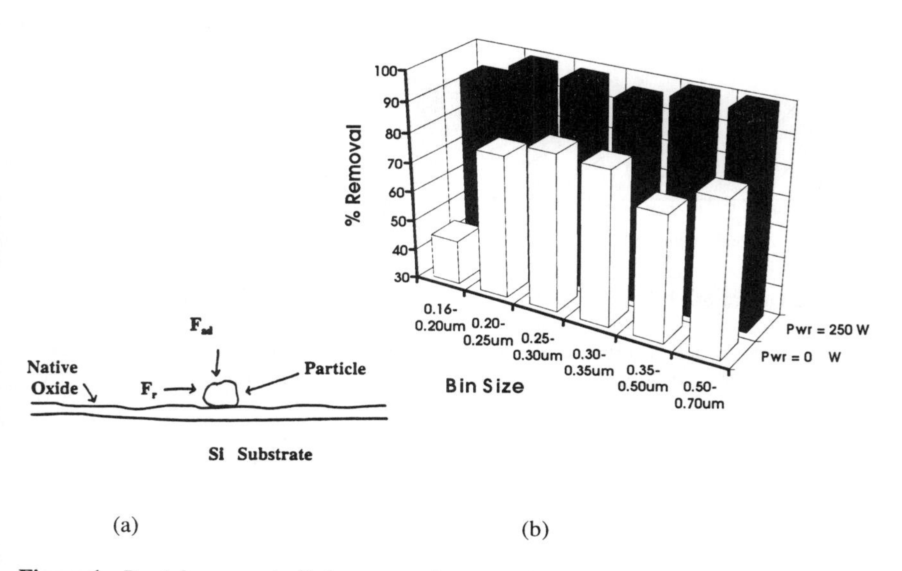

Figure 1: *Particle removal efficiency as a function of particle size in an SC-1 clean. As particle diameter decreases, the ratio of removal force to adhesion force F_r / F_{ad} decreases (a). Wafers processed with megasonic energy show improved particle removal efficiency for all diameters measured (b).*

Improvements in particle removal are also dependant on bath process conditions and chemistry. Both temperature and the $NH_4OH:H_2O_2$ ratio in an SC-1 solution contribute significantly to particle removal efficiency. This is shown in Figure 2 for a standard SC-1 chemistry (ratio = 1.0) and a more dilute chemistry (ratio = 0.1). In each case, megasonic power and process temperature improve removal efficiencies to better than 97% for particles at 0.16 μm and above. In the dilute case, however, there is a larger optimal process window for achieving high particle removal efficiency. This difference becomes significant when one considers cost of ownership (CoO) and bathlife. The dilute chemistry provides the same high particle removal efficiency as the standard chemistry with reduced chemical usage and the potential for longer bath life at lower temperatures. A model for particle removal efficiency over the experimental space was developed with a multiple regression analysis:

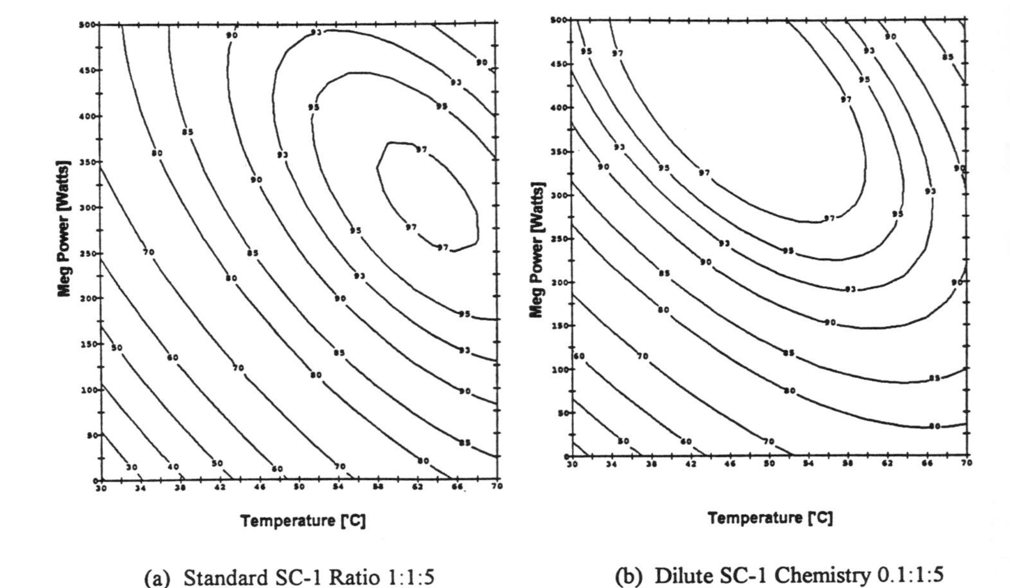

(a) Standard SC-1 Ratio 1:1:5 (b) Dilute SC-1 Chemistry 0.1:1:5

Figure 2: *Response surface for particle removal efficiency. Particle removal efficiencies better than 97% are shown for the standard SC-1 chemistry (a), as well as the dilute case (b).*

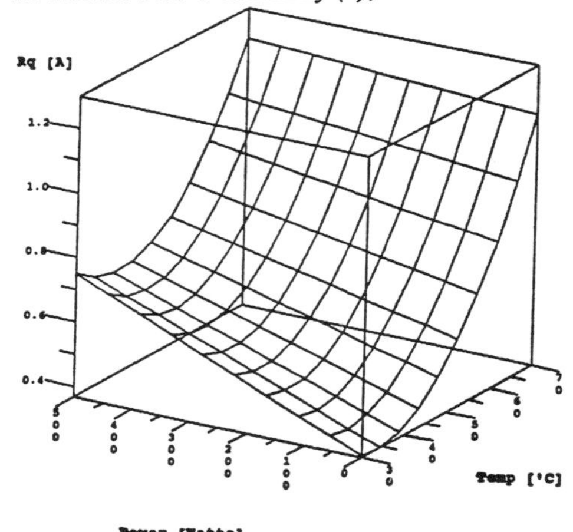

Figure 3: *Response surface for RMS surface roughness with standard (ratio = 1.0) SC-1 chemistry. Surface roughness increases dramatically for bath temperatures above 50 °C regardless of megasonic power levels. High megasonic power levels can contribute to surface roughening.*

$$\mathrm{Eff}(C,T,P) = -11.2T^2 - 13.9P\cdot T - 12.6P^2 + 5.9\alpha\cdot C\cdot T + 12.1T + 15.2P - 2.8\alpha\cdot C + 92.6 \quad (4)$$

where $\alpha =$ 1 for a ratio of 1.0, or
-1 for a ratio of 0.1

In these equations C represents $NH_4OH:H_2O_2$ ratio, T represents the temperature in °C, and P represents the megasonic power in watts. To simplify the analysis, the following transforms were used for temperature and megasonic power:

$$P = (P - 250)/250 \quad (5)$$
$$T = (T - 50)/20 \quad (6)$$

2. Surface Roughness

The roughening of the wafer surface was also studied as a function of megasonic power, bath temperature, and $NH_4OH:H_2O_2$ ratio. Figure 3 shows the response surface developed for rms roughness with a standard SC-1 chemistry. Both temperature and megasonic power significantly influenced surface roughness in this model. As megasonic power increased, surface roughness also increased linearly. The most significant factor affecting surface roughness was temperature. Above 50 °C the surface roughness increased dramatically regardless of megasonic power level. Equation 7 shows the model developed from the regression analysis for rms roughness:

$$R_q(P,T) = 0.23T^2 - 0.08P\cdot T + 0.31T + 0.11P + 0.64 \quad (7)$$

where P and T are as defined in equations 5 and 6 above, and R_q is given in units of Å.

3. Substrate Etching in SC-1

The increase in surface roughness with process temperature appears to be related to the competing oxidation and etching mechanisms in the SC-1 chemistry. These effects are related to the $NH_4OH:H_2O_2$ ratio and bath temperature as shown in Figure 4. Note that as temperature increases above 50 °C for the standard chemistry, the etch rate of thermal oxide increases substantially. This correlates well to surface roughness and the reduction in removal efficiency seen in the particle model at high temperature and megasonic power. It also suggests that the removal efficiency for very small particles (haze) may be influenced by surface roughness.

To understand the relationship between SiO_2 etchrate in SC-1 and particle removal/surface roughness, the models for these responses were plotted as a function of removed oxide. The data is shown in Figure 5. This information suggests that wafers processed without megasonic energy, as compared to those processed with megasonic energy, require more substrate removal to achieve high particle removal efficiency. This is consistent with theory which suggests that particles are etched from the wafer surface and electrically repelled due to the surface and particle zeta potentials in alkaline solutions[7]. It is also interesting to note that beyond 15 Å of etched substrate the removal efficiency saturates. This is approximately the thickness of a native/chemical oxide. In the case of wafers processed with megasonic energy, only a slight etching (< 6Å) of the substrate was required to achieve high particle removal efficiency. Thus, there is a clear distinction in the removal mechanisms for wafers processed with and without megasonic energy. Finally, it can be seen that surface roughness increases substantially after 12 Å to 15 Å of substrate has been removed. This also correlates well to the amount of oxide (native or chemical) that may initially be present on the wafer. If the silicon is exposed during an SC-1 clean, the etch rate can be several times that of an oxide resulting in increased surface roughness[9]. This problem will be compounded if the initial native/chemical oxide is non-uniform to begin with.

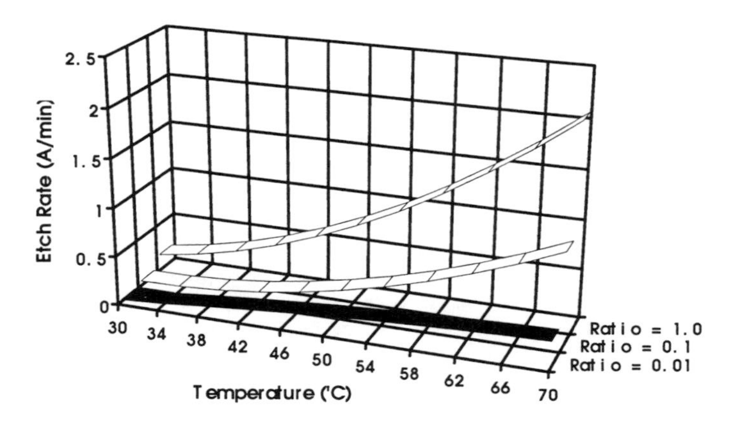

Figure 4: *Etch rate of thermal oxide as a function of SC-1 bath temperature and $NH_4OH:H_2O_2$ ratio. For the standard ratio = 1.0, the etch rate increases dramatically after bath temperatures of 50 °C.*

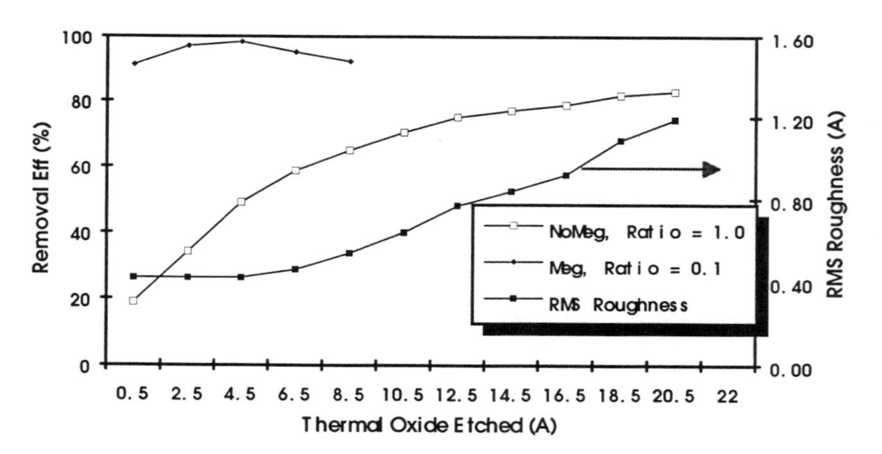

Figure 5: *Particle removal efficiency and surface roughness as a function of etched oxide. Wafers processed in dilute SC-1 chemistries with megasonic power show high particle removal efficiencies for less than 5 Å of substrate removed. Surface roughness increases substantially after 12 Å to 15 Å of etched substrate.*

4. Surface Metallic Contamination

It has been well established that heavy metal impurities (Fe, Zn, Al) can be a problem in alkaline processes such as an SC-1 clean[4,8,9]. The purity of the NH_4OH and H_2O_2 used in the SC-1 solution are critical for controlling these heavy metal impurities[9]. In addition to the purity of the chemical, however, there are also process conditions which can impact the level of heavy metal contamination seen on the wafer surface. Figure 6 shows Fe surface concentration as the $NH_4OH:H_2O_2$ ratio was varied (R:1:5) with temperature. For each ratio considered, the Fe surface contamination increased as temperature was increased. This difference was especially pronounced for temperatures above 50 °C. The mechanisms for increased deposition of Fe at higher temperatures are not clearly understood, although, it is possible that the increase in temperature may shift the reaction rate constants such that dissolution onto the Si surface is promoted. The effect of megasonic power on surface Fe concentration was also examined. At lower temperatures the difference was negligible, however, at higher bath temperatures there was an appreciable increase in Fe contamination levels for wafers processed with megasonic power (high points in Figure 6).

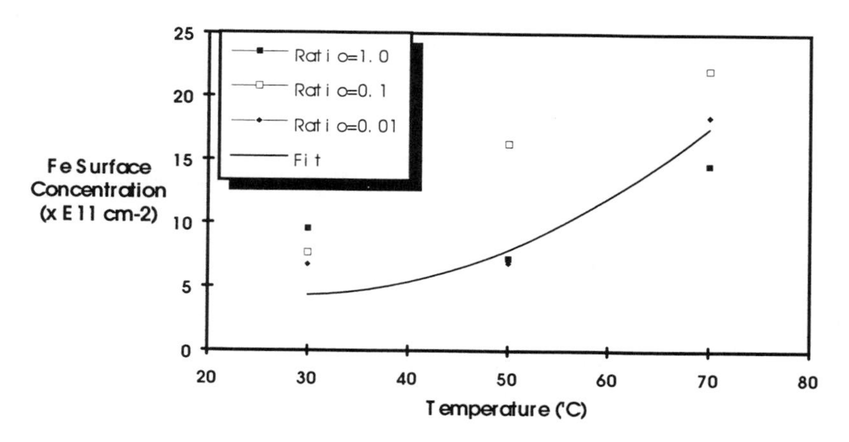

Figure 6: *Surface Fe concentration as bath temperature and $NH_4OH:H_2O_2$ ratio are varied. Fe contamination levels increase with increasing temperature. The high points at each temperature setting were wafers processed with megasonic power.*

CONCLUSIONS AND RECOMMENDATIONS

The efficient removal of particles from the wafer surface in SC-1 solutions requires an understanding of process temperature, bath chemistry, and megasonic power. Under optimized conditions for these process parameters, removal efficiencies better than 97% have been demonstrated for particles down to 0.16 μm in size. It has also been shown that only a slight etching of the wafer substrate (less than 5 Å) is required for high particle removal efficiency, thereby reducing surface roughness. This differs from previously published data which suggest that 20 Å of substrate removal is required for optimal particle removal efficiency[3]. The primary reason for this difference is the use of megasonic energy in the SC-1 chemistry which aids in the particle removal process. Megasonic energy also enables the use of dilute SC-1 chemistries and lower bath temperatures without compromising removal efficiency. This results in a dramatic reduction in CoO due to reduced chemical usage, extended bathlife, and improved defect levels. Surface metallic contamination levels have also been investigated. The results of these tests showed a net reduction (3X) in surface Fe contamination levels for wafers processed below 50 °C as compared to higher temperatures.

Despite the large amount of data that is available on processes involving SC-1 chemistries, there is still much to be learned. For example, the particle removal process in SC-1/megasonic cleaning is not well understood[10]. For this reason, the limitations of this technology are also in question. Process technologies at the 0.25 μm level and below will require the removal of particles less than 0.1 μm in size, and consequently, cleaning technologies must be developed at that level. There is also little work that has been published concerning removal efficiencies on patterned wafers with SC-1/megasonic cleans. The removal efficiency for wafers with extreme topography (ie. 256 Meg DRAM) could be significantly different than those for bare Si wafers. A successful implementation of cleaning technologies for advanced ULSI processing will require a better understanding of our current process technology.

REFERENCES

1. P.J. Resnick et. al., in A Design of Experiments Approach to an Optimized SC-1/Megasonic Clean for Sub-0.15 Micron Particle removal, edited by J. Ruzyllo and R. Novack, (ECS, **PV94-7**, Pennington, NJ, 1994) pp. 450-457.
2. R.M. Hall et. al., in Optimization of SC1-Megasonic Cleaning Efficiency for 200 mm Wafer Processing, (Proceedings: Microcontamination-94, Santa Monica, CA, 1994) pp. 529-537.
3. M.M. Heynes et. al., in New Wet Cleaning Strategies for Obtaining Highly Reliable Thin Oxides, edited by G.S. Higashi, E.A. Irene, and T. Ohmi, (Mater. Res. Soc. Proc. **315**, Pittsburgh, PA,1993) pp. 35-45.
4. M. Berry, G. Depinto, and J. Steinberg, in A Methodology For Continuous Defect Reduction In A High Volume Sub-Micron CMOS Factory, edited by D.N. Schmidt, (ECS, **PV92-21**, Pennington, NJ, 1992) pp. 208-222.
5. L. Jastrzebski, in Surface Photovoltage Monitoring of Heavy metal Contamination in IC Manufacturing, Solid State Technology, (Dec. 1992), pp. 27-31.
6. W. Kern, Handbook of Semiconductor Wafer Cleaning Technology, 1st ed. (Noyes Publications, New Jersey, 1993) pp. 48-50.
7. T. Ohmi, in Future Process Innovation Based on Ultra-Clean Technology, (Proceedings: Santa Clara Plastics 2nd International Symposium, Boise, ID, 1994), pp. 7-20.
8. C.R. Helms, and H. Park, Electrochemical Equilibrium of Fe in Acid/Base/Peroxide Solutions Related to Si Wafer Cleaning, edited by J. Ruzyllo, and R. Novack, (ECS, **PV94-7**, Pennington, NJ, 1994) pp. 27-33.
9. H.F. Schmidt et. al., in Evaluating The Effects of Chemical Purity Within The RCA Wafer-Cleaning Process, Microcontamination, (Sept. 1993), pp. 27-32.
10. D. Zhang, D.B. Kittelson, and B.Y.H. Liu, in An Investigation of Large Scale Acoustic Streaming in Megasonic Cleaning,(Proceedings: Microcontamination-94, Santa Monica, CA, 1994) pp. 215-224.

EFFECTS OF SC-1 DILUTION AND TEMPERATURE VARIATIONS ON ETCH RATE AND SURFACE HAZE

K. K. Christenson and S. M. Smith
FSI International, Inc., Surface Conditioning Division, Chaska, Minnesota

ABSTRACT

The slight etch and accompanying roughening of the silicon by the APM solution are of great concern as device geometrys and gate oxide thicknesses decrease. This report covers the variations in the SiO_2 etch rate and surface roughening of an APM solution as a function of NH_4OH and H_2O_2 concentration and temperature. In general, the etch rate and roughening increased with increasing temperature and NH_4OH concentration but was unaffected by H_2O_2 concentration. Other portions of the larger APM study covering the addition and removal of metals and the removal of particles by APM are published elsewhere.[1, 2, 3] Together, these studies provide a complete set of process response surfaces for the SC-1 chemistry. These surfaces can be used to select the most promising regime of parameter space for any particular process.

INTRODUCTION

The RCA-1 or APM standard clean (NH_4OH:H_2O_2:H_2O), as published by Kern in 1970, has been the primary means of removing particles, trace organics and some metals in pre diffusion cleans for over 20 years.[4, 5] When a silicon wafer is immersed in an APM solution, an equilibrium is established between the oxidation of the silicon by the H_2O_2 and etching of the SiO_2 by the NH_4OH. The etch rate increases with increasing NH_4OH concentrations, decreasing H_2O_2 concentrations and increasing temperature.[6] A critical H_2O_2 concentration of $3x10^{-3}$ M (1:0.002:5 blend ratio) is necessary at 70° C to prevent gross etching of the underlying Si.[7]

For wafers with a thick (>>10 Å) layer of SiO_2, the indiffusion rate of oxygen is negligible and the etching of the surface by NH_4OH is the dominant reaction. Etch rates are determined by measuring the change in thickness of the bulk oxide with interferometric or ellipsometric techniques. Due to the speed, ease and low cost of this technique, this study, and most other studies, report the etch rate of SiO_2 and not of Si. Confusion in the literature exists because this is often not explicitly stated.

Note that there can be substantial differences in etch rate between silicon, thermal SiO_2 and other "glasses." Hossain reports etch rates of 88.5 Å/min for BPSG vs 2.5 Å/min for thermal SiO_2 under the same conditions.[8] Raheem reports variations in the etch rate ratio of Si/SiO_2 of 2.62 to 4.61 as a function of temperature and NH_4OH concentration.[9]

It has been shown by many workers that exposure of bare silicon to APM solutions can induce surface roughening and that this roughening is correlated to a reduction in V_{BD} and Q_{BD} in gate oxides.[10, 11, 12, 13] It has also been reported that roughness alone does not cause break down, but the combination of roughness with metallics.[14] This roughening can be measured directly with an AFM, or indirectly through the measurement of surface haze.[15, 16] Surface haze is a measure of the diffuse scatter of light from a surface. A haze of 1 part per

Mat. Res. Soc. Symp. Proc. Vol. 386

million (ppm) indicates that on average, for every 1 million incident photons, one will be scattered and not specularly reflected. While the exact correlation has not been established, it is very likely that an increase in surface haze is an indicator of increased surface roughness.

When measuring surface roughness, both the amplitude of the roughness and the period or wavelength of the roughness are important. Roughness with periods in the Ångstrom range causes degradation in the electrical properties of the gate oxide. "Roughness" with periods of mm to cm can create problems with focusing in lithography. Optical scattering techniques are sensitive to roughness with periods from Ångstroms to microns. Since the APM mixture is only etching a few tens of Ångstroms, the periods of the APM roughening are likely to be in the Ångstrom range. Increases in haze during exposure to APM will therefore correspond to degradations of gate oxide electrical performance.

EXPERIMENT

The experiment consisted of exposing test wafers to 27 treatments in a 3x3x3 full factorial matrix of H_2O_2 and NH_4OH concentrations and temperatures along with a 28th "rinse only" treatment. The experimental runs were carried out in an FSI MERCURY® MP centrifugal spray processor. A simplified plumbing diagram is shown in Figure 1. In the spray processor, Ashland SEMI grade chemicals and water were blended on-line and sprayed onto the wafers. Chemical flow rates of 23, 92 and 320 cc/min were chosen to give two, approximately 4x, changes in concentration and a blend ratio range of n:m:5 with n and m ranging from to 0.06 to 1.45. Due to a 90 cc/min practical lower limit on the flow systems, the H_2O_2 and NH_4OH were diluted 3:1 in the canister to achieve effective flow rates of 23 cc/min for these chemicals when required. The in-line temperature at the spray post was varied over 20, 60 and 95° C. The following process sequence was used:

- 5 minute DI water rinse to pre-heat the wafers to the steady state process temperature
- 30 second dispense of H_2O and H_2O_2
- 20 minute APM dispense at 60 RPM with 20 second, 500 RPM ramps every two minutes
- 4.5 minute rinse in hot DI water utilizing ramped rinsing[17]
- 6 minute spin dry

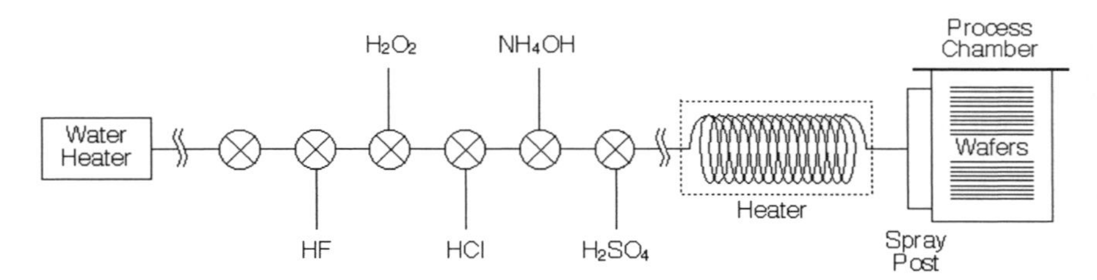

Figure 1: Simplified MERCURY MP Spray Processor Plumbing Diagram

For the runs at elevated temperatures, the in-line temperature of the DI water pre-heat and the APM chemistry was maintained at 60 or 95° C by an in-line infrared chemical heater.[18] Heat loss through processes such as evaporation reduced the on-wafer temperature from 60 and

95° C to approximately 50 and 70° C. A constant total APM flow rate of 1760 cc/min was used for all treatments so that the transient and steady state on-wafer temperatures were the same.[18]

N type, <100>, 150 mm CZ wafers were initially cleaned in the MERCURY MP with a B Clean chemistry sequence (SPM, DHF, APM, HPM). For etch rate measurements, 4,000 Å of oxide was grown in 60 minutes at 1,000 °C with in 37.5% oxygen in H_2 followed by a 10 minute anneal in N_2. Pre and post measurements of the thickness of the oxide were performed on a Prometrix SM 200. 49 points on each wafer were read 4 times before and 4 times after the etch and the readings averaged to determine the thickness of oxide removed.

One oxide wafer and two bare wafers were run in each of the 28 APM treatments. Post measurements of the surface haze were performed on a Tencor 6200 at the Tencor application laboratory in Mountain View, CA. Six wafers per box were not processed in the 20 minute APM and were used as controls to determine the initial haze levels of the wafers.

RESULTS AND DISCUSSION

The thermal oxide etch rate as a function of temperatures and blend ratios are shown in Figure 2. The rate increases rapidly with temperature. The 30 and 20° C rises (20 to 50 and 50 to 70° C) in on-wafer temperatures resulted in 8x and 4x increases in etch rates agreeing well with Arrhenius' rule that reaction rates double with every 10° C increase in temperature. The SiO_2 etch rates near 2 Å/min at 70° C are consistent with those measured by other experimenters under similar conditions.

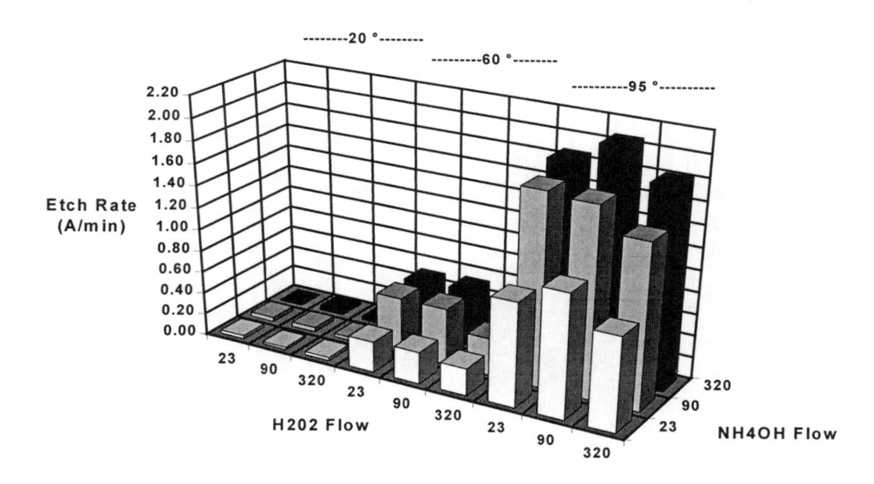

Figure 2: Summary of All Thermal Oxide APM Etch Rate Data
Thermal oxide etch rate of an APM in a FSI MERCURY MP vs NH_4OH and H_2O_2 flow rates and temperatures.

The lack of response to H_2O_2 concentration should be expected. When etching a thick thermal oxide, H_2O_2 is not needed to continuously reoxidize the silicon as it is when etching a "bare" silicon surface. The response for bare silicon has been shown to be strongly dependent on the concentration of H_2O_2.[6]

The oxide etch rate also increases with the concentration of NH_4OH. There is an approximately 2x increase between the 23 and 90 cc/min NH_4OH flow rates and a 20% increase between the 90 and 320 cc/min flow rates. The 2x increase from 23 to 90 cc/min is consistent with the etch rate being proportional to the concentration of OH- ions. For ammonia, a weak base:

$$[NH_4^+]\,[OH^-]/\,[NH_4OH] = 1.79\text{x}10^{-5} \qquad \text{or...}$$

$$[OH^-] = (1.79\text{x}10^{-5} \text{ x } [NH_4OH]\,)^{1/2} \qquad \text{assuming } [NH_4^+] = [OH^-]$$

The 4x increase in $[NH_4OH]$ from 23 to 90 cc/min would correspond to a 2x increase in $[OH^-]$. Since the etch rate scales as $[NH_4OH]^{1/2}$, it would in general be more economical to achieve a chosen target removal by using a more dilute solution and increasing the exposure time or temperature.

The small 20% increase in etch rate from 90 to 320 cc/min is not consistent with this theory. One possible explanation is that a greater fraction of the NH_4OH evaporates from the high concentration solutions than from the low concentration solutions. Further work is necessary to resolve this issue. Further increases in NH_4OH concentration beyond 320 cc/min do not appear likely to significantly increase the etch rate.

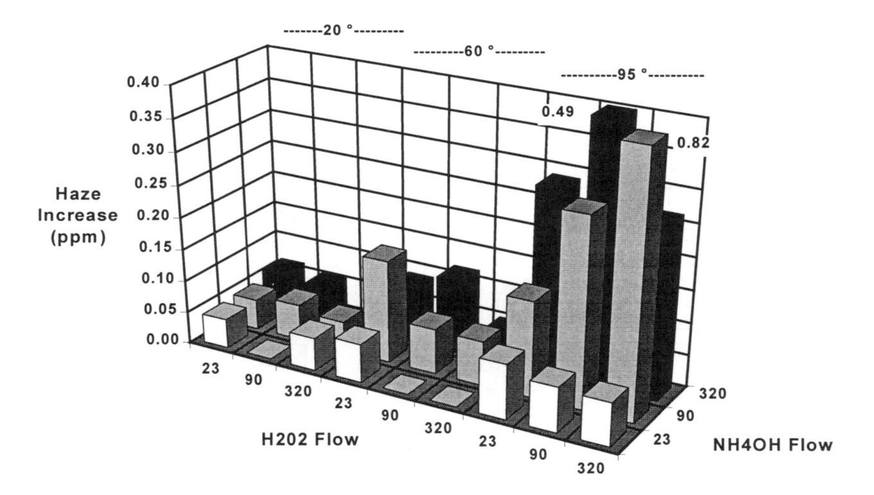

Figure 3: Summary of Haze Increase vs Blend Ratio and Temperature
Increase in surface haze on bare silicon after an APM in a FSI MERCURY MP vs NH_4OH and H_2O_2 flow rates.

The increase in surface haze as a function of temperature and blend ratio is shown in Figure 3. While there is a significant amount of noise in the data, some conclusions can be drawn. First, haze increases very rapidly with temperature. Even after the 20 minute exposure, there was little or no increase in haze at 20° C for any blend ratio. At 95° C, the was a large increase in haze for NH_4OH flows of 90 and 320 cc/min. One obvious trend in the 95° C data is that low NH_4OH concentrations (23 cc/min, 0.07:n:5, $NH_4OH:H_2O_2:H_2O$) gave very little roughening. This may be due to a lack of etching by the NH_4OH; even low levels of H_2O_2 can provide enough oxidization of the silicon to "keep up" with the slow etch.

Figure 4 shows that there is a high correlation between the amount of silicon etched and the roughening of the surface. This data can be used to select the proper parameters for any given clean. For instance, any processes that are sensitive to surface roughening should be run at low temperatures, low NH_4OH concentrations, or both. In particular, the APM pre-gate clean after the sacrificial oxide is removed should be cold, dilute and short. These conditions will also reduce the metal and particle removal efficiency of the SC-1 solution.[1, 2, 3] But at pre gate, the primary concern must be the protection of the gate, even at the expense of other metrics.

The device surfaces that are exposed after the deposition of the gate electrode will probably be much less sensitive to increases in roughness than the gate. Particle removal in post gate cleans could therefore be much more aggressive using high concentrations, exposure times and temperatures. Likewise, there should be no device impact as a result of roughening an oxide such as the sacrificial oxide that will later be stripped.

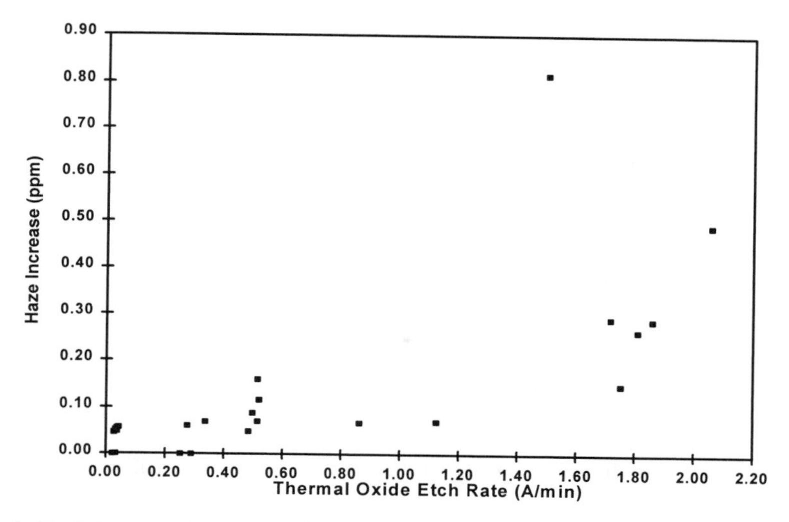

Figure 4: Etch Rate of Thermal Oxide vs Increase in Surface Haze

SUMMARY

The etching rate of thermal oxide depends strongly on temperature with the etch rate doubling every 10° C. The rate also increased with increasing NH_4OH concentration, but appears to reach a plateau. Little increase in rate is seen for NH_4OH flows above 90 cc/min (~ 0.3:0.3:5). The concentration of H_2O_2 had little effect. The surface haze increases rapidly with increasing temperature, particularly at high NH_4OH concentrations. The increase is near or at the detection limit for all blends at 20° C. Any processes that are sensitive to surface etching or roughening should be run at low temperatures, low NH_4OH concentrations, or both.

ACKNOWLEDGMENTS

We would like to acknowledge the generosity of the Motorola Incorporated whose support made this work possible. In particular, we thank Robert Duffin and Kathy McCormack for their involvement in the FSI IRONMAN program. We would also like to thank Brad DeSelms and Eric Persson of ATT, Jeff Glick of AMD, John Grant of Sharp and Scott Becker, Don Deal, Jim Oikari and Don Grant of FSI for suggestions on the design of the experiment and appropriate particle challenges.

---

1. Shelley Smith, K. K. Christenson and Dennis Werho, *Proceedings of the 1995 Semiconductor Pure Water and Chemicals Conference*, edited by M. Balazs (Balazs Analytical Laboratory, Sunnyvale, CA, 1995).
2. K. K. Christenson, Shelley Smith and Dennis Werho, *Proceedings of the 1995 Semiconductor Pure Water and Chemicals Conference*, edited by M. Balazs (Balazs Analytical Laboratory, Sunnyvale, CA, 1995).
3. K. K. Christenson and Shelley Smith, submitted to *Proceedings of the Fourth International Symposium on Cleaning Technology in Semiconductor Device Manufacturing*, edited by J. Ruzyllo and R. Novak (The Electrochemical Society, Pennington, NJ, 1996).
4. W. Kern and D. Puotinen, RCA Review, **31**, 187-206 (1970).
5. W. Kern, *Proceedings of the First International Symposium on Cleaning Technology in Semiconductor Device Manufacturing*, edited by J. Ruzyllo and R. Novak (The Electrochemical Society, **PV 90-9**, Pennington, NJ, 1990) pp. 3-19.
6. Y. Sugihara, S. Shimokawa and Y. Oshida, *Proceedings of the 1993 Semiconductor Pure Water and Chemicals Conference*, edited by M. Balazs (Balazs Analytical Laboratory, Sunnyvale, CA, 1992) pp. 191.
7. J. van den Meeraker and M. van der Straaten, Journal of the Electrochemical Society, **137**, 1239-1243 (1990).
8. S. D. Hossain, *Extended Abstracts of the 1993 Spring Meeting* in Honolulu, Hawaii (The Electrochemical Society, **93-1**, Pennington, NJ, 1993) pp. 787-788.
9. R. Raheem and J. Glick, *Proceedings of the 1994 Semiconductor Pure Water and Chemicals Conference*, edited by M. Balazs (Balazs Analytical Laboratory, Sunnyvale, CA, 1994) pp. 226-241.

10. J. Takano et al., *Proceedings of the 1992 Semiconductor Pure Water and Chemicals Conference*, edited by M. Balazs (Balazs Analytical Laboratory, Sunnyvale, CA, 1992) pp. 199.
11. T. Ohmi, M. Miyashita, M. Itano, T. Imaoka and I. Kawanabe, IEEE Transactions on Electron Devices, **139**, 537 (1992).
12. M. Heyns et al., *Surface Chemical Cleaning and Passivation for Semiconductor Processing*, edited by G. Higashi, E. Irene and T. Ohmi (Mater. Res. Soc., **315**, Pittsburgh, PA, 1993) pp. 35-45.
13. K. Mori, N. Ishikawa, T. Shihoya and A. Yamashita, *Proceedings of the 1992 Semiconductor Pure Water and Chemicals Conference*, edited by M. Balazs (Balazs Analytical Laboratory, Sunnyvale, CA, 1992) pp. 191.
14. M. Heyns, presented at the *1995 Semiconductor Pure Water and Chemicals Conference*, San Jose, CA 1995 (unpublished).
15. Igor J. Malik et al., *Extended Abstracts of the 1993 Spring Meeting* in Honolulu, Hawaii (The Electrochemical Society, **93-1** Pennington, NJ, 1993) pp. 1133-1134.
16. H. F. Schmidt et al., *Proceedings of the Third International Symposium on Cleaning Technology in* Semiconductor *Device Manufacturing* edited by J. Ruzyllo and R. Novak (The Electrochemical Society, **PV 94-7,** Pennington, NJ, 1994) pp. 102-110.
17. K. Christenson, *Proceedings of the Third International Symposium on Cleaning Technology in Semiconductor Device Manufacturing* edited by J. Ruzyllo and R. Novak (The Electrochemical Society, **PV 94-7,** Pennington, NJ, 1994) pp. 153-162.
18. K. K. Christenson, *Proceedings of the Third International Symposium on Cleaning Technology in Semiconductor Device Manufacturing* edited by J. Ruzyllo and R. Novak (The Electrochemical Society, **PV 94-7,** Pennington, NJ, 1994) pp. 474-483.

DIRECT SURFACE ANALYSIS OF ORGANIC CONTAMINATION FOR SEMICONDUCTOR RELATED MATERIALS

J.J. LEE, R.W. ODOM, G.S. STROSSMAN AND P.M. LINDLEY
Charles Evans and Associates, 301 Chesapeake Drive, Redwood City, CA 94063

ABSTRACT

Direct surface analysis of Si wafers and environmental materials such as polymers for wafer carriers and for high purity water systems is important to identify contaminants and their sources. Organic contaminants on the surfaces are difficult to analyze; however, their adverse effects have been cited in recent years. This paper demonstrates the detection and identification of surface contaminants using Time-of-Flight Secondary Ion Mass Spectrometry (TOF-SIMS) on semiconductor related materials. The analytical information provided by TOF-SIMS can be useful for maintaining material cleanliness or for failure analysis.

INTRODUCTION

Surface cleanliness has become a critical material processing issue in modern industry.[1] Contamination in the levels of monolayers or sub-monolayers can drastically alter surface properties such as adhesion, wettability, and optical or electrical characteristics. Low level organic contamination can cause severe yield problems and/or product failures. In the semiconductor industries, for example, many material-related problems have been attributed to possible organic contamination, including slow or irregular oxidation rate, haze formation, changes in surface wettability or electric characteristics, unintended C implantation and poor adhesion.[2] However, identifying surface organic contamination presents an analytical challenge because molecular structures may not survive the harsh processing steps required in manufacturing. In addition, most modern surface analysis techniques fail to provide the specific information to needed to identify complex organic molecules.

Time-of-Flight Secondary Ion Mass Spectrometry (TOF-SIMS) can provide sensitive and specific detection of both inorganic and organic contaminants.[3] Direct surface analysis using TOF-SIMS provides a viable means to verify possible contaminating sources. The data presented in this paper demonstrate the detection and identification of surface organic contamination from materials commonly found in the semiconductor manufacturing environment, *i.e.*, wafer carriers and high purity water transfer systems.

The effects of volatile organics on material surfaces have been attracting increasing attention in recent years.[4] The detection of these contaminants could be difficult for most surface analysis techniques employing vacuum sample chambers. A low temperature analysis indicates that the detection of volatile organics on Si surface is possible using TOF-SIMS with a cold stage.

EXPERIMENTAL

Analyses were performed on a Charles Evans and Associates Time-of-Flight Secondary Ion Mass Spectrometer (TOF-SIMS) equipped with liquid metal (Ga) and Cs sources.

Mat. Res. Soc. Symp. Proc. Vol. 386 © 1995 Materials Research Society

I. Analysis of Si Wafers and Carrier Materials

Si wafers were stored in polypropylene (PP) and polycarbonate (PC) carriers. One set of samples, including each type of carrier materials, were treated at elevated temperature (80° C) for 24 hours. Analyses of Si from center and edge of the wafers as well as of the carrier materials were performed.

II. High Purity Water System Materials

Two poly(vinylidene fluoride), PVDF, -(CH_2-CF_2)-, samples were analyzed to identify possible surface contamination on the materials. One was a molded part and the other was an extruded pipe. About 1 cm x 1 cm pieces were obtained from the samples using a clean saw for TOF-SIMS analysis.

III. Detection of Volatile Organic Molecule using a Cold Stage

Ethylene glycol was deposited on a clean Si wafer surface using a micro-pipette. The sample was analyzed at low temperature (-50° C) using a cold stage equipped in the TOF-SIMS instrument.

RESULTS AND DISCUSSION

I. Analysis of Si Wafers and Carrier Materials

Figure 1 contains a representative positive spectrum obtained from the Si wafer stored in a polypropylene (PP) carrier. The peak for Si^+ is the base peak in the spectrum. Hydrocarbon fragment ions, $C_xH_y^+$, which may arise from adventitious hydrocarbons and organic contaminants on the surface, are observed in the spectrum. The peaks at m/z 149, 279 and 391 are attributed to $C_8H_5O_3^+$, $C_{16}H_{23}O_4^+$ and $C_{24}H_{39}O_4^+$, respectively. The detection of these peaks indicates the presence of dioctylphthalate (DOP) on the wafer surface. DOP is a common plasticizer used in polymer formulations.[5,6] DOP had also been detected in a clean room environment.[5] Another peak at m/z 219 arises from butylated hydroxytoluene (BHT) ions, $C_{15}H_{23}O^+$. BHT ion is a fragment that can be observed from a group of anti-oxidants. The protonated molecular ion for dioctyl adipate is found at m/z 371, $C_{22}H_{43}O_4^+$. Although trace amounts of various organic species are found on the Si surface, the prominent Si^+ peak indicates that the surface contamination is at very low levels, probably in the range of monolayers or less. The adverse effects of these contaminants may increase as they accumulate on the surface in longer time periods or after more handling.

In comparing the spectra from Si wafers stored in polycarbonate (PC) carriers, Figure 2 demonstrates that the protonated as well as the oxidized molecular ions for Irgafos at m/z 647 and 662, respectively, are detected on the wafer edge that was stored in a heat-treated polycarbonate (PC) carrier. These peaks are detected from the PC carrier material at more prominent intensities. This observation suggests that the commonly used antioxidant Irgafos was transferred to the edge of the stored wafer from the PC surface. Analyses by Xray Photoelectron Spectroscopy indicate that the amount of P on the PC surface is less than 0.1 atomic percent and it is lower than the XPS detection limit on the Si surface.

The data from these samples indicate that BHT ions are found on the Si stored in a PP carrier and Irgafos ions are detected from the PC carrier. DOP is found on all the wafers analyzed, which might be due to a different source.

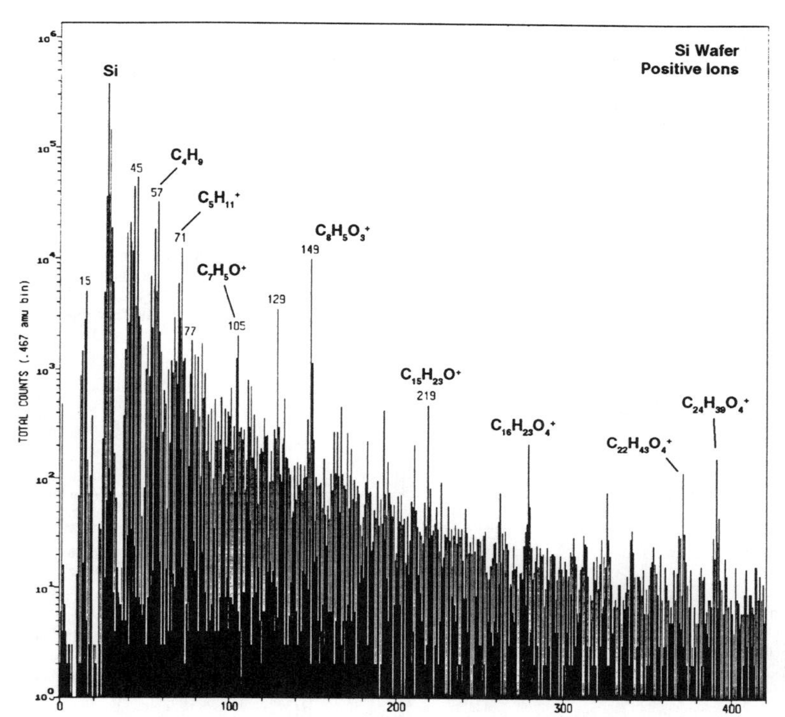

Figure 1. Positive ion spectrum of a Si wafer stored in a polypropylene (PP) carrier.

II. High Purity Water System Materials

Fluorocarbon peaks indicative of poly(vinylidene fluoride), PVDF, are detected in the spectra of both samples as shown in Figure 3. The PVDF peaks are observed at more prominent intensities from the extruded pipe. Contaminant peaks are found in the spectra from the molded sample. The contaminant ion peaks include Na^+, K^+, $C_3H_9Si^+$, $C_5H_{15}Si_2O^+$, Cl^-, $C_{12}H_{25}SO_4^-$, $C_{14}H_{29}SO_4^-$ and $C_{16}H_{31}O_2^-$. The peaks $C_3H_9Si^+$ and $C_5H_{15}Si_2O^+$ result from polydimethyl siloxane (PDMS) which is a common lubricant or a mold release agent. $C_{12}H_{25}SO_4^-$ and $C_{14}H_{29}SO_4^-$ peaks are indicative of sulfate-containing surfactants such as dodecyl sulfate. The presence of surfactant residue on the surface may suggest an incomplete rinse. The fatty acid anion $C_{16}H_{31}O_2^-$ could arise from finger oil or other sources. The extruded pipe is essentially free from these contaminants due to the clean extrusion process.

III. Detection of Volatile Organic Molecule using a Cold Stage

Protonated molecular ions and cluster ions of ethylene glycol are detected as shown in Figure 5. Peaks for DOP are also found in the spectrum. Volatile organics such as ethylene glycol on sample surfaces are easily desorbed in the vacuum chamber. However, using a cold

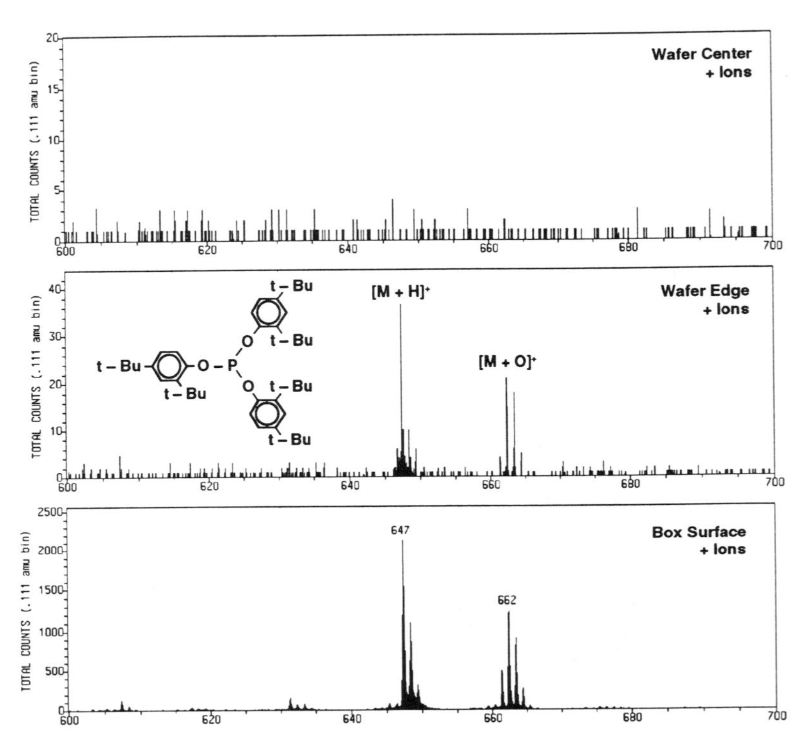

Figure 2. Spectra of wafer center, edge and polycarbonate carrier material. The wafer stored in the carrier was treated at 80°C for 24 hours.

stage operated at low temperature (-50° C) in the TOF-SIMS instrument, the peaks for ethylene glycol are easily detected. Direct surface analysis of more volatile species can be achieved at low temperature.

Conclusion

This paper demonstrates that direct surface analysis using TOF-SIMS can provide useful information about contaminants and their possible sources. Organic contamination on Si wafers was detected using TOF-SIMS. Detection of Irgafos on both the wafer edge and on the polycarbonate carrier material indicates that the antioxidant has migrated from the carrier to the wafer in a heating process.

Metallic cations, Na^+ and K^+, as well as fatty acids, surfactants and PDMS were detected from the surface of a molded part for a high purity water system. Low temperature analysis facilitates the detection of possible volatile contaminants using TOF-SIMS.

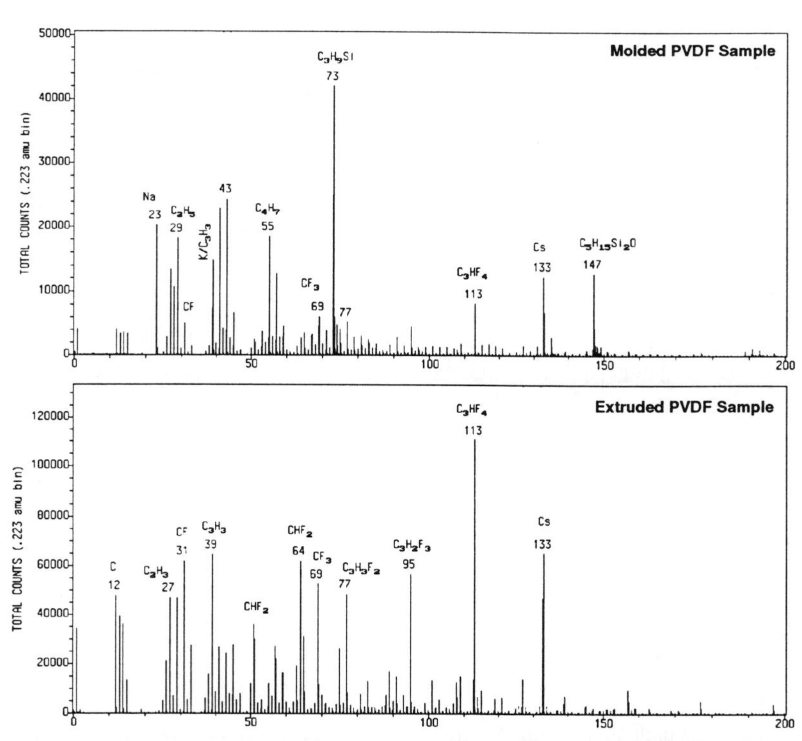

Figure 3. Spectra of molded and extruded poly(vinylidene fluoride), PVDF, parts for high purity water system.

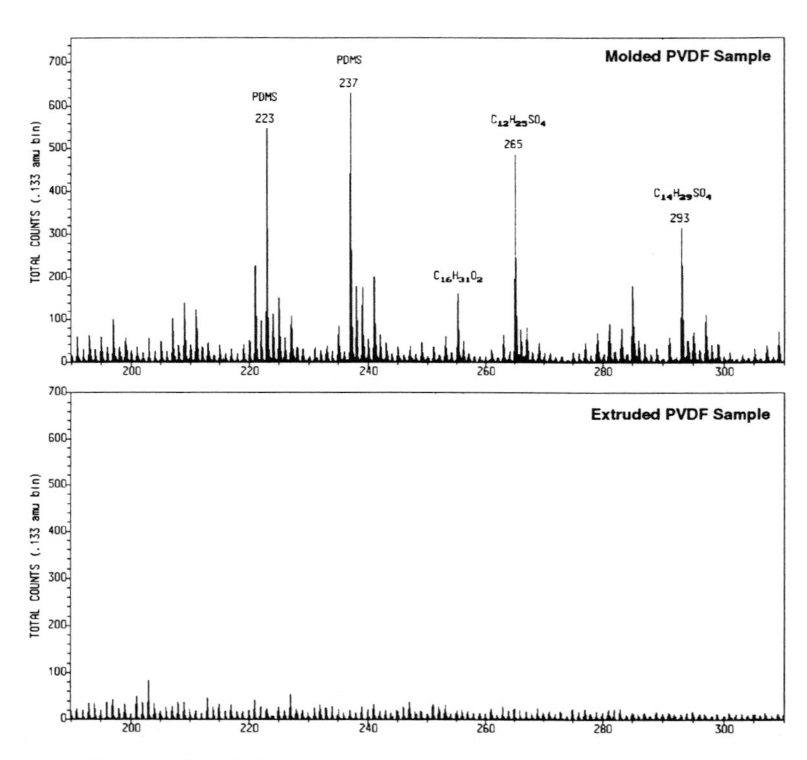

Figure 4. Spectra of molded and extruded PVDF samples.

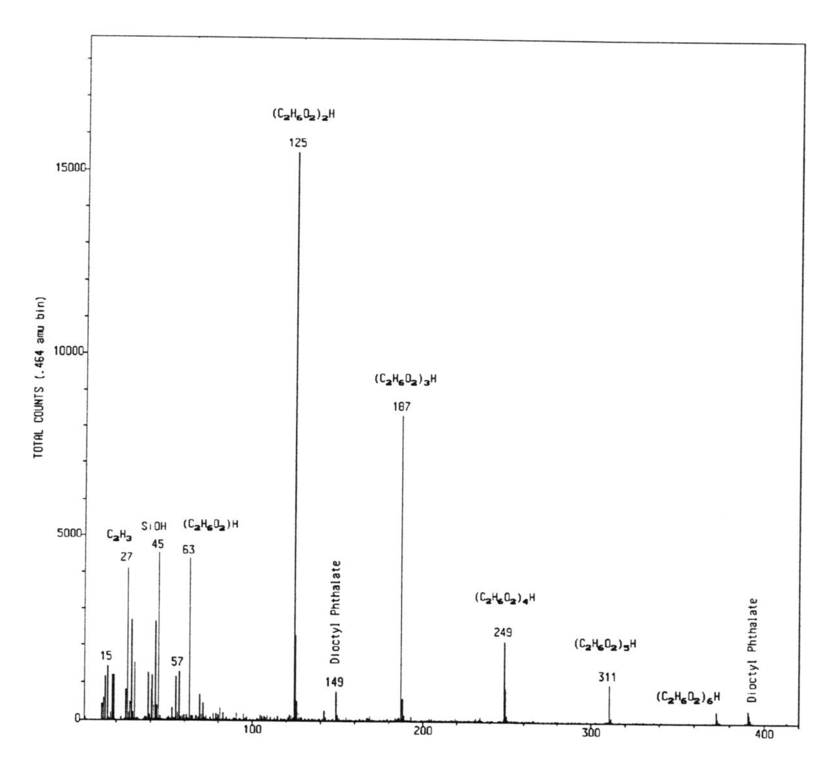

Figure 5. TOF-SIMS spectrum of ethylene glycol on Si using a cold stage.

References

1. P. Burggraaf, Semiconductor Int'l, **10**, 56 (1992)
2. E.J. Mori, K.D. Dowdy and L.W. Shive, Cleanroom Technology Forum, Microntamination 92, 154 (1992).
3. B. Schueler, Microsc. Microanal.Microstruct. **3**, 1 (1992)
4. A.J. Muller, L.A. Psota-Kelty, H.W. Krautter and J.D. Sinclair, Solid State Technology, p61, September, 1994.
5. V.K. Bjola, Microcontamination, p31, Dec. 1993.
6. A. Licciardello, O. Puglisi and S. Pignataro, Appl. Phys. Lett., **48**(1), 41 (1986).

DEPOSITION OF Ca^+, Na^+, Al^+, K^+ ON SINGLE-CRYSTAL Si FROM DIH_20: NH_40H: H_20_2 AND DIH_20: HCL: H_20_2 AS ANALYZED BY ToF-SIMS AND SPV

DAMON DEBUSK*, JOHN LOWELL*, AND FRASER REICH**
*Advanced Micro Devices, 5204 E. Ben White Blvd., Austin, TX 78741
**PHI Evans, 301 Chesapeake Drive, Redwood City, CA 94063

Abstract

The solutions DIH_20: NH_40H: H_20_2 and DIH_20: HCL: H_20_2 have been used extensively in the semiconductor industry for the removal of transition and alkali metals from wafer surfaces. The former, when used in a ratio of approximately 5:1:1 effects the wet oxidation removal of organic surface films and trace metals (Cu, Au, Ag, Ni, etc.) while the latter, in an approximate 6:1:1 ratio dissolves alkali ions and hydroxides of Al^{+3}, Fe^{+3}, and Mg^{+2} desorbing by complexing residual metals. Of concern is that these mixtures are known to plate out charged ionic species, particularly Fe, which remains on the Si surface after exposure to these solution. This effect adds an enhanced surface potential barrier which can be negative or positive depending on the concentrations of species deposited.

While the effects of Fe as deposited by the NH_40H solution is well-known, we are concerned in this study with the deposition of Ca, K, and the lighter ions, Na and Al. Conventional analytical methods such as ICP-MS or AAS or TXRF have been used to detect these ions but are problematic since each has a limited range of detectable species or is destructive to the surface when used with VPD to increase resolution. In this study, we have verified the deposition of these species by a ToF-SIMS (Time-of-Flight Secondary Ion Mass Spectrometry) analysis since it has the ability to detect the complete range of the desired ions. In addition, we have applied high-injection surface photovoltage (SPV) to obtain an average surface charge on each sample. To illustrate the use of these methods, we have applied them to the analysis of various closed acid processing systems. Our interest here is the amount of these selected ions being deposited from these solutions and the influence of using a standard canister or bulk acid supply system. Comparisons were also made to standard and megasonic RCA cleans and hot phosphoric nitride strips. We will discuss ToF-SIMS and how it was applied to analyze the addition of these charged trace ions. This work suggests that the problem of trace ion detection on Si must include the entire range of transition and alkali metals, and that ToF-SIMS is a promising analytical tool for this purpose.

Introduction

During the initial qualification of a new closed acid processing system, identified as Tool #1, various analytical short-loop tests were performed to document contamination levels and

Mat. Res. Soc. Symp. Proc. Vol. 386 © 1995 Materials Research Society

measure the impact of particles generated during wafer laser scribing. Our primary concern was that the new tool behave similarly to the existing tools which were considered to have acceptable performance.

RCA cleans basically consist of two or more chemical mixtures, known as Standard Clean 1 (SC1), Standard Clean 2 (SC2), and in some cases a dilute DIH_2O:HF mixture. The SC1 is made up of $DIH_2O:NH_4OH:H_2O_2$ which is typically at a 5:1:1 ratio and is used to effect wet oxidation removal of organic surface films and expose the surface for desorption of trace metals (Au, Ag, Cu, Ni, Cd, Zn, Co, Cr, etc.). In addition, this mixture continues forming and dissolving hydrous oxide films while having etch rates of approximately 0.9 Å/min. for SiO_2 and 0 Å/min. for Si (20 minutes at 85°C). The SC2, which is made up of $DIH_2O:HCl:H_2O_2$ typically at a 6:1:1 ratio, is used to dissolve alkali ions and hydroxides of Al^{+3}, Fe^{+3} and Mg^{+2} and desorb by complexing residual metals. In addition, this mixture leaves a protective passivation hydrated oxide film on the wafer surface. A 30 second DIH_2O:HF at a ratio of 10:1 is frequently used prior to and/or after the SC1 to strip any impure hydrous oxide surface layers and re-expose the silicon surface. The mixture ratios, etch times and temperatures are typically varied to achieve the desired etch rate or surface microroughness.

One benefit of having a closed acid processing system is that acids are freshly mixed during each clean. This reduces the variation in acid concentration that occurs in immersion tank processing when the H_2O_2 decomposes, and the NH_4OH (SC1) volatilizes.

Another issue of interest is the switch from a standard canister system to a bulk acid supply system for units like our TOOL #1 & #2, in order to reduce down time by eliminating the frequent manual filling of the canisters by material handlers. During this contamination baselining of the closed acid processing systems, we decided to establish a comparison with a standard and megasonic wet hood RCA clean, and Hot Phosphoric/DI which is regularly used to strip nitride. For all of these experiments we took advantage of two relatively new metrologies, namely ToF-SIMS and high-injection surface photovoltage (SPV).

ToF-SIMS is a new technique which we are using for its unique ability to analyze charged ions on silicon surfaces [1]. In particular it is sensitive to the mass analysis of both organic and inorganic species and complexes on semiconductors, and can also be used to evaluate polymer materials or organic residue on treated materials. For comparison, other techniques such as TXRF are only sensitive to metallics. However, Al which is prominent in this cleaning chemistry is typically not considered detectable by this method. ToF-SIMS uses a pulsed primary ion beam (low dose) to desorb and ionize species from the sample. The secondary ions produced are accelerated into a mass spectrometer where they are mass-analyzed by measuring their time-of-flight by the relation $E=1/2mv^2$ from the surface to the detector. The physics in this equation relates masses in time (dispersion) and energy for identification. The mass spectrometer can also estimate the location of the species on the sample and present an image thereof at the detector. The mass spectrum and the resultant image then combine to determine the composition and geometric distribution.

Unlike previous TOF-SIMS techniques, the unit in this work uses stigmatic secondary ion optics to transmit the mass-resolved image for simultaneous data acquisition. Our results on test wafers indicate that the spectrometer is more surface sensitive than XPS, TXRF, VPD/AAS or Auger (>9000 @ 28m/e). With it, Wenner and co-workers have confirmed existing models for the diffusion of Fe into surface oxides and the resultant silicide formation [2]. The principal scientific advantages of TOF-SIMS are (1) mass resolution or the ability to resolve peaks from interfering species, (2) assignment of mass or the exact mass of a spectral peak, and (3) lateral resolution. Like standard SIMS, the technique does use ion milling for depth profiling. However, its depth resolution is on the order of tenths of angstroms which is not typical of SIMS. Here the advantage is to look at species in the near-surface region for contaminant metals or shallow-junctions. In the case of contaminates, one can also find exotic sequences such as CF_2-O-CF_2 (from a perflorinated polyalkylether) which are difficult to find with more traditional techniques. Thus its usefulness as an all-purpose surface analyzer is applicable to bulk and surface contaminants from a wide variety of organic and inorganic sources (photoresists, improper cleans, contaminated solvents, construction materials of process equipment). We believe that like micro-Raman or possibly RBS it could be used to find species on the surface of glasses, polysilicon, and dielectrics with vastly wider mass resolution.

High-injection SPV, hereinafter referred to as Surface Charge Imaging (SCI), principles are as follows: The electrostatic potential on the wafer surface barrier is changed by minority carriers generated by laser illumination. Changes of this potential are sensed by a remote electrode that does not touch the wafer. The generation rate is very high (> 10^{19} photons cm^{-2} sec^{-1}), and minority carriers generated by light at a constant wavelength compensate the charge in the space charge region at the surface. This provides the flat band conditions, which are directly related to surface charge.

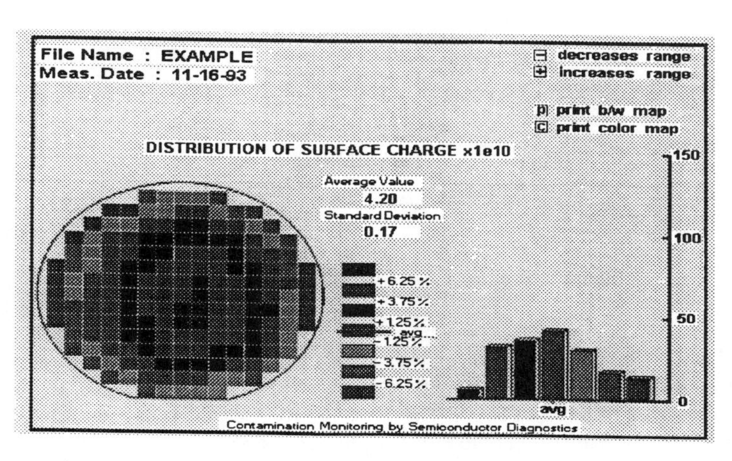

Figure 1. Typical Wafer Map of Post-RCA Clean Surface Charge

Our findings, and the SCI data of others [3], suggests that the addition of charged trace ionic contamination on the wafer surface causes a corresponding measurable change in surface charge. In this case, monitoring wafer cleaning processes with an SCI tool may be very beneficial in ensuring product consistency and aid in diagnosing device problems, such as edge failures. Non contact and non destructive mapping of the wafer surface is possible without fabricating a capacitor and takes approximately 1 second per point (Figure 1). This is in contrast to conventional CV plotting techniques that are not considered real time, typically only test one or two positions on a wafer, are influenced by the preparation variables (i.e. sputtering film quality), and have limited detection limits.

Experiments

For the initial testing, we obtained standard Czochralski (CZ) grown wafers (100 P-type). Additional FZ wafers were added as references in order to eliminate oxygen precipitation effects on SPV diffusion length measurements, which can occur to various degrees during the thermal treatment necessary with this technique. The wafers were then scribed except for one CZ wafer from each group and processed per the following process flow chart (Figure 2). In this paper we will only discuss the portion of these charts wherein the SCI and TOF-SIMS work was done. Additional testing will be referenced but not shown.

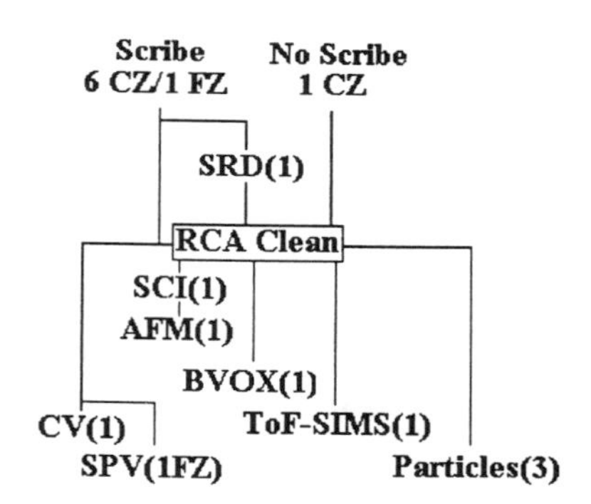

Figure 2. Process Flow Chart for Tool (Reference)

The closed acid system cleaning process was an SC1 for 3 minutes followed by an SC2 for 1.5 minutes, at a rotation speed of 450 RPM. The mixture temperatures at the wafer surfaces is estimated to be 55 to 60°C. The wet hood RCA clean process was an SC1 for 5 minutes and SC2 for 5 minutes at mixture temperatures of 85°C. The wet hood megasonic clean was performed in SC1 only at room temp for approximately 10 minutes, and the wet hood hot phosphoric processing was done at 160° for 35 minutes.

Results

In this experiment (Table I), the levels of ions measured by ToF-SIMS varied considerably between treatments and ranged from of $9X10^8$ to as high as $2X10^{11}$ atoms/cm^2. Due to the lack of correlation with existing techniques, these values are only estimates based on the count frequency of the trace element being analyzed [4]. The only significantly high level of contamination was that of Al, which was detected on the megasonic processed wafer surface. We were surprised to observe that the semiconductor grade phosphoric had relatively low contamination levels. This acid is typically known to have high contamination levels (i.e., some researchers have used hot phosphoric as a controlled method for sodium contaminating of oxides for passivation studies).

Tool	#1 (canister)	#2	Hood
Average Surface Charge (Q/cm^2)	$4.1X10^{10}$	$4.0X10^{10}$	$4.3X10^{10}$
Estimated Na^+ Concentration (atoms/cm^2)	$4X10^{10}$	$3X10^{10}$	$3X10^{10}$
Estimated Al^+ Concentration (atoms/cm^2)	$4X10^{10}$	$2X10^{10}$	$5X10^9$
Estimated K^+ Concentration (atoms/cm^2)	**$2X10^9$**	**$9X10^8$**	**$3X10^9$**
Estimated Ca^+ Concentration (atoms/cm^2)	$<1X10^9$	$4X10^9$	$<1X10^9$

Tool	#1 (bulk)	Mega	Hot P
Average Surface Charge (Q/cm^2)	$5.3X10^{10}$	$5.7X10^{10}$	$4.7X10^{10}$
Estimated Na^+ Concentration (atoms/cm^2)	$8X10^{10}$	$5X10^{10}$	$9X10^{10}$
stimated Al^+ Concentration (atoms/cm^2)	**$2X10^{10}$**	**$2X10^{11}$**	**$9X10^9$**
Estimated K^+ Concentration (atoms/cm^2)	$1X10^9$	$2X10^9$	$2X10^9$
Estimated Ca^+ Concentration (atoms/cm^2)	$2X10^{10}$	$2X10^{10}$	$9X10^{10}$

Table I. TOF-SIMS and SCI Results

Aside from the noted differences, spectra from all of the wafers except for the hot phosphoric wafer are quite similar. Spectra from the hot phosphoric sample contain no B peaks, different intensities of some organic ions, and varying relative intensities of some Si-containing peaks arising from the wafer substrate. As discussed above, aliphatic hydrocarbons ($C_xH_y^+$) species described as "adventitious carbon" are present in all of the spectra. Surface concentrations of transition metals such as Fe, Ni, Cr, Cu and Zn are below the detection limit at less than 10^{10} atoms/cm^2. Surface charge levels remained at their previous levels. This suggests that if surface concentrations are at levels of 10^{10} atoms/cm^2 they do not impact the surface charge or barrier levels significantly. As others have suggested [3], this makes the use of surface charge as an indication of high ionic contamination levels in the process an effective, non-destructive indicator.

Conclusions

For the sake of completeness, we can make the following observations in regard to the system aspects (i.e. the differences between cleaning methods) for the two sets of experiments reported.

1) At the time these tests were performed, the tools were considered to have acceptable contamination levels for the technologies being processed.

2) The Al concentration is the major difference observed between the megasonic and the other tools tested.

3) No major differences exist between Tool #1, Tool #2, and the standard wet hood.

4) The surface charge values reflect the change in concentration of positive ionic species detected by ToF-SIMS.

5) The limited number of wafers (one per treatment) used in this testing is insufficient for determining statistical significance. The data may reflect normal tool to tool, or material variation and is presented only to show feasibility of using the two techniques.

However the main purpose of this study was to reach these conclusions by using the effective methods of SCI and TOF-SIMS. We believe that the SCI can be a reliable metrology for a quick diagnosis of post-clean wafers using a prominent (or SPC-like) threshold of surface charge as a statistic for the unacceptable level of contaminant ions in the chemical or in the tool used. The spectroscopy provided by the TOF-SIMS allows us to see the entire range of singular metallic ions and molecules and suggests that changes in charge will track with concentrations of important species such as Na, Al, K, and Ca. The combination of these methods can be an effective method for future on-wafer diagnosis of contaminant metals.

References

1. R.W. Odom, *et.al.*, Electrochemical Society, Extended Abstracts, Spring Meeting, San Francisco, May 1994, p. 559.
2. V. Wenner, *et.al.*, Advanced Micro Devices Technical Symposium, Santa Clara, February 1994.
3. J. Lowell, V. Wenner and L. Jastrzebski, Materials Research Society Spring Meeting, San Francisco, April 1993
4. Private communication Charles Evans & Associates

SurfaceSIMS, SECONDARY ION MASS SPECTROMETRY USING OXYGEN FLOODING: A POWERFUL TOOL FOR MONITORING SURFACE METAL CONTAMINATION ON SILICON WAFERS.

STEPHEN P. SMITH*, LARRY WANG, JON W. ERICKSON, and VICTOR K.F. CHIA
Charles Evans and Associates, 301 Chesapeake Drive, Redwood City, CA 94063

ABSTRACT

Secondary ion mass spectroscopy (SIMS) coupled with oxygen flooding of the silicon surface during analysis provides an analytical technique capable of detecting $\leq 10^{10}$ atoms/cm^2 of many surface elemental contaminants. Of particular importance to meet the future needs of the semiconductor industry is the current ability to detect Al and Fe contamination at a level of $2x10^9$ atoms/cm^2.

INTRODUCTION

As the critical dimensions of microelectronics devices steadily decrease, concurrent improvement in the cleanliness of starting Si wafers and of IC process steps such as ion implantation and resist strip is vital. Measurement of the surface metal contamination on silicon wafers is an essential part of any such process improvement. Secondary Ion Mass Spectrometry (SIMS) provides such measurements because of the high sensitivity and good depth resolution of the technique.

SIMS with oxygen primary ion bombardment (O-SIMS) has been used in particular to investigate surface cleanliness because of its high sensitivity for light elements such as Li, Na, Mg, Al, and K which cannot be readily detected by X-ray techniques such as Total Reflection X-ray Fluorescence. However, a disadvantage of conventional O-SIMS for surface contamination measurements is the variation in secondary ion yields that occur during the analysis of the first 10nm of a sample. This variation is related to the initial increase toward steady state of the concentration of the implanted primary beam oxygen, and reflects the enhancement of secondary ion yields due to the presence of the reactive oxygen atoms in the substrate.

To provide a more accurate SIMS depth profile of near-surface contaminants, the initial transient changes in secondary ion yields must be minimized. This can be done by increasing the local concentration of oxygen in the sample surface to a level where the secondary ion yields are constant during the primary beam equilibration. It is well known for silicon that this increase can be achieved by exposing the sputtered surface of the sample to a jet of oxygen gas to ensure that the surface is continually saturated with adsorbed oxygen. Combined with other necessary analytical protocols, this moderation of the secondary ion yield from silicon is an essential feature of the measurement referred to as *SurfaceSIMS* at Charles Evans and Associates.

*Corresponding Author

Mat. Res. Soc. Symp. Proc. Vol. 386 © 1995 Materials Research Society

SENSITIVITY AND DETECTION LIMITS

In the absence of any measurement background that would limit the detection of a given impurity, the sensitivity of the *SurfaceSIMS* measurement can be represented by the number of secondary ions detected for a nominal impurity areal density:

$$\textbf{Detected Ions} = \textbf{(Areal Density)} \times \textbf{(Analysis Area)} \times \textbf{(Useful Ion Yield)} \quad (1)$$

The useful ion yield is an empirical yield (ions detected per atom sputtered) that is the absolute ion yield modified by factors including efficiencies of ion collection, transmission, and detection. For a nominal areal density of 10^{10} atoms/cm^2 measured with a high transmission magnetic sector SIMS instrument (e.g. CAMECA IMS 4f), useful ion yields are about 0.1 to 10^{-5} ions/atom sputtered for most elements, and a typical detected area may be approximately 10^{-5} cm^2. Under these conditions, the number of detected ions should range from 10^4 to 1 in the absence of any significant background contribution. Thus a *SurfaceSIMS* detection limit of about 10^{10} atoms/cm^2 or less is expected for many surface-contaminating elements.

Detection limits observed by SurfaceSIMS using magnetic sector SIMS are listed in Table I. Instrumental techniques used to separate interfering ions of the same nominal mass/charge ratio as the desired element include careful selection of the isotope monitored, increased mass resolution of the secondary spectrometer, and the use of kinetic energy filtering ('energy offset').

TABLE I: Practical *SurfaceSIMS* Detection Limits for Surface Impurities on Silicon

Element	Detection Limit 10^9 atoms/cm^2	Element	Detection Limit 10^9 atoms/cm^2
Li	1	**Fe**	2
B	1	**Ni**	2
C	1000	**Cu**	10
Na	1	**As**	10
Mg	1	**Mo**	1
Al	1	**Sn**	1
P	10	**Sb**	1
K	1	**Ta**	10
Ca	1	**W**	2
Cr	1		

The tabulated values are consistent with the detection limits suggested from Eqn. (1). The few exceptions are for atmospheric elements (e.g. C) where the detection limit is influenced primarily by the partial pressure of species in the residual vacuum of the sample chamber.

Current requirements of the semiconductor industry for individual metal contamination on starting Si wafer material are stated as $<10^{11}$ atoms/cm^2 for Na, Al, and Ca, and $<5E10^{10}$ atoms/cm^2 for Fe, Ni, Cu, and Zn[1]. A metrology capability of 10% of the surface contamination level requires detection limits of 1 to 0.5 $x10^{10}$ atoms/cm^2. The required limits are expected to decrease by a factor of four to five in the next six years. The current *SurfaceSIMS* detection limits for Na, Al, Ca, and in particular Fe satisfy these needs. For other species, the detection limits meet the present characterization requirements. Improved *SurfaceSIMS* protocols are being developed to match possible future requirements for lower detection limits.

REPRODUCIBILITY

To provide useful information for IC manufacturing process control improvement, measurements of surface contaminant areal densities should have a relative precision of about 10% (1 sigma) or better. This level of precision will allow reliable identification of samples with contamination levels differing by a factor of two (or indeed by as little as 30%, 3 sigma), which should indicate significant reduction in the contamination associated with a given process or process step.

SIMS measurement precision is usually monitored by repeat measurements of both control samples and unknown samples. A previous study demonstrated that the reproducibility of *SurfaceSIMS* measurements was quite similar to that of conventional depth-profiling SIMS (±10 to 20%)[2]. No significant differences were observed between measurement results obtained on two different SIMS instruments by different analysts using the same analytical protocol. In Table II we have extended the results of that earlier study using a longer series of measurements on standard and control samples extending for a period of over two years. The table lists the mean areal density of each control sample, the number of measurements N, and the percent relative standard deviation (%RSD) of the N measurements.

TABLE II: Long-Term Reproducibility of *SurfaceSIMS* Measurements of Na, Al, and K on Silicon Control Wafers. Areal Density expressed in units of 10^{10} atoms/cm^2. Samples ranked in order of Al concentration

Control Sample	Al			Na			K		
	Mean Areal Density (E10 at/cm^2)	N	% RSD	Mean Areal Density (E10 at/cm^2)	N	% RSD	Mean Areal Density (E10 at/cm^2)	N	% RSD
1	4410	34	8.8%						
2	3600	36	7.6%						
3	1300	18	12%	3	16	36%	2	13	65%
4	107	5	5%	0.8	5	23%	0.5	5	108%
5	33	36	9.8%	120	23	30%	93	19	29%
6	27	10	16%	2.2	9	38%	0.6	7	106%
7	12	5	16%	33	5	6%	24	5	17%
8	5.2	3	7%	0.6	3	38%	3	3	166%
9	3.9	5	22%	9	5	14%	6	5	14%
10	0.5	66	80%	0.6	25	111%	0.2	20	113%
Calibration Standard	*121*			*344*			*285*		

The control samples are calibrated on an 'in-load' basis using a Consensus Reference Standard (as defined by ASTM F 1569) that is a silicon wafer uniformly contaminated with controlled amounts of Na, Al, and K prepared by a spin-coating technique. The Na, Al, and K contamination levels on the standard (Table II) were determined by VPD/AAS (Vapor Phase Decomposition/Atomic Adsorption Spectrometry). The standard sample is measured in the same sample load as the control (and unknown) samples to determine in-load relative sensitivity factors (RSFs). For comparison with the control data, the %RSD values of the mean in-load sensitivity factors for Na, Al, and K measured over the same period of time are 29%, 20%, and 35%, respectively.

For a given contaminant, the RSD values in the table tend to be larger for low areal densities. This trend is due to a combination of the increased statistical variability of measurements at low ion count rates, as well as to an increased likelihood of small-scale unintentional contamination of the surfaces at low concentration levels. Careful inspection of the data in the table shows that among the most heavily contaminated samples, the results for Na and K on control wafer #5 seem to contradict this pattern. This probably reflects some real variation in the intentional contamination across the wafer. The RSD for aluminum on control #3 also seems slightly high compared to data for the other samples, and may also reflect non-uniformity.

In general, the in-load procedure provides an improvement in calibration precision of at least a factor of two compared to the use of a mean sensitivity factor. The fact that the in-load control sample RSDs are substantially reduced means that much of the variance in the calibration sample sensitivity factor measurement is systematic variation due to differences in analytical conditions such as sputter rate.

It is important to realize that the %RSD values in the table combine contributions from any non-uniformity of the calibration standard and of the individual control sample, as well as from random variation in the measurement process. Data presented in Table III provide further information concerning the uniformity of the calibration standard, and on the precision of the measurement process for aluminum.

TABLE III: Short-Term Reproducibility of *SurfaceSIMS* Measurements of Al on the Calibration Standard Wafer. Areal Density expressed in units of 10^{10} atoms/cm^2. Repeat measurements made on twelve sample chips analyzed in a single load.

Sample Holder Window	Mean Areal Density (E10 at/cm^2)	N	% RSD		Sample Holder Window	Mean Areal Density (E10 at/cm^2)	N	% RSD
1	122.5	3	4.72%		7	127.8	2	3.02%
2	118.0	2	1.24%		8	123.7	2	9.15%
3	133.5	2	2.15%		9	112.4	2	3.43%
4	124.9	2	1.60%		10	119.9	2	0.11%
5	109.5	2	0.43%		11	119.7	2	3.06%
6	118.2	2	0.62%		12	121.3	3	0.56%

Mean areal density of 26 individual measurement	**121.0 atoms/cm^2**
%RSD of 26 individual measurements	**5.63%**
Average %RSD of measurements in a single window	**2.51%**

The data in Table III were obtained during a single analysis session from twelve pieces of the calibration sample loaded in one multiple-window sample holder. The precision of repeat measurements on a single sample in any one window is ±2.5% on the average. This number sets a limit on any variation in the Al contamination on the standard on a scale of 1mm (the typical separation of the repeat analyses). The %RSD of the total set of 26 measurements is somewhat larger, ±5.6%. This suggests that there is some systematic variation in the results obtained from the different windows. This may be due to variation in the Al contamination on a centimeter scale (the characteristic dimension of the area on the original wafer from which the samples were cleaved). Alternatively, departures of the faceplate of the sample holder from its ideal flat configuration may also contribute to the increased RSD of the total measurement set. The measured RSD provides an upper limit to the magnitude of either of these effects in this experiment. The still somewhat larger %RSD values (7.6% to 12%) obtained for the long-term measurements of control samples 1 to 3 (Table II) imply that the variability of the Al contamination on these samples is ±5 to 10% (on a 10 centimeter scale).

ACCURACY

The accuracy of *SurfaceSIMS* measurements of Na, Al, and K areal density is estimated to be within a factor of ±2 based on the uncertainty of VPD-AAS measurements used to determine the Na, Al, and K levels on the standard wafer. The accuracy of the *SurfaceSIMS* calibration for other elements is less certain. Relative sensitivity factors for other elements are currently determined from ion implanted or bulk doped standards measured under *SurfaceSIMS* conditions. Most of the impurity element in the implant standards used to determine the RSFs lies below the surface of the sample. In contrast, accurate determination of an areal density of a surface contaminant depends on the first few data points of a concentration depth profile since it is often decreasing rapidly with depth. Although the *SurfaceSIMS* technique minimizes ion yield variations very near the sample surface, it is not clear that the variations are completely eliminated. Therefore it is not certain that the RSF determined from measurements of subsurface implanted impurity atoms is necessarily the appropriate value to use for areal density determinations of surface contaminants. A limited number of comparisons of calibrations for Na, Al, K, Ca, Fe, and Sn based on both surface-contaminated and ion-implanted samples have been made (unpublished results). The results suggest that the *SurfaceSIMS* calibrations based on ion implants are accurate to within a factor of three or four.

FUTURE ADVANCES

SurfaceSIMS detection limits must be lowered to keep pace with the increasingly stringent demands of the semiconductor industry. Some improvement in the precision of the measurements may also be required. Maintaining or improving analysis throughput is desirable, and may provide a constraint on improvement of detection limits in some cases.

If we rearrange equation (1) to solve for the impurity areal density, and define an operational detection limit of 30 integrated ion counts for normal *SurfaceSIMS* measurement conditions, then the minimum detectable areal density, here called **ADL** (for Areal Detection Limit) is given by equation (2) as:

$$\textbf{ADL}\text{ (Areal Detection Limit)} = 30 \times (\textbf{Area} \times \textbf{Useful Ion Yield})^{-1} + (\textbf{Background}) \quad (2)$$

In equation (2) we have added an explicit background term, since for some contaminants (e.g. carbon) this may be the controlling factor in setting current or future detection limits. Detection limits potentially may be improved by increasing the area of analysis, increasing the useful ion yield, or reducing backgrounds.

Increasing the area of analysis is an option for unpatterned wafers, but not necessarily for patterned product wafers. The increase can be direct, by increasing the area from which ions are accepted by the secondary spectrometer, or possibly indirect, by using a pre-concentration technique like vapor phase decomposition. One or more orders of magnitude improvement in detection limit may be gained through analysis area increases.

Increasing useful ion yields may have less potential for improving detection limits except in specific cases. Significant increases in absolute secondary ion yields are not generally possible, except if a reliable post-sputter ionization step can be incorporated. For elements that must be analyzed at high mass resolution to eliminate interferences, an improvement of about an order of magnitude would be possible using a secondary spectrometer of improved design (high transmission at HMR conditions).

Reducing measurement background is important now for certain elements, and probably will become important for many other elements as detection limits are improved by other means. Some improvement can come about through minimizing wafer handling, for example by employing a SIMS sample chamber capable of accepting a full wafer. Other avenues for improvement may come about through instrument modifications (such as improvements in UHV, modification of construction materials, and increasing the primary ion beam purity). Finally, attention can be paid to design of analytical protocols to minimize backgrounds from sources such as memory effects.

SUMMARY

SurfaceSIMS is an analytical technique capable of detecting 10^{10} atoms/cm^2 or fewer of surface contaminants on silicon. These contaminants include low-Z elements such as Li, B, and C, important species such as Na, Al, K, and Ca, cross-contaminants such as P, As, and Sb, transitions metals such as Cr, Fe, Ni, and Cu, and high-Z metals such as Mo, Ta, and W. The precision of the measurements are comparable to those of conventional SIMS ($\pm$10% or better at moderate areal densities, and controlled by ion-counting statistics at low concentrations). The accuracy of the technique depends on the accuracy of standards used for calibration. Under consideration at Charles Evans & Associates are several approaches which may be combined to improve the areal density detection limits by an order of magnitude or more. *SurfaceSIMS* is therefore a powerful tool for analysis of surface metal contamination on silicon.

REFERENCES

1. The National Technology Roadmap for Semiconductors (Semiconductor Industry Association, San Jose, 1994), p.113.

2. Stephen P. Smith, in Secondary Ion Mass Spectrometry (SIMS IX), edited by A. Benninghoven, Y. Nihei, R. Shimizu and H. W. Werner (John Wiley & Sons, Chichester, 1994) pp. 476-479.

Part III

Metallic and Organic Surface Contamination

DETERMINATION OF ORGANIC CONTAMINATION FROM POLYMERIC CONSTRUCTION MATERIALS FOR SEMICONDUCTOR TECHNOLOGY

Klaus J. Budde,

SIEMENS AG, Research Laboratories, ZFE T MR 3
Otto Hahn Ring 6, D 81730 Munich, Germany

Abstract

Volatile organic surface contaminants on silicon wafers lead to strong detrimental impact on semiconductor production yield and product reliability. Our model for the mechanism of impact is introduced and discussed.
Only a few metrology methods are suited for the ultratrace detection of volatiles on surfaces and their identification. The data presented in this paper were achieved by ion mobility spectrometry followed by mass spectrometry (IMS/MS).
Eleven commercial wafer storage and transport boxes were screened by sampling the contaminants onto silicon wafers at room temperature. In a second set of experiments, enhanced stress testing was performed at elevated temperatures for polypropylene, polycarbonate, polytetrafluoro ethylene, perfluoro alkoxy polymer, polyvinylidene fluoride and acrylonitrile-butadiene-styrene copolymer. From the outgassing behaviour of single contaminants, valuable information can be achieved.
Test method E 46 (SEMI) samples the contaminants onto the wafer under the real conditions of use. The data for six virgin minienvironments are shown.

1. Introduction

The structural dimensions of the semiconductor devices have been shrinking with each product generation. This process is strictly connected to increasing problems with contamination. A few years ago it was proposed that for the production of semiconductor generations of the future, not the particles nor the metallic ions will be the true enemy, but the adsorbed molecules on the silicon surface /1/. Todays observations clearly demonstrate that adsorbed, volatile organics on silicon wafers already exhibit a strong detrimental impact to semiconductor yield and product reliability. It was shown that volatile organic surface contaminants cause unintended hydrophobization of surfaces /2-6/, spoil most wet and dry chemical treatments /3,4,6-12/, enhance the deposition of particles /7/, strongly influence oxide growth kinetics /2,13-15/, increase surface roughness /16/, and may lead to silicon carbide formation /4,8,17-19/ and sometimes even to counter doping effects /20,21/. A more detailed overview about the various kinds of impact is given in /22/. In most cases, the sources of the organic contaminants are polymeric materials that are used in cleanrooms, in cleanroom equipment and related tools. The requirement to check all of these materials in order to eliminate contamination sources and to improve the polymeric construction materials is obvious.

Mat. Res. Soc. Symp. Proc. Vol. 386 © 1995 Materials Research Society

2. Mechanisms of impact

In the following, a model mechanism is presented that is able to explain all the observed kinds of impact due to organics. Fig. 1 shows the structural formula of butylated hydroxytoluene (BHT), as it is adsorbed onto a silicon surface. This molecule is a very frequently used antioxidant. We observed it outgassing from most polymeric construction materials of today. Adsorption takes place by a strong Van-der-Waals-bonding of the polar moiety of BHT (-OH, an alcoholic function) to the surface of the substrate, regardless, if this surface in particular is silicon, silicon mono- or dioxide or siliconhydroxide or even hydrogen terminated silicon. The rest of the BHT-molecule contains bulky aliphatic t.-butyl groups that stick out of the surface plane and lead to a local protection against all aqueous based chemical treatments, i.e.: the spot that is covered by BHT will not be wetted in such processing. According to this, any following process step encounters a surface that is different from the desired surface state - defects may be the result. Additionally, the Zeta-potential of the surface is changed in such way that enhanced deposition of particles is supported, sometimes even leading to wafer haze.

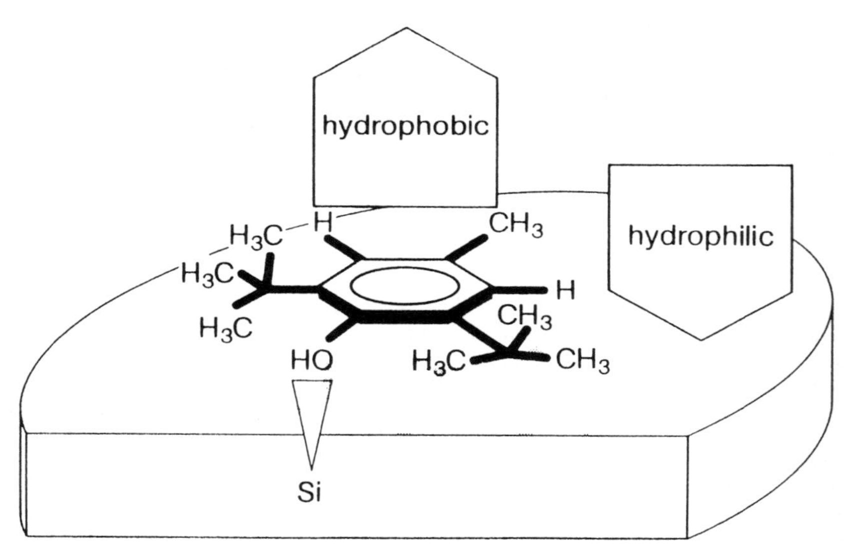

Fig. 1 Adsorption of a typical organic contaminant [2,6 di t.-butyl 4-methyl - hydroxybenzene (=BHT; common antioxidant)] onto silicon

Although the BHT is volatile, it takes some time to break the Van-der-Waals-bonding and to desorb the molecule. If this period of time is not supplied by a subsequent processing step of heating etc., the molecule will fragment on the surface. Those parts of the molecule that remain on the silicon then often turn into silicon carbide, which will cause further defects due to its resistance against the usual chemical treatments. At the latest it is in this stage, that the surface roughness is enhanced and - simultaneously - oxide growth kinetics are spoiled. As a result the gate oxide integrity

(GOI) drops. Further heating of the silicon carbide covered surfaces will induce diffusion of the carbon into the bulk.
Some plasticizers still contain phosphorous, e.g.: tributylphosphate or tri[(di t.butyl) hydroxybenzene] phosphate. In such cases, phosphorous, too, can diffuse into the bulk. The result may be counter doping, which was observed upon storage in wafer boxes.

A different mechanism is chemical carry over: here, usually a wafer carrier takes up some chemical compound because of the existence of micro-pores in the polymeric materials. These are especially present in PTFE. In a first processing step this chemical is incorporated into the carrier and then released in one of the subsequent steps. The result may well be a replica of the wafer carrier, reproduced by the silicon wafer surface. Another modification of chemical carry over implies the chemical reaction of incorporated chemicals within the carrier with a second group of chemicals from a subsequent processing step. Then the chemical reaction product may further react with the silicon surface, again producing imprints.

It seems obvious from the foregoing descriptions that an absolute figure for the maximum tolerable surface covering by organic molecules can not be given in general. Whether or not a given amount of organics will be deleterious, depends - among others - on the particular process step under consideration, on the sticking coefficient(s) of the adsorbed organic substance(s) and on the required yield for this process step. The identical concentration of one particular organic compound can destroy the effect of one process step and may have no impact during a different step. Surface coverings as low as 10^{12} Carbon atoms / cm^2 have been reported to be detrimental, whereas sometimes even 10^{14} at/cm^2 can be tolerated. Here, many more investigations are required to obtain sufficient knowledge of this phenomenon.

3. Metrology methods

The requirements for the appropriate metrology methods can be deduced straight forward from the foregoing description: the analytical tool must be capable to detect volatiles, it has be show an extreme sensitivity towards organics and it should be able to measure on the silicon surface itself. Furthermore it should allow to unambiguously identify the organic compound in order to clearly evaluate the particular source of contamination. From these considerations, the use of the classical analytical methods applied in semiconductor development such as electron microscopy and related techniques (SEM, TEM, EDX, WDX, EELS), Auger electron spectroscopy (AES, SAM), electron spectroscopy for chemical analysis (ESCA, UPS, XPS) and Time-of-Flight secondary ion mass spectrometry (TOF SIMS, STAT SIMS) seems to be limited. All these methods require high vacuum conditions. Under these conditions, most volatiles disappear into the pumping system before they can be detected. Furthermore, except for the TOF-SIMS, the clear identification of the single contaminants is hard to achieve.
Contact angle measurements, wafer bonding and micro balance methods can detect volatiles and are of sufficient sensitivity for most investigations, but these methods completely lack the required identifying power.

Methods that are better suited for the detection and identification of volatile organics are: especially tailored techniques of Infrared spectroscopy (MIR FT-IR, ATR FT-IR), Atmospheric pressure ionization mass spectrometry (APIMS), Ion mobility spectrometry / mass spectrometry (IMS/MS) and Gaschromatography mass spectrometry following thermodesorption (TD GC/MS). A detailed overview about the available metrology methods, including a discussion of the single methods can be found in /23/.

Most of the results shown in this paper were achieved by IMS/MS. Fig. 2 shows the schematics of an ion mobility spectrometer with a quadrupole mass spec serving as the second detector. The IMS itself consists of the sample compartment including the oven and thermodesorption unit, the ion-molecule-reactor with the radioactive source, and the drift cell with the first detector - a Faraday plate. The wafer with the surface contaminants is introduced into the oven and smoothly heated; typically to reach 160°C. By a constant stream of carrier gas (usually zero grade air with 10 ppm water) the desorbed contaminants are swept into the ion-molecule-reactor. Here ionization takes place in two steps: first, the molecules of the carrier gas are ionized and turned into the so called reactant ions. Next, these reactant ions lead to a chemical ionization of the target molecules (contaminants) at atmospheric pressure. This second step often reaches an ionization yield of 100% for trace compounds. Among others, the absolute sensitivity of IMS measurements is depending on the matrix and on the chemical nature of the target molecule; low limits of detection of low femtograms or minimum concentrations in the ppq-range can be achieved. In the drift cell, the different target ions are separated by electrophoresis in the gaseous state and detected by a fast acting electrometer. The result is a chromatogram with typical drift times in the range from 5 ms to 40 ms, an example is shown in fig. 3. If unknown species /samples are encountered, a quadrupole mass spec is used as the second detector. In this case, the sensitivity of the equipment is decreased by approx. two orders of magnitude.
For a detailed description of the method and its applications to semiconductor technology see /22,24/, for theoretical background refer to /25-27/, more general applications are discussed in /28-30/.

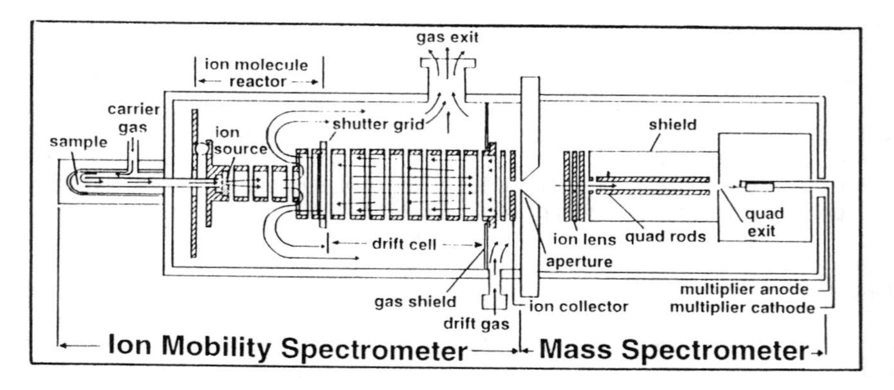

Fig. 2 Schematic drawing on an ion mobility spectrometer - mass spectrometer

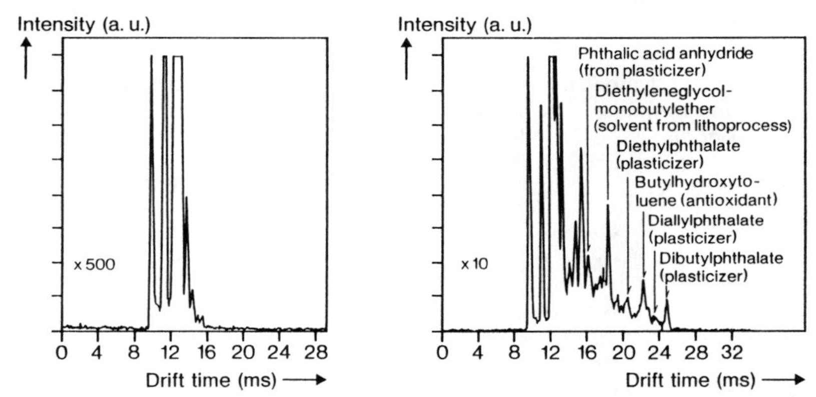

Fig. 3 Wafer contamination due to storage in a polypropylene wafer box; left: clean wafer, right: after 24 hours of storage at 25°C

4. Results

Fig. 3 shows the ion mobility spectrum (positive ions only) of a freshly cleaned wafer in the left picture and the spectrum of the same wafer after 24 hours storage in a polypropylene wafer box in the right picture. It is important to note that for the exposure the wafer first was put into a quartz container, which then was placed into the box. So all the contaminants due to the storage must have diffused through the air, any physical contact between the wafer and the material of the box was strictly prohibited. In the spectrum of the clean wafer almost entirely the reactant ions produce signals, from the left: ammonia, nitrogen oxide, protonated water clusters. The other, small signals are due to acetone and other contaminants of the zero grade air, their concentration is in the very low ppb range. The picture on the right shows a lot of additional, quite intensive signals (note the different scales of vertical enhancement!). Every single peak stands for a different chemical compound. We identified various plasticizers of the phthalate type, the common antioxidant BHT and numerous degradation products from the polymer itself. The presence of diethyleneglycol monobutylether indicates a cross contamination from the photolithographic department of the cleanroom. The total sum of the contaminants may be assumed to be definitely lower than a monolayer of coverage. Much higher (up to a factor of 10^6) amounts of contamination have been found to arise from storage boxes.

Table 1 gives the overview about the first systematic investigation into the contamination from wafer boxes /24/. Every dot stands for one order of magnitude. The data clearly demonstrate that there are good boxes in the market, e.g. no´s 3 and 10 and such ones, that have a very high amount of intrinsic contaminants, see no´s 1, 2, 7 or 11. By the IMS measurements easily the contamination due to the polymeric material itself, to plastic additives or to cross contamination from prior use can be identified.

Table 1: Volatile organic surface contamination from various wafer boxes

BOX	t	Source of contamination										
		polymeric material					cross contamination					
		prepolymers oligomers	polymer additives				solvents				nico-tine	other
			BHT	DEP	DBP	PAA	NMP	CSA	DEG MBE	ben-zene		
1	6	***	***	****	**		***			***		***
2	7		***	***	***	****			***			*
3	320						*	*				*
4	16	***	**	**	*		***	***		**	**	*
5	16				**		***		***			****
6	17		***						**			****
7	17	****	**	***		***	***	**				
8	18				***	***	****				**	*****
9	18		*	*								
10	13		*									
11	3		*****									

t: exposure time of the wafers in days, at T= room temperature;
BHT:butylated hydroxytoluene (2.6 ditert. butyl 4 hydroxytoluene), antioxidant; DEP:dietylene phthalate, plasticizer; DBP: dibutyl phthalate, plasticizer; PAA: phthalic acid anhydride, degradation product of plasticizers of the phthalate type;
NMP: n-methyl pyrrolidone; DEGMBE: diethylene glycol monobutyl ether.

In order to improve the good polymeric materials, e.g. that from box no. 3, enhanced stress tests were performed. It has to be noted that direct outgassing experiments, especially such under enhanced temperature treatment, do not give the same results upon adsorption behaviour onto the wafer than the test described before. But sampling times of almost one year are not acceptable in industrial research. On the other side, the data gained from the outgassing under thermal stress are a valuable information where things have to be improved. Fig. 4 shows the total amount of outgassings from several typical polymeric construction materials as a function of temperature. Quite naturally, PTFE turned out to be the best, followed by PFA. As long as the temperature is kept well below the softening point, PC shows considerable low outgassing rates. The ABS measurements were terminated because of too high outgassing rates. One of the polypropylene samples (PP1) shows an unexpected effect: upon heating above T=100°C, the amount of outgassed material dropped. This behavior clearly indicates a non-perfect production of the polypropylene, which can easy be improved. Additional to the total sum of outgassings, the single volatile components can be identified and their temperature dependence followed, see fig. 5. From this, the most valuable information about possible improvements of the polymer can be drawn.

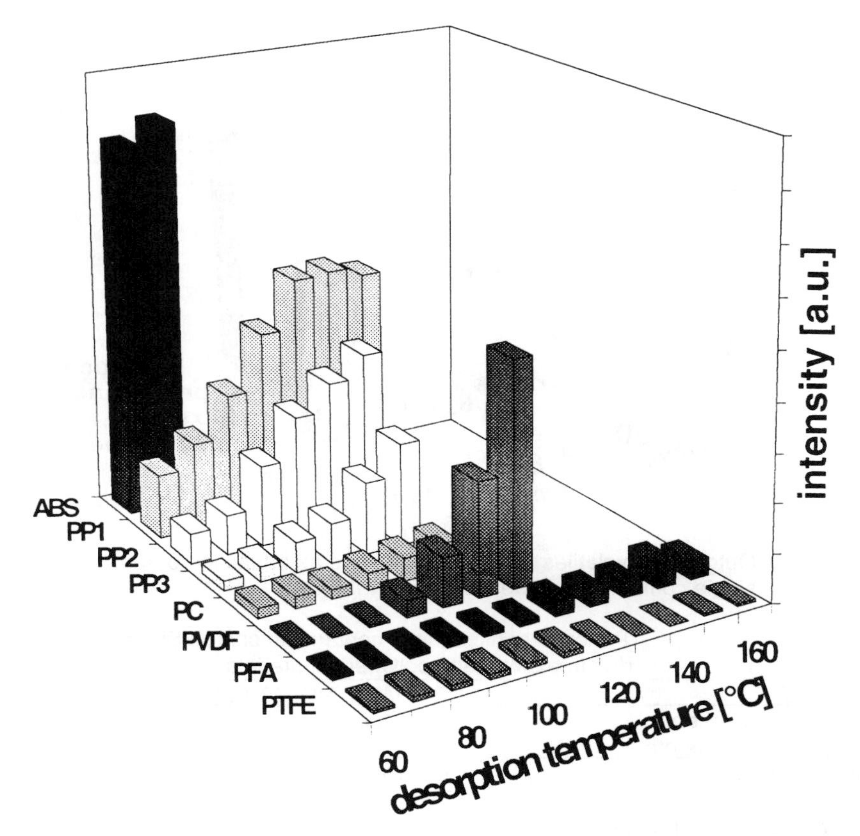

Fig. 4 Total amount of outgassing as a function of temperature for polypropylene, polycarbonate, polytetrafluoro ethylene, polyvinylidene fluoride, perfluoro alkoxy polymer and acrylonitrile-butadiene-styrene copolymer

In a cleanroom there are many more applications for polymers than just wafer boxes and carriers. We found organic volatiles outgassing from a large variety of polymeric materials used in semiconductor processing. We will not show all of the single results here in detail, but give an overview about typical sources, see table 2.

The need for a standardized test method for the detection of volatile organics in order to perform trouble shooting, to improve the wafer processing and for the development of new, outgas-free materials seems obvious after the foregoing. Recently, a new SEMI standard test method for the determination of organic contamination from minienvironments was published /31/. This standard test method is based on IMS measurements and implies the application for testing of polymeric materials as well. As IMS involves chemical ionization of the contaminants at atmospheric pressure, the

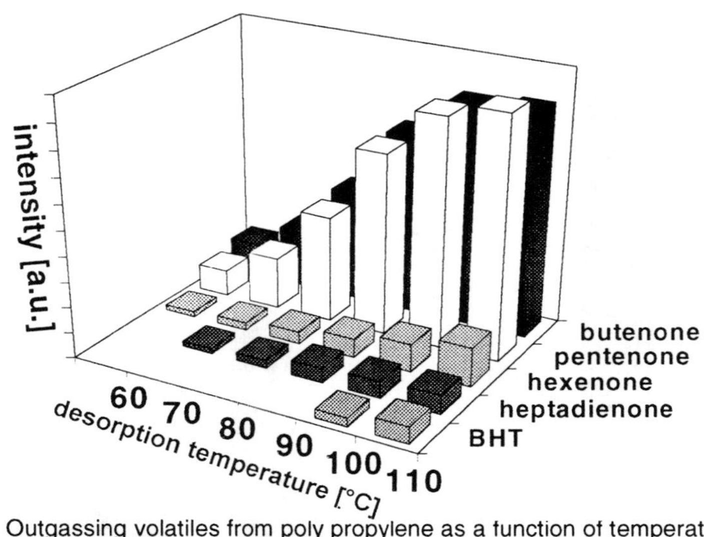

Fig 5 Outgassing volatiles from poly propylene as a function of temperature. The single components were identified by IMS/MS.

Table 2 Typical organic surface contaminations on Silicon wafers due to exposure to polymeric materials

SOURCE	ORGANIC VOLATILES
wafer boxes,carriers, photo plate covers etc.	see Table 1
floor covering	mainly dioctylphthalates and butylated hydroxytoluene
filter frames including the appropriate fixing glue	oligomeric dimethylsiloxanes
silicon rubbers (sealing material)	silicones, oligomeric dimethylsiloxanes
clean room paper	BHT, tributylphosphate, high risk of cross contamination
tubings for the distribution of process media	BHT, polymeric degradation products, their nature depending on the polymer and the process medium
ion exchanging resins for water purification	all kinds of polymeric degradation products according to the particular water cleaning system
sealing foils	BHT, plastizicers, oligomers and polymeric breakdown products
cutting foils (sectioning)	hydroxybenzene, toluene, acetates, plasticizers, sulphuric and chlorinated compounds
photolithography	solvents, plasticizers
epoxy dyes	polymeric degradation products like the bisphenoles A and -F, and their epichlorohydrine derivatives

quantification of single compounds in mixtures is hard to achieve. Anyhow, the discussions during the development of SEMI E 46 showed that the chosen methodology is the best presently available: i) only the adsorbed, i.e. harmful organics are measured, because the wafer itself is used for the sampling, ii) the total amount of these organics is evaluated with sufficiently high precision. SEMI E 46 does not comprise the identification and quantification of single contaminants. This would only be useful, if all correlations between the various contaminants and the different process steps were known.

In order to assure international compatibility and interchangeability of the results, an external reference material is used in the measurement. For a detailed description of the measuring procedure and the evaluation of the "contamination value", CV, see /31/. Table 3 shows the results of the first investigation into the characterization of commercial minienvironments according to the SEMI E 46.

Table 3: Application of SEMI E 46 to six commercial minienvironments

Sample	CV (pos. ions)	saturation	CV (neg. ions)	saturation
A	1.47 ± 0.14	yes	5.268 ± 0.025	yes
B	2.06 ± 0.06	yes	3.63 ± 0.5	yes
C	0.430 ± 0.009	no	0.15 ± 0.05	no
D	0.73 ± 0.09	no	2.38 ± 0.09	yes
E	1.44 ± 0.01	yes	4.1 ± 0.4	yes
F	1.06 ± 0.11	yes	0.62 ± 0.14	no

All samples were virgin ones except for samples E and F. E originally was identical to sample D, but had undergone a cleaning treatment. The results clearly show that this type of cleaning introduced additional organic contamination. Sample F had been used in a different standard test method (SEMI E 45, see /32,33/) prior to our measurements. The data indicate, that at least not much of organic contamination was added to the minienvironment during this test method.

Fig. 6 comprises the data from table 3 together with that of two additional samples in graphical form. Sample H was a fiberglas reinforced duct for the use in cleanrooms. Here the measurement was different from that used for samples A through F, because the neat polymer was tested. Anyway, the results show that the material is very well suited for application in cleanrooms. Sample G passed a cleaning process, then was stored for one week in a quartz container and shipped. The very low value for G demonstrates that handling and storing of silicon wafers is possible without introducing considerable organic contamination.

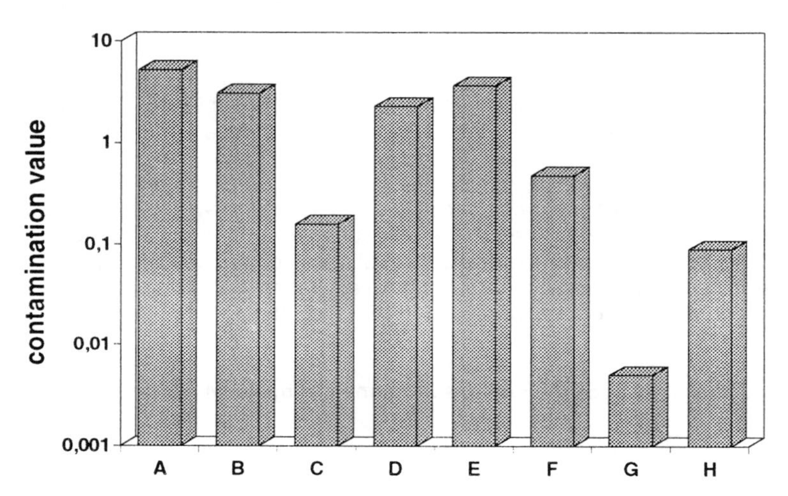

Fig. 6 Application of SEMI E 46 to polymers and minienvironments. Data shown are the "contamination values" (CV) for negative ions.

5. Conclusion

In todays semiconductor processing, volatile organics are encountered everywhere. They produce considerable detrimental impact during a lot of processing steps. Most of the polymeric materials used in cleanroom architecture, equipment and tools have to be improved for their outgassing behavior. This will be supported by the existence of standardized test methods, such as SEMI E 46. On the other side our investigations clearly show that some well-suited polymeric materials already exist and - even more important - that handling and processing without introducing organic contamination is possible.

Acknowledgments

We gratefully acknowledge the contributions of W.J. Holzapfel and M.M. Beyer to the Siemens Research results presented in this paper. The SEMI E 46 was developed by the Environment Contamination Source Task Force of SEMI EUROPE, Brussels, Belgium. For the first measuring program we would like to acknowledge financial support by the following companies: Applied Materials, Santa Clara, CA, USA; Automation in Cleanroom GmbH, Niedereschach, D; Robert Bosch GmbH, Reutlingen, D; GEC Plessey Semiconductors, Plymouth, UK; IBM Deutschland Halbleiter GmbH, Böblingen, D; Jenoptik Technologie GmbH, Jena, D; Kessler & Luch GmbH, Giessen, D; Meissner + Wurst GmbH, Stuttgart, D.

Further we wish to express our gratitude to Fluoroware, Inc., Chaska, MA, USA; SIEMENS AG, München, D; and Wacker Chemitronic GmbH, R&D, Burghausen, D for support in terms of material and services. We further gratefully acknowledge the cooperation of the Fraunhofer Institute for Integrated Circuits in Erlangen, Germany.

References

/1/ T. Ohmi, Microcontamination **6**(1988),16

/2/ A. Licciardello, O. Puglisi, S. Pignataro, Appl. Phys. Lett. **48**(1986),41

/3/ P. Parimi, V. Sundarsingh, J. Electrochem. Soc. **PV 90-9**(1990),260

/4/ M. Grudner, H. Jacob, Appl. Phys. **A 39**(1986),73

/5/ L. A. Zarrera, J.F. Moulder, J. Electrochem. Soc. **136(2)**(1989),484

/6/ M.G. Yang, K.M. Koliwad, G.E. McGuire, J. Electrochem. Soc. **122(5)**(1975),675

/7/ M. Beyer, K. Budde, W. Holzapfel, Appl. Surf. Sci. **63**(1993),88

/8/ R.C. Henderson, J. Electrochem. Soc. **119**(1972),772

/9/ B. Deal, M. McNeilly, D. Kao. Extended abstracts of the 1991 Intern. Conf. on Solid State Devices and Materials, Yokohama, 27.-29.8.1991,496

/10/ D. Burkman, Semiconductor International, July 1981,103

/11/ T. Ohmi, T. Isagawa, M. Kogure, T. Imaoka, J. Electrochem. Soc. **140**(1993),804

/12/ J. Goodman, S. Andrews, Solid State Technology, July 1990,65

/13/ K.B. Kim, Ph. Maillot, A.E. Morgan, J. Appl. Phys. **67**(1990),2176

/14/ I. A. Aizenberg, C. V. Kopetskii, S. V. Nosenko, Doklady Akademii Nauk SSSR, December 1988, **303(5)**(1988),1147

/15/ S. Hossain, C. Panatano, J. Ruzyllo, J. Electrochem. Soc. **137**(1990),3287

/16/ I. Kawanabe, F. W. Kern, M. Itano, M. Miyashita, T. Ohmi, MRS Spring Meeting 1991, Anaheim, CA

/17/ S.R. Kasi, M. Liehr, J. Vac. Sci. Technol. **A 10(4)**(1992),795

/18/ S. Verhaverbeke, doctoral thesis, Leuven, June 1993

/19/ C. J. Sofield et al., Proc. Materials Research Society, Symposium B, Spring Meeting 1992, San Francisco, CA

/20/ V. Lehmann, private communication, 1989

/21/ R. Stengl, private communication, 1988

/22/ K.J. Budde, W.J. Holzapfel, M.M. Beyer, J. Electrochem. Soc. **142**, (1995) 888

/23/ K.J. Budde, W.J. Holzapfel, Proc. of the 38th Annual Technical Meeting of the IES, Nashville, TN, 3.-8.5 1992, pp.483-488

/24/ K.J. Budde in „Analytical Techniques for Semiconductor Materials and Process Characterization", eds. B.O. Kolbesen, D.V. McCaughan, W. Vandervorst (the Electrochemical Society, Pennington, 1990) **PV 90-11**, 215

/25/ E.W. McDaniel, E.A. Mason, "The Mobility and Diffusion of Ions in Gases",John Wiley and sons, New York, 1975

/26/ "Ion Molecule Reactions", Vol. I: Eds. E.W. McDaniel et al., John Wiley and sons, New York, 1970; Vol. II: Ed.: J. Franklin, Butterworths, London, 1972

/27/ „Techniques for the study of ion-molecule-reactions", Eds.: J.M. Farrar and W.J. Saunders,Jr., Wiley Interscience, New York, 1988

/28/ „Instrumentation for Trace Organic Monitoring", Eds.: R.E. Clement, K.W.M. Siu and H.H. Hill, Jr., Lewin Publishers, Boca Raton 1992

/29/ H.H. Hill, Jr. et al., Anal. Chem. **62** (1990) 1201 A

/30/ T.W. Carr, Ed., "Plasmachromatography", Plenum Press, New York, London, 1984

/31/ SEMI E 46 "standard test method for the determination of organic contamination from minienvironments“, yellow ballot 2´94, BOSS volumes 1995

/32/ SEMI E 45 „standard test method for the determination of inorganic contamination from minienvironments“, yellow ballot 2´94, line ballot #2389 in 8`94, BOSS volumes 1995

/33/ N. Streckfuss et al., Proc 41st Annual Technical Meeting of the Institute of Environmental Sciences, Anaheim, CA, 30.4.-5.5.1995, p 147

CHEMICAL STABILITY OF SC1-CLEANED HYDROGEN TERMINATED Si(100) SURFACES

C.H. Bjorkman, H. Nishimura[1], T. Yamazaki[1], J.L. Alay, M. Fukuda[1], and M. Hirose[1]
Research Center for Integrated Systems, Hiroshima University, Higashi-Hiroshima 724
[1] Department of Electrical Engineering, Hiroshima University, Higashi-Hiroshima 724

ABSTRACT

Surface contamination and chemical stability of hydrogen terminated Si(100) surfaces have been studied using Fourier Transform Infrared Attenuated Total Reflection (FT-IR-ATR) spectroscopy and X-ray Photoelectron Spectroscopy (XPS). Hydrogen terminated Si(100) is obtained by removing the chemical oxide, formed in a low-concentration-NH_4OH SC1 clean, in various HF based solutions. Using standard cleaning and loading conditions, we find a direct correlation between surface roughness and the amount of adsorbed C contamination. Oxidation during water rinsing and wafer loading observed by both FT-IR-ATR and XPS indicates that dihydride terminated silicon atoms are preferentially oxidized. Optimizing the water rinse and wafer loading conditions reduces total atomic concentration of C, O, and F surface contamination from 20-30 at.% to less than 6 at.%. These clean surfaces enable XPS-identification of the Si-H_x components of the Si 2p core-level spectra as well as estimation of the relative surface concentration of Si-H_x and contamination species.

INTRODUCTION

As MOS device dimensions are reduced, atomic level control of the morphology and contamination of the Si surface during device processing is becoming increasingly important, especially in the gate oxidation step. Many processing steps are preceded by chemical cleaning of the Si surfaces, such as the SC1 clean. These chemical cleans leave a chemical oxide which is subsequently removed in an HF-based solution. Such hydrogen terminated Si surfaces are further contaminated in clean-room air or in introduction chambers of processing or analysis equipment.

The objective of this study was to determine the chemical stability of Si-H_x species on Si(100) and Si(111) surfaces prepared by ex-situ chemical treatments, thus assuring saturation coverage of hydrogen. However, ex-situ chemical treatments introduce the problem of controlling and limiting the contamination levels from species such as C, O, and F especially during water rinsing and loading into vacuum chambers for deposition, oxidation, or analysis. We use in this study a method that drastically reduces contamination and produces Si(100) and Si(111) surfaces of equally low levels of contamination. These surfaces were found to be sufficiently clean to allow identification of the Si-H_x species by XPS.

EXPERIMENTAL PROCEDURES AND RESULTS

The Si(100) substrates used in this study were cut from 10 Ω-cm p-type CZ wafers. After 10 min cleaning in a low NH_4OH concentration (NH_4OH:H_2O_2:H_2O = 0.15:3:7) SC1 solution at 80°C, the chemically grown oxide was removed in one of the following solutions: dilute HF,

Mat. Res. Soc. Symp. Proc. Vol. 386

$HF:H_2O_2$ mixture or 40 % NH_4F. The treatment time was 5 min for all solutions followed normally by a 5 min rinse in deionized (DI) water.

The FT-IR-ATR measurements were performed using a JEOL JIR 3510 with an attenuated total reflection (ATR) kit, which uses a Ge crystal, whose edges are cut at 60°, as the ATR element. XPS measurements were performed using Scienta Instruments ESCA-300 equipped with an monochromatized Al Kα source and with an acceptance angle of 3.3°. The photoelectron take-off angle was 5° for all spectra in this study in order to enhance the signal from species on the surface.

Standard cleaning and loading conditions

A survey spectrum of a surface cleaned and loaded using standard procedures is displayed in Fig. 1(a). We can easily identify significant amounts of C and O on this surface. The major component of the C 1s peak is that of hydrocarbon, while C-O makes up about 25%. As discussed below, the O 1s peak indicates that the majority of oxygen is bonded to Si as oxide or suboxide. Also, NH_4F-treated Si(100) surfaces exhibited only carbon and oxygen contamination while fluorine was detected on surfaces treated in dilute HF. The spectra in Fig. 1(b)-(d) will be discussed in the following section.

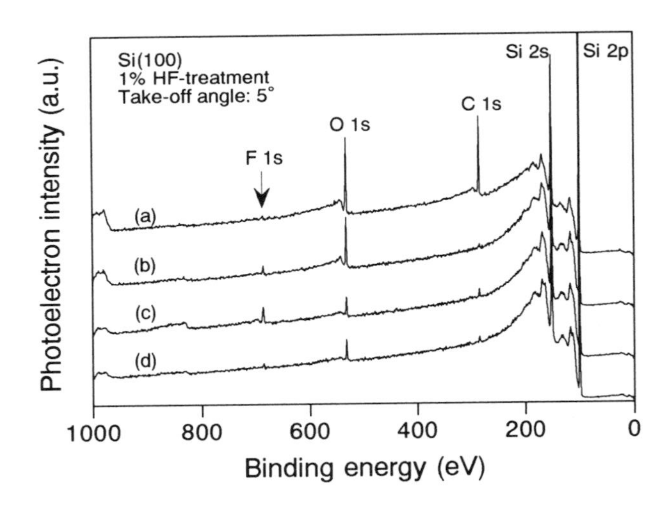

Fig. 1. XPS survey spectra after (a) 5 min DI-water rinse and monitoring the pressure in the introduction chamber by a Penning gauge, (b) 5 min DI-water rinse, (c) no DI-water rinse, and (d) three second DI-water rinse. Each of the Si(100) wafers was subjected to a low-NH_4OH-concentration SC1 clean and the chemical oxide was removed in 1% HF. Note that the Penning gauge was disconnected for the wafers whose spectra are shown in (b), (c), and (d).

Using XPS we studied C contamination on the different surfaces and found that the lowest C concentration (~18 at.%) was obtained on $1:9=HF:H_2O_2$ treated surfaces. This solution has previously been shown to produce flat Si(100) surfaces on which terraces could be observed using scanning probe microscopy.[1] The next lowest C contamination (~27 at.%) was found on 1% HF treated surfaces while the highest (~39 at.%) was found on surfaces treated in NH_4F, a solution which forms (111) facets on the Si(100) surface, i.e., there is a direct correlation between surface roughness and the amount of adsorbed C contamination originating from the clean-room air and rotary pumps. The organic contamination of H_2O_2 is the highest and about 20 ppm. Nevertheless, the surface treated by a HF:H_2O_2 mixture is most clean, indicating that the carbon contamination does not come from the chemicals.

Next we studied the initial oxidation of Si(100) by measuring the FT-IR-ATR spectra before and after XPS measurements. During loading into the XPS system, the pressure gauge produces active oxygen species that oxidize the Si surface. The 1% HF and 1:9=HF:H_2O_2 treated surfaces both exhibit dominant asymmetric dihydride termination before loading into the XPS system,[2] but when ATR spectra were measured after the XPS measurement, the asymmetric dihydride termination was significantly reduced (Fig. 2). We instead observed Si atoms bonded to H and O using both FT-IR-ATR and XPS indicating that dihydride terminated silicon atoms on terraces are preferentially oxidized.

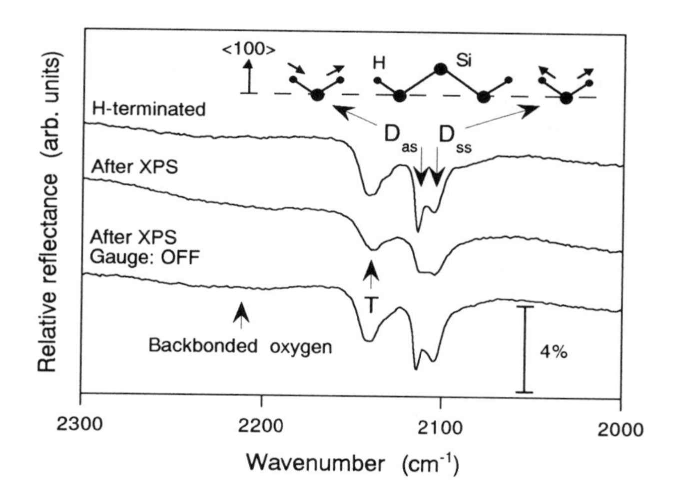

Fig. 2. ATR spectra of the Si-H_x region of H-terminated Si(100). The chemical oxide was removed by a 5 min 1% HF-treatment, and the spectra were obtained before loading into the XPS system (top curve), after 5 min DI-water rinse and monitoring the pressure in the introduction chamber by a Penning gauge (middle), and after a 3 sec DI-water dip and disconnecting the Penning gauge (bottom curve).

Optimized cleaning and loading conditions

As a first attempt to reduce the contamination, a plasma-generating Penning gauge in the introduction chamber was switched off. Also, by monitoring the pumping time, we assured that the ion-gauge in the XPS chamber was not switched on until the pressure reached ~2×10^{-9} Torr. The resulting spectrum is shown in Fig. 1(b). Although this procedure reduced the amount of carbon significantly, most of the oxygen still remains on the surface. However, this result illustrates how important it is to consider contamination due to pressure gauges in surface analysis as well as process equipment.

Since the amount of oxygen on the surface was found to be independent of the HF concentration of the solution used to remove the chemical oxide, we realized that the sources of oxygen were most likely the processing and handling of the wafer after the dilute HF dip. In order to investigate the influence of the water rinse, we eliminated this step which resulted in the spectrum shown in Fig. 1(c). A significant reduction in the oxygen concentration was observed, which clearly illustrates that even overflow rinsing in deionized water with a dissolved oxygen content of about 50 ppb cannot eliminate the growth of suboxide and oxide. Figure 3 shows the reduction and change in shape of the O 1s peak after turning off the Penning gauge and eliminating the DI-water rinse after the 5 min 1% HF treatment. Also indicated are the energies assigned by Alay[2] and Grunthaner et al.[3] to suboxides, stoichiometric SiO_2, and adsorbed oxygen. As all three components are reduced by eliminating the water rinse, the ratio of the suboxide and oxide peaks remain approximately the same, while the component due to adsorbed oxygen exhibit even greater relative reduction. Returning to Fig. 1(c), we note that the elimination of the water rinse, while effective in reducing the amount of oxygen, leaves unwanted fluorine on the surface. We found that this fluorine is quickly removed in deionized water, as illustrated in Fig. 1(d). The spectrum was obtained after a 3 sec water dip and this treatment appears to retain the low oxygen concentration of the spectrum in Fig. 1(c) while at the same time removing most of the unwanted fluorine. The difference between the F 1s peaks of Fig. 1(c) and (d) is shown in greater detail in Fig. 4. As this figure illustrates, the elimination of water rinse yields a F 1s peak made up of two components, i.e., those of adsorbed fluorine at 687.4 eV and F-Si at 685.9 eV.[4] On the other hand, a 3 sec water dip drastically reduces both components, especially adsorbed fluorine.

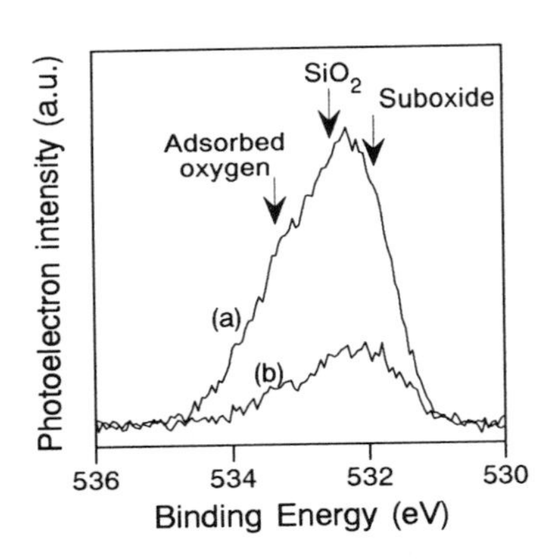

Fig. 3. XPS core-level spectra of the O 1s region of H-terminated Si(100). The chemical oxide was removed by a 5 min 1% HF-treatment and the spectra were obtained (a) after 5 min overflow DI-water rinse and monitoring the pressure in the introduction chamber by a Penning gauge and (b) after a 3 sec DI-water dip and disconnecting the Penning gauge.

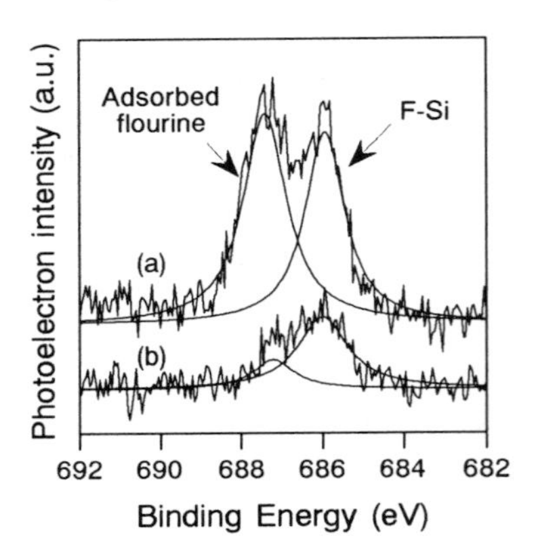

Fig. 4. XPS core-level spectra of the F 1s region of H-terminated Si(100). The chemical oxide was removed by a 5 min 1% HF treatment and the spectra were obtained (a) before and (b) after a subsequent 3 sec DI-water dip.

Before analyzing the Si 2p peaks we must verify that this region doesn't show any evidence of detectable contamination. As shown in Fig. 5, disconnecting the Penning gauge and limiting the DI-water rinse to three seconds resulted in a reduction of the signal over the region of suboxide and SiO_2 oxidation states compared to standard rinsing and loading procedures. As discussed below, the total concentration of O in the form of C-O_x and Si-O is about 3 at.%, while the concentration of Si-F bonds is less than 0.5 at.%. Figure 5 also demonstrates that we as expected obtain more signal from the substrate as the surface contamination is reduced.

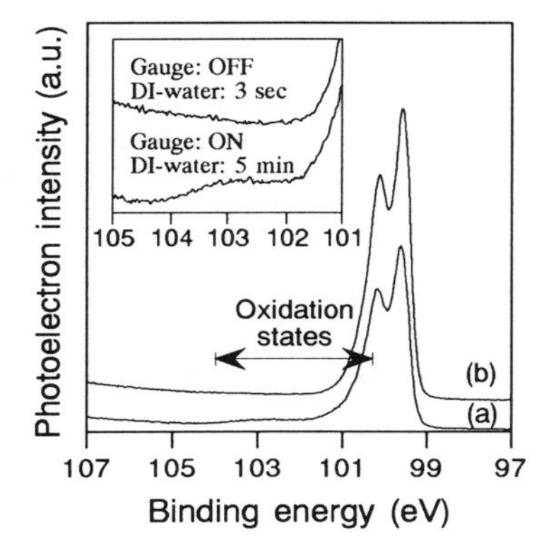

Fig. 5. XPS core-level spectra of the Si 2p region of H-terminated Si(100). The chemical oxide was removed by a 5 min 1% HF-treatment and the spectra were obtained (a) after 5 min DI-water rinse and monitoring the pressure in the introduction chamber by a Penning gauge and (b) after a 3 sec DI-water dip and disconnecting the Penning gauge.

Fig. 6. XPS core-level spectrum of the Si $2p_{3/2}$ region of H-terminated Si(100) treated in 1% HF obtained after deconvolving the corresponding spectrum in Fig. 5 into its $2p_{3/2}$ and $2p_{1/2}$ components.

DISCUSSION

The Si 2p core-level peaks in Fig. 5(b) was deconvolved into its respective Si $2p_{3/2}$ and Si $2p_{1/2}$ components by assuming a spin orbit splitting of 0.61 eV and intensity ratio of 2:1. After subtraction of a Shirley background,[5] the bulk components were fitted with a Voight line shape (i.e., a Lorentzian convoluted with a Gaussian), while the remaining peaks were fitted using pure Gaussian shapes. In addition, the Si $2p_{3/2}$ and Si $2p_{1/2}$ peaks exhibited a slight asymmetry as observed by several authors, among them Himpsel et al.[6] and Mitchell et al.[7] This was accounted for by adding a Gaussian peak of the same FWHM, but with peak area equal to 5% of the main peak and relative peak position shifted 0.34 eV towards higher binding energies.

Figure 6 shows the extracted $2p_{3/2}$ spectra obtained after fitting the spectrum in Fig. 5(b). of Si(100) treated in 1% HF. In addition to the bulk Si component, we also observe Si-H, Si-H_2, and Si-H_3 shifted 0.25, 0.48, and 0.71 eV relative to the bulk peak. After deconvolving the spectra we can now estimate the concentration of the different species observed in the survey spectrum in Fig. 1(d). The results are summarized in Table 1. The first column shows the relative atomic concentration. We can also estimate the surface concentration of the different species by assuming that our surfaces had saturation coverage of hydrogen. After subtracting the contribution from bulk Si and renormalizing the relative concentrations of the surface species, we obtain the results shown in the second column. As seen, total surface contamination is less than

15%, divided almost equally between oxygen and carbon. There is also a small amount of fluorine in addition to the dominant Si-H_x termination.

Table 1. Atomic concentrations of Si and surface contamination for the surface whose survey spectrum is shown in Fig. 1(d).

Species	Total concentration (%)	Surface concentration (%)
Si-bulk	58.9	
Si-H_x	35.2	85.6
C	2.2	5.4
O	3.0	7.3
Adsorbed F	0.26	0.6
F-Si	0.44	1.1

CONCLUSIONS

We have established a method for preparing and loading H-terminated Si surfaces into equipment for analysis as well as process chambers. This method involves using dilute HF for the removal of the chemical oxide followed by a short dip in deionized water. During evacuation of the introduction chamber, we found that a plasma-generating pressure gauge activate mostly carbon species. The resulting carbon contamination shows a direct correlation with the roughness of the Si surface, i.e., increasing surface roughness leads to increasing levels of carbon contamination. However, we obtained a drastic reduction in carbon by disconnecting the pressure gauge and instead monitoring the pumping time. Combining these precautions prior to and during introduction into a high resolution XPS system, we could limit total contamination from C, O, and F to less than 6 at.% at a take-off angle of 5°. The resulting Si 2p spectra show no detectable amount of Si-O bonds, while we can easily detect the Si-H and Si-H_2 components of the Si $2p_{3/2}$ peak. After extracting the Si-H_x components of the Si 2p spectrum, we were able to estimate the surface concentration of the different species by assuming that our surfaces had saturation coverage of hydrogen. In addition to C, O, and adsorbed F, there was also about 1 at.% F bonded to Si on the surface.

REFERENCES

1. C.H. Bjorkman, M. Fukuda, T. Yamazaki, S. Miyazaki, and M. Hirose, Jpn. J. Appl. Phys. **34**, 722 (1995).
2. J.L. Alay, Ph.D. Thesis, Katholieke Universiteit Leuven, Leuven, Belgium, 1993.
3. F.J. Grunthaner and P.J. Grunthaner, Mater. Sci. Rep. **1**, 69 (1986).
4. M. Hirose, T. Ogura, Mater. Res. Soc. Symp. Proc. **75**, 357 (1987).
5. D.A. Shirley, Phys. Rev. B **5**, 4709 (1972).
6. F.J. Himpsel, F.R. McFeely, A. Taleb-Ibrahimi, J.A. Yarmoff, and G. Hollinger, Phys. Rev. B **38**, 6084 (1988).
7. D.F. Mitchell, K.B. Clark, J.A. Bardwell, W.N. Lennard, G.R Massoumi, and I.V. Mitchell, Surf. Interface Anal. **21**, 44 (1994).

OUTPLATING OF METALLIC CONTAMINANTS ON SILICON WAFERS FROM DILUTED ACID SOLUTIONS

A.L.P. ROTONDARO, T.Q. HURD*, H.F. SCHMIDT, I. TEERLINCK, M.M. HEYNS AND C. CLAEYS
IMEC, Kapeldreef 75, Leuven B-3001, Belgium
* assigned to IMEC from Texas Instruments, Dallas TX, USA

ABSTRACT

The outplating behaviour of Fe and Cu was investigated for diluted solutions of HCl and HNO_3. The deposition of the metallic contaminants was found to be strongly dependent on the type of surface that is exposed to the contaminated solution. Cu deposits heavily on bare silicon surfaces, whereas only low levels of Fe deposition are observed. On the other hand, on thermal oxide surfaces, the levels of deposited Fe are consistently higher than the Cu ones. The acid used appears to have no major impact on the deposition process. The pH of the solutions has a major effect on the Cu deposition and a minor effect on the Fe case.

INTRODUCTION

Metallic contamination has a strong effect on reducing the yield of silicon devices [1]. Metals can degrade the dielectric properties of gate oxides causing premature breakdown [2] and can diffuse into the bulk of the silicon material resulting in the increase of junction leakage and the reduction of minority carrier lifetime [3]. It is crucial for ULSI processing that metallic contamination is controlled and kept at negligible levels.

Acid solutions are universally used in cleaning processes that aim for metal removal. Since the introduction of the RCA cleaning sequence [4] the metal removal step (SC2) has been adopted as a standard and it has been widely used without any major changes in its composition or stoichiometry in silicon processing. Nowadays, however, as the semiconductor industry faces the new challenge of reducing the amount of chemical waste and its environmental impact, diluted chemistries appear to be an interesting alternative to the classic SC2 mixture [5]. The use of diluted acid solutions also results in reduced chemical consumption, thus lowering the cost of the cleaning sequence. Another interesting feature of these mixtures is that they provide the possibility of avoiding particle deposition [6] which is the major drawback of the standard SC2.

In this paper, the outplating of metallic contaminants on silicon wafers from diluted acid solutions is studied in detail. The interaction of the metallic contaminants with different wafer surfaces is discussed and the effects of pH and the use of different acids is assessed.

EXPERIMENTAL

Hydrophobic (bare silicon) and hydrophilic (20 nm thermally grown silicon dioxide) wafers were processed in the diluted solutions to investigate the outplating behaviour of the metallic contaminants on different silicon surfaces.

The amount of contamination in solution was varied by spiking the final mixtures with metal impurities in the range of 1 to 100 ppb. Fe and Cu were selected to be studied as they are major contaminants found during processing and are known to be detrimental to silicon devices.

Solutions of HCl and HNO_3 were investigated for dilution ratios of 1:10000 to 1:100 in volume, which corresponds to solutions of 0.0012 M to 0.12 M.

All tests were carried out in ultraclean quartz beakers without recirculation or filtration at a process temperature of 70 °C. The substrates were 125 mm diameter, CZ, <100>, p type, 1-30 Ωcm silicon wafers. The wafers were immersed for 10 minutes in the stagnant solutions and immediately rinsed in DI water in an overflow tank to 18 MΩcm water resistivity.

The surface metal contamination on the wafers was measured with the technique of Vapour-Phase-Decomposition, Droplet-Surface-Etching, Total reflection X-Ray Fluorescence (VPD-DSE-TXRF) [7] which has a detection limit of 5×10^8 at/cm^2.

Mat. Res. Soc. Symp. Proc. Vol. 386

RESULTS

Copper Behaviour

The copper outplating on hydrophobic and hydrophilic surfaces scales with the amount of copper spiked in the solutions (Figure 1). However, the final levels of Cu on the wafers are dependent on the nature of the surface that is exposed to the spiked mixture. On hydrophobic surfaces (bare silicon) the amount of copper deposited is always higher than on the hydrophilic ones (thermal oxide), this effect being more pronounced in heavily contaminated solutions.

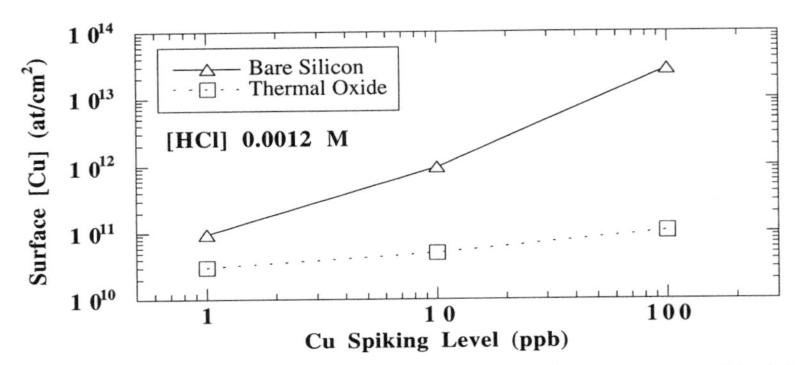

Figure 1: Surface [Cu] on bare silicon and thermal oxide surfaces treated in 0.0012 M HCl mixtures spiked with Cu.

The increase of the HCl concentration minimizes the transfer of Cu contaminants to the wafers on both hydrophobic and hydrophilic surfaces (Figure 2). However, the bare silicon wafers still present higher final levels of surface copper contamination as compared to the thermal oxide ones.

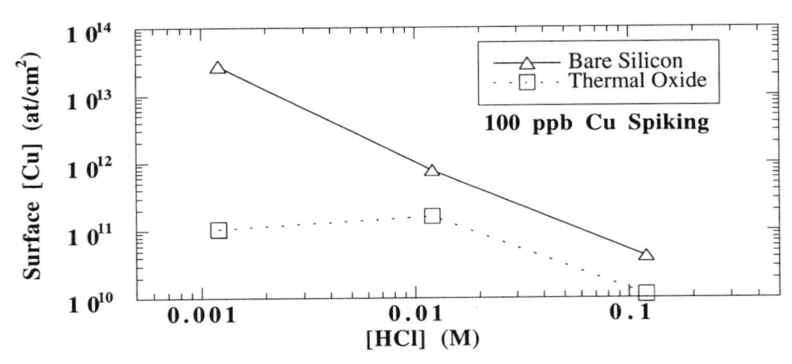

Figure 2: Surface [Cu] on bare silicon and thermal oxide surfaces treated in diluted HCl mixtures spiked with 100 ppb Cu.

The acid used seems to have no significant impact on the outplating behaviour of copper on both surfaces (Figures 3a and 3b). Only a slight improvement is observed in the case of bare silicon surfaces when HNO_3 is used instead of HCl. On both surfaces the major effect on the reduction of Cu outplating is related to the reduction of the pH of the solution, this effect being more evident on bare silicon wafers (Figure 3a).

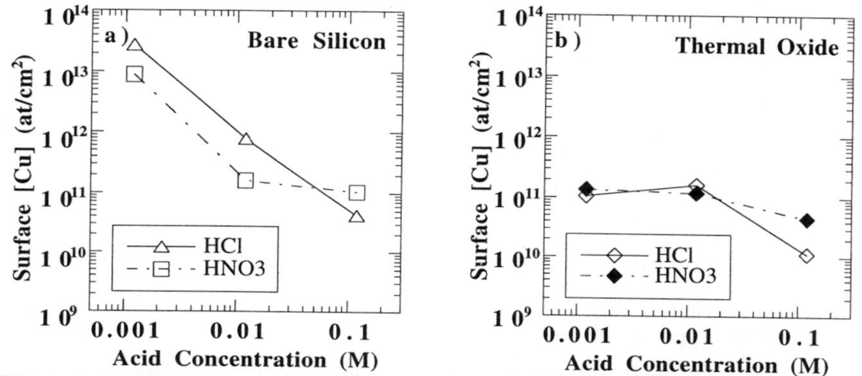

Figure 3: Surface [Cu] on: a) Bare silicon and b) Thermal oxide surfaces treated in diluted HCl or HNO_3 mixtures spiked with 100 ppb Cu.

Iron Behaviour

A strong impact of the wafer surface nature is also found on the Fe outplating. Bare silicon wafers ended up with two orders of magnitude lower Fe contamination when compared to the thermal oxide ones (Figure 4). On the other hand, only an increase of one order of magnitude in the surface Fe concentration is observed when the contamination in solution is varied from 1 ppb to 100 ppb for both surfaces.

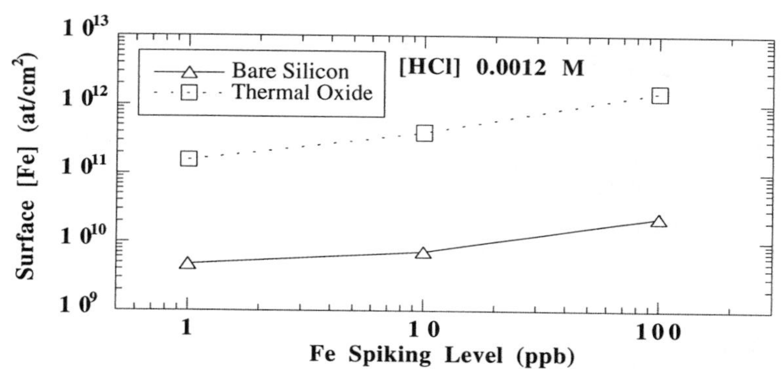

Figure 4: Surface [Fe] on bare silicon and thermal oxide surfaces that have been treated in 0.0012M HCl mixtures spiked with Fe.

The reduction of the pH of the solution from 3 (0.0012M) to 1 (0.12M) has no major effect on the Fe deposition on both surfaces (Figure 5).

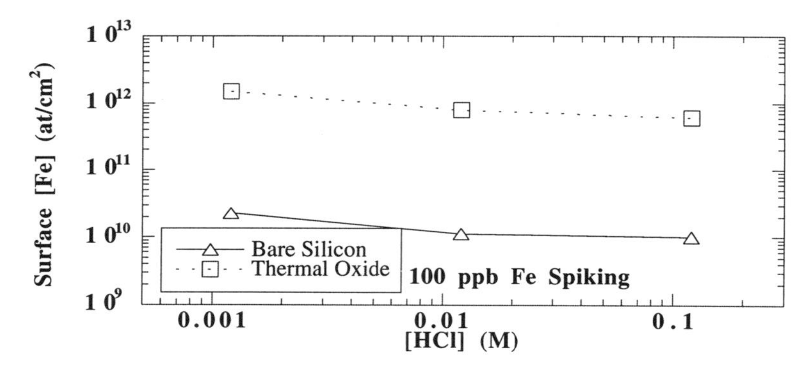

Figure 5: Surface [Fe] on bare silicon and thermal oxide surfaces that have been treated in diluted HCl mixtures spiked with 100 ppb Fe.

As shown on Figures 6a and 6b the outplating behaviour of Fe in solutions of HCl and HNO_3 is similar, presenting only a very small dependence on the pH of the solutions.

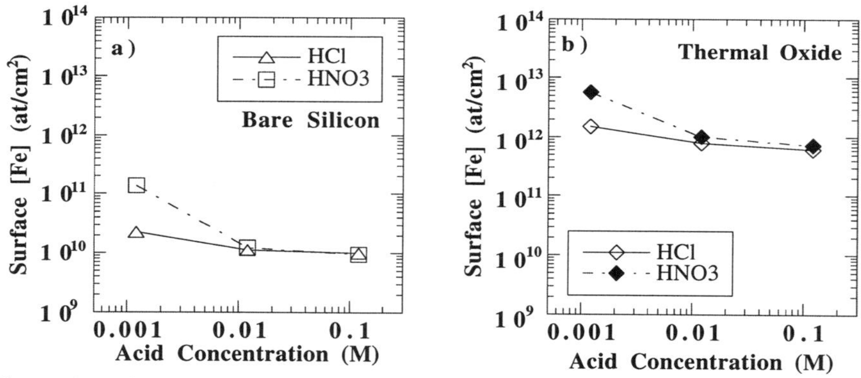

Figure 6: Surface [Fe] on: a) Bare silicon and b) Thermal oxide surfaces treated in diluted HCl or HNO_3 mixtures spiked with 100 ppb Fe.

DISCUSSION

Thermal Oxide Surfaces

The SiO_2 surface in aqueous media is terminated by OH groups which can interact with metal ions. As a result the metal ions become adsorbed onto the oxide surface in exchange for the hydrogen atoms.

Fe is probably present in the diluted solutions in the Fe^{3+} state. In the case of Cu, Cu^{2+} is expected to be the most abundant species in the spiked mixtures. As shown in Figure 7, Fe is more prone to adsorb on the oxide surface than Cu in the range of the investigated pH (1 to 3). This indicates that Fe^{3+} is more effective in coordinating with the surface hydroxyl groups than Cu^{2+}. The reduction of the pH of the solution will lead to the shift of the reaction towards the release of the metal ions from the surface, as can be clearly observed for Cu. In the case of Fe the effect of pH is less evident due to the high affinity of Fe to react with the hydroxyl groups.

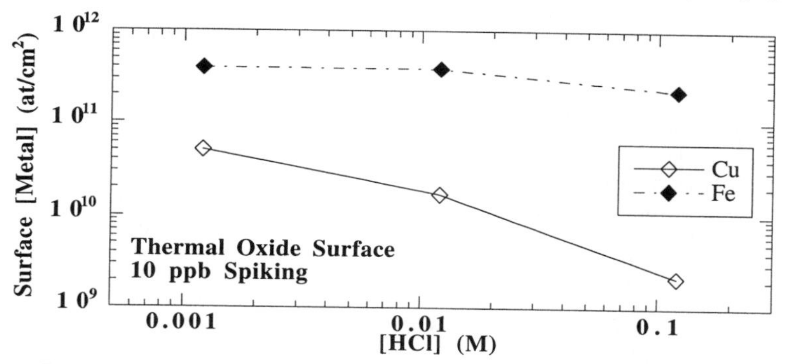

Figure 7: Surface metallic contamination measured on thermal oxide surfaces treated in diluted HCl mixtures spiked with 10 ppb Cu or Fe.

Bare Silicon Surfaces

The hydrophobic surfaces are hydrogen passivated and electroless plating is the main mechanism for metal deposition to occur. As Cu has a standard reduction potential higher than hydrogen and silicon, it deposits on the silicon surface. On the other hand, although having a standard reduction potential higher than silicon, the Fe potential is lower than the hydrogen one. This means that hydrogen will prevail at the expense of Fe in bonding with Si. This is in good agreement with the results of Figure 8. Surface Cu contamination is orders of magnitude higher than Fe. Also the Cu deposition is strongly influenced by the pH of the solution indicating that as the acid concentration in solution increases, the Cu plating is reduced.

The bare silicon surface is slowly hydrolized in the diluted solutions. This phenomena creates hydroxil sites on the surface that are very reactive with Fe. As the hydrolization of a hydrogen passivated surface is a slow process, just a very low amount of reacting sites are available for Fe to attach to the surface. This explains the low levels of Fe observed on the surface after treatment (Figure 8). The minor effect of the pH on the Fe deposition agrees with the fact that Fe has a high affinity to react with the hydroxil group, as seen on thermal oxide surfaces.

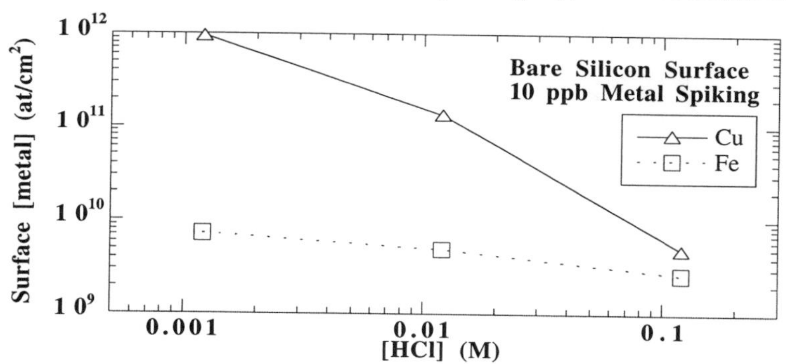

Figure 8: Surface metallic contamination measured on bare silicon surfaces treated in diluted HCl mixtures spiked with 10 ppb Cu or Fe.

CONCLUSIONS

The type of silicon surface that is treated in diluted acid solutions has a major impact on the outplating properties of the metallic contaminants. Cu primarily deposits on bare silicon and Fe is more easily adsorbed on thermal oxide surfaces.

Changing the pH of the solutions from 3 to 1 (0.0012 M to 0.12 M) significantly reduces the Cu contamination transferred to the wafer surface. However, the same variation in pH has just a minor impact on the Fe concentration measured on the surface of the wafers after treatment.

The acid used has no major effect on the outplating behaviour of Fe and Cu as similar contamination levels are observed after immersion in HCl and HNO_3 solutions.

If the goal of having $< 10^{10}$ at/cm^2 of metallic contaminants on the wafer surface after cleaning is pursued for the next generation of ULSI processing, it is possible to define the contamination level allowed in solution for the different metals. This means that 0.1 M solutions can receive as much as 10 ppb of Cu contamination without leading to "over-the-limit" contamination on the wafers. The same limits are valid for Fe contamination when bare silicon surfaces are considered but for thermal oxide surfaces the allowed Fe contamination must be reduced to levels below 1 ppb.

ACKNOWLEDGEMENTS

A.L.P. Rotondaro would like to thank CNPq (Conselho Nacional de Desenvolvimento Científico e Tecnológico) Brazil for financial support. The authors would like to acknowledge Christian Wilms for processing the samples.

REFERENCES

1. A. Hiraiwa and T. Itoga, IEEE Trans. on Semicond. Manufact. **7** (1), 60 (1994).
2. P.W. Mertens, M. Meuris, S. Verhaverbeke, M.M. Heyns, A. Schnegg, D. Gräf and A. Philipossian, Proceedings of the 38th Annual Tech. Meet. of the IES, Nashville TN, 475 (1992).
3. P.J. Ward, J. Electrochem. Soc. **129** (11), 2573 (1982).
4. W. Kern and D. Poutien, RCA Rev. **31**, 187 (1970).
5. T.Q. Hurd, P.W. Mertens, H.F. Schmidt, D. Ditter, L.H. Hall, M. Meuris and M.M. Heyns, Proceedings of the 40th annual Tech. Meet. of the IES, Chicago IL, 218 (1994).
6. T.Q. Hurd, P.W. Mertens, L.H. Hall and M.M. Heyns, Proceedings of the 2nd Internat. Symp. on Ultra-clean Process. of Silicon Surf. (UCPSS'94), Brugge Belgium, 41 (1994).
7. C. Neumann and P. Eichinger, Spectroch. Acta **46B** (10), 1369 (1991).

SUBMICRON, NOBLE METAL PARTICLE REFERENCE STANDARDS: A PROPOSAL

R. S. HOCKETT, ANGELA Y. CRAIG, and DIEM LE
Charles Evans & Associates, 301 Chesapeake Drive, Redwood City, CA 94063

ABSTRACT

There is a need in the semiconductor industry to develop new techniques and instrumentation for the elemental and chemical analysis of submicron, particularly <0.2 μm, particles. The development of these techniques and instrumentation could be assisted by submicron particle reference standards. We propose that high number-density, noble metal (Cu, Ag, Au) particles on silicon, with controlled diameters in the range of 0.02 μm to 0.10 μm, be developed and used as particle reference standards.

INTRODUCTION

The National Technology Roadmap for Semiconductors[1] (NTRS) indicates there will be the need to control particle contamination in semiconductor processing at the sizes shown in Table I.

Table I. Particle Size Requirements

Year	1995	1998	2001	2004	2007	2010
IC Design Rule	0.35 μm	0.25 μm	0.18 μm	0.13 μm	0.10 μm	0.07 μm
Particle Size	0.12 μm	0.08 μm	0.06 μm	0.04 μm	0.03 μm	0.02 μm

The primary analytical technology for submicron particle analysis has been Secondary Electron Microscopy (SEM) equipped with an Energy Dispersive Spectrometer (EDS). Standard SEM/EDS systems lose much of their analytical capability below the size range of 0.2 μm to 0.5 μm, depending upon the particle composition and the support matrix. New technology which may successfully analyze smaller submicron particles include: Field Emission Auger Electron Spectroscopy (FE-AES) and Time-of-Flight Secondary Ion Mass Spectrometry (TOF-SIMS). Atomic Force Microscopy (AFM) can provide some morphology information on submicron particles. A very new technique, Near Field Optical Microscopy (NFOM), may be successful in identifying submicron particle morphology as well. A longer term technique, Synchrotron x-Ray, micro X-ray Absorption Near Edge Spectroscopy (SR-μXANES), may be able to provide the chemistry as well as composition of sub-0.1 μm particles.

In order to facilitate the development of instrumentation and analytical techniques to determine the composition and/or morphology of these contaminant particles, reference materials are needed. The use of polymeric particle reference materials, common in the development of particle detectors using optical laser scanners, is not sufficient to meet this need. We anticipate that many different methods will be developed in the near future for preparing submicron particle reference materials.

Recently, pre-formed gold clusters (diameters of 5 nm, or 0.005 μm) on silicon were prepared to test AFM tips and therefore AFM resolution.[2] The gold clusters were formed using

Mat. Res. Soc. Symp. Proc. Vol. 386 ©1995 Materials Research Society a Multiple Expansion Cluster Source for vapor deposition onto the silicon substrate. In another approach, particle deposition of aluminum and of alumina onto silicon substrates was done at NIST.[3] The purpose was to prepare reference materials for comparing the analysis capability of SEM/EDS, FE-AES, and TOF-SIMS for these particles with sizes down to 0.05 µm. The preparation method used particle suspension in isoproponal or water. Lastly, Morinaga et al[4] recently reported on the preparation of ultrafine gold particles (order of 10 nm, or 0.01 µm) on silicon using a gas deposition method. Their purpose was to study the wet chemistry removal of such ultrafine particles. In the next section we propose another preparation method.

NOBLE METAL PARTICLE DEPOSITION FROM HF OR NH4/HF

Recently, there has been a surge of published work on the formation of very small (<0.1 µm) noble metal particles on silicon surfaces immersed in HF or buffered HF.[5-13] The intent of this research has been to identify the mechanism of the particle formation and growth, and to develop chemical cleaning processes to either remove the particles or to keep them from forming at all. Our proposal is likewise to understand the particle formation and growth. However, in contrast to the reported research, we propose the idea of intentionally creating controlled particle sizes, morphology and number densities on silicon surfaces as surface particle reference materials.

Formation Mechanism

The mechanism of noble metal particle formation and growth onto the silicon surface from HF or buffered HF (NHF/HF) has been mostly studied for copper,[5-8,10-13] less for silver[9,10,12] and inferred for gold.[10] The formation mechanism appears to be as follows. The noble metal in the solution is electrochemically reduced at silicon surface nuclei sites, which, once fully occupied across the wafer, become the locations for particle growth up to some maximum size. The noble metal particle is not formed in the solution and then deposited onto the surface. Particle formation and growth does not occur on silicon oxide.[10-11] One report gave the maximum number density of copper particles as $10^{10}/cm^2$ for 2 ohm-cm p-type silicon (100) after 60 seconds in solution; initial number densities were about 2x lower.[7] There can be an effect from dopant type and level, because these affect the number of surface states available for electrochemical nuclei.[13] The particle number densities can be varied from $10^6/cm^2$ to $10^{10}/cm^2$, by varying the surface doping density and type.[13]

Growth Mechanism[10]

After the metal from the solution is reduced at the silicon surface and all surface sites for nucleation are filled, the metal nuclei, having a higher electronegativity than Si, attract electrons from Si to become negatively charged. Other metal ions in solution obtain electrons from the surface metal and deposit with the metal nuclei. In this way the surface metal particle grows. The silicon surface underneath the metal particle releases as many electrons as required by the metal ions to be charged, while SiO_2 is formed. A pit is formed by the removal of the underlying SiO_2 by etching. The resultant particle is metallic.[10,12] The growth process is diffusion-limited and a function of $t^{1/2}$, where t is time.[6] One reported[10] particle shape, as inferred by the chemical removal of the particle followed by the AFM measurement of the resultant pit, was 0.1 µm in width and 0.008 µm in depth (into the silicon). Thus the particles may be more like saucers rather than spheres.

Reported copper particle diameters by different research groups have been 0.018 µm in 100 sec[7], 0.050 µm in 180 sec[10], and 0.040 µm in 25 min[12]. Although not all researchers used the same conditions, the maximum size diameter for copper has been near 0.1 µm, as reported to-date, and it is not known if this represents some fundamental limit. In a comparative study between Ag and Cu, the maximum reported size for silver particles was about two times larger than the copper particles (0.085 μm for Ag and 0.040 μm for Cu).[12] There have been no reports on gold particle sizes using this chemical method.

If the proposed growth mechanism is correct, then with an infinite source of metal in solution and infinite time, what is limiting the particle size? A possibility is that the SiO_2 underneath the metal particle cannot be effectively etched away as the particle saucer shape increases to some size, so that the SiO_2 becomes an insulator and impedes the transfer of electrons from the silicon to the metal copper and thus to the metal ions in solution. If the metal type is changed to a higher redox potential (e.g., Cu to Ag) compared to Si, then the electron transfer takes a thicker silicon oxide underneath to stop the growth, and therefore it may grow larger. We note that the maximum size of the Ag particles is larger than that for Cu. It would be instructive to determine, if possible, if the underlying oxide is thicker for the Ag particles compared to the Cu particles. It would also be instructive to determine if Au can form larger particles than Ag. One puzzle however is why the particle cannot continue growing laterally.

EXPERIMENTAL EQUIPMENT AND PROCEDURE

All reported work in this research area has used chemical baths where the noble metal contaminant source was "infinite" from a diffusion standpoint. In our experiments we have used a spin coater (Laurell WS-200-8T2) to supply the HF solution containing noble metal contaminants onto the silicon surface. In this case, the metal source is not "infinite" and the kinetics will not be fully diffusion limited after some time period.

For our experiments we have used Cu, Ag and Au in dilute HF. Impurity solution levels were: 1 ppm, 10 ppm, 100 ppm Cu; 1 ppm, 10 ppm, 100 ppm Ag; and 0.3 ppm, 1 ppm Au. The experimental procedure for forming the particles is listed below. The polished silicon wafers were 100 μm in diameter. Analysis of the particles was done using the following instruments: JEOL 5300 SEM/EDS, JEOL 6400F FE-SEM, PHI™ 670 FE-AES, PHI™ 5500 ESCA (after sputter etch), TECHNOS TREX 610T TXRF, and Digital Instruments Nanoscope III AFM (tapping mode). There was no study of pits after chemical removal of the particles.

*SC1 soak (H_2O_2:NH_4OH:H_2O @1:1:5) for 10min
*Spin at 3500 RPM for 30 s
*Ultrapure water (UPW) soak, 3 times for 3.5 min each
*Spin at 3500 RPM for 30 s after each water soak
*0.5% HF for 10 min
*Spin at 3500 RPM for 30 s
*UPW soak, 3 times for 3.5 min each
*Spin at 3500 RPM for 30 s after each water soak
*0.5% HF spiked with $CuCl_2$, $AgNO_3$, or $AuCl_3$ (soak for 2 min, except 100 ppm Cu which was for 10 min)
*Spin 3500 RPM for 30 s
*UPW soak, 3 times for 3.5 min each
*Spin at 3500 RPM for 30 s after each soak
*Air dry for 5 min
*Store in fluoroware

RESULTS AND DISCUSSION

Table II summarizes the data. The amounts of Cu or Ag are similar to the other reports. ESCA demonstrated that the Cu and Ag bonds are metal, after some sputter etching, as was reported by others for Cu. The particle sizes are not singular across a wafer, i.e., there is a distribution of sizes. Some of the morphology seen under SEM or FE-SEM and under AFM indicate the morphology is not always like a sphere or saucer, but sometimes like a moon crater. In addition, FE-AES was used to evaluate the Cu or Ag level between particles. For the 10 ppm Cu and Ag samples, the FE-AES results showed Cu in the 0.1% atomic range between particles, but no detectable Ag between particles. The results of the Au spin coating were not clear in that we have not shown that the apparent particles (<0.1 μm) are Au, although the Au level on the wafers is in the $5E13/cm^2$ and $2E13/cm^2$ for the 1 ppm and 0.3 ppm Au respectively.

TABLE II. Analysis Summary

Solution	SEM/EDS	FE-SEM	TXRF	ESCA	AFM
CU	μm	μm	at/cm2	Bond	μm
1 ppm	0.05-0.1		7e13		0.02-0.1
10 ppm	0.05-0.1		9E14		0.05-0.1
100 ppm	0.1-0.5		>E15	metal	0.1-0.5
AG					
1 ppm	0.01-0.05		7E14		0.05-0.1
10 ppm	0.05-0.1	0.01,0.1	9E14		0.05-0.1
100 ppm	0.1-0.5		2E15	metal	0.1-0.5
AU					
0.3 ppm			5E13		
1 ppm			2E13		

CONCLUSION

Our work has shown that a spin coat dilute HF process can form Cu and Ag particles on silicon wafer surfaces in the size range of 0.01 μm to 0.5 μm, however the size distribution and morphology are not constant across the wafer. A chemical bath process may be able to better control the size distribution and morphology, if a constant size distribution and morphology is desired. These types of reference samples should be useful for assisting the development of new analytical technology for submicron particle analysis.

ACKNOWLEDGEMENT

The authors would like to thank Jenny Metz, Jim Vitarelli, Jeff Kingsley, Dave Harris, and Rob Brigham of Charles Evans & Associates for the TXRF, SEM/EDS, FE-SEM, FE-AES, and AFM analyses.

References

1. National Technology Roadmap for Semiconductors, Semiconductor Industry Association, (San Jose, CA) 1994.

2. Donald Chernoff, "Using Cluster Materials for High Resolution Tools in AFM," Extended Abstracts of the Industrial Applications of Scanned Probe Microscopy, A Workshop Sponsored by NIST, SEMATECH, ASTM E-42, and the AVS, (NIST, Gaithersburg, MD) March 24-25, 1994, pp. 64-65.

3. Alain C. Diebold, George Mulholland, Pat Lindley, Kent Childs, to be published.

4. H. Morinaga, T. Futatsuki, T. Ohmi, E. Fuchita, M. Oda, and C. Hayashi, J. Electrochem. Soc. 142, 96 (1995).

5. M. Suyama, H. Morinaga, M. Nose, T. Ohmi, and S. Verhaverbeke, 1994 Semiconductor Pure Water and Chemicals Conference Proceedings (Balazs Analytical Laboratory, Santa Clara, CA) pp. 93-109 (1994).

6. J. A. Sees, L. H. Hall, O. M. R. Chyan, J.-J. Chen, H. Y. Chien, "Kinetics and morphology of Cu deposition on silicon surfaces," Proceedings of the Second International Symposium on Ultra Clean Processing of Silicon Surfaces, Bruges, Belgium, Sept. 19-21, 1994 (Uitgeverij Acco, Leuven, Belgium, 1994), pp. 147-150.

7. Oliver M. R. Chyan, J-J. Chen, and H. Y. Chien, L. Hall, J. Sees, "Electrochemical aspects of noble metals related to Si wafer cleaning," ibid., pp. 213-216.

8. H. Morinaga, M. Suyama, M. Nose, S. Verhaverbeke and T. Ohmi, "Metallic particle growth and metal induced pitting (MIP) on silicon surfaces in wet processing and its prevention," ibid., pp. 217-220.

9. D. Levy, P. Patruno, L. Mouche and F. Tardiff, "Pitting on wafers by Ag trace in dilute HF," ibid., pp. 293-296.

10. H. Morinaga, M. Suyama, and T. Ohmi, J. Electrochem. Soc. 141, 2834 (1994).

11. K. K. Yoneshige, H. G. Parks, S. Raghavan, J. B. Hiskey, and P. J. Resnik. J. Electrochem. Soc. 142, 671 (1995).

12. K. K. Yoneshige, H. G. Parks, C. R. Helms, S. Halapete, and S. Fang, "The deposition of noble metals from buffered oxide etchants onto silicon surfaces - a morphological study," Extended Abstracts Vol. 95-1, (The Electrochemical Society, Pennington, NJ), Abstract 174, pp. 280-281 (1995).

13. J. E. Turner, K. Miyazaki, Y. Fukazawa, and H. Muraoka, "Copper adsorption at a Si surface from HF solutions," ibid., Abstract 342, pp. 518-519 (1995).

ADSORPTION AND DESORPTION OF CONTAMINANT METALS ON Si WAFER SURFACE IN SC1 SOLUTION

G.MAEDA, I.TAKAHASHI, H.KONDO, J.RYUTA AND T.SHINGYOUJI
Mitsubishi Materials Corporation, 297, Kitabukuro-cho,
Omiya-shi, Saitama-ken, Japan

ABSTRACT

Variation in the surface concentration of Fe, Ni, Cu and Zn on Si wafers due to treatment in $NH_4OH/H_2O_2/H_2O$ solution called SC1 is investigated. The metal concentration on the wafer surface depends on the initial surface concentration, concentration in the solution, adsorption probability, desorption rate constant and the treatment time. The surface metal concentration behavior is explained by taking into account the effects of these parameters. The variation in the desorption rate constant with the metal species, the concentration in the solution, treatment temperature and mixing ratio of SC1 is discussed.

INTRODUCTION

The requirement for clean wafers is becoming more and more urgent in manufacturing high-quality LSIs. The specification of concentration of metallic impurities on wafers for a 16Mbit dynamic random access memory (DRAM) is 10^{10} atoms·cm $^{-2}$. Much effort has been made to improve the cleanliness of wafers by adjusting the cleaning conditions, such as the RCA cleaning process [1] is widely used.

It is known that even a slight contaminat in the cleaning solutions can degrade the cleanliness of the wafer surface. Not only desorption, but also adsorption of metals, can occur in the solution. Therefore clarification of the mechanisms of adsorption and desorption of metallic impurities in the solutions is important to improve the cleaning efficiency of wafers.

Recently, Hiratsuka et al. reported the adsorption mechanism of metallic impurities on the basis of the reaction model of the frontier orbital [2]. Kern et al. also reported the adsorption mechanism by taking into account the effects of the half-cell potential [3]. However these mechanisms can be applied only to cases of adsorption at equilibrium conditions. Thus the kinetics of the adsorption of metallic impurities has never solved.

In this paper, variation in the surface concentration of the metallic impurities with treatment time is investigated. Based on such an investigation, the kinetics of the adsorption of metallic impurities in SC1 solution is discussed, and the application of the Langmuir theory to metal adsorption on Si wafers in the SC1 solution is carried out.

EXPERIMENT

Variation in the surface concentration of Fe, Ni, Cu and Zn due to the SC1 treatment was examined. Polished (100) Si wafers, 6inches in diameter and P-type, 10 Ω ·cm were used in the present experiment. The SC1 solution was prepared by mixing de ionized water with H_2O_2

Mat. Res. Soc. Symp. Proc. Vol. 386 © 1995 Materials Research Society

solution (30%) and NH_4OH solution (30%). The SC1 treatment was followed by rinsing with de ionized water for 10min.

In order to examine the adsorption of metals, the standard solutions for the atomic absorption spectrochemical analysis containing metallic impurities were intentionally added in the SC1 solution. Wafers preliminary cleaned by the RCA cleaning process (referred to as "clean" wafer) were used in this adsorption experiment. The analysis of the surface concentration of the metals were made by the total reflection X-ray fluorescence method using TREX 610. The detection limit for Fe, Ni, Cu and Zn is about 10^{10} atoms·cm^{-2}. From "clean" wafers, no characteristic X-ray of metallic impurities were detected.

RESULTS AND DISCUSSION

The influence of impurity concentration in SC1 solution on the surface cleanliness of wafers was first examined for Fe, Ni, Cu and Zn. The "clean" wafers were treated in SC1 solution which mixing ratio of NH_4OH:H_2O_2:H_2O was 0.25:1:5 at volume. The treatment temperature was 80 ℃, and cleaning time was 10min. The result is shown in Fig.1; The abscissa indicates the impurity concentrations in the solution.

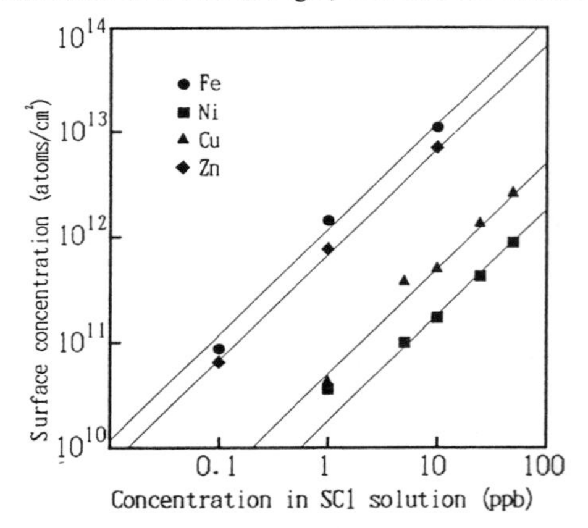

Fig. 1. Effect of impurity concentrations in intentionally contaminated SC1 solution on the surface concentration. SC1 (0.25:1:5) 80℃, 600sec.

The authors also investigated that the influence of the initial surface contamination levels on the surface concentration after treatment in the clean SC1 solution. The result was the surface concentration, irrespective of initial surface concentration, was determined by the final treatment condition. When the final treatment is performed with a clean solution, impurities on initially contaminated wafers are removed to a concentration in equilibrium with the low impurity concentration in the final solution. These results indicate that the adsorption and desorption of metallic impurities occur simultaneously at some rate, depending on the surface concentration and concentration in the SC1 solution.

Ryuta et al. proposed the application of Langmuir theory to describe the phenomenon of adsorption and desorption [4]. The adsorption and desorption were assumed to be first-order reactions. Thus the adsorption rate per unit area, Ra, is express as follows:

$$Ra = N \cdot p_1(1 - \theta) \quad (1)$$

Here, N is the incident frequency of impurities per unit area, that is thought to be proportional to the impurity concentration in the solution, p_1 is the adsorption probability and θ is the surface coverage of metallic impurities. In the present experiment, surface concentration of metallic impurities were less than 2×10^{13} atoms·cm^{-2}, whereas the surface concentration of Si atoms in the (100) monolayer is 6.8×10^{14} atoms·cm^{-2}. Therefore, the approximation of $1 - \theta = 1$ was taken.

The desorption rate per unit area, Rd, is expressed as follows:

$$Rd = C \cdot p_2 \quad (2)$$

Here, C is the surface concentration and p_2 is the desorption rate constant. The value p_2 corresponds to the reciprocal of the residence time of metallic impurities on the surface.

Variation in the surface concentration with treatment time should be the difference of the adsorption rate and desorption rate. Therefore,

$$dC/dt = Ra - Rd \fallingdotseq N \cdot p_1 - C \cdot p_2 \quad (3)$$

Here, p_1 and p_2 are independent of the other parameters. By solving this differential equation,

$$C = (Co - N \cdot p_1/p_2)\exp(-p_2 \cdot t) + N \cdot p_1/p_2 \quad (4)$$

where Co is the initial surface concentration. Equation (4) indicates that the surface concentration changes, with treatment time to the saturated surface concentration.

Figure 2 shows a result of surface concentration on initially "clean" wafers in intentionally contaminated solutions. In this case, the solution contains Fe to a concentration of 1ppb. The mixing ratio of the SC1 solution used was 0.25:1:5. The treatment temperature was varied from 28 to 80 ℃. The surface concentration increases first and after a certain treatment time, the rate of increase becomes sluggish and the surface concentration reaches a maximum value depends on the treatment temperature. The solid curves indicate the surface concentration expected from eq.(4); fitting to the measured values was made by assuming suitable values for parameters in the equation. In Table Ⅰ, the parameters chosen here are listed. From Fig.2, it is clear that equation (3) and (4) represent very well the data of Fe contamination. This indicates that the desorption is thought to be a first-order reaction.

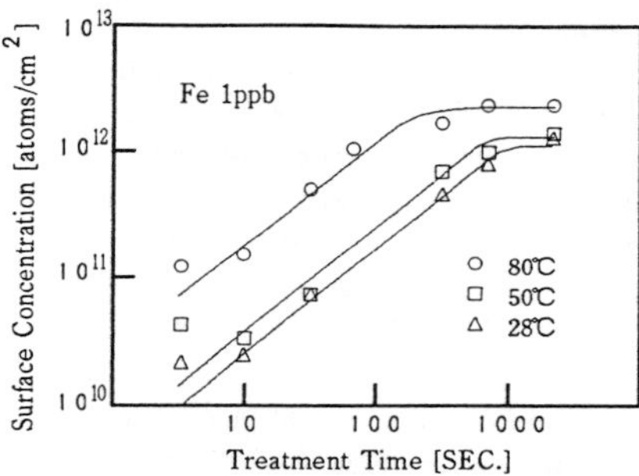

Fig.2. Variation with the treatment time in surface concentration of Fe on initially clean wafers.

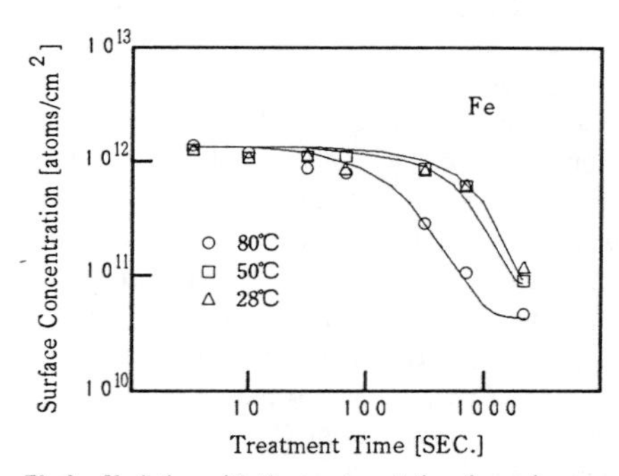

Fig.3. Variation with the treatment time in surfane concentration of Fe on initially contaminated wafers.

Table Ⅰ. The values of $Np1$ and $p2$ adoped in Figs. 2 and 3

	$Np1$		$p2$	
	Fig. 2	Fig. 3	Fig. 2	Fig. 3
28℃	2.0E9	3.5E7	1.5E−3	1.5E−3
50℃	3.1E9	5.3E7	1.8E−3	1.9E−3
80℃	1.4E10	2.8E8	5.5E−3	5.5E−3

The cleaning process, using clean solution, of initially contaminated wafers is shown in Fig.3 for the case of Fe. The fit of eq.(4) was also determined and the results are shown as the solid curves in the figure. The correlation is very good again. The values of the parameters chosen here are listed in Table Ⅰ. It should be noted that the values of the desorption rate constant, p_2, chosen from Figs.2 and 3 are very close to each other. This indicates that the desorption rate constant, p_2, is independent of the initial surface concentration, Co, the saturated surface concentration, $N \cdot p_1/p_2$, and the impurity concentration in the solution. Here, the impurity concentration in the solution were not more than 100ppb in the present experiment.

Table II. Mixing ratio of SC1 solution for experiment. The indicated values are ratio of volume each solusion.

	NH_4OH	H_2O_2	H_2O
No. 1	1	1	5
No. 2	0.1	1	5.9
No. 3	1	0.1	5.9

The cases of contamination of the other metallic impurities were also examined. As in the case of Fe contamination, the measured data can be fitted well by eq.(4). For the results, the saturated surface concentration, $N \cdot p_1/p_2$, and the incident frequency of impurities, N, are linearly proportional to the impurity concentration, but the adsorption probability, p_1, seems to be indepenent on the concentration, as assumed to derive eq.(4). This suggest that the adsorption and desorption of metallic impurities on Si wafers in the SC1 solution seems to be a first-order reaction.

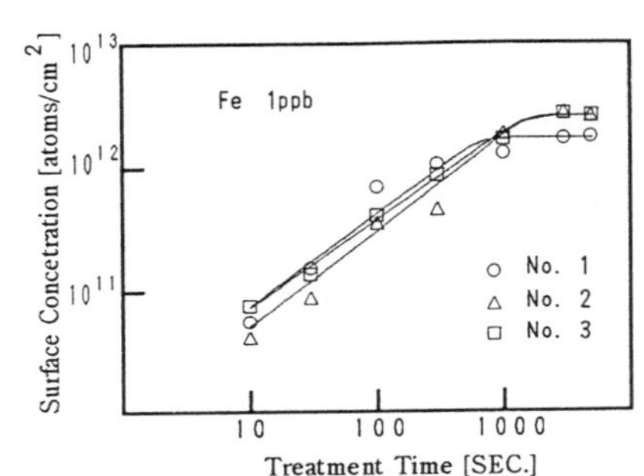

Fig. 4 Variation with the treatment time in surface concentration of Fe for each mixing ratio of SC1 discripted Table II.

Figures 4 and 5 show results of surface concentration on initially "clean" wafers in intentionally contaminated solutions. In these cases, the solutions contain Fe or Cu. In this experiment, mixing ratios of SC1 solutions are varied as in Table Ⅱ and treatment temprature is 50 ℃. As shown in Figs.4 and 5, The surface concentration of Cu vary with the mixing ratio of SC1 solution, but those of Fe do not vary so much. Here Langmuir theory applies again to obtain the kinetic parameters, the values of adsorption rate and desorption rate constant. The solid curves indicate the surface concentration expected from eq.(4); fitting to the measured values are made by assuming suitable values for parameters in the equation.

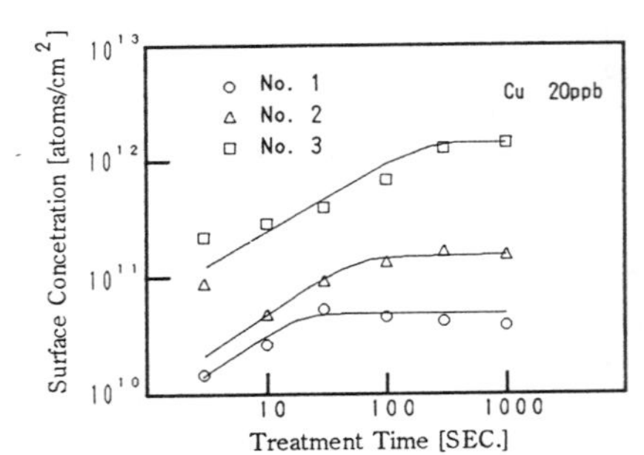

Fig. 5 Variation with the treatment time in surface concentration of Cu for each mixing ratio of SC1 discripted Table II.

The values of the adsorption rate, $N \cdot p_1$, and the desorption rate constant, p_2,chosen here are shown in Figs. 6. and 7. It is clearly seen that the adsorption rate, $N \cdot p_1$, is not varied with the mixing ratio of SC1 solution. When the concentration of NH_4OH is varied, the coordination number of ligand of complex of metallic ions might change. Cu can have so much four

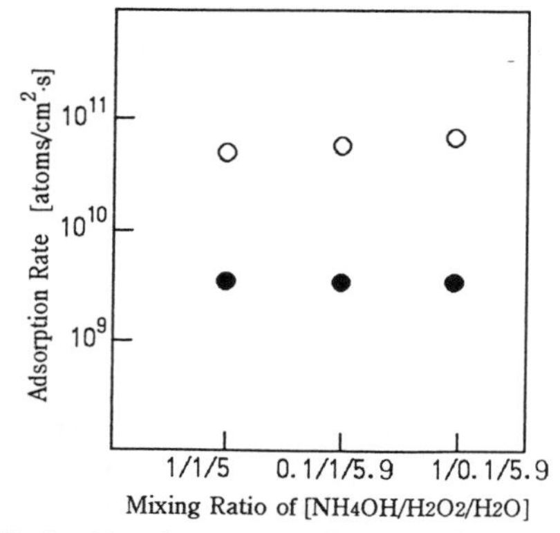

Fig.6 Adsorption rate of Fe [○] and Cu [●] for each mixing ratio of SC1 discribed Table II.

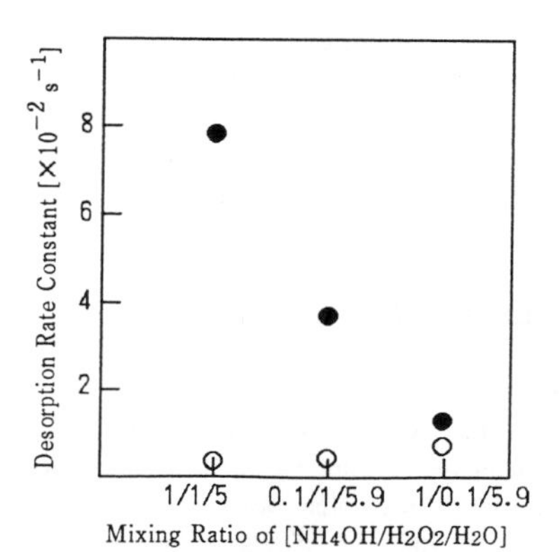

Fig.7 Desorption rate constant of Fe [○] and Cu [●] for each constitusion of SC1 discribed Table II.

anmonia ligands, and each concentrations are determined by their stability constants. The estimated values of the complex concentrations for this experiment are described in Table III. From the result, $[Cu^{2+}]$, $[Cu(NH_3)]^{2+}$ and $[Cu(NH_3)_2]^{2+}$ are very low concentrations, so that they are negligible to adsorption behaivor for this experiment. So the $[Cu(NH_3)_3]^{2+}$ and $[Cu(NH_3)_4]^{2+}$ are discussed to adsorption rate . These adsorption species might have to free their ligands when they adsorb on Si surface, so that it is thought the complex having more ligands is later adsorption rate. In this experiment, however, the adsorption rates of each SC1 solution any change. The difference among these complex concentrations might have no influence of the adsorption rate.

Table III. Concentration of ammine complex of cupper estimated by stability constant for each mixing ratio of SC1. [mol/l]

	$[Cu^{2+}]$	$[Cu(NH_3)^{2+}]$	$[Cu(NH_3)_2^{2+}]$	$[Cu(NH_3)_3^{2+}]$	$[Cu(NH_3)_4^{2+}]$
1/1/5	1.52E-18	1.13E-14	1.61E-11	5.11E-9	3.09E-7
0.1/1/5.9	1.32E-14	9.83E-12	1.40E-9	4.43E-8	2.68E-7

Table IV. Etching rate of Si surface for each mixing ratio of SC1 solution.

	1/1/5	0.1/1/5.9	1/0.1/5.9
Etching Rate [Å/min]	5	3	12

The mixing ratio of SC1 changes etching rate of Si. The etching rate of each mixing ratio SC1 for this experiment are described in Table IV. The surface adsorption site would be changed by etching reaction of Si. Because it was thought that the site which had negative charge increased with desolved Si. However, as the saturated surface consentration was more than 10^{14} atoms·cm^{-2} (shown Fig. 1), the adsorption site would be much more than adsorption metals in this experiment.

As mentioned above, metal complex concentration and etching rate of Si had no effects on adsorption rate in this experiment.

As shown in Fig. 7, the desorption rate constant, p_2 of Fe do not vary. When the mixing ratio of SC1 solution was varied, the etching rate also shown in table IV , The desorption rate constant of Cu varied with the mixing ratio. The desorption behavior has been considered the influence of etching rate. However, desorption rate constant didn't relate to etching rate described Table IV. So numbers of Si which desolves by etching reaction per unit area in one second time could be estimated by etching rate. If adsorption metals on surface were 10^{11} atoms·cm^{-2}, surface concentration of metals on (100) surface after Si desolved were estimated to 9.91 × 10^{10} atoms·cm^{-2} for 1/0.1/5.9 and 9.96 × 10^{10} atoms·cm^{-2} for 1/1/5. Afterall the desorption amount of metal in the etching reaction of Si were a few. It may be considered that the desorption rate constant of Cu varied with NH_4OH and H_2O_2 concentration. This discussion wouldn't be mentioned only these parameter. We are investigating how these elements work to desorption rate constant, now.

CONCLUSIONS

Adsorption and desorption of Fe, Ni, Cu and Zn impurities on Si wafers in the SC1 solution were investigated. It was clarified that adsorption and desorption in the SC1 solution take place simultaneously. On the basis of the variation in the surface concentration with the treatment time, the kinetics of adsorption and desorption was discussed. The application of the Langmuir theory to the analyses of adsorption and desorption of metallic impurities on Si wafers in the SC1 solution was carried out for the first time. The surface concentration of metallic impurities was formulated as a function of the initial surface concentration, concentration in the solution, adsorption probability, desorption rate constant and the treatment time. The theoretical equation involving these parameters was found to represent the measured surface concentration well. We recognized that the adsorption rate is changed by metal concentration and temperature. As the desorption rate constant, especially Cu, it has influensed of mixing ratio of SC1.

The kinetic analysis can be used to clarify the mechanism of adsorption and desorption, and to improve the cleanliness of wafers by adjusting the cleaning conditions.

ACKNOWLEGEMENTS

The authers would like to thank Kazushige Takaishi of Mitsubishi Materials Silicon Co. for his advice.

REFERENCES

1. W.Kern and D.A.Puotinen, RCA Rev. 31, 207, (1970)
2. H. Hiratsuka, M. Tanaka, T. Tada, R. Yoshimura and Y. Matsushita, Proc. 11th Workshop on ULSI Ultra Clean Technology, Tokyo, 1991, Ultra Clean Soc., p.5.
3. F. W. Kern Jr., M. Itano, I. Kawanabe, M. Miyashita and T. Ohmi, Proc. 11th Workshop on ULSI Ultra Clean Technology, Tokyo, 1991, Ultra Clean Soc., p.23.
4. J. Ryuta, T. Yoshimi, H. Kondo, H. Okuda and Y. Shimanuki, Jpn. J. Appl. Phys. 31 (1992) p.2338.

IRON DEPOSITION FROM SC-1 ON SILICON WAFER SURFACES

S. DHANDA*, C. R. HELMS*, P.("KIM") GUPTA**, B. B. TRIPLETT** AND M. TRAN**
* Solid State Electronics Laboratory, Stanford University, Stanford, CA 94305
** Intel Corporation, Santa Clara, CA

ABSTRACT

This work examines the extent of the deposition of iron on the wafer from (iron) contaminated SC-1 solutions on silicon wafer surfaces, models this effect, and also predicts the chemical state of the iron thus deposited on the wafer surface. The deposition of iron from SC-1 on three different wafer surface terminations was studied. The surfaces were characterized by: (i) the presence of ~10 Å of native oxide, (ii) by relatively little native oxide and (iii) by a thick thermal oxide. Experiments were performed at room temperature using a 1:1:5 SC-1 (NH_4OH-H_2O_2-H_2O) solution, and also at 80°C with a more dilute composition (0.25:0.5:5). We found that irrespective of the initial surface termination, the amount of iron deposited on the silicon surface from SC-1 exhibited remarkably little deviation over a wide range of spiking levels, leading to the conclusion that in all cases an initial rapid oxidation of the silicon took place, followed by the preferential oxidation of the iron and its inclusion as the oxide into the oxide film. Finally, the model developed predicts that lower temperatures and more concentrated chemistries are more effective in keeping the iron in solution.

1. INTRODUCTION

The deposition of metals from cleaning solutions onto wafer surfaces during various cleaning steps has been identified as a key issue for current and future technology needs. Considerable evidence supporting these effects has been published [1-3], and work continues to document and understand these effects. In particular, the presence of transition elements such as iron results in a reduction of the measured lifetimes in silicon[4], an increase in surface recombination velocity[5], reduction in the breakdown voltage[6] and other such deleterious effects.

This work investigates the deposition of iron on the wafer from spiked SC-1 solutions on silicon wafer surfaces. Based on the data obtained from the experiments performed, and that available in the literature, a model for the depositon of iron from SC-1 onto silicon is proposed.

2. EXPERIMENT

SC-1 (NH_4OH, H_2O_2, H_2O mixture) was chosen as a medium for depositing iron on the wafer surface. (Such a scenario would be of particular relevance in a real fabrication process where metallic microcontamination occurs from cleaning solutions during the course of a pre-gate oxidation clean). The experiment investigated the effect of different wafer surface terminations on the extent of iron deposition. Deposition on three different wafer surface terminations was studied. These included surfaces cleaned using : (i) a SPM (H_2SO_4, H_2O_2 mixture) /HF/SC-1/SC-2 (HCl, H_2O_2, H_2O mixture) sequence, (ii) a SPM/HF/SC-1/SC-2/HF sequence surface, and (iii) a SPM/HF/SC-1/SC-2 sequence, followed by the growth of a thermal oxide, 64 Å thick. These cleans were performed at room temperature using 1:1:5 SC-1 (NH_4OH-H_2O_2-H_2O), and 1:1:5 SC-2 (HCl-H_2O_2-H_2O) solutions. Surface (i) is characterized by the presence of ~10 Å of native oxide, (ii) by relatively little native oxide and (iii) by a thick thermal oxide. The contamination was performed at room temperature using a 1:1:5 SC-1 solution, and also at 80°C with a more dilute composition (0.25:0.5:5).

Mat. Res. Soc. Symp. Proc. Vol. 386 © 1995 Materials Research Society

3. RESULTS AND DISCUSSION

We found that irrespective of the surface clean performed prior to the Fe/SC-1 treatment (or equivalently, the surface termination), the amount of iron deposited on the silicon surface from SC-1 exhibited remarkably little deviation over a wide range of concentrations (0.1 ppb-10 ppm) as shown in Fig.1. In particular, the wafer surface levels for iron deposited from a 1:1:5 SC-1 solution, at 28°C, are the same, regardless of whether the initial surface termination is of type (i), (ii) or (iii). Similarly, the wafer surface levels for iron deposited from a 0.25:0.5:5 SC-1 solution, at 80°C, are the same, regardless of whether the initial surface termination is of type (ii) or (iii).

Since an identical amount of iron is deposited on the wafer surface regardless of the wafer surface termination, the mechanism of iron deposition is evidently independent of the starting silicon surface. The iron is deposited onto/into the silicon dioxide, which in turn is continuously grown/etched in the SC-1 solution[8-10]. However, it is not easy to predict the exact form(s) of the Fe participating in the reaction, due to the presence of different forms of the iron (Fe^{2+}, Fe^{3+}, $HFeO2^-$ etc.) at pH values characteristic of the SC-1 solution[7]. A reaction that is consistent with the preceding dicsussion (and also consistent with the predictions of the model discussed later) is:

$$HFeO_2^- + 1/2\ H_2O_2 = 1/2\ Fe_2O_3 + 1/2\ H_2O + OH^- \quad (1)$$

The reaction is, however, inconsistent with a mechanism where iron is deposited in its elemental form, bonded to the silicon surface. Thus if the iron was indeed preferentially deposited in its reduced form, it would manifest itself as a difference in the deposition characteristics for the HF terminated surfaces on the one hand, and the native oxide (SC-2)/ thermal oxide terminated surfaces on the other hand, since deposition on a HF terminated

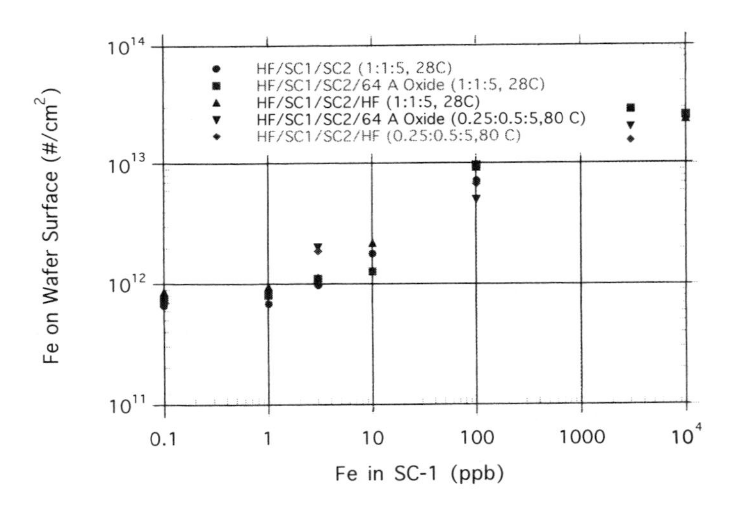

Fig.1 Iron Deposition from SC-1 on Si with varying surface terminations

surface would require a reaction like:

$$12OH^- + 4Fe^{3+} + 3Si = 3H_2SiO_3 + 3H_2O + 4Fe, \quad (2)$$

where the Si is oxidized by the Fe^{3+} ion.

Thus, the fact that the deposition characteristics are the same seems to suggest that in all cases an initial rapid oxidation of the silicon wafer surface takes place first, followed by preferential oxidation of the iron and its inclusion as the oxide into the oxide film.

At solution concentrations below ~3ppb the flat region is due to the iron initially present in the bath containing the (unspiked) SC-1. This is probably because similar experiments have repeatedly been performed in the wetbench in use. Hence the elevated baseline (8X1011 cm^{-2}), at the low end of the spiking range. Also worthy of note is the fact that there is a limit on the absolute maximum in terms of the amount of iron that may be deposited on the wafer surface irrespective of the amount of iron in solution, the solution chemistry, or the temperature. This observed maximum has a value of ~ $3X10^{13}$ cm^{-2}.

4. MODELING THE RESULTS

A major thrust of this work was the development of a mathematical model for the deposition of Fe onto Si surfaces from SC-1 solutions, that can be implemented for computer aided analysis. Ideally, this would also be based on an understanding of metal deposition, which is complicated by the solution chemistry, electrochemical effects, the nature of the Si surface, and etching of the surface by some of the solutions. Consideration of the chemistry of the process and the data lead to the following kinetic model for Fe deposition:

$$d\sigma/dt = k_1\rho(1- \sigma/\sigma_0) - k_2\sigma \quad (3)$$

where ρ is the surface concentration and σ the solution concentration. The solution of this equation for an arbitrary initial surface concentration, σ_1, is:

$$\sigma = k_1\rho\tau\{1-\exp(-t/\tau)\} + \sigma_1\exp(-t/\tau), \quad (4)$$

where $\tau = (k_1\rho/\sigma_0+k_2)^{-1}$. k_1 is a rate constant relating to the rate of adsorption of Fe onto the surface , and k_2 is the desorption rate constant, presumably, at least partially associated with the etching in an SC-1 solution. Measurements of the adsorption of Fe in the limit of long contact times are shown in Fig.2[3] ; long contact times imply steady state. Examination of Eqn.1 reveals that for low solution concentrations, the deposition (build-up) rate on the wafer surface is equal to $k_1\rho$. Hence, the initial slope on the curve(s) in Fig.2 may be used to determine k_1 at different temperatures. In the absence of a desorption mechanism, adsorption of iron onto the surface would lead to a continuous buildup until the solution was depleted of iron. The fact that there is an absolute maximum on the amount of iron that can be deposited irrespective of the solution concentration, is indicative of a competing desorption/etching mechanism which arises due to the presence of NH_4OH in the solution. (The mechanism for this saturation is not clear; it may be related to a density of a particular chemical entity on/in the native oxide present). Hence, an SC-1 clean also involves etching of the Si leading to a redissolution of some of the iron. As stated earlier, k_2 is the rate constant relating to the rate of desorption of Fe from the surface into the solution. An examination of Eqn.4 indicates that in the limit of low solution concentration the steady state surface concentration ($t \to \infty$) is equal to $k_1\rho\tau \sim k_1\rho/k_2$. (Care must be taken while defining a "low" solution concentration, which depends on the solution temperature and the chemistry under consideration). k_2 can thus been determined, once k_1 has been ascertained. Fig.3 shows the "leaching" of Fe from a surface in a pure SC-1 solution[3]. In principle, these curves may also be used to determine k_2. Note that k_1 and k_2 vary with the temperature and

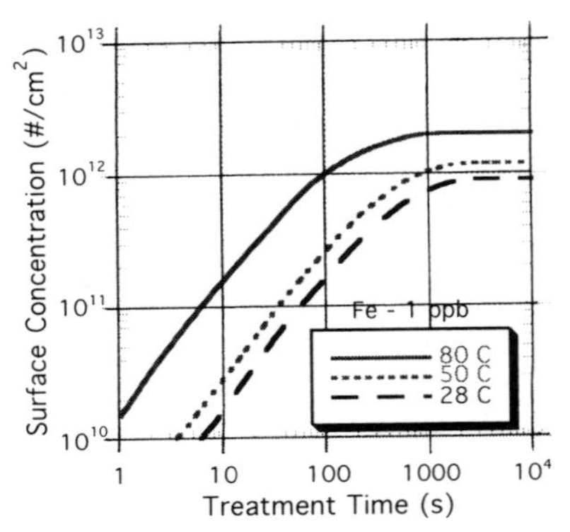

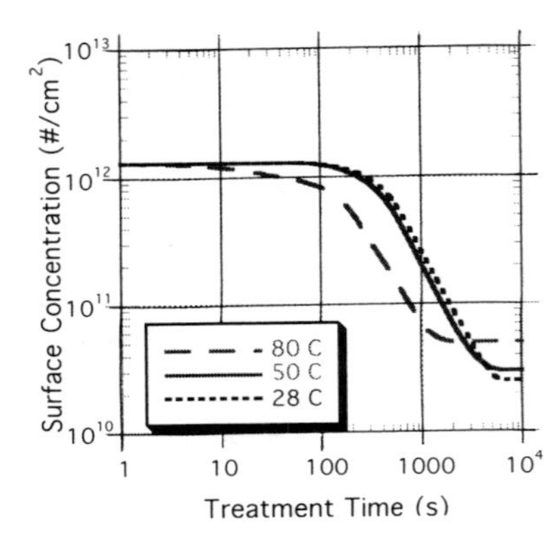

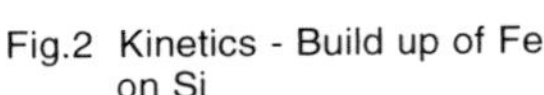
Fig.2 Kinetics - Build up of Fe on Si

Fig.3 Leaching of Fe from surface in pure SC-1.

the solution chemistry under consideration. Thus a "complete" model would need to incorporate these dependencies too.

As a first pass we decided to consider the effect of temperature (only) on k_1 and k_2, at a particular solution chemistry. Examination of the values of k_1 at different temperatures, for a 0.25:1:5 chemistry, (from Fig.2) yielded an approximately exponential dependence on the temperature. The exponential fit resulted in an activation energy associated with this process equal to 0.32 eV. k_1 at a particular solution chemistry may thus be modeled by the following equation:

$$k_1=12.93 \exp(-0.32/k_bT), \tag{5}$$

where k_b is the Boltzmann constant. Inclusion of the effect of solution chemistry proves to be more complex and is modeled by modifying the pre-exponential factor as follows:

$$k_1=6.47\{[H_2O_2]/[NH_4OH]\}^{1/2}\exp(-0.32/k_bT) \tag{6}$$

The modification was based on our experimental data (which involved kinetic studies, similar to that in Fig.2, at different contamination levels, chemistries and temperatures), as well as that available in the literature[3].

A justification for this dependence of k_1 on solution chemistry was obtained by closely examining Eqn.1, which clearly indicates that the adsorption rate, k_1,is proportional to to $[H_2O_2]^{1/2}/[OH^-]$, which is in turn proportional to $\{[H_2O_2]/[NH_4OH]\}^{1/2}$. This is in exact accordance with the form of Eqn.6.

Determination of the temperature dependence of k2 was done in a similar fashion (using the curves in Fig.3). k_2 was also found to have an approximately exponential dependence on the temperature . The activation energy associated with k_2 was found to be 0.265 eV. k_2, at a particular solution chemistry may hence be modeled by the following equation:

$$k_2=23.6 \exp(-0.265/k_bT), \tag{7}$$

As in the case of k_1, the dependence of k_2 on solution chemistry is fairly complex and is as given below:

$$k_2=188.6\{[NH_4OH]/[H_2O_2]\}^{3/2}\exp(-0.265/k_bT) \quad (8)$$

While the exact power dependence of k_2 on the solution chemistry is hard to explain, the direct proportionality of the desorption rate on the concentration of the NH_4OH in solution is consistent with the fact that the etch rate increases with an increase in the amount of NH_4OH in the solution.

Fig.4 shows that the predictions of the model are in good agreement with the experimental data. Finally, predictions of the model, indicating the manner in which the deposition of iron varies with the SC-1 chemistry and temperature, are shown in Fig.5.

7. CONCLUSIONS

The experimental data supports the theory that the iron deposited from SC-1 on the silicon surface is incorporated as the oxide into the native oxide film, contrary to the belief held by some that it is predominantly present in its elemental (reduced) form, bonded to the silicon surface. The model developed accurately predicts the deposition characteristics for iron from (SC-1) solution, for varying chemistries, and at different temperatures. Based on the predictions of the model (Fig.5), a clean optimized for low iron levels on the wafer surface would require lower temperatures and more concentrated chemistries (ie. a higher ratio of the concentration of the NH_4OH to H_2O_2). Note however, that the model is based on a relatively limited set of data points, and merits verification by performing further experiments at different chemistries and temperatures.

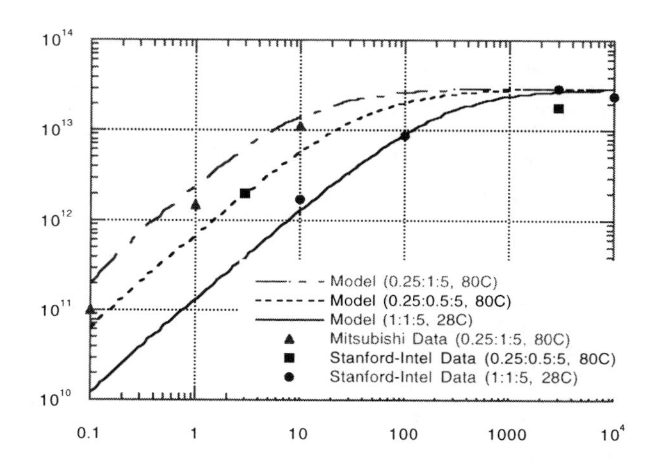

Fig.4 Predictions of the model and comparision with experimental data for different solution chemistries and temperatures.

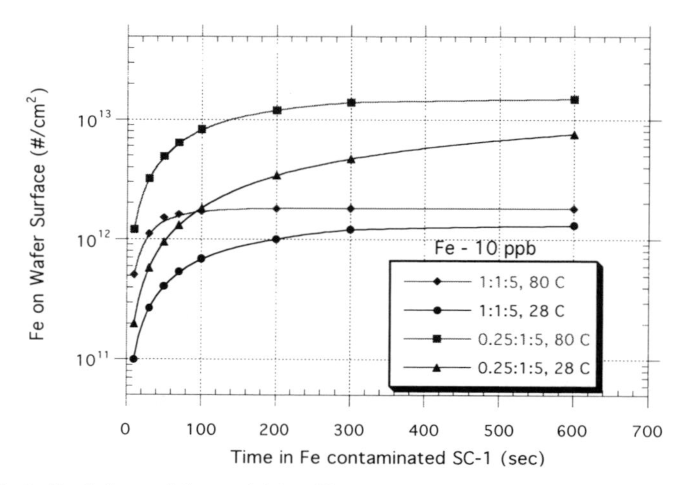

Fig.5 Predictions of the model for different solution chemistries and temperatures.

References

1. C. R. Helms, H.-S. Park and S. Dhanda, in Fundamental metallic issues for ultraclean wafer surfaces from aqueous solutions, (Proceedings of the Second International Symposium on Ultra-clean Processing of Silicon Surfaces, Acco, 1994), p. 205.
2. P. Gupta, S. H. Tan, Z. Pourmotamed, F. Cristobal, N. Oshiro and B. McDonald, in Novel methods for trace metal analysis in process chemicals & DI water and on silicon surfaces, (Proceedings of the Symposium on Contamination Control and Defect Reduction in Semiconductor Manufacturing III), p. 200, (1994).
3. J. Ryuta, T. Yoshimi, H. Kondo, H. Okuda, and Y. Shimanuki, Jpn. J. Appl. Phys., 31, 2338 (1992).
4. Y. Hayamizu, T. Hamaguchi, S. Ushio, T. Ab and F. Shimura, J. Appl. Phys., 69, 3097 (1991).
5. S. M. Sze, Physics of Semiconductor Devices, (John Wiley and Sons, NY,1981), p. 35-38 .
6. K. Honda, T. Nakanishi, A Ohsawa and N. Toyokura, in Metal impurities at the SiO2-Si interface, (Proceedings of Microsc. Semicond. Mater. Conf., Oxford,1987) p 403.
7. M. Pourbaix, Atlas of Electrochemical Equilibria in Aqueous Solutions, (Pergamon Press, NY, 1961), p. 309.
8. H. Kobayashi, J. Ryuta, T. Shingyouji and Y. Shimanuki, Jpn. J. Appl. Phys., 32, L45 (1993).
9. J. E. A. M. van den Meerakker and M. H. M van der Straaten, J. Electrochem. Soc., 137, 1239 (1990).
10. H. Kaigawa, K. Yamamoto and Y. Shigematsu, Jpn. J. Appl. Phys., 33, 4080 (1994).

MEASURING DIFFUSION LENGTHS IN EPITAXIAL SILICON BY SURFACE PHOTOVOLTAGE

JOHN LOWELL, VALERIE WENNER, AND DAMON DEBUSK
Advanced Micro Devices, 5204 E. Ben White Blvd., Austin, TX 78741

Abstract

In CMOS, the use of epitaxial layers for prevention of latch-up in logic technologies is well-known and pervasive. One of the crucial parameters is the amount of metallic contamination due to transition metals such as Fe in the epi since this phenomena effects both device performance and quality. However, the ability to measure this parameter on product material is not generally available due to inherent problems with most known methods. The limitation of traditional surface photovoltage is that the deep optical penetration of over a hundred microns is well-beyond the depth of most epitaxial layers and does not accurately profile the epitaxial region [1]. In this paper we report on the application of optical surface photovoltage (SPV) using a set of ultra-shallow optical filters to both quantify and qualify as-grown epitaxial layers on CZ P-type silicon. We believe that a non-contact, SPV measurement of Fe concentration and diffusion lengths within an epitaxial region has not been previously reported.

Introduction

Most VLSI logic technologies currently in production use epitaxial silicon starting wafers. Despite the increased cost over single crystal wafers, their inherent latch-up prevention and defect gettering properties are considered invaluable to product yield and reliability. Since semiconductor manufacturers choose to purchase wafers with the epi regions as-grown, maintaining the integrity of this layer is important to wafer vendors worldwide as well. It is also known that as dependence on epi has increased so has the cost making the issue of epi quality an important and vital issue. One of the crucial epi parameters is the amount of metallic contamination due to transition metals (such as Fe) in the epi since this phenomena is known to degrade both device performance and quality. Currently, passive measurement of this parameter on epitaxial wafers is not possible at incoming inspection or anywhere in-line since contact to the epi surface or destructive bulk spectroscopy is required. Therefore, process control and IC quality require a way to extract directly information *nondestructively* on the diffusion length and metallic properties in this layer.

Mat. Res. Soc. Symp. Proc. Vol. 386 © 1995 Materials Research Society

Here we report on the application of optical surface photovoltage (SPV) to both quantify and qualify as-grown epitaxial layers on CZ P-type silicon. SPV methodology passively determines minority carrier diffusion length from the spectral dependence of a surface photovoltage on the optical penetration depth [1]. Sweeping the wavelength through a series of optical filters as shown in Table I, the penetration depth is varied from 11 to 145 μm [2]. The traditional problem with this approach is that one must use an extrapolation of at least three measurement points to obtain reliable estimates of diffusion length for accurate Fe bulk concentration calculation. Taking the three lowest penetration depths of 11.3-38.2 μm (shown in Table I) one observes that the deepest point is well-beyond the depth of most epitaxial layers. Thus the lifetime measurements are dominated by single-crystal properties, and the results do not accurately profile the epitaxial region. To date this problem exists in most optical lifetime characterization techniques. Recent work with electrolytic metal tracer methods have shown capability in this area using a single optical penetration depth of 4 μm [3]. While a distinct improvement, a single depth prevents ranging within the epi, and may have problems with very thin layers (< 5 μm). Hence the analytical problem still remains.

Wheel Position	Wavelength (micron)	Penetration Depth (micron) at 21°C
1,8	0.800	11.3
2	1.004	157
3	0.991	121
4	0.975	90.9
5	0.950	59.5
6	0.915	38.2
7	0.870	22.7

Table I. Standard specifications for optical filters

To address this problem within the framework of a standard tool, work was begun leading to our development of an entirely new set of optical filters compatible with the system developed by the SPV manufacturer [2]. The parameters of this new set are given in Table II and are optimized for shallow depths. Unlike those shown in Table I, these filters attain a minimum depth of 1.1 μm in depth and a *maximum* of 38.2 μm. Between them are seven additional filters with a *selectable* resolution of 1.6-22.7 μm. The depth of 1.16 μm is the smallest currently allowed due to optical manufacturing constraints on the filters. To date we believe this is the shallowest depth resolution of SPV ever reported.

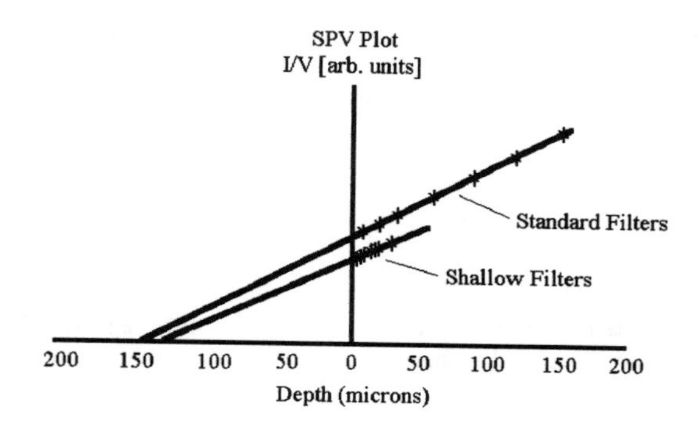

Figure 1. Example plot of the standard and shallow filters

Wheel Position	Wavelength (micron)	Penetration Depth (micron) at 21°C
1,10	0.500	1.16
2	0.550	1.66
3	0.600	2.38
4	0.650	3.43
5	0.700	5.02
6	0.750	7.50
7	0.800	11.5
8	0.870	22.7
9	0.915	38.7

Table II. Specifications for shallow-depth optical filters

Experiments

Initial tests were performed on different samples of unprocessed P/P+ epitaxial wafers with an specified epi thickness of 17 μm. Each wafer was measured by SPV and later subdivided for additional tests. Portions of each were passed on for analysis by deep-level transient spectroscopy (DLTS) or time-of-flight SIMS (ToF-SIMS) for additional estimates for iron. DLTS biasing was set for a space charge region of 4 microns [5]. For these tests the range of optical depth penetration was set to between 5.02 and 22.7 μm measuring nine points per sample. This practice eliminates the aforementioned problem of interference from the single crystal layer in the measurements. Since the epi/single crystal junction allows the formation of a transition region at the P/P+ interface, the depth of 22.7 μm is allowable for a 17 μm thick layer to accurately reflect the epi character. Dissociation of the FeB pair (i.e. activation of Fe_i) was done by thermal treatment at 250° C/2 minutes for DLTS, and by optical exposure for SPV. Here, Fe_i refers to non-bonded, interstitial Fe.

Results

Table III lists estimated Fe concentrations by our new "epi SPV" method which are compared to those calculated from DLTS. The standard formula for this estimate is shown below [1,4]:

$$N_{Fe} = 1.05 \times 10^{16} (L_{pre}^{-2} - L_{pos}^{-2}) \quad \text{in cm}^3 \tag{1}$$

Diffusion length value L_{pre} is prior to Fe activation and L_{pos} is post-activation. All L values are in microns.

In all measurements, the SPV correlation factor was above the threshold of 0.98 indicating no errors in the measurement and assuring the integrity of the samples. Concentration correlation of the SPV data to the measured ToF-SIMS data was closer than that estimated by DLTS (for Fe_i). Indeed the DLTS measurements underestimate the concentrations shown by ToF-SIMS and SPV. This is noteworthy since its epi penetration (4 μm) is greater than the ToF-SIMS and much less than the SPV. To date on material used in this test (and other samples) results have shown that diffusion lengths of 5X the epitaxial thickness (17 μm) can been measured.

Sample I.D.	Avg. N(Fe) DLTS	Avg. N(Fe) SPV
E2-A	0.09E+12	1.3E+12
E2-B	0.12E+12	.63E+12
G0-B	0.25E+12	1.2E+12
G0-C	0.21E+12	2.0E+12

Table III: Concentration of Fe in the epitaxial layer as estimated by DLTS and high-resolution surface photovoltage. ToF-SIMS analysis showed an average concentration of 2.0E12 a/cm^2 on all samples.

Next we present in Table IV pre and post-activation values obtained by our new SPV method and their respective recombination lifetime values, tau_r. In relation to diffusion lengths obtained on CZ wafers, we observe that for the epitaxial layer (1) the Fe concentration is high and (2) the diffusion lengths are low. While it is probable that most of the Fe in the epi would diffuse to the single layer during the first thermal cycle in the process line, these levels are probably above the limits for starting material for most logic manufacturers. Nevertheless we have shown that isolated, passive measurements of metallic contamination within the epi layer are possible and well-correlated to actual concentration.

Sample I.D.	Diff. Length pre-act (micron)	Diff. Length post-act (micron)	Tau_r pre-act (microsec)	Tau_r post-act (microsec)
E2-A	81.6	66.1	2.19	1.44
E2-B	78.9	66.9	2.05	1.47
G0-A	39.2	35.7	0.51	0.42
G0-B	38.7	35.4	0.49	0.41
G0-C	46.3	38.6	0.71	0.49
G0-D	46.5	39.0	0.71	0.50

Table IV. Shallow optics results on epitaxial samples. The results shown are averaged over five points on the sample surface.

Conclusion

In this work our results report for the first time the application of high-resolution, optical filters optimized for short-diffusion length SPV measurements. Using this system one can set the range of measurement to be within most specified epitaxial regions wherein the recombination lifetime and estimated Fe concentration reflect only the epi component with little contribution from the bulk Si wafer. In addition our optics are compatible with currently available SPV machines, and inexpensive to produce. Like standard SPV, this new system is fast, non-contact, and non-destructive. Its output is a complete map or set of distributed points of distributed lifetime measurements.

Since the ability to qualify epitaxial layers on product wafers particularly for metallic contamination is important to both wafer vendors and IC manufacturers, this work suggests that it can be monitored and measured directly by passive methods. We also believe that this optical system may also be useful for characterizing any shallow region beyond 1 micron for its minority carrier diffusion length properties. We also observe that our obtained values of diffusion length are much smaller than the predicted values in epitaxial material. While more work in this area is clearly needed, there maybe definite differences between theoretical diffusion lengths and actual measured values particularly for small epitaxial thicknesses.

References

1. J. Lagowski, *et.al.*, Semicond. Sci. Tech., **7**, A185 (1992).
2. SPV Tools User's Manual (Semiconductor Diagnostics Inc., Tampa, FL, 1994), p.26.
3. Robert Falster, Diagnostic System for Metal Contamination in Wafer Manufacturing (Esprit Project, 1993).
4. G. Zoth, Tech. Proc. SEMICON Europa '90 (Zurich), p.24 (1990)
5. Dieter Schroder, Semiconductor Materials and Device Characterization, (Wiley, New York, 1990), p. 319.

Part IV

Pre-Cleaning Impact on Gate-Oxides and Silicidation

INTERFACIAL DEFECTS INDUCED BY SILICIDATION AND EFFECTS OF H-TERMINATION AT METAL/SILICON CONTACTS

SHIGEAKI ZAIMA AND YUKIO YASUDA
Department of Crystalline Materials Science, School of Engineering, Nagoya University, Furo-cho, Chikusa-ku, Nagoya 464-01, Japan

ABSTRACT

We have investigated crystallographic structures and electrical properties at the interfaces of transition metals such as Ti, Zr, Hf and V and Si(100), from the viewpoint of an application to ohmic contacts in future ULSI's with low contact resistivity and high reliability. We have achieved very low contact resistivities of $3\text{-}5\times10^{-8}$ Ωcm^2 for Zr/ and Hf/n^+-Si(100) contacts and $1\text{-}2\times10^{-7}$ Ωcm^2 for p^+-Si at 400-600°C. It is found that the silicidation reaction in this temperature range brings about the formation of deep levels associated with vacancies in Hf/Si and with metal atoms in V/Si. This difference is considered to be related to that in the silicide formation process between these systems. Furthermore, an anomalous electrical characteristics observed for Hf/p-Si contacts has been found to be markedly improved by using a H-termination treatment of Si surfaces.

INTRODUCTION

Formation of ohmic contacts with low contact resistance and high reliability is one of the key issues in realizing future ultra-large scale integrated circuits (ULSI's). Because the contact resistance is drastically increased with decreasing device dimensions [1], the current driving capability of small metal-oxide-semiconductor field-effect transistors (MOSFET's) is reduced by the parasitic series resistance of ohmic contacts. In order to accomplish improved device performance by miniaturization, it is essential to develop new materials and process technologies to realize ohmic contacts with resistivities as low as 10^{-8} Ωcm^2. Furthermore, interfacial silicidation reaction may lead to the redistribution and segregation of impurity atoms resulting in the destruction of very shallow pn junctions in source and drain regions. Also the reduction of contact areas would bring about new problems related to size effects and local stress. Therefore, understanding interfacial phenomena related to solid-phase reactions are necessary in order to form highly-reliable ohmic contacts.

Transition metals in the group IVa, Va and VIa and their silicides can be expected to be low-resistivity contact materials for ULSI's, because of their thermal and chemical stability and Schottky barrier heights as low as the half of the band gap of Si. We have investigated crystallographic structures and electrical characteristics of the interfaces between these metals and Si(100) substrates from the viewpoint of applications to ohmic contacts of ULSI's. Contact resistivities of $3\text{-}5\times10^{-8}$ Ωcm^2 in Zr/ and Hf/n^+-Si(100) at annealing temperatures of 400-600°C [2-4] have been achieved. We have also observed a bilayer structure composed of an amorphous and a crystalline silicide interlayer at the metal/Si interfaces in this annealing

Mat. Res. Soc. Symp. Proc. Vol. 386 © 1995 Materials Research Society

temperature range [4, 5]. The silicidation reaction at the interface also induces the formation of defects such as vacancies and diffused metal atoms. Since these defects are thought to cause the increase in leakage current and to impede the precise control of impurity profiles of very shallow pn junctions, it is important to clarify the relation between interfacial reactions and electrical characteristics of these contact systems.

In the present study, we have examined the formation of interfacial defects by the silicidation reaction and the effect of H-termination of Si(100) surfaces on the electrical properties. The termination of dangling bonds with H atoms makes Si surfaces passive against oxygen adsorption even in atmosphere [7]. The H-termination of Si surfaces is promising as a new technology for surface passivation to control the formation of native oxide. A marked improvement of the electrical characteristics in p-Si Schottky diodes has been found upon H-termination treatments [8].

EXPERIMENTAL

Substrates used were n- and p-type Si(100) wafers with a resistivity of 0.5-0.7 and 1-3 Ωcm, respectively. Si wafers were chemically cleaned and dipped in a diluted HF solution ($HF:H_2O=1:50$) to remove thin oxide layers and rinsed in deionized (DI) water for 10 min before placing in an ultrahigh-vacuum (UHV) evaporation chamber. After introducing the substrates into the evaporation chamber, with a base pressure less than 5×10^{-10} Torr, transition metal films were deposited using an electron-beam evaporator and then the samples were annealed at 400 and 600°C for 30 min in the chamber. In this paper, the effects of H-termination of Si surfaces on electrical characteristics are also presented, using the procedure described below. The substrate surface treated with these procedures was examined by high-resolution electron energy loss spectroscopy (HREELS) and x-ray photoelectron spectroscopy (XPS) using MgKα radiation.

In order to examine electrical properties of metal/Si interfaces, Schottky diodes were formed on the Si(100) substrates. The electrode area of Schottky diodes was 1 mm^2. Ohmic contacts with substrates were formed by Al electrodes through heavily-doped diffused layers on the backside of substrates. Current-voltage (I-V) characteristics of Schottky diodes were measured in the temperature range of 77 to 300 K. Schottky barrier heights were determined from the temperature dependence of I-V characteristics and also capacitance-voltage (C-V) characteristics at 1 MHz. Furthermore, deep levels at the metal/Si interface were examined by deep-level transient spectroscopy (DLTS). The contact resistivities were measured by a four-terminal probe method using Kelvin patterns [9] with a contact area of 6.21 μm^2. Phosphorus and boron concentrations in n^+- and p^+-Si diffused layers, N_D and N_A, in the Kelvin patterns were 2×10^{20} and 1×10^{20} cm^{-3}, respectively.

RESULTS AND DISCUSSIONS

Contact Resistivity

Contact resistivities of Ti, Zr, Hf and V to n^+- and p^+-Si, ρ_{cn} and ρ_{cp}, have similar dependencies on the annealing temperature for all metals and also for both ρ_{cn} and ρ_{cp} [3]. Typical temperature dependence of ρ_{cn} and ρ_{cp} are shown in Fig. 1 for Hf/Si(100) contacts [4]. The values of ρ_{cn} and ρ_{cp} decrease with increasing the annealing temperature and reach their minimum values at temperatures ranging from 400 to 600°C. Further increase in annealing temperature leads to a drastic increase in both ρ_{cn} and ρ_{cp}. The minimum values and the corresponding annealing temperatures for ρ_{cn} and ρ_{cp} are listed in Table I [5] for Ti, Zr, Hf, V and Cr contacts with Si. Very low contact resistivities of 3-5x10^{-8} Ωcm^2 can be obtained for Zr/ and Hf/n^+-Si. It is noted that the minimum values of ρ_{cp} for these metals are 2-3x10^{-7} Ωcm^2, higher than those of ρ_{cn}. In the case of Cr/p^+-Si, a very low value of ρ_{cp} is observed in as-deposited contacts, however, the thermal annealing leads to an increase in ρ_{cp}.

The current flowing at the interface between metals and heavily-doped semiconductors is dominated by the tunneling process of carriers through the Schottky barriers. The contact resistivity in this case is given by [10]

$$\rho_{cn} \propto \exp\left[\frac{2\sqrt{\varepsilon_s m^*}}{\hbar}\left(\frac{\phi_{Bn}}{\sqrt{N_D}}\right)\right], \quad (1)$$

where ε_s is the permittivity, m^* the effective mass of carriers and N_D the doping concentration. The contact resistivity of an ideal contact system, therefore, is governed by both the Schottky barrier height and the doping concentration in semiconductor substrates. For n-Si, it has been found that the Schottky barrier heights also show minimum values at the annealing temperature range of 400-600°C and increase above 600°C, which means that the changes of ρ_{cn} by thermal annealing correspond to changes in Schottky barrier height [3]. On the other hand, the correlation between ρ_{cp}'s and Schottky barrier heights is not clear for p-Si, because the Schottky diodes of metal/p-Si, especially for the group IVa metals, exhibit nonideal characteristics [3, 11]. The main features of the nonideal characteristics are explained by the presence of additional series capacitance and of large excess current components. One of the possible origins of the additional series capacitance is dielectric-like layers formed at the metal/p-Si interface [11], and the large excess current components are related with interfacial defects. In the following sections, we discuss the formation of defects due to silicidation reaction and the improvement of electrical characteristics in Hf/p-Si by a H-terminated treatment of Si surfaces.

Formation of Defects

Figure 2 shows DLTS spectra of Hf/n-Si(100) diodes with a conventional treatment at various annealing temperatures. We can observe several peaks in the DLTS spectrum for the as-deposited sample, and it should be noted that these peak intensities are drastically reduced by thermal annealing. This fact is consistent with the growth of epitaxial silicide at the

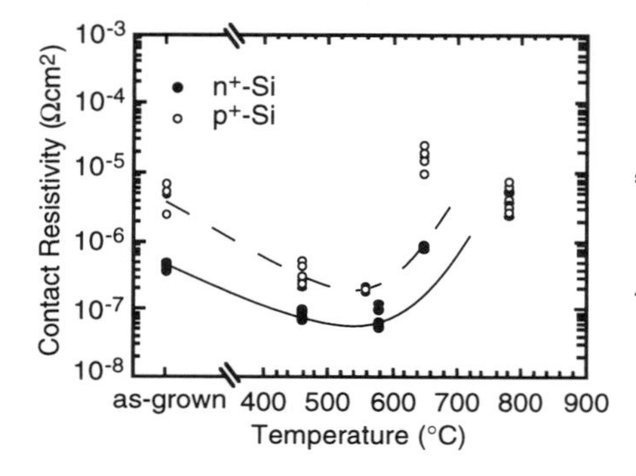

Fig. 1. Specific contact resistivity of Hf/n$^+$- and p$^+$-Si(100) as a function of annealing temperature.

Table I. Minimum contact resistivities and the corresponding annealing temperatures for Ti, Zr, Hf, V and Cr/n$^+$- and p$^+$-Si(100) contacts.

Metal	ρ_{cn} (Ωcm^2)	Annealing Temp. (°C)	ρ_{cp} (Ωcm^2)	Annealing Temp. (°C)
Ti	2.4×10^{-7}	520	2.8×10^{-7}	580
Zr	3.4×10^{-8}	420	1.8×10^{-7}	560
Hf	5.4×10^{-8}	580	1.8×10^{-7}	560
V	1.3×10^{-7}	550	2.1×10^{-7}	550
Cr	2.2×10^{-7}	350	2.1×10^{-7} ($< 2 \times 10^{-8}$	450 as-grown)

Hf/Si(100) interface, as reported previously [4]. After annealing at 460 and 580°C, a bilayer structure consisting of an amorphous layer and an epitaxially-grown layer of Hf_3Si_2 is formed between Hf films and Si substrates. The epitaxial relationship of Hf_3Si_2 layers is Hf_3Si_2(110)//Si(010) and Hf_3Si_2[110]//Si[100] with Si(100) [4]. Therefore, it can be considered that the growth of epitaxial Hf_3Si_2 layers leads to a decrease in trap densities. The minimum values of ρ_c are obtained at 580°C in this system.

The energy levels of electron traps denoted by E1, E2 and E4 in Fig. 2 are determined by Arrhenius plots as E_c-E_t=0.15±0.01, 0.20±0.05 and 0.43±0.03 eV. The trap E1 can be assigned to a complex of Si vacancies and O atoms (0.17 eV [12]), which is not observed for samples with H-termination as shown in Figs. 8 and 10. The trap energies of E2 and E4 are very close to those of divacancies (0.23 and 0.43 eV [12-14]) and a complex of vacancies and P atoms (0.44 eV [12, 14]). The average trap densities of E1, E2, and E4 are about 2×10^{13}, 3×10^{12} and 1×10^{14} cm^{-3}, respectively. The densities of these vacancy-related traps are reduced by thermal annealing, e. g., values of E1, E2 and E4 for the sample annealed at 460°C are 1/3-1/10 times smaller than those for as-deposited one. After annealing at 580°C, these vacancy-related traps vanish, which means that the densities are less than 10^{11} cm^{-3}. The DLTS signals of E3 (0.29±0.01 eV) and E5, the origins of which are not clear yet, were less than the detection limit of DLTS measurements at 460°C. A new trap, labeled E6, is produced at 580°C, which has an activation energy and an average density of 0.29±0.02 eV and 1×10^{12} cm^{-3}, respectively. The formation of vacancy-related traps observed in Fig. 2 is related with the interfacial reaction, as discussed below.

In the case of V/n-Si diodes, however, we do not observe any vacancy-related traps. Figure 3 shows the DLTS spectra of V/n-Si(100) at various annealing temperatures [15]. In the as-grown sample, DLTS signals are very small. By annealing up to 550°C we can observe a marked increase in DLTS signals with increasing annealing temperature. In this temperature

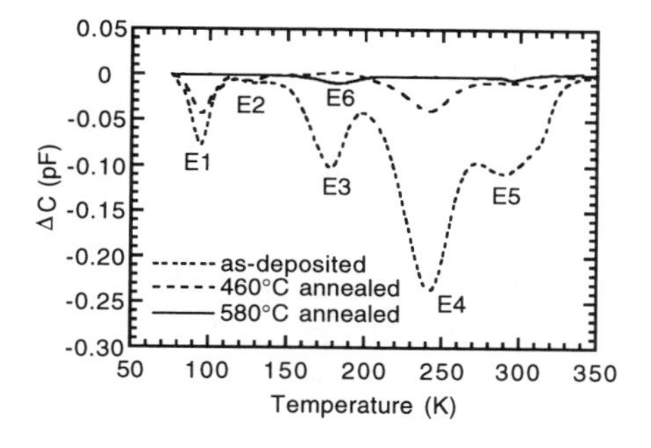

Fig. 2. DLTS spectra of Hf/n-Si(100) with various annealing temperatures.

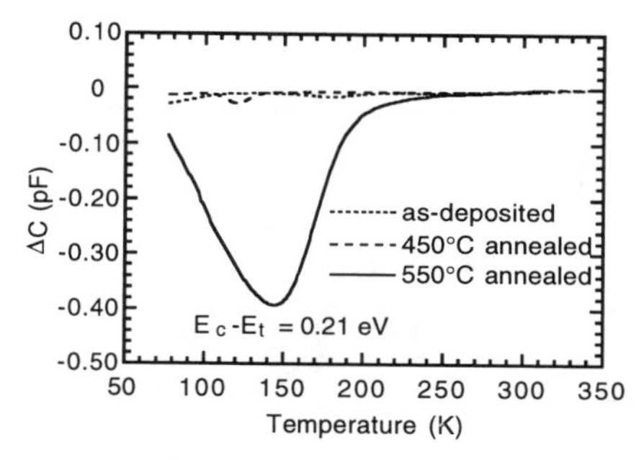

Fig. 3. DLTS spectra of V/n-Si(100) with various annealing temperatures.

range, the energy level of an electron trap is determined to be E_c-E_t=0.21±0.02 eV, the concentration of which is about $8x10^{12}$ and $4x10^{14}$ cm^{-3} at 450 and 550°C, respectively. Vanadium atoms are reported to be fast diffusers in Si substrates [16] and this trap can be identical with an acceptor level of interstitial vanadium atoms [17]. The metal atom diffusion into Si substrates is also observed for Ti/Si systems in this temperature range [18].

The formation of different types of traps between Hf and V by annealing is considered to be closely connected with silicidation reaction at the interface. The silicidation reaction at the Hf/Si interface has been reported to be limited by diffusion of Si atoms in silicide layers [19]. In addition, it is noted that the interfacial reaction is initiated by the existence of SiO_2 on Si surfaces and the silicidation takes place even for as-deposited samples [4, 20]. Therefore, the vacancy-related traps observed in Hf/Si are thought to be formed as a result of the silicidation reaction. In the case of V/Si(100) systems, on the other hand, the density of traps related with interstitial V atoms is markedly increased with an increase in annealing temperature. It has been reported that the diffusion of V atoms into the Si substrate is dominated in the silicidation process below 650°C [15]. Furthermore, the relatively good Schottky characteristics can be obtained for both V/n- and V/p-Si even by the conventional treatment. In V/p-Si, the influence of additional capacitance observed in Hf/p-Si is insignificantly small and the n-values are also close to unity [15]. These facts suggest that not only the defect formation but the electrical characteristics at metal/Si interfaces are significantly affected by the silicide formation process.

H-Termination Treatments

It has been reported that H-terminated Si(100) surfaces can be obtained by dipping in a HF solution and then rinsing in DI water for a very short time, and that the native oxide thickness on Si surfaces increases with increasing rinsing time [21, 22]. The interfacial reaction is complicated by external factors such as native oxide and impurities. Therefore, the

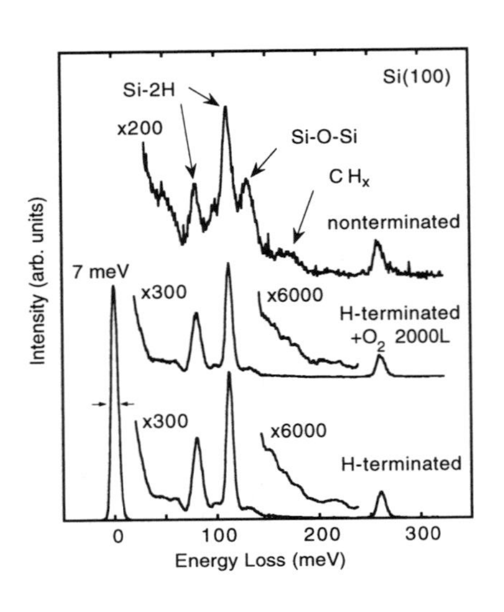

Fig. 4. HREELS Spectra of H-terminated Si(100) surfaces with and without 2000-L-O_2 exposure and a nonterminated surface.

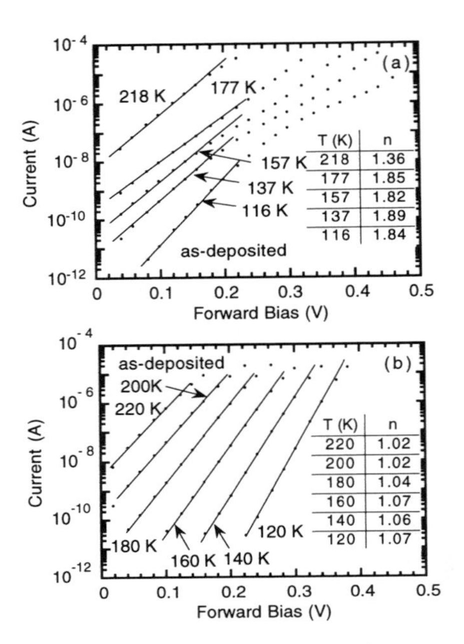

Fig. 5. Forward current-voltage characteristics of as-deposited Hf/p-Si(100) formed on (a) nonterminated and (b) H-terminated Si surfaces.

development of surface cleaning and/or passivation processes is also essential in contact technologies. In this experiment, Si substrates were rinsed in DI water for 2-5 s to obtain H-terminated Si(100) surfaces and then introduced into the UHV chamber while paying strict attention to oxygen and carbon contamination. Figure 4 shows HREELS spectra of the H-terminated p-Si(100) surface as-treated and exposed to 2000 L of O_2 at room temperature [8] and the conventionally-treated (nonterminated) surface. The primary energy of electron beam was 7.5 eV and specularly scattered electrons were measured. Energy loss peaks are observed at 61, 78, 112 and 259 meV for the as-treated surface, which can be assigned as those of rocking, wagging, scissor and stretching modes of dihydride (Si-2H) groups [23], respectively. This fact indicates that a dihydride phase is formed on the Si(100) surface by the H-terminated treatment. We can also see a small peak at 125 meV, which is considered to be due to a B_1 mode of Si-O-Si bonds [24] or a stretching mode of C-O bonds [25]. The extremely small peaks at 140-200 meV are vibrational modes of adsorbed hydrocarbon [26]. Impurity concentrations of H-terminated surfaces determined by XPS were less than a few percent of a monolayer for both oxygen and carbon, but no Si2p XPS peak of SiO_2 can be detected for repeated experiments. In spite of the presence of these impurities on the surface, the spectrum hardly changes after being exposed to 2000 L of O_2, as seen in Fig. 4. This fact indicates that the H-terminated surface is very stable for O_2 adsorption. For nonterminated surfaces, on the other hand, the intensity of an elastic peak is reduced by about 1/10 compared with that for H-terminated ones and we can clearly observe two peaks of A_1 and B_1 modes of Si-O-Si bonds at about 50 and 125 meV. In XPS spectra, the Si2p peak of SiO_2 was observed and the average thicknesses of SiO_2 were estimated to be less than 1 Å, which is consistent with the values reported by Gräf et al. [21]. Thus we confirmed that the Si substrate surfaces before the evaporation of Hf films are well passivated against native oxidation and oxygen adsorption by this procedure.

H-Termination Effects on Electrical Characteristics

Figures 5(a) and 5(b) show forward I-V characteristics of as-deposited Hf/p-Si(100) diodes formed on the nonterminated and the H-terminated surface, respectively [8]. The I-V characteristic of Schottky diodes is generally evaluated by an ideality factor. The ideality factor, n, is defined as [27]

$$n = \frac{q}{k_B T} \frac{\partial V}{\partial (\ln I)}, \tag{2}$$

where I is the forward current, k_B the Boltzmann constant, T the absolute temperature in measurements, q the electric charge, and V the forward bias voltage. When the thermionic emission current dominates the I-V characteristic, the n-value is equal to unity. In Fig. 5, the n-values obtained from the linear part in log$I$ versus V plots are also listed. In the case of nonterminated diodes, the n-value deviates significantly from unity and the I-V characteristics are far from the ideal thermionic emission characteristic, which is considered to be affected by the interfacial oxide. On the other hand, the current is found to be dominated by the thermionic emission and the n-value of H-terminated diodes is very close to unity. The results in Fig. 5 clearly show that the I-V characteristics of Hf/p-Si diodes are markedly improved by

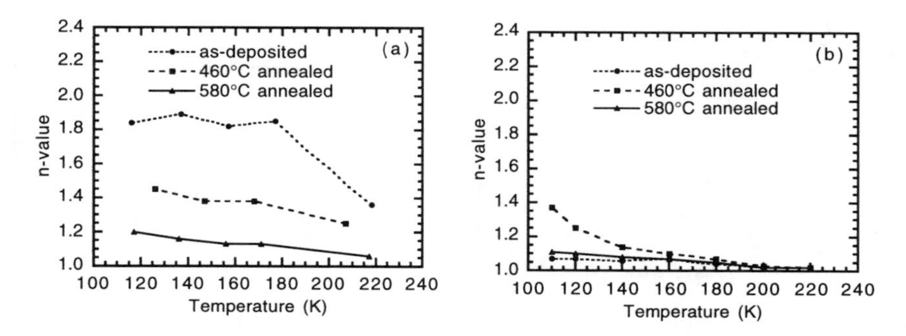

Fig. 6. Changes in n-values of as-deposited, 460°C-annealed and 580°C-annealed Hf/p-Si(100) diodes formed on (a) nonterminated and (b) H-terminated Si surfaces, as a function of measurement temperature.

the H-termination of Si surfaces even for as-deposited diodes.

In order to clarify the effect of thermal treatments, the n-values of annealed diodes are plotted in Fig. 6, as a function of measurement temperature [8]. Figures 6(a) and 6(b) are for the nonterminated and the H-terminated diodes, respectively. It is evident for nonterminated diodes that the thermal annealing is very effective in improving the I-V characteristics, and the n-values approach unity upon annealing at 580°C. In the case of the H-terminated sample, very low n-values are obtained not only for as-deposited diodes but also for annealed ones, although the annealing tends to cause slight degradation of n-values. The degradation of n-values by thermal annealing can be explained by the formation of traps due to interfacial reaction, as mentioned below.

It is also found that the H-terminated treatment improves the reverse I-V characteristics of diodes. Figure 7 shows reverse I-V characteristics of as-deposited Hf/p-Si diodes at measurement temperatures of 110, 220 and 300 K, as a function of reverse bias voltage [8]. The reverse current of as-deposited diodes with H-termination is about 10^{-1}-10^{-2} times smaller that without H-termination at low measurement temperatures. In the case of nonterminated diodes the reverse current is not greatly diminished by thermal annealing. On the other hand, the H-terminated diodes show a large reduction of the reverse current upon annealing. After annealing at 460 and 580°C, the reverse current of H-terminated diodes is less than 10^{-3} of that of nonterminated ones. These facts suggest that the I-V characteristics of nonterminated diodes are governed by defects and that the H-terminated treatment is very effective in forming the metal/Si interface with a very few defects.

Measured DLTS spectra of H-terminated Hf/p-Si(100) diodes are shown in Fig. 8 [8]. For nonterminated diodes, DLTS spectra could not be measured because of very large reverse current. For as-deposited diodes with H-termination, the DLTS peaks of both electron and hole traps can be observed. Unfortunately, the trap energies cannot be determined, because of

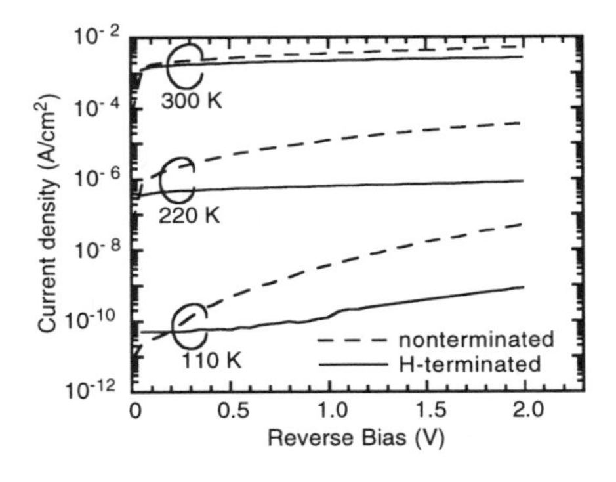

Fig. 7. Reverse current density-voltage characteristics of as-deposited Hf/p-Si diodes formed on nonterminated and H-terminated Si surfaces at measurement temperatures of 300, 220 and 110 K.

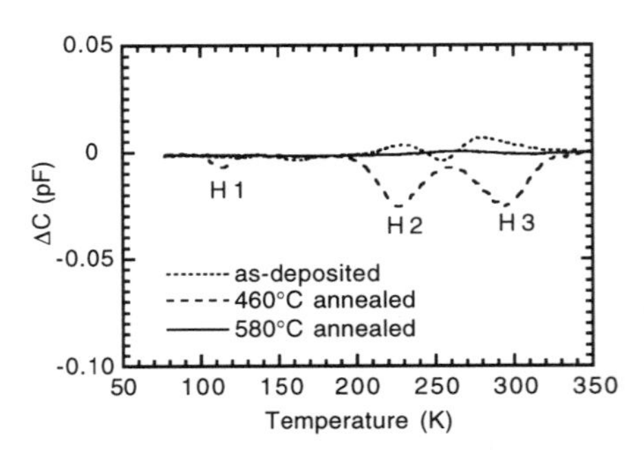

Fig. 8. DLTS spectra for as-deposited, 460°C-annealed and 580°C-annealed Hf/p-Si(100) diodes with H-termination.

an overlap of these peaks. After the annealing at 460°C, three kinds of hole traps denoted by H1, H2 and H3 are produced, the energy levels of which are E_t-E_v=0.14±0.01, 0.43±0.02 and 0.52±0.02 eV, respectively. The origin of H1 is considered to be due to Si vacancies (0.13 eV [28]) and those of H2 and H3 are complexes of B atoms with vacancies or interstitial Si atoms [29], whose average densities are $4x10^{12}$, $1x10^{13}$ and $1x10^{13}$ cm^{-3}, respectively. These vacancy/interstitial-related traps are considered to be formed by the silicidation at 460°C, which indicates that the silicidation reaction at the interface with H-termination takes place at higher temperatures than that without H-termination, as mentioned above. These traps are annealed out and only a small electron trap signal can be observed by annealing at 580°C.

The Schottky barrier heights and the n-values obtained for H-terminated and nonterminated Hf/p-Si(100) diodes are summarized in Table II [8]. The Schottky barrier height from I-V characteristics was determined from the temperature dependence of forward current, and the ideality factor is an average for the measured temperature range. The characteristic features of nonterminated p-Si diodes are that the n-values are much larger than unity and, moreover, that anomalous large barrier heights are obtained from C-V characteristics. A possible explanation for the anomaly observed in nonterminated diodes has been proposed to be the formation of a dielectric-like layer at the interface [3, 11]. The dielectric layer is considered to act as a series capacitance in C-V characteristics. Since the native oxide layer of Si is reduced by Hf and the silicidation takes place even at the as-deposited interface [4], the dielectric-like layer is considered to be formed as a result of the solid-phase reaction.

As seen in Table II, the H-termination not only improves the I-V characteristic but also the C-V characteristics. The effective SiO_2-conversion thicknesses of dielectric-like layers, d_{ox}, are also listed for comparison in Table II. These values are evaluated from C-V measurement using the barrier heights obtained from nonterminated Hf/n-Si [4]. The values of d_{ox} in the H-terminated diodes are 1/4-1/3 times smaller than those in the nonterminated ones. Since the H-termination protects Si surfaces from native oxidation and adsorption of O atoms, the main reason of the anomalous characteristics observed at nonterminated p-Si interfaces is thought to be due to native oxide. However, a discrepancy between the barrier heights obtained from I-V and C-V characteristics in H-terminated diodes is still observed. The

Table II. Schottky barrier heights, $q\phi_{Bp}$, n-values and effective SiO_2-conversion thicknesses of dielectric layers, d_{ox}, of Hf/p-Si diodes formed on nonterminated and H-terminated surfaces.

Annealing temperature	from I-V characteristics				from C-V characteristics			
	H-terminated		nonterminated		H-terminated		nonterminated	
	$q\phi_{Bp}$ (eV)	n-value	$q\phi_{Bp}$ (eV)	n-value	$q\phi_{Bp}$ (eV)	d_{ox} (nm)	$q\phi_{Bp}$ (eV)	d_{ox} (nm)
as-deposited	0.56	1.04	--	1.73	0.69	0.16	--	--
460°C	0.53	1.06	0.50	1.34	0.75	0.26	(1.76)	1.1
580°C	0.51	1.04	0.51	1.12	0.74	0.28	(1.67)	0.95

barrier heights of nonterminated Hf/n-Si obtained from these procedures diodes are identical in an experimental error after annealing [4]. Therefore, this fact suggests that there exists some substantial factor causing the anomaly observed at Hf/p-Si interfaces.

As compared with p-Si, H-termination is not very effective for improving the electrical characteristics of n-Si diodes. Observed temperature-independent current components at low measurement temperatures and the H-termination leads to an increase in the current component, which is a distinctive feature of the H-termination of n-Si surfaces. This current component becomes negligibly small by annealing at 580°C. The H-termination of n-Si surfaces was also confirmed by XPS. The typical temperature dependence of the current at a forward bias of 0.15 V is shown in Fig. 9. It is found that this current component has an activation energy of 0.05-0.1 eV, which suggests that the electrical conduction is governed by the multi-step tunneling through defects existing at the interface [30, 31].

Corresponding to the presence of the large multi-step tunneling component, the n-value is so large that DLTS measurements of as-deposited diodes with H-termination were not successful. Figure 10 shows DLTS spectra of 460°C-annealed Hf/n-Si(100) with and without H-termination. The spectrum of the nonterminated sample is the same as that in Fig. 2. These two spectra are very different from each other. In the case of H-termination, the traps of vacancy-O complexes (E1) cannot be detected, which also suggests that the H-terminated surface is realized by this treatment. There is also observed a large background over the wide temperature range. Small signals from traps are observed at about 120 and 190 K. The origin of the former is divacancies (E2) and the energy levels of the latter cannot be determined. The large background suggests the presence of many kinds of traps with different energy levels, which would be a reason for large multi-step tunneling current components in the H-terminated interface. Similar tendencies are also observed for other metals [18].

The effects of H-termination on contact resistivity are still under investigation. Further

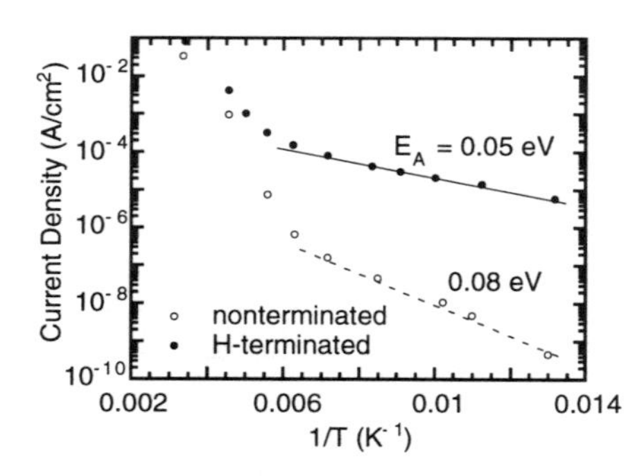

Fig. 9. Temperature dependences of forward current densities of as-deposited H/n-Si(100) with and without H-termination at a bias voltages of 0.15 V.

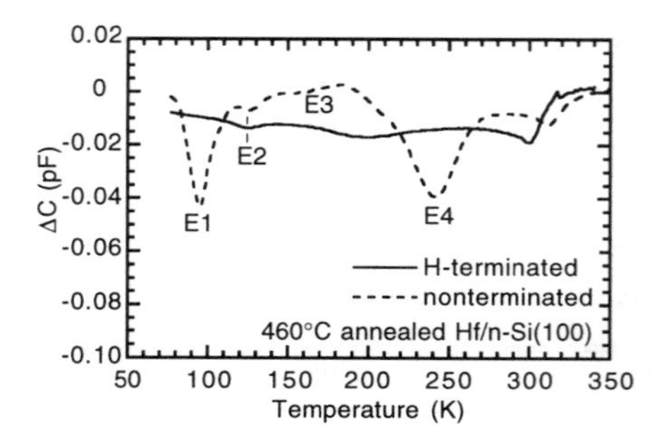

Fig. 10. DLTS spectra of Hf/n-Si(100) with and without H-termination at a annealing temperature of 460°C.

studies are necessary to clarify the H-termination effect on electrical characteristics of transition-metal/n-Si contacts.

Conclusions

We have investigated the defect formation by silicidation reaction and the H-termination effect on electrical characteristics of transition-metal/Si(100) interfaces. It is found that the silicidation reaction at annealing temperatures of 400-600°C yields vacancy/interstitial-related traps at the Hf/Si interface and interstitial metal atom-related traps at V/Si. The silicide formation process is considered to have a close connection with the production of these traps.

The H-terminated Si(100) surface is confirmed to be very stable for the formation of native oxide and the adsorption of oxygen by HREELS and XPS. From I-V and C-V measurements, it can be concluded that the electrical characteristics of Hf/p-Si diodes are markedly improved by the H-termination. The effective dielectric layer thicknesses of H-terminated diodes are 1/4-1/3 times smaller than those in the nonterminated ones, and the n-values of I-V characteristics are close to unity. It is also found that the H-terminated Hf/p-Si interface has a very low density of defects. Therefore, we can conclude that the H-termination of Si surfaces is very effective in controlling and improving the metal/p-Si interfacial properties. On the other hand, the very different results are found for the H-termination of n-Si.

Acknowledgment

This work is partly supported by the Murata Science Foundation.

References

1. R. H. Dennard, F. H. Gaensslen, H. N. Yu, V. L. Rideout, E. Bassous and A. R. LeBlanc, IEEE J. Solid-State Circuits, **SC-9**, 256 (1974).
2. T. Yamauchi, S. Zaima, K. Mizuno, H. Kitamura, Y. Koide, and Y. Yasuda, Appl. Phys. Lett., **57**, 1105 (1990).
3. S. Zaima, T. Yamauchi, Y. Koide, and Y. Yasuda, Appl. Surf. Sci., **70/71**, 624 (1993).
4. S. Zaima, N. Wakai, T. Yamauchi, and Y. Yasuda, J. Appl. Phys., **74**, 6703 (1993).
5. Y. Yasuda and S. Zaima, in Proc. Advanced Metallization for ULSI Applications in 1993, San Diego, edited by D. P. Favreau, Y. S. Diamand and Y. Horiike, pp. 191, Materials Research Society (1994).
6. T. Yamauchi, S. Zaima, K. Mizuno, H. Kitamura, Y. Koide, and Y. Yasuda, J. Appl. Phys., **69**, 7050 (1991).
7. T. Takahagi, I. Nagai, A. Ishitani, H. Kuroda, Y. Nagasawa, J. Appl. Phys., **64**, 3516 (1988).
8. S. Zaima, J. Kojima, M. Hayashi, H. Ikeda, H. Iwano and Y. Yasuda, Jpn. J. Appl. Phys., **34**, 742 (1995).
9. S. J. Proctor and L. W. Linholm, IEEE Electron Device Lett., **EDL-3**, 294 (1982).

10. C. Y. Chang and S. M. Sze, Solid-State Electron., **13**, 727 (1970).
11. T. Yamauchi, S. Zaima, M. Kataoka, Y. Koide, and Y. Yasuda, Appl. Surf. Sci., **56-58**, 545 (1992).
12. S. D. Brotherton and P. Bradley, J. Appl. Phys., **53**, 5720 (1982).
13. S. D. Brotherton, G. J. Parker and A. Gill, J. Appl. Phys., **54**, 5112 (1983).
14. A. Chantre and L. C. Kimerling, Appl. Phys. Lett., **48**, 1000 (1986).
15. S. Zaima, M. Kosaka, S. Tomioka and Y. Yasuda, in Proc. 1st Int. Symp. Control of Semiconductor Interfaces, edited by I. Ohdomari, M. Oshima and A. Hiraki, pp. 57, Elsevier Science B. V. (1994).
16. E. R. Weber, Appl. Phys. A, **30**, 1 (1983).
17. T. Sadoh and H. Nakashima, Appl. Phys. Lett., **58**, 1653 (1991).
18. M. Hayashi, J. Kojima, S. Zaima, H. Ikeda and Y. Yasuda (unpublished).
19. K. N. Tu, and J. W. Mayer, in Thin films - Interdiffusion and Reactions, edited by J. M. Poate, K. N. Tu and J. W. Mayer, pp. 359, John Wiley & Sons (1978).
20. T. Yamauchi, H. Kitamura, N. Wakai, S. Zaima, Y. Koide and Y. Yasuda, J. Vac. Sci. Technol. A, **11**, 2619 (1993).
21. D. Gräf, M. Grundner and R. Schulz, J. Vac. Sci. & Technol. A, **7**, 808 (1989).
22. T. Takahagi, A. Ishitani, H. Kuroda and Y. Nagasawa, J. Appl. Phys., **69**, 803 (1991).
23. J. A. Schaefer, F. Stucki, J. A. Anderson, G. J. Lapeyre and W. Göpel, Surf. Sci., **140**, 207 (1984).
24. J. A. Schaefer and W. Göpel, Surf. Sci., **155**, 535 (1984).
25. J. A. Stroscio, S. R. Bare and W. Ho, Surf. Sci., **154**, 35 (1985).
26. M. Nishijima, J. Yoshinobu, H. Tsuda and M. Onchi, Surf. Sci., **192**, 383 (1987).
27. S. M. Sze, Physics of Semiconductor Devices, 2nd ed., Chap. 5 (John Wiley & Sons, New York, 1981).
28. S. K. Bains and P. C. Banbury, J. Phys. C: Solid State Phys., **18**, 1109 (1985).
29. A. Mitic, T. Sato, H. Nishi and H.Hashimoto, Appl. Phys. Lett., **37**, 727 (1980).
30. J. P. Donelly and A. G. Milnes, Proc. IEEE, **113**, 1469 (1966).
31. A. R. Riben and D. L. Feucht, Solid-State Electron., **9**, 1055 (1966).

TIME-DEPENDENT DIELECTRIC BREAKDOWN IN THIN INTRINSIC SiO_2 FILMS

J. S. SUEHLE[†] and P. CHAPARALA[‡]

† Semiconductor Electronics Division, National Institute of Standards and Technology, Gaithersburg, MD 20899

‡ Center for Reliability Engineering, University of Maryland, College Park, MD 20742

ABSTRACT

Time-Dependent Dielectric Breakdown studies were performed on 6.5-, 9-, 15-, 20-, and 22.5-nm thick SiO_2 films over a wide range of stress temperatures and electric fields. Very high temperatures (400 °C) were used to accelerate breakdown so that stress tests could be performed at low electric fields close to those used for device operating conditions. The results indicate that the dependence of TDDB on electric field and temperature is different from that reported in earlier studies. Specifically, the electric-field-acceleration parameter is independent of temperature and the thermal activation energy was determined to be between 0.7 and 0.9 eV for stress fields below 7.0 MV/cm.

Failure distributions of high-quality current-generation oxide films are shown to be of single mode and have dispersions that are not sensitive to stress electric field or temperature, unlike distributions observed for oxides examined in earlier studies. These results have implications on the choice of the correct physical model to describe TDDB in thin films. The data also demonstrate for the first time the reliability of silicon dioxide films at very high temperatures.

INTRODUCTION

The physical process of Time-Dependent Dielectric Breakdown, TDDB, in thin dielectric films has been the subject of controversy for nearly two decades and is still not completely understood. Earlier studies have reported conflicting electric field and temperature dependencies for this process, and there is no general agreement on the exact dependencies. Characterization and modeling of TDDB involves accelerating dielectric failure of test samples using high electric fields and temperatures. These results are used to predict life at-use electric fields and temperatures via an acceleration model. This model must be carefully chosen because lifetime at-use conditions are usually obtained by extrapolating orders of magnitude in time from lifetime measured at accelerated stress test conditions.

Two popular field acceleration models have been developed to predict dielectric lifetime from accelerated testing conditions. One predicts a linear dependence of $\log(t_{50})$ with electric field, while the other predicts a reciprocal dependence. The linear field model was observed by Crook [1], Berman [2], and by McPherson and Baglee [3]. McPherson and Baglee assumed that TDDB was a thermodynamic process and can be described by an Eyring model. They observed that the slope of the $\log(t_{50})$ versus electric-field characteristic or field-acceleration parameter was a function of temperature, and the thermal activation energy, a function of electric field.

Mat. Res. Soc. Symp. Proc. Vol. 386 © 1995 Materials Research Society

The reciprocal-field model developed by Chen [4], Lee [5], and Moazzami [6] is based on Fowler-Nordheim electron injection into the oxide, and it is assumed that holes are injected or produced via impact ionization. These holes are trapped and alter the cathode electric field, leading to enhanced electron injection, hole generation, and eventual dielectric breakdown. The field-acceleration parameter is also temperature dependent, and the thermal activation energy is assumed to be dependent on electric field as in the case of the linear field model.

The results presented provide evidence that current-generation thin, intrinsic SiO_2 films have similar and predictable electric field and temperature dependencies for TDDB stress conditions. Results from 6.5-to 22.5-nm-thick oxides show that the $\log(t_{50})$ is linearly dependent on electric-field, with a field-acceleration parameter approximately equal to 1.0 decade/MV/cm. The field acceleration parameter is not sensitive to temperature. All of the films exhibited a thermal activation energy of approximately 0.7 to 0.9 eV at stress electric fields at or below 9 MV/cm.

A description of the constant field stress test methodology is provided in Section II. Time-to-failure data obtained over a wide range of stress temperatures and electric fields are presented and discussed in Section III. Conclusions are given in Section IV.

EXPERIMENTAL

Wafer-level TDDB data were collected from polysilicon gate test capacitors fabricated by several different technologies. The capacitors were fabricated on a *p*-type substrate and have an area of 0.001 cm^2. Oxide thickness ranged from 6.5 to 22.5 nm. The effect of positive and negative gate bias was also studied. An n^+ guard ring surrounded the devices to facilitate inversion testing.

High-temperature stressing was implemented by using a specially designed wafer-level probe station capable of performing tests at temperatures as high as 400 °C [7, 8]. This station features a water-cooled probe card that utilizes a water jacket attached to a standard 10-cm (4-in.) probe card.

The test devices are connected through a current-limiting series resistor to a constant voltage source. The voltage is monitored by scanning each device sequentially with a computer-controlled relay matrix system. The test system consists of equipment connected to a microcomputer via an IEEE interface bus. The system is capable of detecting a failure event in 100 ms. Stress temperatures varied from 60 to 400 °C, and electric fields were in the range from 5.5 to 12 MV/cm.

RESULTS AND DISCUSSION

Figure 1 shows Fowler-Nordheim tunneling characteristics measured over the temperature range from 23 to 400 °C. Even at a temperature of 400 °C, the tunneling current is still qualitatively described by Fowler-Nordheim indicating that the high temperatures used to accelerate breakdown did not induce a failure mode that was significantly different than that which occurs at lower temperatures.

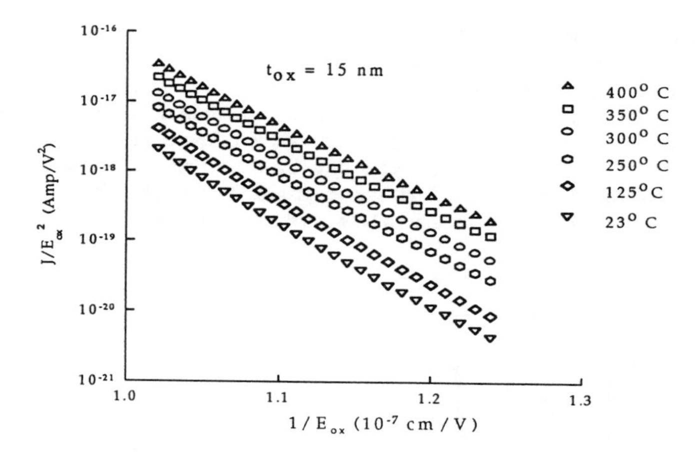

Figure 1 Fowler-Nordheim tunneling characteristics measured on 15-nm oxide devices at temperatures between 23 and 400 °C.

All of the failure distributions examined are best described by log-normal statistics and are of single mode with no apparent extrinsic population. Figures 2 and 3 show typical failure distributions obtained from the 15-nm oxide samples. The distributions shown in Figure 2 are for a stress electric field of 10 MV/cm at 125, 200, 350 and 400 °C, and Figure 3 shows distributions for stress electric fields from 5.5 to 10 MV/cm at 400 °C. Unlike failure distributions for oxides reported in earlier studies [3], the slopes (commonly referred to as sigma) of the distributions do not vary significantly as a function of temperature or electric field.

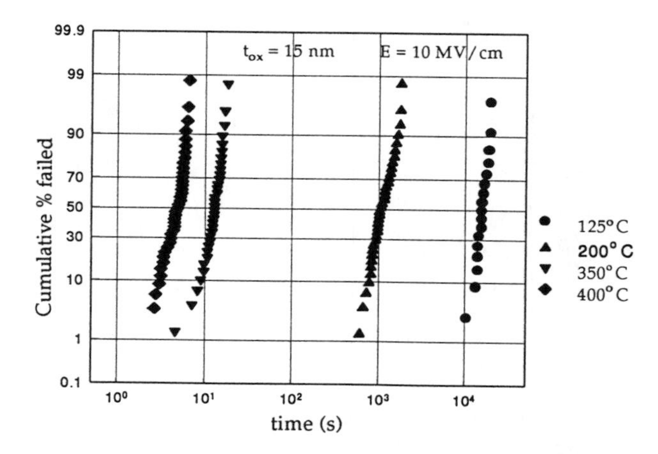

Figure 2 TDDB failure distributions obtained at a stress electric field of 10 MV/cm. Distributions are best described by log-normal statistics and are shown for 125, 200, 350, and 400 °C.

Electric field dependence of TDDB was examined by plotting $\log(t_{50})$ plotted against reciprocal electric field (left plot of Figure 4) and linear electric field (right plot of Figure 4) for the 6.5-, 9-, 15-, and 22.5-nm oxides. All of the data were collected at a stress temperature of 400 °C. The data shown on the left plot should fall on a straight line if the $\log(t_{50})$ values fit the reciprocal field model. This is not evident for data collected at stress electric fields below 8 MV/cm. The right plot of Figure 4 shows that all of the data exhibit a better fit to the linear field model. Note that a good fit can result from either model for electric fields greater than 8 MV/cm. Therefore, it is necessary to obtain data at lower electric fields to distinguish between the two models.

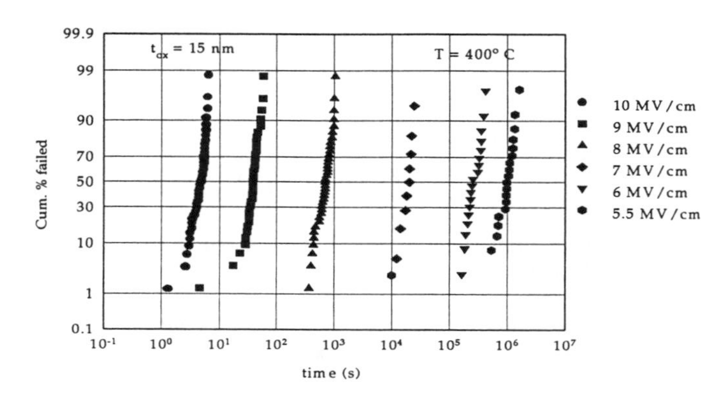

Figure 3 TDDB failure distributions obtained at a stress temperature of 400 °C. Distributions are shown for 5.5, 6, 7, 8, 9, and 10 MV/cm.

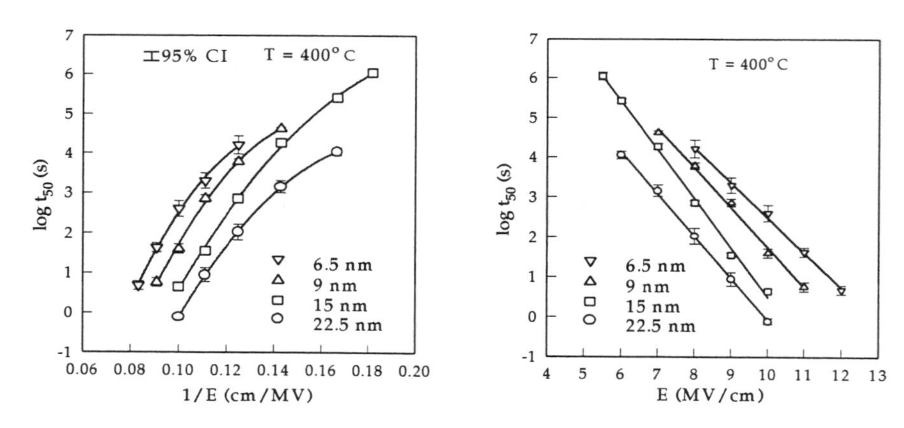

Figure 4 $\log(t_{50})$ plotted as a function of reciprocal electric field (left plot) and linear electric field (right plot) for 6.5-, 9-, 15-, and 22.5-nm oxides. Note that results obtained at fields below 7 MV/cm deviate away from the reciprocal field dependence and are well described by the linear field model.

Figure 5 shows the $\log(t_{50})$ plotted against linear electric field for the 15-nm oxide over a wide range of stress temperatures. Note that the field acceleration parameter, γ, defined as the slope of the $\log(t_{50})$ vs. E curve does not change with stress temperature in the range from 60 to 400 °C. This is contrary to earlier studies where γ was found to be dependent on temperature. The temperature independence of γ is clearly shown in Figure 6 for 6.5-, 9-, 15-, and 22.5-nm oxides. Note the value of γ is between 0.85 and 1.25, exhibiting a slight trend of lower values for thinner oxides. A similar trend was noted in [9, 10]. The values of γ shown in Figure 6 are in contrast to larger values between 3 to 7 decades/MV/cm reported for earlier generation oxide technologies [1-3].

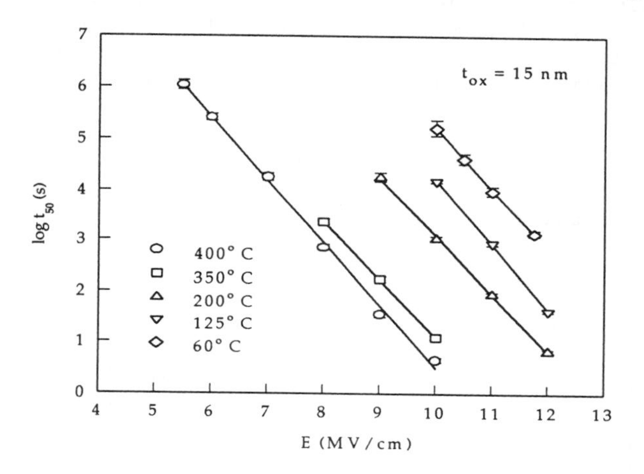

Figure 5 Log(t_{50}) plotted as a function of linear electric field for the 15-nm oxide. The curves are for stress temperatures of 60, 125, 200, 350, and 400 °C.

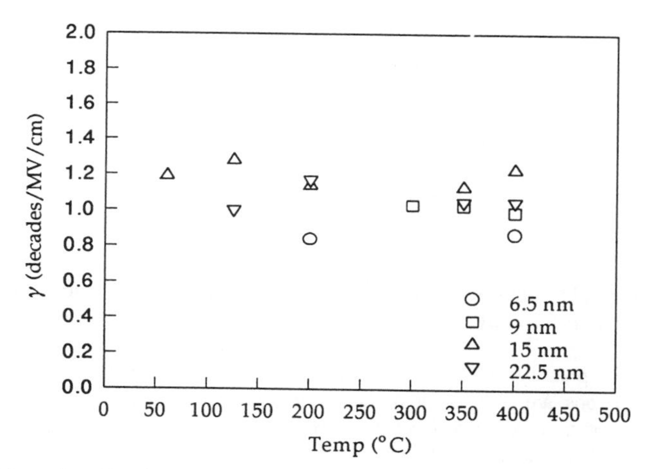

Figure 6 Field acceleration factor, γ, versus temperature for 6.5-, 9-, 15-, and 22.5-nm oxides. Note that γ is insensitive to temperature and exhibits a slight decrease in value for decreasing oxide thickness.

Thermal activation energies were calculated from the Arrhenius plots similar to those shown in Figure 7 for the 15-nm oxide. The plot indicates that the thermal activation energy (defined as the slope of the curves) is dependent on electric field. This dependence has also been observed in earlier studies. However, the 22.5-nm oxide did not exhibit a change in E_a for the range of electric fields from 7 to 10 MV/cm. The dependence of thermal activation energy on electric field is shown in Figure 8 for the 6.5-, 9-, 15-, and 22.5-nm oxides. Note that E_a for the 15-nm oxide has a low value of approximately 0.39 eV at 12 MV/cm and increases to 0.8 eV at 9 MV/cm. The value does not change for lower electric fields. Boyko and Gerlach [11] also observed a saturation of the thermal activation energy value at fields below 6.0 MV/cm for extrinsic oxide failures. It is important to realize that the value of E_a must be determined for the electric field range of interest. Our results suggest that E_a remains constant at electric fields below 7 MV/cm. Therefore, it is only necessary to determine the value at or near this electric field for most actual device applications that are operating at lower electric fields.

Similar observations of a temperature-independent field acceleration parameter and a thermal activation energy that was not dependent on electric field were reported by Shiono and Itsumi [12] for a smaller range of temperatures and electric fields.

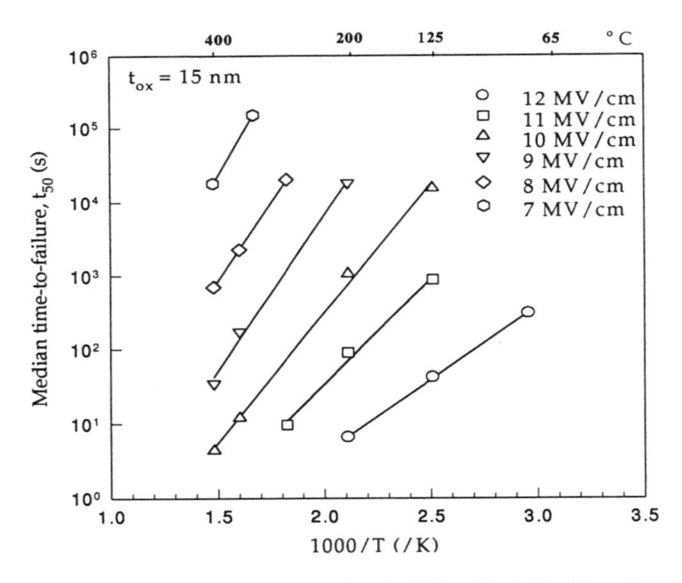

Figure 7 Arrhenius plots obtained at stress electric fields of 8, 9, 10, 11, and 12 MV/cm for the 15-nm oxide. Note that the activation energy exhibits a dependence with electric field.

The effect of bias polarity on TDDB is shown in Figure 9 for 20-nm oxide samples [13]. Under positive bias, electrons are injected into the oxide from the silicon-silicon dioxide interface. Electrons are injected into the oxide from the polysilicon-silicon dioxide interface under negative bias. Differences in dielectric life due to bias are believed to be caused by the efficiency differences of electron injection from each interface. It has been observed that the polysilicon-silicon dioxide interface can be mechanically rougher than the silicon-silicon dioxide interface. Such an interface provides localized regions of high electric field that enhance electron injection. The figure shows that oxide life under positive bias lifetime is significantly longer than that under negative bias. Also note that the slope or the field-acceleration factor is greater for the positive bias case.

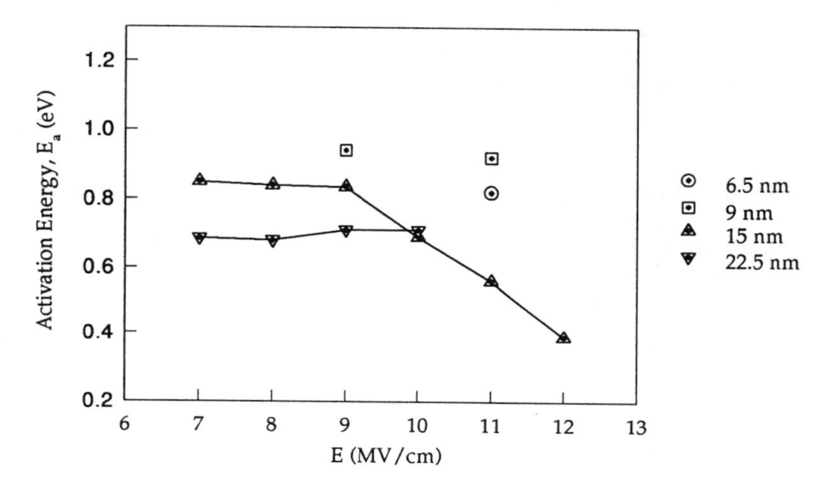

Figure 8 Activation energy plotted as a function of electric field for the 6.5-, 9-, 15-, and 22.5-nm oxides. The activation energies have values between 0.7 and 0.92 eV at electric fields below 9 MV/cm.

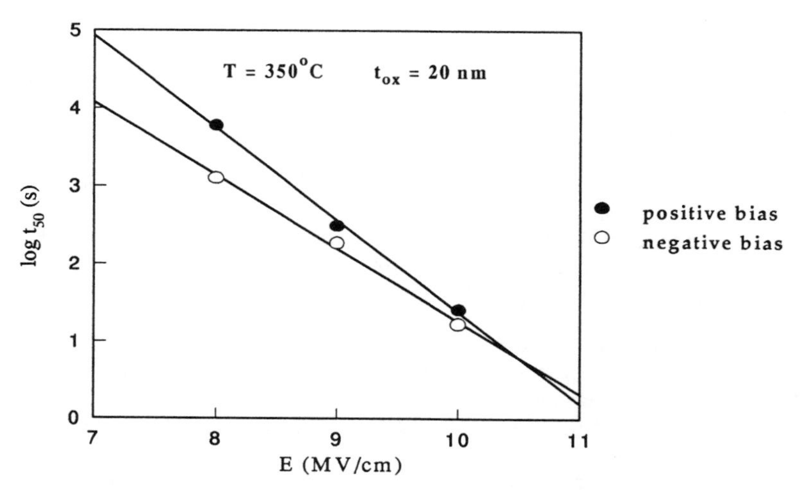

Figure 9 Log(t_{50}) plotted as a function of linear electric field for the 20-nm oxide at a stress temperature of 350 °C. Data obtained from positive and negative gate bias are shown. Note the greater lifetimes obtained under positive bias.

CONCLUSIONS

Time-dependent dielectric breakdown data are presented for 6.5-, 9-, 15-, 20-, and 22.5-nm-thick intrinsic SiO_2 films collected over a wide range of stress temperatures and electric fields. The logarithm of the median-test-time-to failure, $\log(t_{50})$, is described by a linear electric field dependence. This field dependence can be verified only if stress electric fields at or below 7.0 MV/cm are used to acquire the data set. Stress temperatures as high as 400 °C were used to accelerate dielectric breakdown at low electric fields.

Contrary to reports in earlier studies, the field acceleration parameter, γ, is observed to be insensitive to temperature and has a value of approximately 1 decade/MV/cm for the range of oxide thicknesses studied. There is a slight trend of decreasing γ with decreasing oxide thickness.

The thermal activation, E_a, was observed to be a function of electric field for the 15-nm oxide. Its value ranged from 0.39 eV for 12 MV/cm to 0.85 for 7 MV/cm and lower. E_a was independent of electric field for the 22.5-nm oxide. However, for all oxide thicknesses, E_a ranged between 0.7 to 0.95 eV for electric fields below 9 MV/cm. These results provide evidence that current-generation thin, intrinsic SiO_2 films have similar and predictable electric field and temperature dependencies for TDDB stress conditions.

ACKNOWLEDGMENTS

The authors wish to acknowledge Cleston Messick of National Semiconductor for providing the thin SiO_2 samples.

REFERENCES

[1] D. L. Crook, "Method of Determining Reliability Screens for Time Dependent Dielectric Breakdown," Proc. Rel. Phys. Symp., 17, p. 1 (1979).
[2] A. Berman, "Time Zero Dielectric Reliability Test by a Ramp Method," Proc. Rel. Phys. Symp., 19, p. 204 (1981).
[3] J. W. McPherson and D. A. Baglee, "Acceleration Factors for Thin Gate Oxide Stressing," Proc. Rel. Phys. Symp., 23, p. 1 (1985).
[4] I. C. Chen, S. E. Holland, and C. Hu, "Electrical Breakdown in Thin Gate and Tunneling Oxides," IEEE Trans. on Electron Devices, Vol. ED-32, No. 2, p. 413 (1985).
[5] J. C. Lee, I. C. Chen, and C. Hu, "Statistical Modeling of Silicon Dioxide Reliability," Proc. Rel. Phys. Symp., 26, p. 131 (1988).
[6] R. Moazzami, J. C. Lee, and C. Hu, "Temperature Acceleration of Time-Dependent Dielectric Breakdown," IEEE Trans. on Electron Devices, Vol. ED-36, No. 11, p. 2462 (1989).
[7] J. S. Suehle, P. Chaparala, and C. Messick, "High Temperature Reliability of Thin Film SiO_2," Second International High Temperature Electronics Conference, Charlotte, NC, June 5-10, p. VIII-15 (1994).
[8] J. S. Suehle, P. Chaparala, C. Messick, W. M. Miller, and K. C. Boyko, "Field and Temperature Acceleration of Time-Dependent Dielectric Breakdown in Intrinsic Thin SiO_2," Proc. IRPS, No. 32 p. 120, 1994.
[9] K. Yamabe and K. Taniguchi, "Time-Dependent Dielectric Breakdown of Thin Thermally Grown SiO_2 Films," IEEE Trans. on Electron Devices, Vol. ED-32, No. 2, p. 427 (1985).

[10] C. Monserie and P. Mortini, "Breakdown Characteristics of Gate and Tunnel Oxides Versus Field and Temperature," Quality and Reliability International, Vol. 9, John Wiley & Sons, p. 321 (1993).
[11] K. C. Boyko and D. L. Gerlach, "Time Dependent Dielectric Breakdown of 210 A Oxides," Proc. Rel. Phys. Symp., 27, p. 1 (1989).
[12] N. Shiono and M. Itsumi, "A Lifetime Projection Method Using Series Model and Acceleration Factors for TDDB Failures of Thin Gate Oxide," Proc. Rel. Phys. Symp., 31, p. 1 (1993).
[13] J. Prendergast, J. S. Suehle, P. Chaparala, E. Murphy, and M. Stephenson, "TDDB Characterization of Thin SiO_2 Films With BiModal Failure Populations," Proc. IRPS, No. 33, p. (1995).

NEW METHOD TO CHARACTERIZE THIN OXIDE RELIABILITY

HENG-CHIH LIN, J. P. SNYDER, AND C. R. HELMS
Department of Electrical Engineering, Stanford University, Stanford, CA 94305

ABSTRACT

Next generation ULSI devices will require ultra thin gate insulators where degradation due to contamination or surface microroughness is an even more important problem. Tunneling and breakdown characteristics are critical electrical testing methods, but unfortunately obtaining meaningful oxide integrity information on the one hand and tunneling IV's on the other is a tedious and time consuming process.

In this research, we report on a new method to measure meaningful IV's, Q_{bd}'s, and V_{bd}'s at the same time. This method uses a linear current ramp strategy where a voltage ramp to between 8-10 MV/cm is applied first followed by a linear current ramp until breakdown is reached. There are several advantages of this new method: The linear voltage ramp quickly and easily identifies low breakdown devices, whereas switching to a linear current ramp provides for nearly constant field stressing to obtain meaningful IV and Q_{bd}.

INTRODUCTION

The drive toward next generation ULSI devices implies shrinking device dimensions and, thus, requires ultra thin gate insulators where degradation due to contamination or surface microroughness is an even more important problem[1]. Tunneling and breakdown characteristics[2-6] are critical electrical testing methods. For efficient studies of gate oxide quality and reliability, we need a fast oxide breakdown test without sacrificing desired information. Conventional methods use constant voltage [7] or constant current stress[7, 9] to measure Charge-to-Breakdown (Q_{bd}) of the gate dielectric. However, measurement time is a great concern here and useful I-V data cannot be obtained through this measurement. Some authors propose the Exponentially Ramped Current Stress (ERCS) [10] but lose the resolution of intrinsic Q_{bd} due to the dramatically increased current stress in that region.

In this paper we propose the use of a linear current ramp stress method. There are several advantages of this new method: First, the linear ramp rate can be chosen so that it is universal for oxides with certain range of thickness with reasonable measurement time. Second, we first ramp the voltage to obtain 8-10 MV/cm fields and then ramp the current so that both intrinsic and low Q_{bd} can be measured accurately. Third, I-V data is measured automatically. This method has the potential to become a wafer-level oxide evaluation procedure to obtain information both on intrinsic breakdown and on low Q_{bd} devices.

Mat. Res. Soc. Symp. Proc. Vol. 386 © 1995 Materials Research Society

MEASUREMENT METHODOLOGY

In this measurement, a PC is used to automatically control the voltage output of HP-4140B and record measured current information. In order to form the linear current ramp, a program using a feedback loop control was written to sense the measured current and adjust the output voltage level. The initial voltage ramp rate and current ramp rate are given by the user as two parameters. After starting the program, a stepwise increasing linear voltage ramp is applied across the oxide until it reaches a cross over point when the stressing is switched to a linear current ramp until oxide breakdown occurs. The cross over point is defined by the point when the measured current level is equal to the defined current ramp rate times the measured time from the beginning.

Figure 1(a-c) shows voltage vs. time, current vs. time, and current vs. voltage plots of this measurement, respectively. From Figure 1(a, b), we can see that the current level associated with linear voltage ramp is very low in the beginning and grows exponentially thereafter. The advantages of this characteristic are twofold: First, it effectively increases the resolution for early breakdown devices; second, the soft turn on current prevents voltage spikes which could potentially damage the gate oxide. After the cross over point, the linear current ramp provides a compromise between the measurement time and resolution in intrinsic breakdown. Figure 1(c) shows a typical Fowler-Nordheim tunneling characteristic of MOS capacitors. This I-V characteristic is a by-product of this method and can be used as a data base for further analysis.

Generally speaking, Q_{bd} is a function of the chosen voltage ramp and current ramp rates. However, due to the fact that there is only a small amount of charge passing through oxide in the initial voltage ramp, the effect of this voltage ramp rate is negligible. On the other hand, the current ramp rate must be carefully chosen to obtain a reasonable Q_{bd} value. With a good choice of the voltage ramp and current ramp rates, we can greatly reduce the measurement time to a reasonable level while retaining the sensitivity to low and intrinsic breakdown regions. Furthermore, we can use this universal chosen ramp rate to measure oxide with certain range of thickness and get the comparable results. In this case, we choose the linear voltage ramp as 1V/sec and linear current ramp as 0.15 μA/sec $(6 mA / cm^2 - \text{sec})$ to test $50 \mu m \times 50 \mu m$ 138 $\AA$-thick capacitors.

EXPERIMENTS AND DISCUSSIONS

To demonstrate this technique, a simple MOS capacitor was used and fabricated as follows:

(1) The starting material is 4" $<100>$, 20 ohm-cm, p-type CZ wafer. After cleaning using standard Stanford cleaning process, a 2350 $\AA$ field oxide is grown at $1000^{\circ}C$.
(2) Pattern then etch active region using 20:1 BOE.
(3) Split to 5 groups, each group using a different preoxidation cleaning process.
(4) Grow 138 $\AA$ gate oxide.
(5) Deposit, dope, then pattern n^+ poly gate.
(6) Deposit 100% Al on wafer backsize; forming gas annealing 45 min.

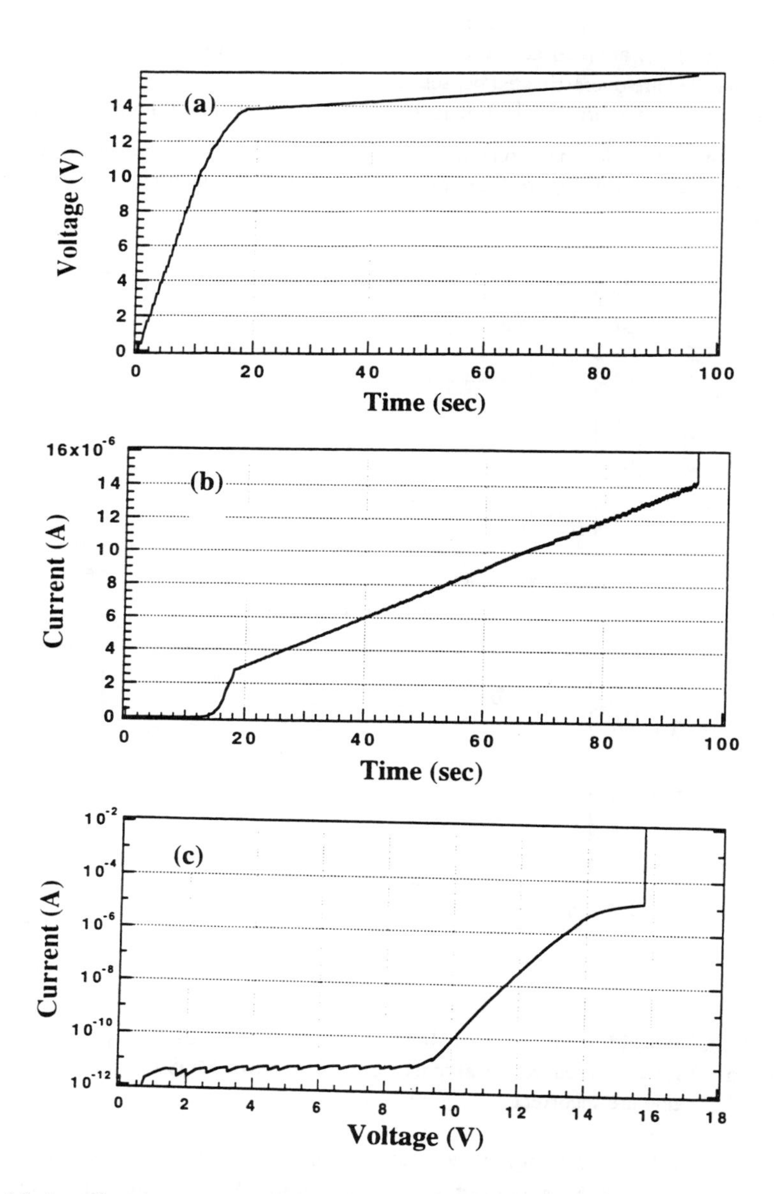

Figure 1: (a) A typical V-t characteristic; (b) A typical I-t characteristic; (c) A typical I-V characteristic of MOS capacitor in our measurement. Q_{bd} can be obtained by calculating the area under current curve in (b).

Figure 2 shows the typical IV curves for early breakdown MOS capacitors. In Figure 2(a), correct Q_{bd} and V_{bd} are recorded according to our program. However, in Figure 2(b), the Q_{bd} and V_{bd} recorded are overestimated because this capacitor has already been damaged before our program "detects" the breakdown point, which is defined as the point when current increase rapidly to over $10^{-2}\,\mathring{A}$. Inspecting the IV curve is necessary otherwise we conclude that capacitor associated with Figure 2(a) has higher Q_{bd} and V_{bd} than capacitor associated with Figure 2(b).

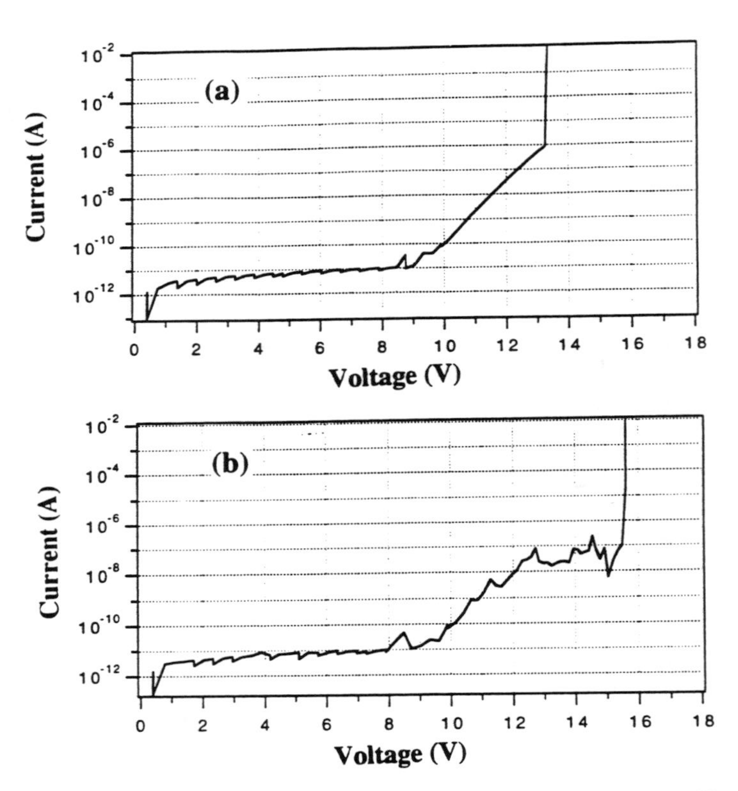

Figure 2: (a). Typical I-V curve of early breakdown MOS capacitor. The Q_{bd} and V_{bd} value are current; (b). Typical I-V curve for another early breakdown MOS capacitor. Notice that the Q_{bd} and V_{bd} value are incorrect.

This can further been seen if we look at plots of Q_{bd} vs. V_{bd} as shown in Figure 3. Fig. 3(a). shows the Q_{bd} vs. E_{bd} plot prior to inspection of the IV curves of each sample. The Q_{bd} and E_{bd} are almost non-correlated, which is inconsistent with our expectation. After carefully examining the IV curve of each capacitor and removing data similar to that of Fig. 2(b), the replot shown in Figure 3(b) shows a positive correlation between Q_{bd} and E_{bd}, as expected.

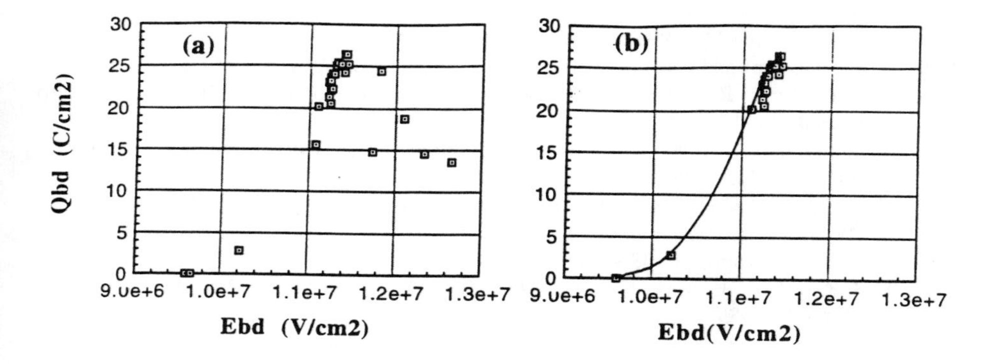

Figure 3: (a). Q_{bd} vs. E_{bd} including questionable IV's. Notice that there is little correlation between Q_{bd} and E_{bd}; (b). Replot of Q_{bd} vs. E_{bd} after removing questionable data, a positive correction is shown between Q_{bd} and E_{bd}.

SUMMARY

A linear current ramp stress method is proposed to investigate gate oxide integrity. It has several advantages over other methods:

1. It is a soft-turn on, time saving test without sacrificing the resolution for low and intrinsic breakdown regions.
2. Fowler-Nordheim characteristics are obtained from the measurement and can be saved for further investigation.

This method therefore is a potential tool for economical wafer-level screening and will be used in the context of our overall program to investigate the effect of microroughness and organic contamination on HF-last cleaned gate oxides.

ACKNOWLEDGMENTS

This research is supported by the SRC under contract SJ-350.

REFERENCE

1. J. D. Meindl, V.K. De and B. Agrawal, Proc. IEEE Intern. Solid-State Circuits Conf. - ISSCC '93, pp.124 (1993).
2. K.F. Schuegraf and Chenming Hu, Proc. of IEEE VLSI Tech. Symp., pp. 43 (1993).
3. D.J. Dumin, J.R. Maddux, R.S. Scott and R. Subramoniam, IEEE Trans. Electron Devices **41**(9), 1570 (1994).
4. R. Subramoniam, R.S. Scott, and D.J. Dumin, Proc. IEEE Intern. Electron Devices Meeting, pp. 135 (1992).
5. P. P. Apte and K.C. Saraswat, IEEE Electron Device Letters **14**(11), 512 (1993).
6. W. W. Abadeer, R. P. Vollertsen, R. J. Bolam, D. J. DiMaria and E. Cartier, Proc. 1994 VLSI Tech. Symp., pp. 43 (1994).
7. D.B. Kao, J.M. deLarios, C.R. Helms nad B.E. Deal, 27th Annual Proc. Reliability Phys. 1989, pp. 9 (1989).
8. P.P. Apte, T. Kubota and K.C. Saraswat, J. Electrochem. Soc. **140**(5), 770 (1993).
9. T. Ohmi, K. Nakamura and K. Makihara, Proc. of 1994 IEEE Intern. Reliability Phy. Symp., pp. 161 (1994).
10. P. Cappelletti, P. Ghezzi, F. Pio, and C. Riva, Proc. IEEE 1991 Int. Conf. on Microelectronic Test Structure, **4**(1), pp.81,(1991).

CONTROLLED NITROGEN-ATOM INCORPORATION AT Si - SiO_2 INTERFACES IN MIS DEVICES

DAVID R. LEE, CHRISTOPHER G. PARKER, JOHN R. HAUSER AND GERALD LUCOVSKY

Departments of Materials Science and Engineering, Physics, and Electrical and Computer Engineering
North Carolina State University
Raleigh, NC 27695-8202

ABSTRACT

We have developed a new pre-deposition, remote N_2O plasma oxidation treatment for forming nitrided SiO_2 films and report here the quality and reliability of devices fabricated with these films. The Si-dielectric heterostructure process has been separated into three independently-controlled steps: i) final Si surface cleaning, and Si - SiO_2 interface formation by plasma-assisted oxidation/nitridation at 300 °C; ii) remote plasma-enhanced chemical vapor deposition of nitrided dielectrics, also at 300 °C; and iii) optional post-deposition rapid thermal annealing. This paper focuses on the first step in which the oxidation has been performed with O_2, N_2O or N_2O / O_2 mixtures to control the amount of N-atoms at the interface, N_{int}. We show that the incorporation of up to ~ 10^{15} N-atoms/cm^2 at the Si - SiO_2 interface by this process has no effect on threshold voltage, V_t, or peak channel transconductance, $g_{m,max}$, but does improve high-field g_m and transistor drive current. Improved resistance to V_t and $g_{m,max}$ degradation during hot-carrier stressing of sub-micron devices is also discussed.

INTRODUCTION

Nitrogen-atom incorporation in SiO_2 films for gate dielectrics in MIS devices has been suggested as a way of maintaining acceptable device reliability as gate lengths are pushed below 0.35 μm with corresponding oxide thickness < 7 nm [1]. N-atoms, particularly when they are located near the Si - SiO_2 interface, are reported to result in improved resistance to trap-generation by hot-carriers. The different nitridation processes used and the difficulty in accurately profiling the N-atom concentration in the oxide film has made it difficult to develop any consensus as to the optimum N-atom concentration and its distribution through the oxide film.

We have previously reported the development of a new process for incorporating N-atoms at Si - SiO_2 interfaces at low temperatures, 300 °C, using a novel pre-deposition, remote N_2O plasma oxidation/nitridation technique [2]. This process, performed *in situ* as a pre-deposition surface treatment prior to SiO_2 deposition by Remote Plasma-Enhanced Chemical Vapor Deposition (Remote PECVD), was found to serve three purposes: 1) reduction of residual C contamination into the low 10^{12} cm^{-2} regime, 2) plasma-assisted growth of ~ 0.5 nm of SiO_2 passivating the Si surface, and 3) incorporation of up to ~ 1 x 10^{15} N-atoms/cm^2 near the Si - SiO_2 interface. This "interface engineering" approach is a continuation of the technique proposed by Yasuda *et al.* for separately controlling interface and bulk film formation in deposited gate oxides [3].

In this paper, we report the transfer of this pre-deposition N_2O oxidation/nitridation technique to a cluster-tool processing test bed at the National Science Foundation Engineering Research Center for Advanced Electronic Materials Processing at N.C. State University. Using the Remote Plasma Processing module of this cluster tool for the gate oxide formation and the

Mat. Res. Soc. Symp. Proc. Vol. 386 © 1995 Materials Research Society

Rapid Thermal Processing (RTP) module for Rapid Thermal Chemical Vapor Deposition (RTCVD) of poly-Si gate electrodes, n-channel field-effect transistors and capacitors were fabricated. In this paper, we discuss how interfacial N-atom concentrations affect the electrical quality and reliability of these devices.

EXPERIMENTAL METHODS

Four-inch, p-type (B), Si (100) wafers with resistivities of 0.05 - 0.10 Ω·cm ($N_A = 4x10^{17}$ cm^{-3}) were used for the fabrication of Metal-Oxide-Silicon (MOS) Field-Effect Transistors (FET's) and capacitors. A four-mask process was used that included thermally-grown field oxide; $POCl_3$ diffusion doping (900 °C) of the gate, source, and drain; wet chemical etching of poly-Si and SiO_2, and liftoff Al/Ti contact metallization all performed by standard methods. After field oxidation and patterning, the gate area was prepared in the following manner: 1) removal of all but 10-15 nm of the field SiO_2 by etching in 10:1 dilute HF, 2) removal of photoresist and standard RCA cleaning, 3) removal of remaining SiO_2 in gate area by etching in 0.5% dilute HF (etch rate ~ 1.2 nm/min), and 4) 5 min rinse in deionized H_2O. After this ex-situ preparation, the wafers were loaded into the load lock of the cluster tool.

The gate oxides were formed by remote plasma processing. All remote plasma processing steps discussed here were performed at 300 °C wafer temperature, 300 mTorr chamber pressure, and 400 W rf (13.56 MHz) plasma power. Mixtures of He (300 sccm), N_2O (30 sccm), and O_2 (5.2 - 15.0 sccm) were directly excited in a glow discharge in a region approximately 28 cm upstream from the heated Si wafer. The O_2 flow rate was adjusted to control the N-atom concentration at the Si-SiO_2 interface, N_{int} [4]. Immediately following the pre-deposition surface treatment, the bulk SiO_2 film was deposited by Remote PECVD using He (300 sccm) and O_2 (15 sccm) upstream in the plasma tube and 10% SiH_4 in Ar (10 sccm) injected through a gas dispersal ring located 2 cm above the Si wafer. Gate oxide thicknesses were approximately 5.5 nm. Following gate oxide deposition, the wafer was moved to the RTP module where 120 nm of undoped poly-Si was deposited by RTCVD at 685 °C and 5 Torr with 300 sccm of 10% SiH_4 in Ar flowing into the chamber. Total processing time for the gate oxide and poly-Si electrode film was just over 2 min (not including wafer transport).

Electrical measurements included high-frequency (HF, 1 MHz) capacitance-voltage (C-V) tests on MOS capacitors (100 x 100 μm) from which oxide capacitance (C_{ox}), equivalent oxide thickness (T_{ox}), effective substrate doping (N_{sub}), and flatband voltage (V_{FB}) were extracted using the common techniques [5]. The drain current versus drain voltage (I_d - V_{ds}) characteristics were measured at different gate voltages (V_{gs}) with the source and substrate grounded. The I_d V_{gs} characteristics in the linear regime (V_{ds} = 50 mV) were also measured and the channel transconductance calculated from derivative of this data. Hot-carrier degradation of short-channel devices was evaluated using devices with a nominal gate length and width of 1.4 and 10 μm, respectively. Control over the gate length was inexact due to single-wafer wet etching of the poly-Si gate electrode, so the effective channel length, L_{eff}, was estimated by measuring the saturation drain current for devices with different nominal gate lengths and extrapolating a plot of $1/I_{d,sat}$ vs L to find the average channel length offset for each sample. All of the devices discussed in this paper had an $L_{eff} \approx 0.7$ μm. The I-V properties of the devices were measured as described above both before and after hot-carrier stressing at 10000 s. For hot-carrier stressing, the source and drain were inverted relative to the measurement condition in order to maximize sensitivity to damage created by hot-carriers. V_{ds} = 5 V. The substrate current, I_{sub}, was measured as a function of V_{gs} for each wafer, and the gate bias corresponding to the peak I_{sub} condition was used as the stress condition since this corresponds approximately to the maximum dc stress condition [6].

RESULTS AND DISCUSSIONS

We have previously established that pre-deposition, remote-plasma oxidation in N_2O resulted in N-atom incorporation at $Si-SiO_2$ interfaces and that the N-atom concentration could be varied by using mixtures of N_2O and O_2 for this oxidation step [4]. Figure 1 includes data that was discussed in Ref. 4 and is included here as a representation of the N_{int} values that are achievable using this technique. These N_{int} values are areal densities obtained from Secondary Ion Mass Spectrometry depth profiles of $Si-SiO_2$ heterostructures. Also in Fig. 1 are plotted data from Green *et al.* [7] who have reported the N_{int} values for nitrided oxide films grown by Rapid Thermal Oxidation (RTO) in N_2O. In the RTO process, temperature was varied between 800-1200 °C in order to control N_{int}. The remote plasma process we discuss in this paper achieves the same range of N_{int} at a wafer temperature of 300 °C by simply varying the relative amounts of N_2O and O_2 supplied during the plasma-assisted oxidation step. In addition to the obvious benefit in terms of the thermal budget of the process, the remote plasma process we describe allows greater process latitude since in an RTO process like the one described in Ref. 7, one has to balance processing temperature and time with both the desired N_{int} and oxide thickness.

The I_d - V_{ds} characteristics of 100 x 100 μm n-MOSFET's with gate dielectrics formed by different remote plasma techniques are shown in Figure 2. For reference, data from an n-MOSFET with a conventional thermal oxide (5.5 nm grown at 1000 °C) is also shown. The devices fabricated with the remote plasma processed gate oxides show good turn-on characteristics and current saturation for both the pure SiO_2 and the nitrided SiO_2 gate dielectrics. The drive current of the device fabricated with the N_2O pre-deposition oxidation was more than 9% higher at $V_{gs} = 4$ V than the drive current of the device fabricated with the O_2 pre-deposition oxidation. The I_d - V_{ds} characteristics of the devices made with mixed N_2O/O_2 pre-deposition oxidation are not shown in Fig. 2, but the drive current increased with the increasing percentage of N_2O used in the pre-deposition oxidation step.

The I_d - V_{gs} characteristics at $V_{ds} = 0.05$ V were also measured, but are not shown here. The device threshold voltage was measured by extracting the linear portion of the I-V curve back to the V_{gs} - axis. The threshold voltages for the 0% (O_2 only), 75%, 85%, and 100% N_2O oxidations were all 0.39 ± 0.02 V and showed no dependence on processing conditions. The channel transconductance, g_m, calculated from the first derivative of the I_d - V_{gs} curves is shown in Fig. 3 for the same devices as in Fig. 2. Peak $g_m = 18$ cm²/s for the pure and nitrided plasma oxides and for the thermally grown oxide. At high effective transverse (perpendicular to direction of channel current flow) fields, however, the g_m of devices with the nitrided gate dielectric were higher than for the devices with the pure plasma oxides. Higher g_m at high transverse fields can be due to smoother $Si-SiO_2$ interfaces. It may also be explained by the reduction of strain, either mechanical or chemical or both, at the $Si-SiO_2$ interface. Chemical strain is reduced in the presence of N-atoms since because they are less electronegative than O-atoms, they withdraw less charge from adjacent Si-Si bonds. In the case of the thermal oxide, mechanical strain was reduced due to the high growth temperature (1000 °C).

Changes in the threshold voltage, ΔV_t, and peak channel transconductance, $\Delta g_{m,max}$, after 10^4 s hot-carrier stressing are shown in Fig. 4 as a function of the percentage of N_2O in the pre-deposition, remote plasma-assisted oxidation. Devices fabricated with the new N_2O process exhibited smaller ΔV_t and $\Delta g_{m,max}$ than the devices fabricated with conventional thermal oxides. The reduction in ΔV_t and $\Delta g_{m,max}$ was a function of the percentage of N_2O used in the pre-deposition process indicating that the incorporation of N-atoms was the mechanism which caused this improvement in device reliability. Reductions in $g_{m,max}$ are attributed to the

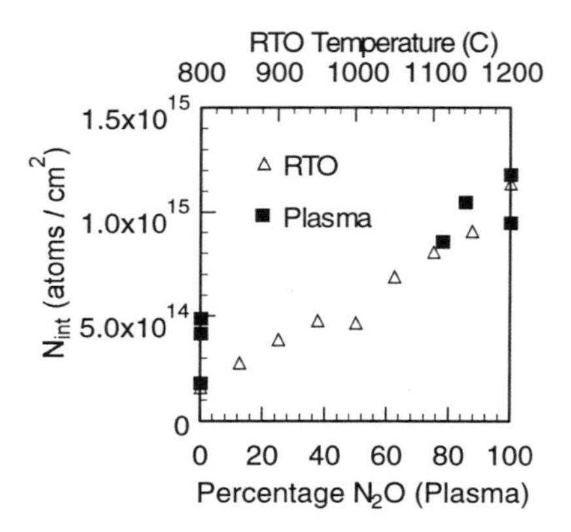

Figure 1 N-atom concentrations, N_{int}, at Si-SiO_2 interfaces formed by remote plasma oxidation at 300 °C in mixtures of N_2O and O_2 (bottom axis) [4] and for SiO_2 films grown in N_2O by Rapid Thermal Oxidation at different temperatures (top axis) [7].

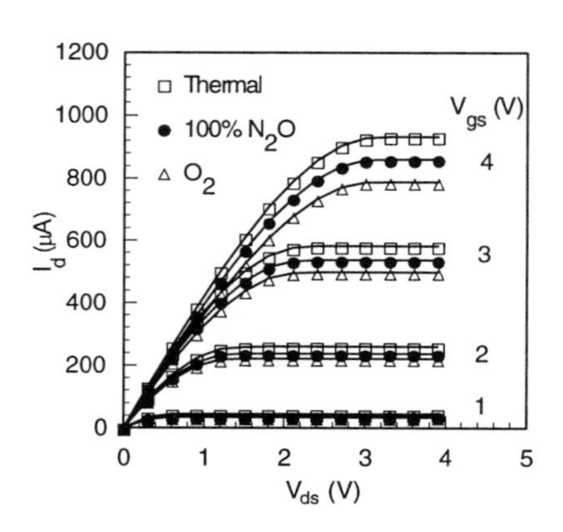

Figure 2 I_d-V_{ds} characteristics for n-MOSFET's with different gate dielectrics. Device dimensions were 100 x 100 μm with 5.5 nm thick gate dielectrics. Measurements were made with the source and substrate at ground.

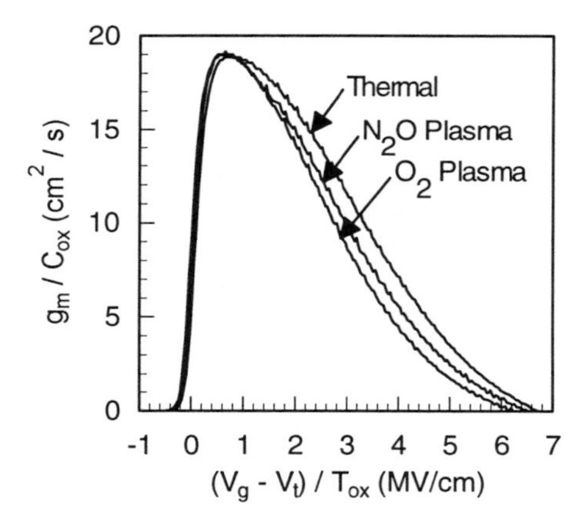

Figure 3 Channel transconductance, g_m, of devices with different gate dielectrics as a function of effective transverse field. Measurements were made on 100x100 μm devices with $T_{ox} \approx 5.5$ nm and $V_{ds} = 50$ mV. Each curve represents an average of data from eight devices.

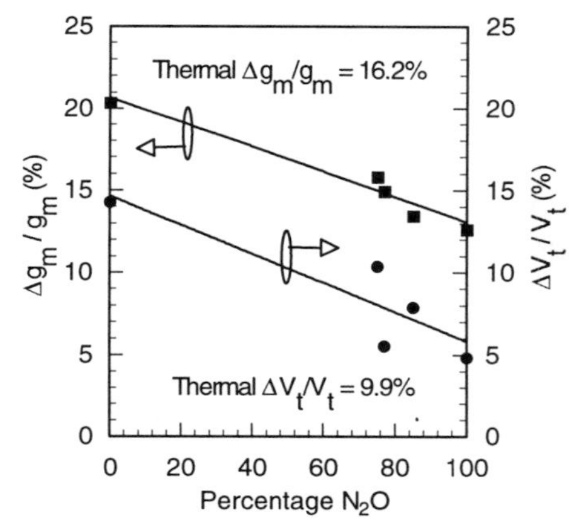

Figure 4 Degradation in $g_{m,max}$ and V_t after hot carrier stressing for 10^4 s for devices fabricated with a remote plasma, pre-deposition oxidation in mixtures of N_2O and O_2. Data for a thermal oxide of the same thickness is shown for reference.

generation of traps at the $Si-SiO_2$ interface. As discussed above, N-atom incorporation at the $Si-SiO_2$ interface reduces chemical strain there; and as a result, strengthens adjacent Si-Si bonds as compared to when O-atoms are bonded at the interface. Therefore, the generation of interface traps under hot-carrier stress is suppressed when N-atoms are bonded at the interface.

CONCLUSIONS

We have demonstrated that a new remote plasma technique for forming thin nitrided gate oxides at 300 °C can be used for the fabrication of n-MOSFET's with initial device qualities (V_t and g_m, but not drive current) similar to and device reliability (ΔV_t and $\Delta g_{m,max,}$) better than when conventional, thermally-grown gate oxides are used. We have shown that the incorporation of up to ~ 10^{15} N-atoms/cm^2 at the $Si-SiO_2$ interface by this process has no effect on V_t or $g_{m,max}$ but does improve high-field g_m and transistor drive current as compared to plasma oxides formed with an O_2 pre-deposition oxidation. Devices fabricated with the nitrided gate oxides were more resistant to degradation of V_t and $g_{m,max}$ during hot-carrier stressing. Improvements in device qualities and reliability were approximately proportional to the expected N_{int}. A model was proposed that explains these improved device properties in terms of a reduction in chemical strain at the $Si-SiO_2$ interface when N-atoms are bonded there rather than O-atoms.

The important result of this work is the contribution of a new, low-temperature alternative process for incorporating N-atoms at $Si-SiO_2$ interfaces in order to improve the quality and reliability of sub-micron MOS devices. Unlike thermally-activated nitridation processes, changes in critical processing parameters such as time and temperature are not required and no special post-processing annealing (other than exposure to subsequent device fabrication steps) was used. This allowed us to change *only* a single source-gas (O_2 replaced with N_2O) in order to realize improved device properties through the use of nitrided gate oxides.

ACKNOWLEDGMENTS

The authors acknowledge the assistance and support of AEMP participants, especially the staff of the microelectronics laboratory who assisted with some of the device processing steps and offered valuable processing advice. Thanks also to Amr Bayoumi for supplying the devices fabricated with thermal oxides. This work was supported by the NSF Engineering Research Centers Program through the Center for Advanced Electronic Materials Processing (Grant CDR 8721505) and the SRC / SCOE program at NCSU (SRC contract #94-MC-509).

REFERENCES

1. T. Ito, T. Nakamura, and H. Ishikawa, IEEE Trans. Electron Devices **29**, 498 (1982).
2. D.R. Lee, G. Lucovsky, M. Denker, and C. Magee, J. Vac. Sci. Technol. A, (1995), in press.
3. T. Yasuda, Y. Ma, S. Habermehl, and G. Lucovsky, Appl. Phys. Lett. **60**, 434 (1992).
4. D.R. Lee, C.G. Parker, J.R. Hauser, and G. Lucovsky, J. Vac. Sci. Technol. B, (1995), in press.

5. E.H. Nicollian and J.R. Brews, *MOS Physics and Technology*, (John Wiley and Sons, New York, 1982) pp. 71-174.
6. C. Hu, S.C. Tam, F.-C. Hsu, P.-K. Ko, T.-Y. Chan, and K.W. Terrill, IEEE Trans. Electron Devices **32**, 375 (1985).
7. M. Green, D. Brasen, K.W. Evans-Lutterodt, L.C. Feldman, K. Krisch, W. Lennard, H.-T. Tang, L. Manchanda, and M.-T. Tang, Appl. Phys. Lett. **65**, 848 (1994).

VALENCE BAND ALIGNMENT OF ULTRA-THIN SiO_2/Si INTERFACES AS DETERMINED BY HIGH RESOLUTION X-RAY PHOTOELECTRON SPECTROSCOPY

J.L. ALAY*, M. FUKUDA**, C.H. BJORKMAN*, K. NAKAGAWA**, S. SASAKI*[1], S. YOKOYAMA*, AND M. HIROSE**

*Research Center for Integrated Systems, Hiroshima University, Higashi-Hiroshima 724, Japan

**Department of Electrical Engineering, Hiroshima University, Higashi-Hiroshima 724, Japan

ABSTRACT

Ultra-thin SiO_2/Si(111) interfaces have been studied by high resolution x-ray photoelectron spectroscopy. The deconvolution of the Si 2p core-level peak reveals the presence of the suboxide states Si^{3+} and Si^{1+} and the nearly complete absence of Si^{2+}. The energy shifts found in the Si 2p and O 1s core-level peaks arising from charging effects are carefully corrected. The valence band density of states for ultra-thin (1.8 - 3.7 nm thick) SiO_2 is obtained by subtracting the bulk Si contribution from the measured spectrum and by taking into account the charging effect of SiO_2 and bulk Si. Thus obtained valence band alignment of ultra-thin SiO_2/Si(111) interfaces is found to be 4.36 ± 0.10 eV regardless of oxide thickness.

INTRODUCTION

Thickness and uniformity control of gate oxides are becoming very important for manufacturing advanced MOS structures. An accurate description of valence and conduction band profiles at the ultra-thin SiO_2/Si interfaces are needed to calculate the tunneling current through oxides thinner than 5.0 nm[1]. Dressendorfer *et al.*[2] reported that the barrier height at the SiO_2/Si interface has a constant value of about 3.0 eV for oxides thicker than 4.0 nm by employing an internal photoemission technique. Horiguchi *et al.*[3] estimated the barrier height for 1.5 to 4.0 nm thick oxides by fitting the experimental results obtained for the Fowler-Nordheim tunneling regime, showing that the barrier height is lower than 1.8 eV for oxides thinner than 3.1 nm and about 3.0 eV for those thicker than 3.6 nm. Also, Heike *et al.*[4] reported a constant barrier height of 2.75 eV for oxide thicknesses 1.8 to 4.5 nm by using an electron beam assisted scanning tunneling microscope technique.

---

[1]Permanent address : Seiko Instruments, 36-1 Takenoshita, Oyama-cho, Sunto-gun, Shizuoka 410-13, Japan.

Mat. Res. Soc. Symp. Proc. Vol. 386

The aim of this paper is to determine the valence band density of states (VBDOS) for ultra-thin (< 4.0 nm) oxides thermally grown on hydrogen-terminated, atomically-flat Si(111) surfaces by high resolution x-ray photoelectron spectroscopy (XPS), and thus the value for the valence band alignment or the conduction band barrier height can be derived.

EXPERIMENTAL

High resolution XPS measurements were performed with an ESCA-300 (Scienta Instruments AB), using monochromatic Al Kα radiation (1486.6 eV). The base pressure during measurements was maintained in the 10^{-10} Torr range. The Si 2p, O 1s, and C 1s core-level peaks were measured at photoelectron take-off angles of θ = 35, 60, and 90 degrees. The valence band spectra were acquired at a take-off angle of 35 degrees in order to enhance the signal from the ultra-thin SiO_2/Si interfaces. The amount of carbon found on the SiO_2 surface was negligible. The background of the various photoelectron peaks was subtracted by using the Shirley method[5], and the Si 2p core-level was deconvoluted into the Si $2p_{3/2}$ and Si $2p_{1/2}$ spin-orbit doublet which is separated by 0.61 eV with an intensity ratio of 0.5 for each of the chemical states such as substrate Si^0, suboxide Si^{1+}, Si^{2+}, Si^{3+}, and oxide Si^{4+} based on the respective energy shifts reported in a previous work[6]. By using the angle-resolved XPS data, the thicknesses of the ultra-thin oxides were calculated from the Si 2p core-level spectra by assuming that the Si 2p escape depths in Si and SiO_2 are 2.7 and 3.4 nm, respectively[7].

Ultra-thin (1.8 - 3.7 nm thick) oxides and a 40 nm thick reference SiO_2 were grown at a temperature of 1000 °C in 2 % dry O_2 diluted with N_2 on atomically flat, hydrogen-terminated p-type Si(111) substrates (10 Ωcm). Si(111) substrates were cleaned with an RCA solution followed by a chemical treatment in a 40 % NH_4F solution[8] which guarantees the atomically -flat, hydrogen-terminated surface. The oxidation time was 0, 1, 4, 7, or 10 min. 0 min stands for the oxidation in which a substrate was loaded into an oxidation zone held at 1000 °C and immediately unloaded to minimise the oxide growth.

RESULTS AND DISCUSSION

The Si 2p spectra measured for ultra-thin oxides indicate the presence of Si^{1+} and Si^{3+} suboxide states together with Si^0 (bulk) and Si^{4+} (oxide) states as illustrated in Fig. 1. The nearly complete absence of Si^{2+} signal in the XPS spectra (deep valley between the $Si^{1+}2p_{1/2}$ and $Si^{3+}2p_{3/2}$ components) could be understood by a layer-by-layer oxidation model which implies atomically flat interfaces mainly composed of Si^{1+} suboxides in the case of Si(111) substrates.

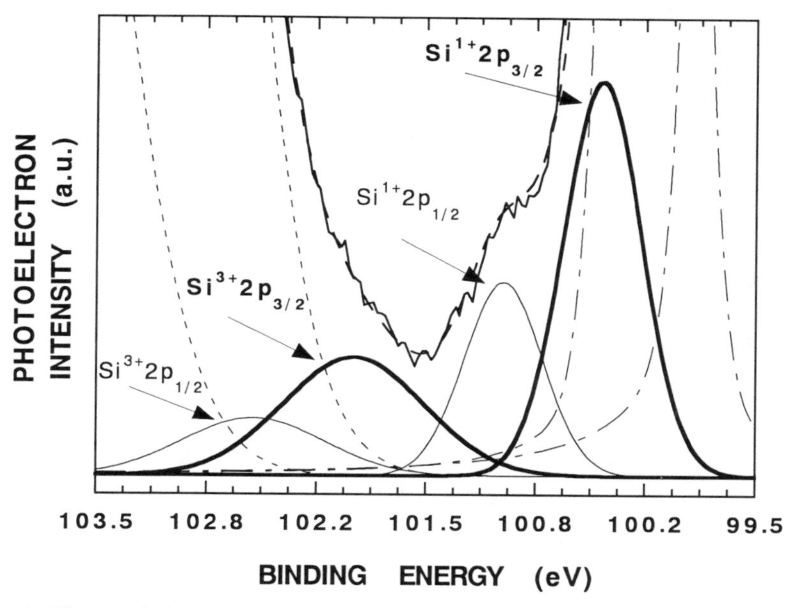

Figure 1 : High resolution XPS spectrum of Si 2p for 1.8 nm SiO_2/Si(111) in the energy range corresponding to suboxide states, Si^{1+} and Si^{3+}.

The oxide thickness, which has been calculated from angle-resolved XPS data, is plotted as a function of oxidation time in Fig. 2. The oxide growth is controlled by a logarithmic law which is typical for the initial oxidation process. It should be noted that the Si $2p_{3/2}$ binding energies for the various chemical states measured at a take-off angle of 35 degrees exhibit shifts towards higher binding energies when increasing the oxide thickness. The observed peak shifts for $Si^{4+}2p_{3/2}$ and O 1s peak are at most 350 and 320 meV, respectively, whereas the $Si^{0}2p3/2$ peak shift is at most 70 meV, a factor of five smaller. This different energy shift is explained by the charging effects for the Si 2p peaks originating from the oxide layer and bulk Si[9].

The VBDOS spectra measured for the various thin oxides are shown in Fig. 3 together with that of the H-terminated Si(111) reference and the 40.0 nm thick SiO_2 reference. The binding energy corresponding to the O 1s or $Si^{4+}2p_{3/2}$ peak for the thinnest (1.8 nm) oxide is taken as the energy reference for all other oxides since charging effect is minimum for this oxide. Therefore, the valence band edge energies have been corrected by shifting each spectrum with the value given by the difference between the respective O 1s or $Si^{4+}2p_{3/2}$ peak position and that of the 1.8 nm oxide to correct the oxide charging contribution.

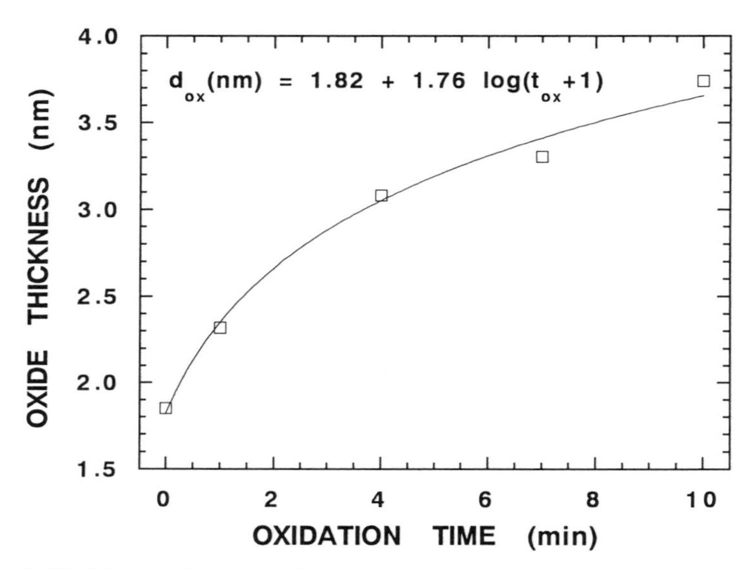

Figure 2 : Oxide thickness plotted as a function of oxidation time. The solid line refers to the logarithmic law indicated in the figure.

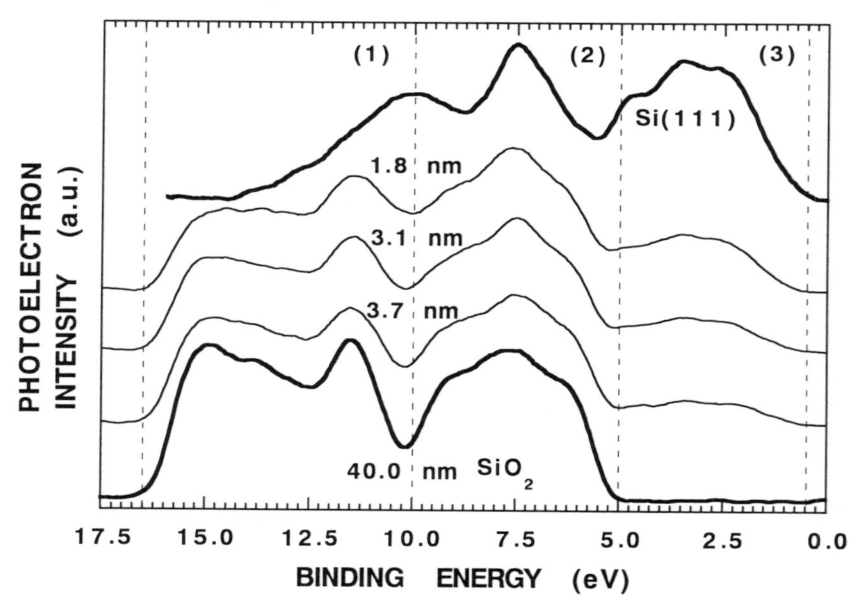

Figure 3 : XPS Valence Band Density of States (VBDOS) or ultra-thin SiO_2/Si(111), thick oxide reference, and H-terminated flat Si(111) reference.

Three energy regions can be distinguished in the VBDOS of the ultra-thin SiO_2/Si interfaces as follows : (1) In the energy range from 16.5 to 10.0 eV, an SiO_2 VBDOS[10] is predominant. (2) From 10.0 to 5.0 eV, the spectra are composed of a mixture of SiO_2 and bulk Si VBDOS[10]. (3) From 5.0 eV to top of the valence band, a pure Si band forms the observed VBDOS. The VBDOS of the ultra-thin SiO_2 layer without any contribution from the bulk Si substrate is obtained by subtracting the valence band spectrum of the hydrogen-terminated, atomically-flat Si(111) reference from the observed VBDOS of the SiO_2/Si(111) interface valence band spectra shown in Fig. 3. The resulting spectra represent the valence band of the ultra-thin oxide layers without any contribution from the underlying substrate as shown in Fig. 4. Furthermore, the top of the oxide valence band obtained for all thin oxides coincides with the top of the thick oxide valence band, indicating that the valence band alignment at the SiO_2/Si interface remains unchanged irrespective of oxide thickness. By taking into account also the small substrate charging effect (maximum 70 meV), the energy difference between the top of the thin SiO_2 valence band and the top of the bulk Si valence band is found to be 4.36 ± 0.10 eV for all oxides.

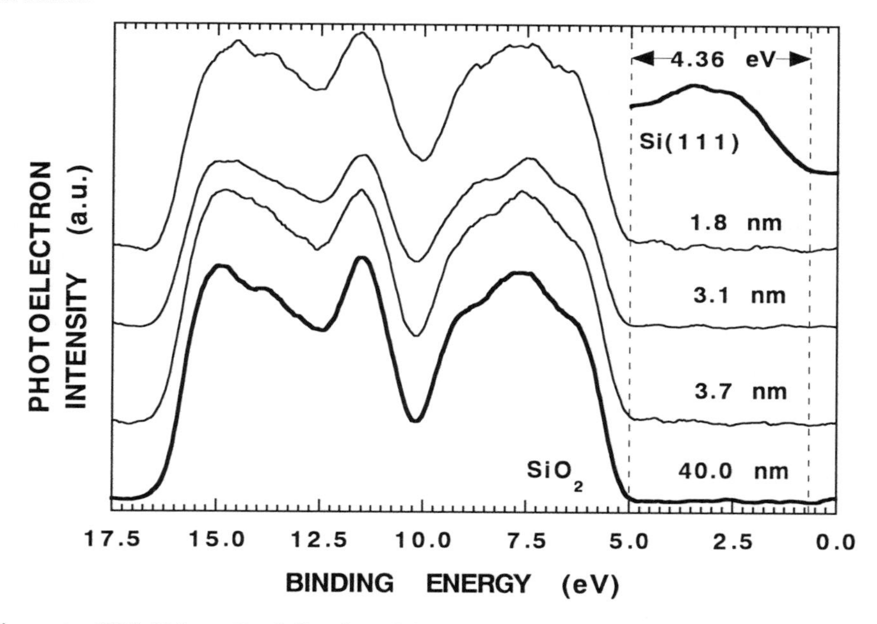

Figure 4 : XPS Valence Band Density of States (VBDOS) for ultra-thin SiO_2 obtained by subtracting the Si contribution from the measured VBDOS shown in Fig.3.

By assuming an SiO_2 band gap of 8.80 eV as obtained for thick oxides[11], a constant barrier height or conduction band alignment of 3.32 eV is estimated for ultra-thin and thick oxides, in agreement with the barrier height value reported from internal photoemission experiments for thick oxides[11].

CONCLUSIONS

The valence band density of states for ultra-thin oxides remains unchanged as compared to that of a thick oxide. The SiO_2/Si band alignment defined as the energy difference between the SiO_2 valence band top and the Si one after correcting for the differential charging effect is obtained to be 4.36 ± 0.10 eV regardless of the oxide thicknesses from 1.8 to 3.7 nm.

ACKNOWLEDGEMENT

One of the authors (Josep L. Alay) would like to thank the Japan Society for the Promotion of Science (JSPS) for the support offered to carry out this research work.

REFERENCES

1) M. Hiroshima, T. Yasaka, S. Miyazaki, and M. Hirose, Jpn. J. Appl. Phys. **33** B, 395 (1994).
2) P.V. Dressendorfer and R.C. Barker, Appl. Phys. Lett. **36**, 933 (1980).
3) S. Horiguchi and H. Yoshino, J. Appl. Phys. **58**, 1597 (1985).
4) S. Heike, Y. Wada, S. Kondo, M. Lutwyche, K. Murayama, and H. Kuroda, Extended Abstracts of the 1994 International Conference on Solid State Devices and Materials (Yokohama, 1994) p. 40.
5) D.A. Shirley, Phys. Rev. B **5**, 453 (1972).
6) F.J. Himpsel, F.R. McFeely, A. Taleb-Ibrahimi, and J.A. Yarmoff, Phys. Rev. B **38**, 6084 (1988).
7) N. Terada, T. Haga, N. Miyata, K. Moriki, M. Fujisawa, M. Morita, T. Ohmi, and T. Hattori, Phys. Rev. B **46**, 2312 (1992).
8) G.S. Higashi, Y.J. Chabal, G.W. Trucks, and K. Raghavachari, Appl. Phys. Lett. **56**, 656 (1990).
9) J. L. Alay, M. Fukuda, C. H. Bjorkman, K. Nakagawa, S. Sasaki, S. Yokoyama, and M. Hirose, (unpublished, submitted to Jpn. J. Appl. Phys. Lett.).
10) J. L. Alay and W. Vandervorst, Phys. Rev. B **50**, 15015 (1994).
11) H. Lefevre and M. Schulz, in *The Si-SiO2 System*, ed. by P. Falk, Elsevier, Amsterdam, 1988.

PROPERTIES OF GATE-QUALITY SiO_2 FILMS PREPARED BY ELECTRON CYCLOTRON RESONANCE CHEMICAL VAPOUR DEPOSITION IN AN ULTRA-HIGH VACUUM PROCESSING SYSTEM

Y. TAO, D. LANDHEER, J.E. HULSE, D.-X. XU, and T. QUANCE
Institute for Microstructural Sciences, National Research Council of Canada, Ottawa, Ontario, Canada, K1A 0R6

ABSTRACT

We have prepared thin SiO_2 layers on Si(100) wafers by electron cyclotron resonance chemical vapour deposition (ECR-CVD) in a multi-chamber ultra-high vacuum (UHV) processing system. The oxides were characterized *in-situ* by single wavelength ellipsometry (SWE) and x-ray photoelectron spectroscopy (XPS) and *ex-situ* by Fourier transform infra-red spectroscopy (FTIR), spectroscopic ellipsometry (SE) and capacitance-voltage (CV) electrical measurements. Films deposited at higher pressures, low powers and low silane flow rates had excellent physical and electrical properties. Films deposited at 400 °C had better physical properties than those of thermal oxides grown in dry oxygen at 700 °C. A 1 minute anneal at 950 °C reduced the fast interface state density from 1.2×10^{11} to 7×10^{10} $eV^{-1}cm^{-2}$.

INTRODUCTION

High quality silicon dioxide films have been produced by the plasma oxidation of silicon using an oxygen electron cyclotron resonance (ECR) plasma[1-3] and the kinetics of this process has been investigated in some detail[2,4]. Films with good physical and bulk electrical properties have also been produced by plasma-enhanced chemical-vapour deposition using an oxygen plasma with silane but less effort has been expended on using this technique to produce gate-quality oxide-semiconductor interfaces[5]. In this work we assess the physical and electrical properties of silicon dioxide films produced in an ECR-CVD system which is part of a UHV multi-chamber processing system with facilities for *in-situ* SWE and XPS. *Ex-situ* physical analysis was done using SE, SWE and FTIR coupled with etch-back of the oxide films. Electrical measurements were made on Si/SiO_2 capacitors with polysilicon gate electrodes.

EXPERIMENTAL

Silicon (100) wafers were given an HF-last RCA clean consisting of immersion in 0.25:1:5 $NH_4OH:H_2O_2:H_2O$ for 10 min. at 80 °C, a deionized water (DI) rinse, immersion for 2 min. in a 1% HF solution, another DI rinse, immersion in 1:1:6 $HCl:H_2O_2:H_2O$ for 10 min. at 80 °C, another DI rinse, and another 2 min. immersion in 1% HF solution. After a final 5 minute rinse in DI water and a nitrogen blow dry the samples were inserted into the vacuum system load-lock.

The multi-chamber vacuum system consists of loadlock, ECR-CVD, metal deposition, XPS and other chambers which communicate via a trolley in a common UHV transfer tunnel. Plasma and loadlock chambers are isolated from the tunnel by buffer chambers. The loadlock entry port opens into a small class 1000 clean room which has a class 10 laminar flow wet bench for wafer cleaning.

Mat. Res. Soc. Symp. Proc. Vol. 386 © 1995 Materials Research Society

The background pressure in the tunnel and buffer and analysis chambers was maintained below $2x10^{-10}$ Torr by a combination of ion- and titanium-sublimation-pumping. The partial pressures of background gases were monitored by an array of quadrupole mass spectrometers. The primary contaminant in the UHV system was hydrogen and the partial pressure of carbon containing species (mainly mass numbers 28, 44, 16 and 12) was $<3x10^{-11}$ Torr (<0.1 monolayer/hr for unit sticking coefficient). The loadlock and the plasma deposition chambers were evacuated by turbo pumps with oil-free backing pumps.

XPS measurements were performed both before and after oxide deposition using a Mg K_α source without a monochromator. The base pressure in the XPS chamber was better than $4x10^{-11}$ Torr and integrated exposure experiments showed an accumulation rate of carbon contamination on bare silicon wafers of ~0.0007 monolayer/hour. Movement of the trolley causes a transient increase in the concentration of hydrocarbon-bearing species with an integrated exposure during transfer measured to be about $8x10^{-8}$ Torr-s (.08 monolayers for unit sticking coefficient). This can account for the contamination on the wafer surface when it reaches the XPS chamber. Titanium-coated pedestals regularly replenished by titanium evaporation on the trolley will reduce this contamination by effectively pumping the tiny volume between pedestal and wafer during trolley motions. XPS measurements were also used to measure the thickness of thin oxide films (<15 nm) using a well-established method[6] which also showed that the stoichiometry of the ECR-CVD samples was $SiO_{2.0\pm0.1}$ for a wide range of deposition conditions.

The carbon contamination level on wet chemically cleaned Si wafers was typically around 0.04±0.01 monolayer after transfer to the XPS chamber. Samples taken out of the chamber and analyzed by secondary ion mass spectrometry coupled with sputter depth-profiling could not detect any carbon at the Si/SiO_2 interface, suggesting that the carbon was removed by the oxygen plasma. However, the sensitivity of this measurement is no better than about 0.001 monolayer.

After oxide deposition, the wafers were transferred directly to the XPS chamber through the UHV tunnel. The carbon contamination on thick (>15 nm) as-deposited sample surfaces was typically around 0.02±0.01 monolayer. There was a large post-deposition pressure transient in the ECR-CVD chamber, whose constituents have not been fully resolved but the carbon contamination levels are consistent with the hydrocarbon backgrounds during the transfer to the XPS system.

An Astex AX4400 single-magnet ECR source was used to create the oxygen plasma for the oxide depositions. The sample stage was situated 43 cm from the source and a 50 cm diameter shaping magnet was placed 10 cm behind it to reduce the divergence of the magnetic field lines. With 150-250 W of microwave power at 2.54 GHz, Langmuir probe measurements at a point 10 cm from the stage indicated an electron temperature in the range of 1-2 eV at 4 mTorr, increasing with decreasing pressure to about 4 eV at 1 mTorr. To keep electron temperatures low most films were deposited with 100 sccm of oxygen flow at a chamber pressure of 3.9 mTorr. This resulted in plasmas 10 cm from the stage with plasma potentials of approximately -2 V and electron densities of 2×10^{12} cm^{-3}. A 13 cm diameter gas distribution ring introduced 25% silane diluted by helium at a plane 15 cm from the sample. The best films were produced with low flows of this mixture and a flow of 0.5 sccm was used for this study. The sample temperature was maintained at 400 °C with a quartz-halogen lamp heater. Further details of the operating conditions will be published elsewhere.

A series of depositions of 10 nm thick films was done with 150 and 210 W of RF power. The stage bias was held at 5 V for these depositions which corresponds to a <8 V accelerating potential between plasma and substrate. Some wafers were given a 1 minute post-oxide-deposition anneal (POA) in a rapid thermal annealer in flowing nitrogen before the polysilicon deposition or before analysis.

The deposition of the oxides was monitored with an *in-situ* single-wavelength ellipsometer operating at 633 nm (Rudolph model i1000). *Ex-situ* ellipsometric measurements on the oxides were performed with a spectroscopic ellipsometer (J. A. Woollam Co. WVASE) and another SWE (Rudolph Auto El II).

FTIR measurements in the range 4000-400 cm^{-1} were made with a Nicolet 550 FTIR spectrometer. They were performed in dry, flowing N_2 using 8 cm^{-1} resolution at normal incidence. The backsides of the wafers were etched clean in dilute HF just before measurement and background corrections were made by subtracting the spectra obtained after the deposited film was etched off.

RESULTS OF PHYSICAL CHARACTERIZATION

Figure 1 shows a typical SWE trace taken during the growth of films deposited with 150 and 210 W of (absorbed) microwave power. The thickness was calculated using the known optical constants of silicon[8] assuming a single layer film with a refractive index of 1.454 at 633 nm as determined by SE. The deposition rate increases with power and there is a rapid initial increase in the thickness that is similar to that observed during the plasma oxidation experiments of Hu *et al.*[4] Also shown in Figure 1 are the results for a 210 W film where the silane was shut off so that only oxidation took place. The similarity of the initial rise with and without silane flow indicates that the initial rapid rise in film thickness in our CVD films can be attributed to plasma oxidation. It should be pointed out the data of Figure 1 is approximate only since the *in-situ* SWE measurements do not fit a simple single layer model as well as the *ex-situ* SWE or SE measurements described below. The reason for this discrepancy is not fully understood and is the subject of further investigation. It could be related to errors in the high-temperature optical constants of the substrate or to a relaxation of the film that takes place during deposition.

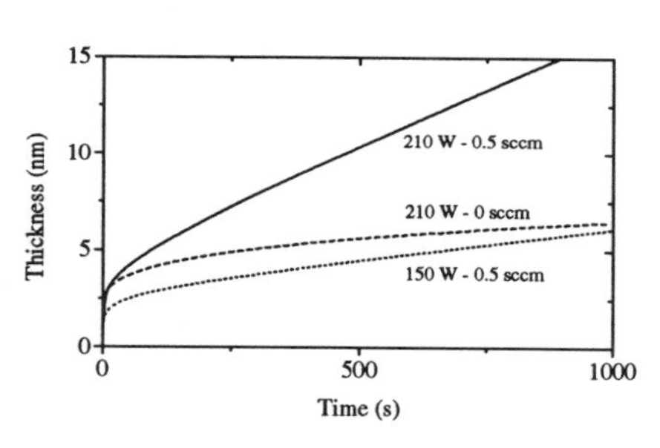

Figure 1. Plot of thickness versus time deduced from the *in-situ* ellipsometer data for films deposited with 150 and 210 W of power and 0.5 sccm of 25% silane. Also shown is the data for the plasma-oxidized film formed with 210 W of power when the silane is turned off.

The *ex-situ* SWE was used to measure the thickness of a series of films made by etch-back using a 1% HF solution. The data for a sample deposited with 150 W of microwave power is shown in Figure 2(a) while the data for another sample taken from the same wafer and annealed for 1 minute at 1100 °C is shown in Figure 2(b). The continuous curves were calculated assuming a single layer of fused SiO_2 with a refractive index of 1.454 at 633 nm. The data of Jellison[9] were used for the silicon substrate. The fit is indicative of the uniformity of the film and excellent

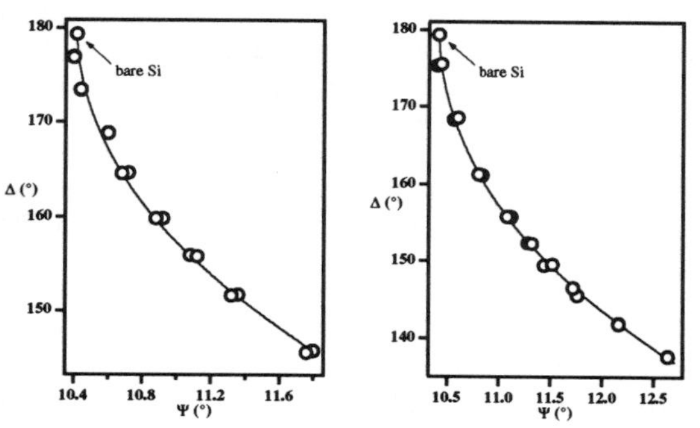

Figure 2. Plot of Δ vs ψ for etch-backs of a) as-deposited and b) 1100°C annealed samples deposited with 150 W of power. The solid curves were calculated assuming a single layer with a refractive index of 1.454.

fits were obtained for the annealed and etched-back sample.

It is of interest to note that the final etches in Figures 2(a) and 2(b) (data point closest to bare Si) produced a surface which was hydrophobic, the usual end-point for etching in HF. The fact that the oxide is apparently not all gone is an indication of the inability of the reputedly H-terminated surface to resist atmospheric degradation. This observation warrants further investigation of the etching process.

SE measurements were carried out to determine the refractive index used above, to check the oxide uniformity and to identify the presence of an interface layer between the silicon substrate and the oxide. The data was fitted over the photon energy range of 1.6 to 4.9 eV by a simple 3-phase model (ambient-film-substrate). The optical constants for the film layer calculated with the Bruggeman effective medium approximation[10] (EMA) assuming the film is a mixture of fused silica and voids. The optical constants for fused silica were taken from Malitson[11] and for the substrate silicon from Jellison[9]. The refractive index at 633 nm n_{633} was determined to be 1.454±0.003 for the sample deposited with 150 W of power, slightly less than that of fused silica. The unbiased estimator δ used by Hu *et al.*[4] was 1.8×10^{-3} for the fit. For the sample of the same wafer annealed for 1 minute at 1100°C the best fit was found for n_{633}=1.469±0.003 with $\delta=2.1\times10^{-3}$. The fit was not improved by using the two layer models of Ref. 4.

The data indicates that a thin boundary layer at the Si/SiO_2 interface is not significant for these films. This is in contrast to the PECVD oxides reported previously[12] which were found to have a thin layer of increased density next to the silicon substrate. It also contrasts with the results of Hu *et al.*[4] who observed a layer next to the silicon up to 2 nm thick that was modelled as having up to a 75% amorphous silicon constituent.

No water related peaks were found in the FTIR spectra taken from 40 nm thick SiO_2 films prepared by ECR-CVD for the deposition conditions used for this study. Using the formula given by Pliskin *et al.*[13], the upper limit of hydroxyl present as SiOH in the films was estimated to be less than 0.5 wt.%.

The frequency (ν_s) and full width at half-maximum (FWHM) of the asymmetric stretching band at 1060~1080 cm^{-1} was also used to characterize the films. It is well known that ν_s is sensitive to the film quality, structure, and composition. The peak position and FWHM for as-deposited ECR-CVD films ~15 nm thick and deposited with 150 W of power were found to be 1065±1 cm^{-1} and 78±1cm^{-1} respectively. In Figure 3, the frequency and FWHM of the Si-O-Si asymmetric stretching band are plotted for this sample as a function of annealing temperature for 1 minute anneals. The films annealed at higher temperature have higher stretching frequency and smaller widths. For comparison, we also plot a group of data obtained from thermal oxide films. These oxide films have comparable thickness as those of ECR-CVD oxide films. They were thermally grown at 700 °C in purified oxygen and annealed for 5 minutes. The Si-O-Si stretching frequency and FWHM of our ECR-CVD SiO_2 films are better than those of the thermal oxide films grown at 700 °C.

In order to investigate the depth dependence of ν_s, FTIR measurements were carried out on the same series of films of different thickness prepared by etch-back and analyzed by SWE in Figure 2. Figure 4 displays the ν_s and FWHM as a function of the film thickness for the as-deposited film and the film annealed at 1100 °C for one minute. There is evidence of a 2 nm thick layer next to the substrate with reduced stretch frequency and increased FWHM; however, this is not conclusive due to the weak signals for these very thin layers. An anneal at 1100 °C for one minute in flowing nitrogen greatly improves the quality of the film as demonstrated by the increased ν_s and decreased half-width. Also shown are the results of a single measurement on a 6.5 nm thick dry thermal oxide grown at 1000 °C . They are close to those of our as-deposited film.

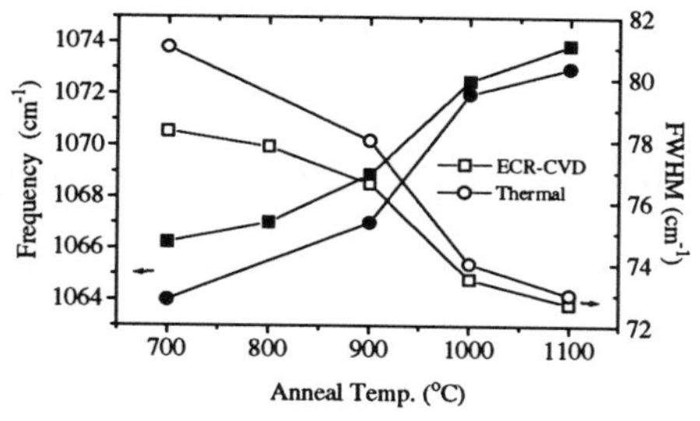

Figure 3. Frequency and FWHM for the Si-O-Si stretch peak as a function of annealing temperature for samples deposited with 150 W of power and annealed for 1 minute. Also shown are the results for a dry thermal oxide grown at 700 °C and annealed for 5 minutes.

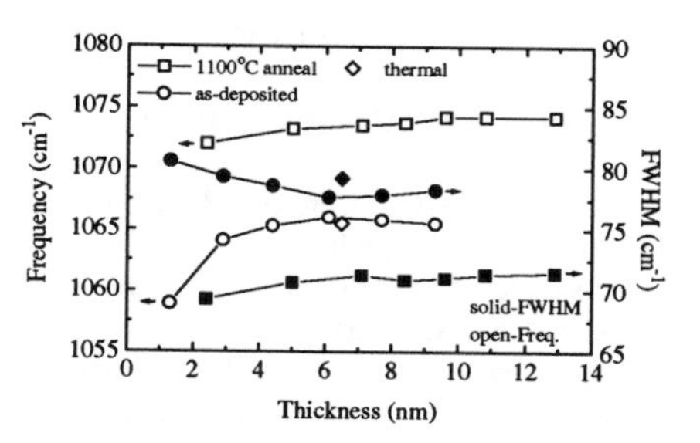

Figure 4. Frequency and FWHM of the Si-O-Si stretch peak as a function of etched-back thickness for as-deposited and 1100 °C annealed sample deposited with 150 W of power. Also shown are the results of a single measurement on a 6.5 nm thick dry thermal oxide grown at 1000 °C.

ELECTRICAL MEASUREMENTS

Poly capacitors were fabricated by opening windows in field oxide on 3-5 Ω-cm *n*-type substrates, followed by the HF-last RCA clean, oxide deposition and removal from the vacuum system for n+ *in-situ*-doped polysilicon deposition at 630 °C. The top electrodes were patterned by wet etching and the capacitors were annealed in forming gas for 30 min at 450 °C. Back contacts were made with InGa eutectic paste.

Interface state densities D_{it} were obtained by the high-low method[14] from quasi-static (QS) CV and 100 KHz (HF) CV data. For the QS-CV measurements the voltage was ramped from inversion to accumulation at 0.1 V/s while for the HF-CV measurements the voltages were stepped at 0.1 V/s. The interface-trapped charge Q_{it} was calculated from the flatband voltage shift ∇V_{fb} using the relation $Q_{it} = \varepsilon\varepsilon_0 \nabla V_{fb}/d_{ox}$, where d_{ox} is the oxide thickness and $\varepsilon\varepsilon_0$ its permittivity.

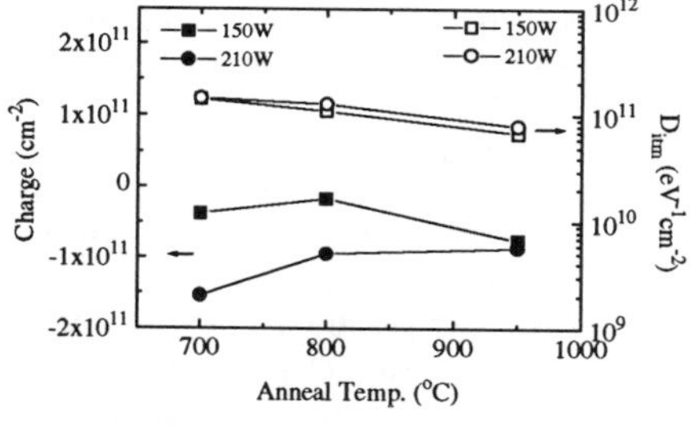

Figure 5. Interface trapped charge and midgap interface state density as a function of annealing temperature for samples deposited with 150 and 210 W of power and annealed for 1 minute.

Figure 5 shows the midgap interface state density D_{itm} and Q_{it} as a function of annealing temperature and power. D_{itm} does not change with deposition power in this range but there is a relatively high density of unpassivated interface states that decreases with increasing annealing temperature. In contrast the interface trapped charge changes with both power and annealing temperature. There is an initial decrease in the negative charge trapped at the interface with increasing annealing temperatures, but at the highest temperatures Q_{it} becomes independent of deposition power. The Q_{it} calculated in this way includes the fixed charge at the interface and a contribution from the charge trapped in the bulk of the oxide. It also contains a contribution from the empty donor interface states above the Fermi energy, but with a uniform interface state density of 10^{11} $eV^{-1}cm^{-2}$, the positive charge due to the donors would be less than 3×10^{10} cm^{-2}. The negative charge may be the result of electrons trapped at hydrogen-related[15] or plasma-created[16] defects during the deposition process.

DISCUSSION AND CONCLUSIONS

We have shown that the ECR-CVD oxide can be a good potential candidate for gate-quality oxide films. By working with low powers, low silane flow rates and low deposition rates films have been deposited at 400 °C that have better physical properties than those of thermal oxides grown in dry oxygen at 700 °C. Initial growth is dominated by plasma oxidation which produces a 1-2 nm thick layer at the Si/SiO_2 interface. This layer is indistinguishable in the *ex-situ* ellipsometric measurements from the bulk oxide. A 1 minute anneal at 950 °C reduced the fast interface state density from $1.2x10^{11}$ to $7x10^{10}$ $eV^{-1}cm^{-2}$.

ACKNOWLEDGEMENTS

We thank A. Golshan, M. Tomlinson and P. Chow-Chong for technical assistance and G.I. Sproule for the secondary ion mass spectrometry. This research is part of a collaborative bilateral agreement between the Telecom Microelectronics Centre of Northern Telecom and the National Research Council of Canada. The authors are grateful to S.-P. Tay and J. Ellul at NT for providing silicon wafers with thermal oxide and for helpful discussions.

REFERENCES

1. G.T. Salbert, D.K. Reinhard, and J. Asmussen, J.Vac. Sci. Technol. A, **8**, 2919 (1990).
2. D.A. Karl, D.W. Hess, M.A. Lieberman, T.D.D. Nguyen, and R. Gronsky, J. Appl. Phys., **70**, 3301 (1991).
3. K.T. Sung and S.W. Pang, J. Vac Sci. Technol. B, **10**, 2211 (1992).
4. Y.Z. Hu, J. Joseph, and E.A. Irene, Appl. Phys. Lett. **59**, 1353 (1991); J. Joseph, Y.Z. Hu, and E.A. Irene, J. Vac. Sci. Technol. B ,**10** , 611 (1992).
5. T.V. Herak, T.T. Chau, D.J. Thomson, S.R. Mejia, D.A. Buchanan, and K.C. Kao, J. Appl. Phys. **65**, 2457 (1989); T.T. Chau, T.V. Herak, D.J. Thomson, S.R. Mejia, D.A. Buchanan, R.D. McCleod, and K.C. Kao, IEEE Trans. Elec. Insul. **25**, 593 (1990); T.T. Chau, S.R. Meija, and K.C. Kao, Mat. Res. Soc. Symp. Proc. Vol. **223**, 69, (1991).
6. M.F. Hochella, Jr. and A. H. Carim, Surf. Sci. Lett. **197**, L260 (1988); F. J. Himpsel, F. R. McFeely, A. Taleb-Ibrahimi, J. A. Yarmoff, and G. Hollinger, Phys. Rev. B, **38**, 6084 (1988).
7. D. Landheer, Y. Tao, J. Hulse, T. Quance, and D.-X. Xu, to be published
8. G.E. Jellison, Jr. and F.A. Modine, Phys. Rev. B, 27, 7466 (1983).
9. G. E. Jellison, Jr., Optical Materials **1**, 41 (1992).
10. D. A. G. Bruggeman, Ann. Phys. (Leipzig) **24**, 636 (1935).
11. I. H. Malitson, J. Opt. Soc. Am. **55**, 1205 (1965).
12. Y. Tao, D. Landheer, J.-M. Baribeau, J. E. Hulse, D.-X. Xu and M. J. Graham, Mat. Res. Soc. Symp. Proc. Vol. **338**, 57 (1994).
13. W. A. Pliskin, J. Vac. Sci. Technol. **14**, 1064(1977).
14. R. Castagne and A. Vapaille, Surface Sci. **28**, 557(1971).
15. F.J. Feigl, D.R. Young, D.J. DiMaria, S.K. Lai, and J. Calise, J. Appl. Phys. **52**, 5665 (1981).
16. J.F. Zhang, S. Taylor, and W. Ecclestone, J. Appl. Phys. **72**, 1429 (1992).

ADVANCED TECHNOLOGY PROCESSING AND STATE-OF-THE-ART SOLUTIONS TO CLEANING, CONTAMINATION CONTROL, AND INTEGRATION PROBLEMS

W. W. Abadeer and A. LeBlanc
IBM Microelectronics Division, 1000 River St.
Essex Junction, VT 05452

A. Kluwe*, P. Schulz*, R. Vollertsen and V. Penka
Siemens Components, Inc.
c/o IBM Microelectronics Division
Essex Junction, VT 05452
* Hopewell Junction, NY 12533

S. Nadahara
Toshiba Corp.
c/o IBM Microelectronics Division
Hopewell Junction, NY 12533

A. Antreasyan, D. Cote, W. Cote, M. Levy
IBM Microelectronics Division
Hopewell Junction, NY 12533

H. Akatsu and R. Ludeke
IBM T. J. Watson Research Center
Yorktown Heights, NY 10598

ABSTRACT

Ultraclean semiconductor processes were developed for ULSI technologies to significantly reduce metallics, foreign materials, and to preserve silicon surface morphology. State-of-the-art detection techniques (Elymat, TXRF, VPD, and SPV) were implemented in all critical process sectors. Acceptable levels of metallic contamination were derived from previous experience and from published work on this subject relating metallic contamination levels to gate oxide reliability and retention time.

A comprehensive diagnostic/characterization system was utilized in evaluating a wide variety of test structures emulating key device process interactions. Advanced techniques were employed for measurement of reliability and surface morphology. Also root causes for all integration problems are identified.

Mat. Res. Soc. Symp. Proc. Vol. 386 © 1995 Materials Research Society

INTRODUCTION

Manufacturing ULSI products depends on solving all integration problems and reducing defects which leads to very competitive yield and reliability. This presentation highlights key challenges and options for optimization of cleaning steps and defect reduction for manufacturing 64Mb and 256Mb DRAM technologies as well as logic applications. Key to manufacturability are the high levels of quality and reliability for all interlevel and intralevel dielectrics involved in the integrated processing, with the thin gate dielectric being the most important. For thin gate dielectric, the key challenges are with silicon-surface preparation and cleaning steps. State-of-the-art electrical characterization procedures using charge-to-breakdown (Qbd) and time-dependent dielectric breakdown testing with both constant current and constant voltage techniques are essential for evaluating thin gate oxide yield and reliability. Part of the overall picture is the utilization of key test structures to test the quality of thin gate oxide and all interlevel and intralevel dielectrics. Test structures based on actual product cell and circuit layouts are also important. This paper highlights key issues in the following areas:

- Silicon-surface cleaning and preparation
- Defect reduction for gate oxide
- Metallic contamination
- Cleaning steps with CMP (chemical mechanical polishing)
- Other integration problems affecting gate oxide

CONTROL OF METALLIC CONTAMINATION

Horizontal shorts (metal line to metal line and gate conductor to gate conductor), vertical shorts (gate conductor to diffusion contact and metal to gate), gate dielectric fails and junction leakage (degradation of DRAM retention time) are all metallic-contamination problems that have been identified through physical and electrical characterization of product and test structures.

The sources identified for metallic contamination are very dependent upon the type of semiconductor process utilized, RIE tools, ion implanters, wet cleans, and deposition tools. Control targets for metal particles are set based on the monitoring procedures and prior art [1-2]. For horizontal and vertical shorts KLA, SEM and EDX are used. The monitoring level for KLA is <0.5 particles/cm^2 with a particle size of ≥ 0.3um. As vertical and horizontal dimensions shrink, the vertical and horizontal shorts become much more prominent as the major yield and reliability detractor, especially in the early phases of a program. For dielectrics, wafer monitors at the RIE, ion implant, deposition and wet-clean tools are utilized. The control targets are strongly dependent on dielectric thickness. For 10nm oxide, a control limit of $\leq 10^{11}$/cm^3 (Fig. 4, Ref. 2) have been utilized. Normally, dielectric fails are not nearly as troublesome as horizontal and vertical shorts. With proper monitoring and detection, metallic particles in gate oxide are easy to track. For junction leakage we have utilized wafer monitors at RIE, ion implant, deposition and wet cleaning tools. Junction leakage is strongly dependent on crystal defects and requires reduction of metallics as well as crystal defects. In DRAMs, junction leakage will show up as an extended tail on the low side of the retention time distribution. Generally, junction leakage is difficult to track due to the interaction of crystal defects with metallic particles. This is also a major problem in early phases of a program, especially for DRAM process development.

For monitoring, several schemes are utilized: VPD, TXRF, SPV and ELYMAT. The detection limits are as follow: $5x10^9$ atoms/cm^2 for TXRF (Ni), 10^{11} atoms/cm^2 for SPV (Fe), and $2x10^{10}$ atoms/cm^2 for ELYMAT (Fe).

CLEANING PROCESSES AND SURFACE MORPHOLOGY

Preserving surface morphology is important for achieving a robust gate dielectric. Achieving a high level of quality for the silicon surface strongly depends on surface cleaning. We experimented with different cleaning and etching solutions and compared the resulting silicon surface roughness and gate oxide quality (3). Bare Si wafers are Ar annealed. The Si(100) samples were divided into two groups in which one group had 10nm dry oxide grown at 850°C (preclean is S/P HF SC-1 SC-2 in a spray processor). Each wafer was broken into pieces with each piece subjected to one of the following etching solutions: 9:1 BHF, 9:1 DHF, and 200:1 DHF. A one-minute DIW rinse was used prior to UHV STM measurement. Figure 1 demonstrates the better surface quality obtained with the DHF solution vs. BHF (4-7). Figure 2 shows that a smoother surface resulted with 200:1 DHF on bare silicon wafers compared to thermal oxidized wafers (8-11). Figure 3 shows no significant difference between 9:1 DHF and 200:1 DHF. Figures 4A and 4B show cumulative percent failure vs. injected current density for 5mm MOS dots, using two different cleaning procedures. Figure 4A the solution is BHF followed by S/P HF SC-1 SC-2 and in Figure 4B an extended (long) BHF followed by S/P HF SC-1 SC-2 was used (5, 7). At an injected current density of $1x10^{-4}$ atoms/cm^2 (breakdown voltage <8mV/cm), much higher cumulative failures resulted with the extended BHF followed by S/P HF SC-1 SC-2. As shown in Figure 5, an excessive oxide overetch in BHF can not only pit Si surfaces, but also increase final foreign material (FM) count. It was also found that by adding SC-1 or SC-2 following the sulfuric peroxide clean significantly reduced both large and small FM count post resist strip (Figure 6) (12-13). Another key process for reducing FM is the incorporation of megasonic energy to the cleaning process, as shown in Fig. 7 (14-15).

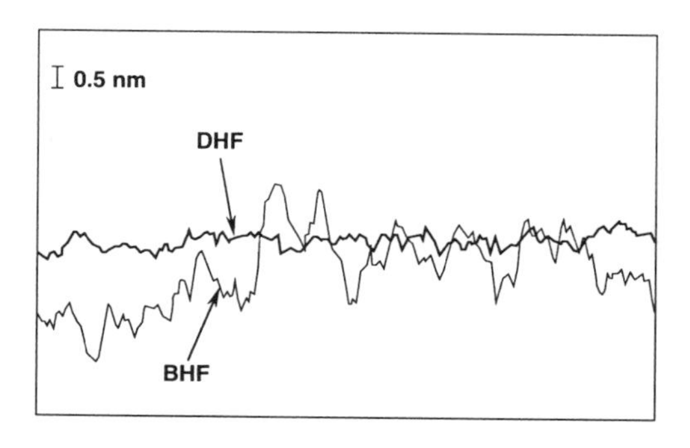

Figure 1. Surface morphology of BHF and DHF.

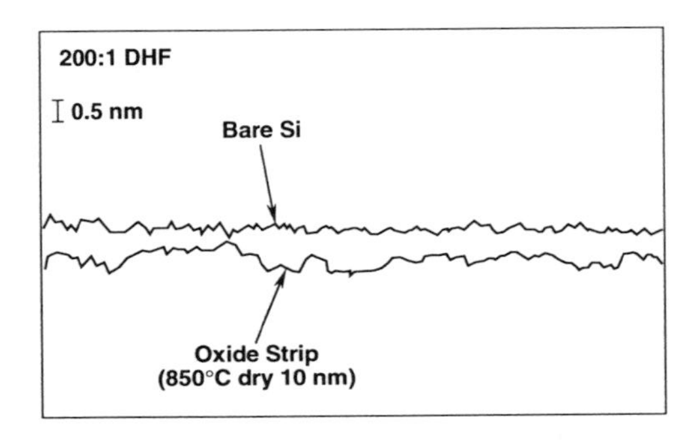

Figure 2. Surface morphology of DHF on bare silicon and after oxide strip.

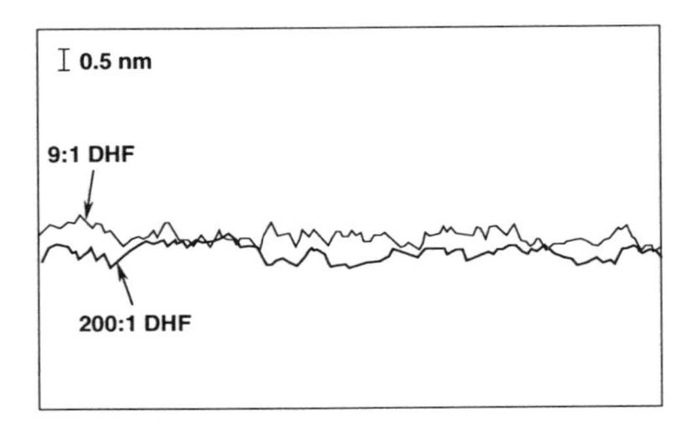

Figure 3. Surface morphology of 9:1 DHF and 200:1 DHF.

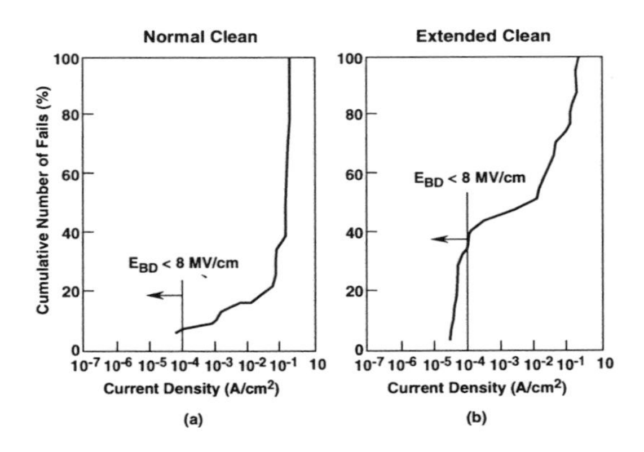

Figure 4. Comparison between BHF and extended BHF.

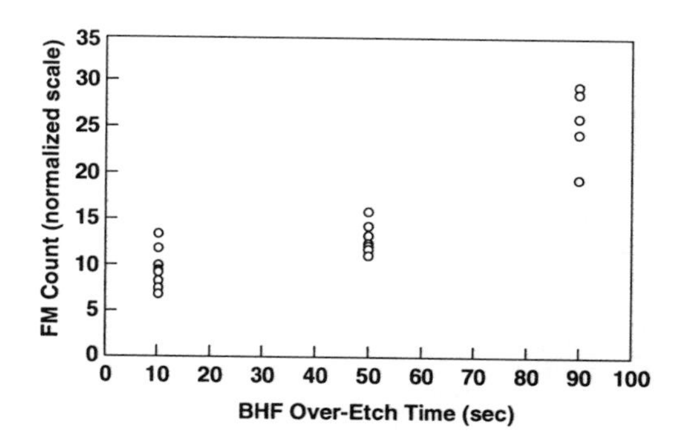

Figure 5. Effect of BHF overetch on FM count.

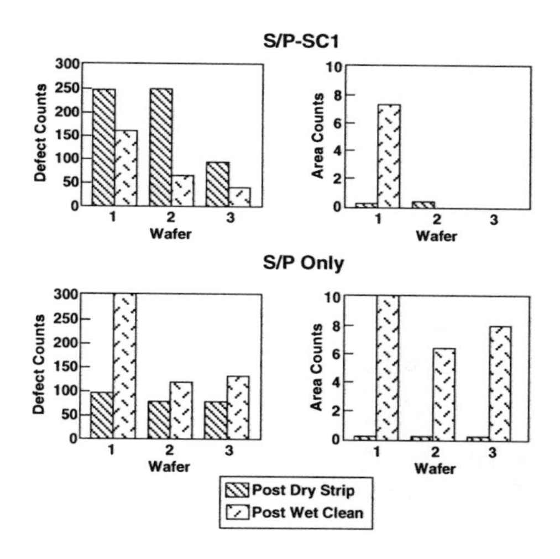

Figure 6. SC-1: reduced FM count.

CLEANING PROCESSES WITH CMP

Chemical-mechanical polishing requires optimized post polish cleaning processes to preserve surface quality and to reduce F M [16]. As shown in Figure 8, optimizing the brush clean is very important for reducing dielectric reliability fails. An improved brush clean in alkaline solution is superior compared to a brush clean in acidic solution . The type of polishing pad is another important issue for CMP. Figure 9 shows the effect on FM count due from three different pad types. Also shown in Figure 9 is the difference between the standard and improved brush cleans. CMP cleaning was also investigated using the 64Mb DRAM chip. The life test was conducted on a personalized version of the 64Mb chip (Figure 10), where direct DC access to all word lines and bit lines could be made. A schematic for the DRAM array test structure is shown in Figure 11.

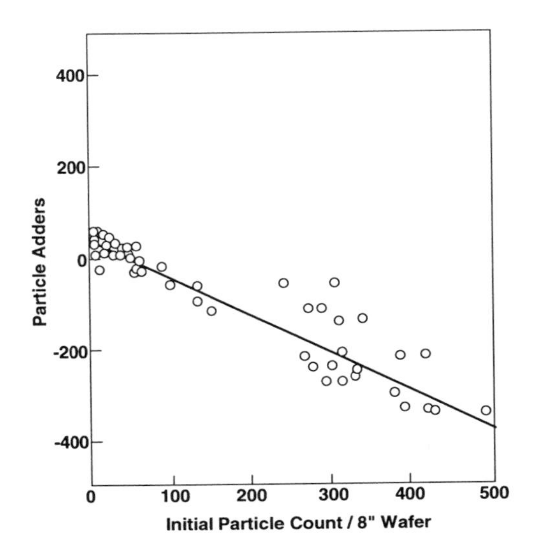

Figure 7. Megasonics: reduced FM.

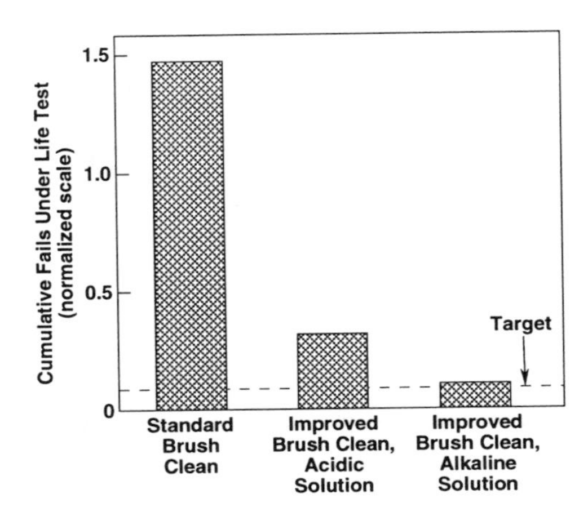

Figure 8. Effect on reliability due to optimization of CMP-brush clean.

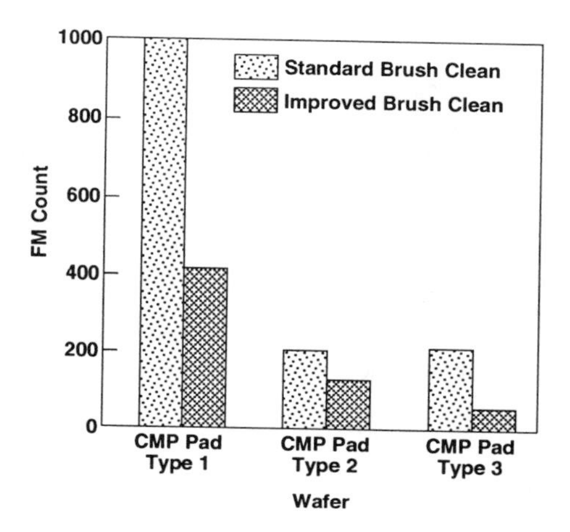

Figure 9. Effect on CMP count due to optimization of pad type.

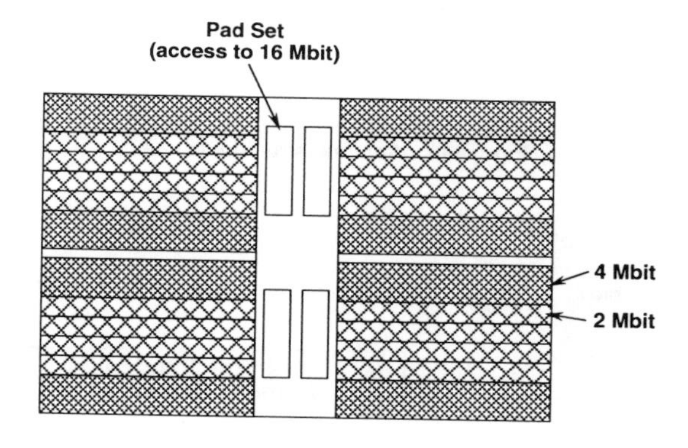

Figure 10. Top layout view of 64-Mbit personalized chip for reliability testing.

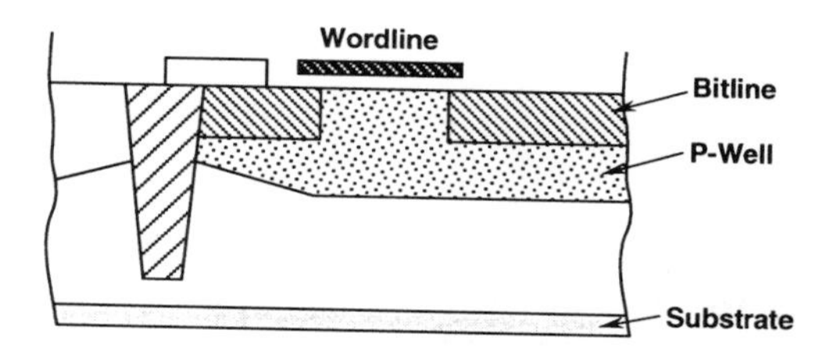

Figure 11. Array schematic for reliability testing.

IN-LINE CHARGING

A comprehensive and complete effort to reduce contaminants and improve yield and reliability also includes process in-line charging issues. A detailed study of this area is beyond the scope of this paper but for completeness, we list a few key issues. Very important to this subject are the types of test structures used for electrical/physical characterization; we have used antenna structures with ratios ranging from 300 to 10 x 10^6. The layout of the antenna as shown in Fig. 12, consists of a dielectric capacitance with a relatively small 1um^2 thin oxide area covered by a gate conductor of much larger area that acts as an antenna. The ratio of the gate conductor area to the thin oxide area determines the antenna ratio. We investigated the impact of plasma-charging at key sectors and successfully optimized the process by using the antenna structures. Table 1 shows the significant improvement in yield and reliability due to optimization of the process.

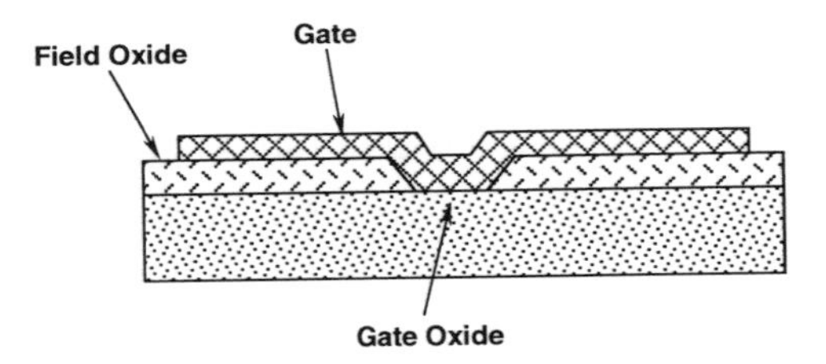

Figure 12. Antenna-structure layout.

Table I. Reduced in-line charging by process optimization.

	Yield for Different Antenna Ratios (%)		
	1 : 2,000	1 : 8,000	1 : 100,000
Old Process	3	0	0
Optimization 1	75	52	8
Optimization 2	100	100	100

CONCLUSIONS

Manufacturability of ULSI products requires continuous efforts for improving yield and reliability. Key issues associated with metallic-particle reduction, foreign materials, and preserving silicon surface morphology have been given.

ACKNOWLEDGEMENTS

The authors thank the following people for their valuable contributions: Susan Cohen of IBM Yorktown Research Center, Michael Liehr of IBM Yorktown Research Center, Dragan Podlesnik of AME Corp., and Russ Randt of IBM Microelectronics Division, Hopewell Junction, New York.

REFERENCES

1. The SIA National Technology Roadmap for Semiconductors, pp. 115-116
2. W. Henley, L. Jastrzebski, and N. Haddad, Proc. Int. Reliab. Phys. Symp., 22 (1993)
3. H. Akatsu, S. Nadahara, R. Ludeke, and S. Cohen, to be submitted
4. G. Higashi, Y. Chabal, G. Trucks, and K. Raghavachari, Appl. Phys. Lett., 56(7), 656 (1990)
5. D. Graf, M. Brahl, S. Bauer-Mayer, A. Ehlert, P. Wagner, and A. Schnegg, 1992 MRS Spring Meeting, San Francisco, CA, 1992
6. P. Dumas, and Y. Chabal, Chemical Physics Letters, 181(6), 537 (1991).
7. M. Dohmen, R. Wijburg, and R. Girsch, Ultra Clean Processes for Semiconductor Surfaces, Bruges, Belgium, Septembter, 1994
8. A. Ogure, J. Electrochemical Society, 138(3), 807 (1991)
9. Mau-Tsu Tang, K. Evans-Lutterodt, M. Green, D. Brasen, K. Kirsch, L. Manchandra, G. Higashi, and T. Boone, Appl. Phys. Lett., 64(6), 748 (1994)
10. Mau-Tsu Tang, K. Evans-Lutterodt, G. Higashi, and T. Boone, Appl. Phys. Lett., 62(24), 3144 (1993)
11. P. Jakob, P. Dumas, and Y. Chabal, Appl. Phys. Lett., 59(23), 2968 (1991)
12. A. Rotendaro, H. Schmidt, M. Meuris, M. Heyens, C. Claeys, and J. Mulready, Ultra Clean Processes for Semiconductor Surfaces, Bruges, Belgium, September, 1994
13. M. Meuris, M. Heyns, P. Mertens, S. Verhaverbeke, and A. Philipossian, Microcontamination, 31 (1992)
14. W. Syverson, M. Fleming, and P. Schubring, J. Electrochemical Society, 10 (1991)
15. P. Resnick, C. Adkins, P. Clews, E. Thoams, and S. Cannaday, Electrochemical Society, Symposium on Cleaning Tech. in Semi. Device Mfg., 184th ECS Mtg, New Orleans, LA, October 10-15, 1993
16. W. Krusell and J. Pollick, Ultra Clean Processes for Semiconductor Surfaces, Bruges, Belgium, September, 1994

Part V

Gas-Phase Cleaning

REDUCED-PRESSURE UV PHOTO-OXIDATION OF ORGANIC CONTAMINANTS ON Si SURFACES

SATISH BEDGE AND H. HENRY LAMB
Department of Chemical Engineering and Department of Materials Science and Engineering, North Carolina State University, Raleigh, NC 27695-7905

ABSTRACT

A reduced-pressure ultraviolet (UV) photo-oxidation process is described for the removal of organic contamination from Si surfaces with concomitant growth of an ultra-thin oxide passivation layer. As *in situ* photo-generation of ozone (O_3) from dioxygen (O_2) is impractical at millitorr pressures, a 2% O_3/O_2 mixture from an ozone generator is fed to the surface conditioning chamber. UV/O_3 surface conditioning employing simultaneous 254/185 nm UV irradiation results in essentially complete removal of carbon contamination in 180 s at 100 mTorr and 100-200°C. Kinetics studies employing cyclohexane-contaminated Si(100) surfaces suggest that carbon removal occurs via two consecutive first-order processes: initial fast photo-oxidation of adsorbed cyclohexane to readily desorbed products and slower photo-oxidation of adsorbed carbonyl intermediates to CO_2 and H_2O. The activation energies for both processes are 2-3 kcal/mol, consistent with the involvement of photo-generated atomic oxygen species. Self-limiting growth of an ultra-thin oxide passivation layer occurs concomitantly with carbon removal. At saturation, the oxide layer is only 3-4 Å thick, and the growth kinetics are described by a first-order Langmuirian adsorption model.

INTRODUCTION

The characteristics of the Si/SiO_2 interface are critical to metal-oxide-semiconductor (MOS) device performance. Consequently, surface chemical purity, microroughness, and defect generation must be carefully controlled during interface formation via either high- or low-temperature processes. For example, the presence of organic contaminants on hydrogen-terminated Si surfaces prior to thermal oxidation can result in a significant reduction of the average breakdown field of MOS capacitors.[1] A similar degradation of MOS device quality occurs when clean, hydrogen-terminated Si is heated to temperature in ultra-pure Ar (containing ~1 ppm of H_2O and O_2) prior to thermal oxidation.[2] This effect was traced to interfacial microroughness that was generated by Si etching via SiO desorption during heating. Si etching is eliminated by providing a stable oxide passivation layer using either the RCA chemistry[3] or atmospheric-pressure ultraviolet (UV)/O_2 surface conditioning.

Hydrogen-terminated Si surfaces resulting from etching of native or chemical oxide layers by dilute HF are hydrophobic and highly susceptible to organic contamination. Typically, Si wafers that are subjected to an RCA clean with a final HF dip and then inserted into a vacuum processing chamber acquire organic contamination either from the atmosphere or from the vacuum ambient. Thus, an *in situ* surface conditioning process is needed to remove organic contamination prior to oxide growth or deposition. Plasma processes have been used for this application, including remote hydrogen and oxygen plasma exposures. An oxygen plasma process is preferred, as Si etching and sub-surface defect generation associated with hydrogen plasma exposure[4,5] are avoided, and a device-quality Si/SiO_2 interface is produced.[6] Although remote plasma oxidation works well, it suffers the disadvantage of subjecting the surface to energetic ion bombardment. In contrast, UV photo-oxidation processes involve only reactive neutral species, such as ozone (O_3) and atomic oxygen.[7]

UV/air and related photo-oxidation processes have been applied *ex situ* to remove organic contamination from Si surfaces and grow a thin oxide passivation layer.[8,9,10] In this paper, we describe a reduced-pressure UV/O_3 surface conditioning process that can be applied *in situ* prior to deposition of device-quality SiO_2 films. A reduced-pressure UV/O_3 process has significant advantages in cleanliness, speed and ease of process integration over an atmospheric pressure UV/O_2 process; the most obvious being obviation of a pump-down cycle from atmospheric

Mat. Res. Soc. Symp. Proc. Vol. 386 © 1995 Materials Research Society

pressure. To determine the kinetics of carbon removal and oxide layer growth, reduced-pressure UV/O_3 conditioning was applied to hydrogen-terminated Si(100) surfaces that were intentionally contaminated with cyclohexane.

EXPERIMENTAL METHODS

Samples (15 x 25 mm^2) taken from a 125-mm *p*-type Si(100) wafer (ShinEtsu) were subjected to UV/air treatment followed by a 60-s etch in dilute HF and a 2-min deionized (DI) water rinse. HF solutions were prepared by diluting 49% HF (Fisher electronic grade) 1:50 with 18 MΩ-cm DI water from a Barnstead Nanopure unit. After drying in flowing N_2, the resultant hydrophobic Si surface was contaminated by applying a self-spread cyclohexane film. Excess cyclohexane was evaporated into flowing N_2 at room temperature. The residual carbon concentration is highly variable, ranging from 25-40 at%, as determined by Auger electron spectroscopy (AES).

Samples were introduced to the UV/plasma surface conditioning chamber via a loadlock that is evacuated by an oil-free Drytel 31 membrane/molecular drag pump and a Varian HV-4 cryopump. The plasma cleaning capabilities, which were not used in the present experiments, will be described elsewhere.[11] The chamber is pumped by an Edwards (Seiko) STPH-200C turbomolecular-drag pump with a magnetically levitated rotor. The base pressure after bakeout is $2x10^{-8}$ Torr. Mass flow controllers (Unit Instruments) are used to regulate the flows of He and O_3/O_2 to the gas-dispersal ring. The O_3/O_2 mixture is produced using a Polymetrics (T-818) ozone generator which is fed with Research-grade O_2. Iodine titrations revealed that under the conditions used in these experiments, the effluent from the ozone generator contained approximately 2% (v/v) O_3 in O_2. The processing pressure is measured by an MKS 127A Baratron gauge and controlled by an Edwards model 1800 feedback controller and butterfly throttling valve.

The Si surface is irradiated through a UV-grade quartz window mounted in a 4.5" Conflat flange (Insulator Seal Inc.). A low-pressure Hg lamp (BHK, Inc.) that emits 254 and 185 nm photons with a total intensity of 15 mWatts/cm^2 at 2.54 cm serves as the UV source. The lamp is enclosed in an Al housing which is purged with dry N_2 to minimize UV absorption by atmospheric O_2. The Si substrate is positioned ~2" from the UV window at the center of a gas-dispersal ring through which process gases are introduced into the chamber. The sample holder/heater assembly is mounted off-axis on a rotary/z-motion feedthrough. The sample is heated radiatively from the backside using a quartz halogen lamp powered by an Eurotherm Model 831 power supply and controlled with a Yokogawa UP25 controller. The wafer temperature is inferred from the temperature measured by a type K thermocouple affixed to the exposed face of a dummy wafer mounted symmetrically below the plane defined by the lamp.

After surface conditioning, samples were transferred via the loadlock chamber to a multi-technique UHV surface analysis chamber. Surface atomic concentrations were determined by AES using a PHI Model PC 3017 AES subsystem comprising a 10-155 single-pass cylindrical mirror analyzer, coaxial electron gun and associated electronics. N(E) spectra were recorded using a primary beam energy of 3 keV, and tabulated AES sensitivity factors were applied.[12] The carbon fraction-remaining after processing was determined from the final-to-initial *CKLL* area ratio.

RESULTS AND DISCUSSION

Exposure of a cyclohexane-contaminated Si(100) surface to O_2 and UV light at 100 mTorr and 25°C for 5 min did not result in carbon removal nor did a similar treatment at 100°C, as evidenced by AES (Table I). We conclude that at 100 mTorr the concentration of *in situ* photo-generated O_3 is too low for effective cleaning. The kinetics of photochemical O_3 generation are strongly pressure dependent, as the mechanism involves a termolecular reaction:[13,14]

$$O(^3P) + O_2(^3\Sigma_g^-) + M \rightarrow O_3 + M \quad (1)$$

in which triplet-state atomic oxygen reacts with ground-state molecular oxygen; a collision partner M is necessary to absorb the energy released by the reaction. These experiments also demonstrate that desorption of neither cyclohexane nor its UV photo-products occurs at 100°C and 100 mTorr.

Table I. Results of reduced-pressure UV/O_2 and UV/O_3 treatments.

Feed Gas Composition	T (°C)	Pressure (mTorr)	UV Radiation λ (nm)	% C Removal effected*
O_2	40	100	185/254	0
O_2	100	100	-	0
O_2	100	100	185/254	0
2% O_3/O_2	100	100	-	14
2% O_3/O_2	40	100	185/254	68
2% O_3/O_2	100	100	185/254	95
2% O_3/O_2	100	10	185/254	83
2% O_3/O_2	100	1	185/254	68

*Treatments were 5 min in duration.

A 5-min exposure to a 2% (v/v) O_3/O_2 mixture at 100 mTorr and 100°C with simultaneous UV irradiation of the chamber contents resulted in 95% carbon removal (Table I). In contrast, exposure to the O_3/O_2 mixture in the absence of UV irradiation resulted in only a 14% decrease in carbon content. The dramatic enhancement of cleaning efficacy by UV radiation suggests that photo-generated singlet atomic oxygen and hydroxyl radical are the principal oxidizing agents. Absorption of a 254-nm photon by O_3 results in dissociation to form singlet atomic oxygen $O(^1D)$ and singlet molecular oxygen $O_2(^1\Delta_g)$:

$$O_3 + h\nu_{(254\ nm)} \rightarrow O_2(^1\Delta_g) + O(^1D) \tag{2}$$

Hydroxyl radical (OH•) is formed by subsequent reaction of $O(^1D)$ with H_2O. O_3 and $O_2(^1\Delta_g)$ are relatively weak oxidizing agents for saturated hydrocarbons, such as cyclohexane, although these species exhibit greater reactivity toward olefins.[15,16] By comparison, the bimolecular rate constants for the reactions of $O(^1D)$ and OH• with hydrocarbons are several orders of magnitude larger. The relative contributions of $O(^1D)$ and OH• to the photo-oxidation rate will depend on the H_2O partial pressure. On the basis of our data, an effect of direct photolysis of adsorbed organics cannot be excluded.

UV/O_3 surface conditioning (using *ex situ* generated O_3) is effective at removing carbon contamination at 100°C and 1 mTorr pressure, despite the low O_3 partial pressure. Cleaning efficacy decreases with pressure in the 1-100 mTorr range (Table I), but the decrease is less than might be anticipated. A possible explanation is more efficient transport of photo-generated oxidant species to the surface due to an increase of the mean free path.

The kinetics of photo-oxidative carbon removal were investigated at 100 mTorr and substrate temperatures of 40, 100 and 200°C. At each temperature, a cyclohexane-contaminated sample was subjected to a series of UV/O_3 treatments. The series was interrupted after each treatment for *in situ* off-line analysis by AES. The results (Figure 1) evidence that the surface carbon concentration does not decrease monotonically with UV/O_3 exposure time. Instead, it decreases rapidly during the initial 40 s of exposure, increases slightly between 40 and 70 s, and decreases slowly thereafter. This temporal behavior is more evident at lower substrate temperatures. Furthermore, carbon removal is incomplete even after a cumulative 5-min UV/O_3 exposure at 40°C, but nearly complete carbon removal is achieved in 180 s at the higher substrate temperatures.

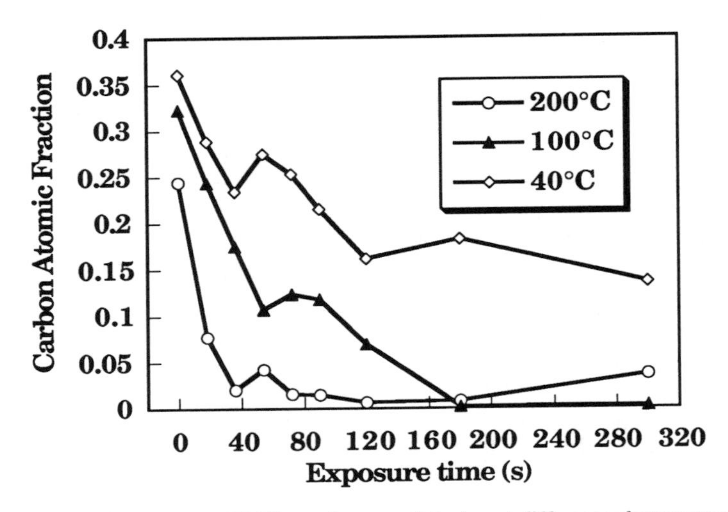

Figure 1. Carbon removal by UV/O_3 surface conditioning at different substrate temperatures.

We associate the initial "fast" kinetics regime with oxidation of cyclohexane to form readily desorbed products and the latter "slow" kinetics regime with oxidation of adsorbed carbonyl intermediates to CO_2 and H_2O. In each regime, the carbon removal rate exhibits a first-order dependence on the adsorbed carbon concentration, and the apparent activation energy is 2-3 kcal/mol. The transition from the fast kinetics regime to the slow kinetics regime coincides with the accumulation of carbonyl intermediates on the surface. Previously, Kasi and Liehr demonstrated by XPS that adsorbed carbonyl species are produced by atmospheric-pressure UV/O_2 exposure at room temperature.[17] The slight increase in the carbon atomic fraction can be attributed to spreading of adsorbed intermediates over the nascent Si oxide film (*vide infra*), as the polar carbonyl compounds are expected to interact more strongly with an oxide surface than with a hydrophobic hydrogen-terminated Si surface. As expected, the concentration of adsorbed intermediates increases with decreasing substrate temperature. Consequently, at 40°C the slow oxidation of adsorbed carbonyl species is responsible for the limited removal of carbon contamination.

The temporal variation of the surface oxygen concentration during UV/O_3 conditioning is shown in Figure 2. Auger electron escape depth analysis using the Si-Si (92 eV) and Si-O (76 eV) *LVV* transitions and an electron escape depth of 6.5 Å[18] indicates that at saturation the oxide layers are 3-4 Å thick, which is consistent with insertion of O atoms into Si-Si bonds within the first 2 atomic layers. The growth rate of the photochemical oxide layer was modeled using first-order Langmuirian adsorption kinetics:

$$R_a = \frac{d\theta}{dt} = k_a(1-\theta) \tag{3}$$

$$\theta = \frac{\text{atomic fraction O}}{\text{atomic fraction O at saturation}} = \frac{C_O}{C_O^S} \tag{4}$$

where k_a is the adsorption rate constant. The values of k_a and C_O^S which are required to fit the data are given in Table II. The curves in Figure 2 illustrate the goodness of fit. Engstrom et al.[19] report that adsorption of $O(^3P)$ on clean Si(100) 2x1 surfaces follows first-order Langmuirian kinetics for the adsorption of 3-5 monolayers. The initial sticking coefficient is near unity and independent of temperature indicating that adsorption of atomic O on clean Si(100) is unactivated. From our data, we estimate the apparent activation energy for $O(^1D)$ adsorption on hydrogen-terminated Si(100) as 2-3 kcal/mol. The existence of a small activation barrier is consistent with hydrogen-saturation of Si dangling bonds.

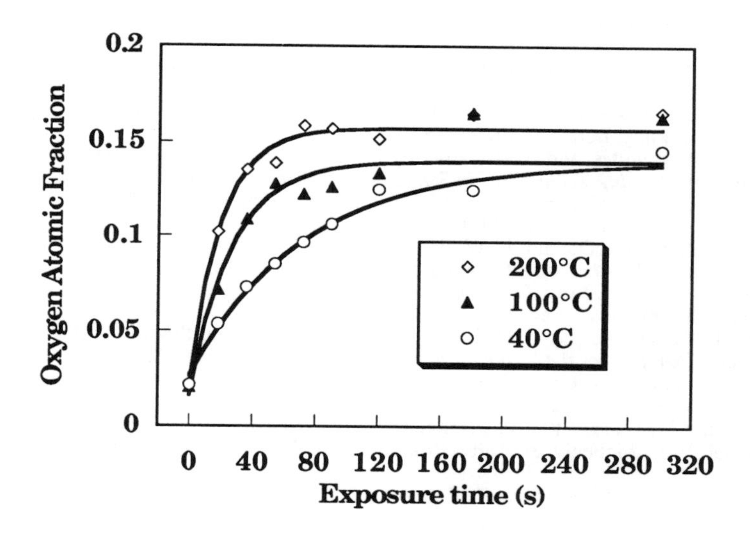

Figure 2. Kinetics of ultra-thin oxide growth by UV/O_3 surface conditioning at different temperatures. The curve fits were obtained using first-order Langmuirian adsorption kinetics.

Table II. Saturation coverages and adsorption rate constants for oxygen on hydrogen-terminated Si(100) using a first-order Langmuirian kinetics model.

Treatment Temperature (°C)	Saturation O coverage (C_O^S)	Adsorption rate constant k_a (s^{-1})
40	0.140	0.0135
100	0.140	0.0371
200	0.157	0.0510

ACKNOWLEDGMENT

This work was supported by the NSF Engineering Research Centers Program through the Center for Advanced Electronic Materials Processing at NCSU (CDR 8721505).

REFERENCES

1. S. R. Kasi and M. Liehr, J. Vac. Sci. Technol. A, **10** (4), 795 (1992).
2 M. Offenberg, M. Liehr, and G. W. Rubloff, J. Vac. Sci. Technol. A, **9** (3), 1058 (1991).
3 W. Kern and D. A. Puotinen, RCA Rev. **31**, 187 (1970).
4 T. P. Schneider, B. L. Bernhard, Y. L. Chen, and R. J. Nemanich, Mat. Res. Soc. Symp. Proc. **259**, 213 (1992).
5 H. X. Liu, T. P. Schneider, J. Montgomery, Y. L. Chen, A. Buczkowski, F. Shimura, R. J. Nemanich, D. M. Maher, D. Korzec and J. Engemann, Mat. Res. Soc. Symp. Proc. **315**, 231 (1993).
6 Y. Ma, T. Yasuda, S. Habermehl, S. S. He, D. J. Stephens and G. Lucovsky, Mat. Res. Soc. Symp. Proc. **259**, 69 (1992).
7 J. R. Vig, in *Treatise on Clean Surface Technology*; edited by K. L. Mittal (Plenum Press, New York, 1987) pp 1-26.
8 J. Ruzyllo, G. Duranko, and A. Hoff, J. Electrochem. Soc. 134, 2052 (1987).
9 M. Tabe, Appl. Phys. Lett. **45**, 1073 (1984).
10 S. Bedge, G. A. Ruggles, and H. H. Lamb in *Proceedings of the Second International Symposium on Cleaning Technology in Semiconductor Device Manufacturing*, edited by J. Ruzyllo and R. Novak (The Electrochemical Society, Pennington, NJ, 1991) pp. 112-121.
11 S. Bedge and H. H. Lamb, to be published.
12 L. E. Davis, N. C. MacDonald, P. W. Palmberg, G. E. Riach, and R. E. Weber, *Handbook of Auger Electron Spectroscopy*; (Physical Electronics, Eden Prairie, MN, 1976).
13 B. Finlayson-Pitts and J. Pitts, *Atmospheric Chemistry: Fundamentals and Experimental Techniques*; (Wiley Press, New York, 1986).
14 S. Bedge, J. McFadyen, and H. H. Lamb, Mat. Res. Soc. Symp. Proc. **259**, 207 (1992).
15 P. Warneck, *Chemistry of the Natural Atmosphere*; (Academic Press, San Diego, 1987).
16 R. P. Wayne, in *Singlet Oxygen: Physical-Chemical Aspects*; edited by A. A. Frimer (CRC Press, Boca Raton, FL, 1985).
17 S. R. Kasi and M. Liehr, Appl. Phys. Lett. **57**, 2095 (1990).
18 F. J. Himsel, F. R. McFeely, A. Taleb-Ibrahimi, J. A. Yarmoff, Phys. Rev. **38**, 6084 (1988).
19 J. R. Engstrom, D. J. Bonser, and T. Engel, Surf. Sci. **268**, 238 (1992).

CORRELATION OF ROUGHNESS AND DEVICE PROPERTIES FOR HYDROGEN PLASMA CLEANING OF Si(100) PRIOR TO GATE OXIDATION

J.S. MONTGOMERY, J.P. BARNAK, C. SILVESTRE, J.R. HAUSER, AND R.J. NEMANICH
Center for Advanced Electronic Materials Processing
North Carolina State University, Raleigh, NC 27695

ABSTRACT

Hydrogen plasma treatment was used as a cleaning and conditioning step prior to gate oxide deposition in the fabrication of cluster-based MOS field effect transistors. Surface roughness was measured by atomic force microscopy and compared to current-voltage characteristics of the MOSFET devices. The MOSFET devices were evaluated on the basis of threshold voltage, peak mobility, interface scattering, and surface roughness coefficient. Following a 10 minute H-plasma exposure at a substrate temperature of 150°C the rms roughness increased from 1.1±0.3 Å to 17±9 Å. The rms roughness for samples treated for 10 minutes at 700°C was 4±1 Å. Analysis of the MOSFET devices treated in the low temperature range (200°C) show significant degradation due to the H-plasma interaction. Threshold voltage for the devices exposed to a 2 minute H-plasma at a temperature of 200°C was 0.72±0.02 V. In contrast the threshold voltage for the 600°C, 2 minute plasma exposure was 0.86±0.03 V. The peak mobility for those devices was 370 $cm^2/V{\cdot}s$. Further device analysis was accomplished from the current-voltage measurements to extract a value of interface scattering and surface roughness scattering for each device. Interface scattering and surface roughness scattering do not increase for H-plasma process temperatures of 450 - 700°C. An H-plasma treatment for 2 minutes at 500°C also resulted in no observable increase in rms roughness, a threshold voltage of 0.92±0.03 V, a peak mobility of 410 $cm^2/V{\cdot}s$, and no increase in interface scattering and surface roughness scattering.

INTRODUCTION

The preparation of silicon surfaces has emerged as one of the limiting factors in the production of reliable small-geometry electronic devices.[1-3] Preparation of silicon surfaces involves removing unwanted surface species, maintaining or creating an atomically flat surface, and providing a surface layer or termination consistent with the following steps of device fabrication. The incorporation of contaminants or an increase in surface roughness during cleaning steps prior to deposition of the gate oxide in MOSFET devices has been shown to degrade device properties such as threshold voltage, breakdown voltage, and peak mobility.[4, 5] Therefore, the pre-gate cleaning steps must be carefully controlled. Plasma cleaning appears to be a reasonable alternative to address these issues of surface preparation. *In situ* processing allows for multi-step processing without exposure to ambient contaminants.[6-9] Previous studies have demonstrated that H-plasma processing can remove surface carbon while reducing levels of residual oxygen.[10-13] The H-plasma treatment has also been shown to effectively terminate the silicon surface with hydrogen which can easily be displaced in subsequent processing steps. In addition, plasma cleaning uses significantly less chemicals than equivalent wet bench cleaning steps. Although much research has been performed in analyzing hydrogen/silicon surface interactions, the implications of these analyses in the operation of fully fabricated devices has yet to be completely realized. In this study rms roughness was compared to threshold voltage, peak

Mat. Res. Soc. Symp. Proc. Vol. 386 © 1995 Materials Research Society

mobility, interface scattering, and surface roughness scattering which were derived from the current-voltage drive characteristics of the fabricated MOSFET.

EXPERIMENTATION

The approach for this experiment was divided into device fabrication, surface analysis, and finally correlations between MOSFET performance and surface characterization. Two sets of wafers were subjected to identical conditions of *ex situ* wet chemistry and *in situ* plasma processing. Following plasma processing, one set of wafers underwent the subsequent steps for MOSFET production; the second set of wafers was analyzed to determine the rms roughness of the surface. The wafers used for the experiment were 100 mm dia., boron-doped, (100) silicon with a nominal resistivity of 0.05 - 0.10 Ω·cm. *Ex situ* wet cleaning consisted of an SC1/SC2/HF series of chemical baths which resulted in an hydrogen terminated surface prior to *in situ* processing. The SC1 solution consisted of $H_20:NH_4OH:H_2O_2$ (5:1:1). The SC2 solution consisted of $H_20:HCl:H_2O_2$ (5:1:1). Both SC1 and SC2 were held at 80°C. The "HF last" was a 0.5% solution of HF in H_20. The sequence of the *ex situ* preparation was *i*) SC1/10 minutes, *ii*) DI rinse/5 minutes, *iii*) HF dip/≈ 20 seconds, *iv*) DI dip-rinse, *v*) SC2/10 minutes, *vi*) DI rinse/5 minutes, *vii*) HF dip/≈ 10 seconds, *viii*) DI dip-rinse, *ix*) spin dry. The samples were then transferred (< 2 min.) to the processing chambers for *in situ* treatments.

In situ processing was performed in a proto-type cluster system.[8] This experiment utilized two of the available modules on the 3-station cluster which includes a remote plasma cleaning module (RPC) for cleaning and conditioning, a remote plasma enhanced chemical vapor deposition module (RPECVD) for deposition of a low temperature gate oxide, and a rapid thermal chemical vapor deposition module (RTP) for deposition of oxides and poly-silicon. In addition to the process modules there is an entry/exit station which can store up to 25 wafers.

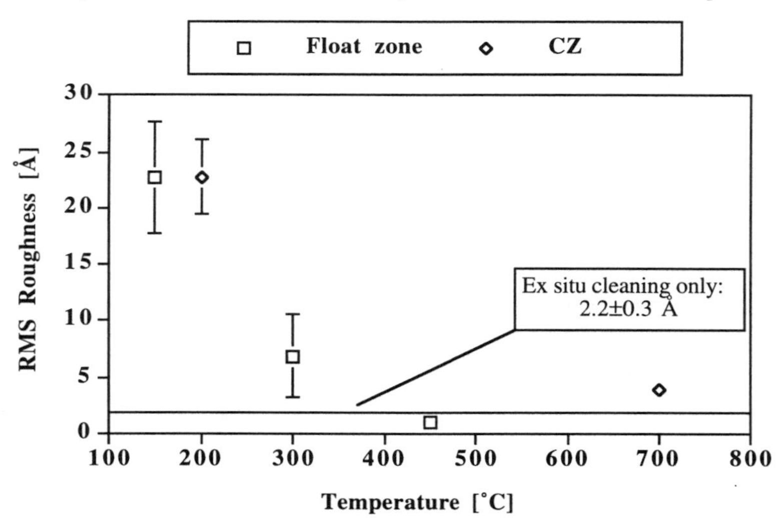

Figure 1.RMS roughness versus substrate temperature following H-plasma processing. Surface roughness prior to H-plasma treated is indicated by the horizontal line at 2.1 Å.

Samples designated for surface analysis were transferred to the cleaning module and subjected to an H-plasma exposure. Samples designated for MOSFET fabrication were subjected to an H-plasma exposure and then transferred to the RTP for a 100 Å gate oxide deposition and then a 1500 Å layer of poly-silicon. The RTP module operation and deposition have been described elsewhere.[14]

The RPC module is a stainless steel UHV system with a nominal base pressure of 5×10^{-9} Torr. During processing a single gas line directed hydrogen to a quartz tube coupled to the chamber. An RF power generator at 13.56 MHz was attached to a coil encircling the quartz tube. Molecular hydrogen dissociates into atomic hydrogen and affects the surface of the silicon wafer positioned ≈ 10 cm below the downstream end of the quartz tube. Samples were treated with combinations of temperature and exposure duration ranging from 200 - 700°C and 2 - 60 minutes, respectively. Process pressure was 15 mTorr with a hydrogen flow of 85 sccm and was maintained by a throttling valve/pressure transducer feed-back loop. The power used in generating the plasma was held constant at 20 W.

Evaluation of the effects of H-plasma cleaning on the surface morphology was accomplished with an ambient 10 μm AFM by Park Scientific, Inc. The cantilevers used in this experiment had silicon nitride pyramidal tips with an aspect ratio of 3:1 and a nominal tip radius of 100 Å. Scan sizes ranged from 1 - 5 μm at 512 points/trace. Scan rates ranged from 0.5 - 2 Hz. RMS values were obtained from no less than 3 separate scans per wafer. No determination was made of the force on the tip during scanning.

Final steps in the fabrication of the MOSFET devices was performed in separate facilities. The transistors evaluated in this experiment had gate areas of $(100\ \mu m)^2$, $(300\ \mu m)^2$, and $(500\ \mu m)^2$ with a gate width/length ratio of 1. Electrical characterization was conducted on a dedicated I-V test station which utilizes a computer controlling an array of Keithley Source

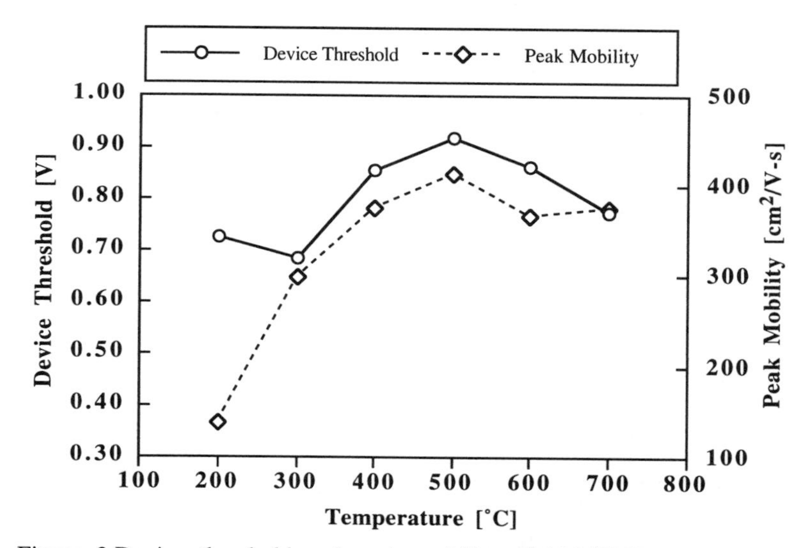

Figure 2.Device threshold and peak mobility of MOSFETs vs. substrate temperature during H-plasma processing. Plasma duration was 2 min. The theoretical threshold voltage for these devices was 0.85 V.

Measuring Units. For each wafer 5 transistors per 3 different chip locations were tested at a drain voltage (V_d) of 0.1 V. The semi-empirical mobility model put forth by Shin, et al.[15-17] was used to evaluate I-V data from the MOSFET devices. Using determined values for the gate oxide thickness and substrate doping, a non-linear least squares fit was obtained from the experimental current-voltage data. From the fitted curve, we then extract, for each device, values for threshold voltage, peak mobility, interface scattering, and roughness scattering coefficient.

RESULTS

The silicon substrates have been analyzed to determine rms roughness following exposure to an H-plasma. The control surface for these studies was only subjected to the standard *ex situ* wet cleaning described previously and was determined to have an rms roughness of 1.1±0.3 Å. The rms roughness as a function of the substrate temperature during a 10 minute plasma exposure is shown in Figure 1. Samples treated at 150°C show a significant increase in atomic scale roughness. The rms roughness for these samples was 23±5 Å. The samples processed with a substrate temperature of 450°C show no significant increase in the rms roughness as compared to the control surface prior to processing. The rms roughness obtained for these samples was 1.0±0.1 Å. In addition, for samples treated at substrate temperatures from 450 - 700°C, no significant increase in surface roughness was observed for processing durations up to 10 minutes.

Devices were analyzed based on threshold voltage, peak channel mobility, interface scattering, and the surface roughness coefficient using a previously published model for current-voltage drive characteristics of MOSFET devices[15]. Figure 2. shows the temperature dependence of threshold voltage and peak mobility for 2 minute exposures. Based on device dimensions and material properties, a theoretical threshold voltage of 0.85 V was determined for

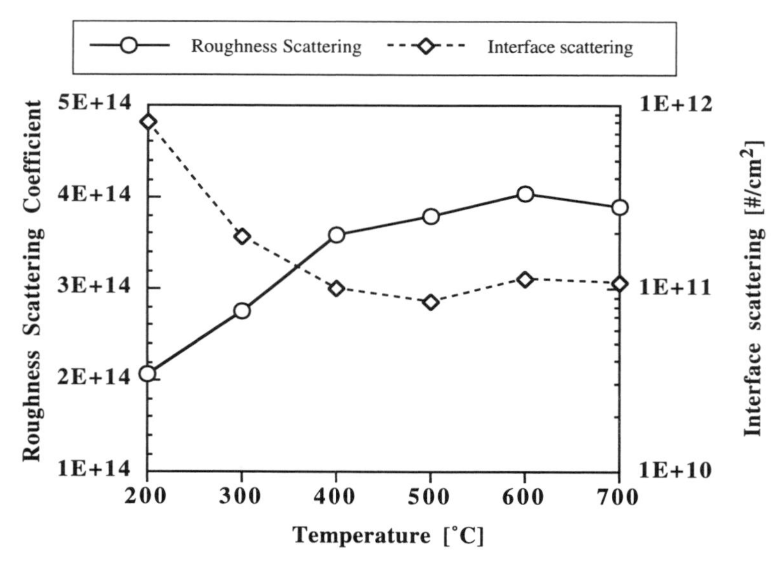

Figure 3. Roughness scattering coefficient and interface scattering vs. substrate temperature. The H-plasma exposure was 2 minutes in all cases.

these devices. The peak mobility also shows a similar trend (Figure 2.) with samples treated < 400°C suffering a significant drop in electron mobility. In Figure 3 values of interface scattering and the surface roughness coefficient are plotted as a function of substrate temperature. A reduction in the roughness scattering mechanisms is seen in the samples treated in the high temperature region (450 - 700°C). The MOSFET devices tested show consistent variations of threshold voltage, peak mobility, interface scattering, and surface roughness scattering as a function of substrate temperature during processing.

CONCLUDING REMARKS

The results display a consistent picture of the effect of the H-plasma treatment on silicon surfaces prior to device fabrication. In the low temperature region (≈ 200°C) the hydrogen is involved in an etching reaction which increases surface roughness, near surface damage, and surface defects. AFM measurements indicate an increase in surface roughness from 2.0±0.3Å before plasma processing to 23±3Å following the plasma exposure. These results along with other studies involving residual gas analysis and transmission electron microscopy suggests that the low temperature processing promotes an etching reaction.[18] In addition, studies conducted by this group and other researchers have shown an increase in the level of the interface roughness and the density of platelet defects following H-plasma cleaning at 150°C.[19, 20]

In the high temperature regime (450 - 700°C) the etching mechanism does not appear to be active. Samples treated with an H-plasma for 10 minutes at 700°C had an rms roughness of 4±1Å which is only a slight increase in roughness compared to the starting surface. To the limits of our AFM analysis, the samples treated with an H-plasma for both the 2 and 10 minute exposures exhibit no increase in surface roughness.

With an increase in interface roughness as observed by AFM, the threshold voltage, electron mobility, and scattering mechanisms in the inversion layer should also suffer. For samples treated with a 2 minute exposure at temperatures ≤ 300°C, the threshold voltage decreases by as much as 0.2 V; and the peak mobility drops to below 200 cm^2/V-s. In the high temperature regime, the threshold voltage is within experimental error of the theoretical value of 0.85 V; and the peak mobility reaches its maximum value of 200 cm^2/V-s. As seen in Figure 3. the interface scattering and roughness scattering also follow the temperature trend with no significant changes observed for samples treated at substrate temperatures ≥ 500°C. This compares well with the AFM observations at substrate temperatures ≥ 450°C which show very little increase in surface roughness.

ACKNOWLEDGMENTS

This work has been partially supported by the National Science Foundation Engineering Research Centers Program through the Center for Advanced Electronic Materials Processing (Grant CDR 8721505). The authors would like to expressly thank the following people for the hours of processing, device testing, and helpful discussions: T.P. Schneider, R.J. Carter, J.B. Flanigan III, C.G. Parker, J. McEwan.

REFERENCES

1. R.B. Fair, Proc. IEEE, **78**, 1687 (1990).
2. R. Iscoff, Semiconductor International, **16**, 58 (1993).
3. P.H. Singer, Semiconductor International, **15**, 36-39 (1992).
4. T. Ohmi, Proc. IEEE, **81**, 716 (1993).
5. P.O. Hahn, M. Grundner, A. Schnegg and H. Jacob, Appl. Surf. Sci., **39**, 436 (1989).
6. R.D. Compton, Semiconductor International, **16**, 11 (1993).
7. J.R. Hauser, N.A. Masnari and M.A. Littlejohn in *Rapid Thermal Annealing/Chemical Vapor Deposition and Integrated Processing*, edited by D. Hodul (Mater. Res. Soc. Proc., **146**, Pittsburgh, PA, 1989) p. 15-26.
8. N.A. Masnari, J.R. Hauser, G. Lucovsky, D.M. Maher, R.J. Markunas, M.C. Ozturk and J.J. Wortman, Proc. IEEE, **81**, 42-58 (1993).
9. P. Singer, Semiconductor International, **16**, 46-51 (1993).
10. S. Banerjee, A. Tasch, T. Hsu, R. Qian, K. Kinosky, J. Irby, A. Mahajan and S. Thomas in *Chemical Surface Preparation, Passivation, and Cleaning for Semiconductor Growth and Processing*, edited by R.J. Nemanich (Mater. Res. Soc. Proc., **259**, Pittsburgh, PA, 1992) p. 43.
11. T.P. Schneider, J.S. Montgomery, H. Ying, J.P. Barnak, Y.L. Chen, D.M. Maher and R.J. Nemanich, in *Third International Symposium on Cleaning Technology in Semiconductor Device Manufacturing*, edited by J. Ruzyllo and R. Novak (Electrochemical Society, New Orleans, 1993) p. 329-338.
12. K. Ueda, Appl. Surf. Sci., **60/61**, 178 (1992).
13. R.E. Thomas, M.J. Mantini, R.A. Rudder, D.P. Malta, S.V. Hattengady and R.J. Markunas, J. Vac. Sci. Technol. A, **10**, 817 (1992).
14. A. Bayoumi and J.R. Hauser, in *Proceedings of the 2nd International RTP Conference*, edited by R. Fair and B. Logek (RTP '94, Monterey, CA, 1994) p. 130.
15. H. Shin, G.M. Yeric, A.F. Tasch and C.M. Mazair, Solid-State Electron., **34**, 545-552 (1991).
16. D.S. Jeon and D.E. Burk, IEEE Trans. Electron Devices, **36**, 1456-1463 (1989).
17. J.T. Watt and J.D. Plummer, Digital Technology Papers, 81 (1987).
18. R.J. Carter, T.P. Schneider, J.S. Montgomery and R.J. Nemanich, J. Electrochem. Soc., **141**, 3136 (1994).
19. H.X. Liu, T.P. Schneider, J.S. Montgomery, Y.L. Chen, A. Buczkowski, F. Shimura, R.J. Nemanich and D.M. Maher in *Surface Chemical Cleaning and Passivation for Semiconductor Processing*, edited by G.S. Higashi, E.A. Irene and T. Ohmi (Mater. Res. Soc. Proc., **315**, Pittsburgh, PA, 1993) p. 231.
20. N.M. Johnson, F.A. Ponce, R.A. Street and R.J. Nemanich, Phys. Rev. B, **35**, 4166 (1987).

RIE PASSIVATION LAYER REMOVAL BY REMOTE H-PLASMA AND H_2/SiH_4 PLASMA PROCESSING

Hong Ying, J.P. Barnak*, Y.L. Chen*, and R.J. Nemanich**
Department of Electrical and Computer Engineering,
*Department of Materials Science and Engineering,
**Department of Physics, North Carolina State University, Raleigh, NC 27695-8202

ABSTRACT

Remote H-plasma and H_2/SiH_4 plasma processes were studied as potential dry cleaning processes following reactive ion etching (RIE). The processes were compared to a process of UV/ozone followed by an HF dip. The native oxide from Si(100) substrates was removed with an RIE etch of CHF_3/Ar. The RIE process produced ~150Å of a continuous fluorocarbon (CF_x) passivation layer on the Si surface. For the post-RIE-cleaning three approaches were studied and compared including (1) uv-ozone exposure followed by an HF dip, (2) remote H-plasma exposure, and (3) remote H_2/SiH_4 plasma exposure. Auger electron spectroscopy (AES) was used to investigate the surface chemical composition, and AFM was used to measure changes in surface roughness. All three processes showed substantial removal of the passivation layer. The CF_x polymer was completely removed in less than 1 min for samples exposed to a 100W remote H-plasma at 15mTorr and 450°C. With the addition of ~0.1% of SiH_4, the remote H_2/SiH_4 plasma also showed increased removal of residual oxygen contamination. The surface roughness of the plasma processed surfaces increased slightly.

INTRODUCTION

RIE processes are becoming more important in VLSI fabrication as device features become smaller. The advantages of RIE include high anisotropy and compatibility with in-situ, single-wafer processing.

Anisotropic SiO_2/Si selective etching has been studied[1-10] since it is an important step prior to producing contact holes[1] or to Si epitaxial growth[2]. Anisotropy is largely characteristic of the RIE technique, but selectivity is dependent upon the etch gases. The use of CF_4, CF_4/x% H_2, and CHF_3 results in different SiO_2:Si selectivity[2-6]. It has been suggested that the deposition of a fluorocarbon (CF_x) film is crucial to achieve the SiO_2/Si etch selectivity[5,6]. The CF_x film is formed on the Si surface which prevents fluorine from attacking the Si, but not on the SiO_2. Therefore the Si etch rate decreases, while the SiO_2 etch rate remains relatively constant. The CF_x film must be removed prior to the next processing step. Processes that have demonstrated success as the post RIE treatment include thermally grown sacrificial oxide film followed by the oxide film removal[2], O_2 plasma/buffered HF dip[3,6,7], UV/O_2 exposures followed by diluted HF dip[8]. In each case, however, a wet etch step is involved.

It would be advantageous to use a dry cleaning process after RIE for the following reasons: (1) to be compatible with in-situ, single-wafer processing, (2) to reduce chemical waste, (3) to obtain a more effective clean than wet chemical cleaning as device features approach 0.1µm. A cleaning procedure based on hydrogen plasma has been reported by J.P. Simko *et al*[9]. Their results indicate that a 50Å fluorocarbon film is reduced to a 10Å

Mat. Res. Soc. Symp. Proc. Vol. 386

carbon film which contains little fluorine. Post RIE treatment in an NF_3/Ar plasma has been studied by H. Cerva *et al*[10]. The CF_x polymer film is completely removed as well as the defect zone, but the surface roughness increases (up to 5nm).

In previous studies from this laboratory, it has been shown that the surface morphologies are dependent on substrate temperature and pressure during H-plasma cleaning[11,12]. The 450°C, 15mTorr, 20W H-plasma exposure of Si(100) substrates leads to a H-passivated surface, and the surface is smooth with no observable subsurface defects. AFM measurements indicated no discernible increase in the RMS roughness for these processes. Furthermore, the 450°C H-plasma process did not induce the near surface platelet defects often observed for H-plasma exposure with wafer temperatures less than 300°C[12]. In this study, the same H-plasma process was utilized as a post RIE cleaning process. The surface chemical composition and surface roughness were studied.

EXPERIMENTAL

Reactive Ion Etching Process

The substrates used in this study were Si(100), 25mm diameter, boron doped (p-type), with a resistivity of 0.8-1.2Ω-cm. An CHF_3/Ar RIE was used to selectively etch the Si native oxide from the Si substrates. RF power was 85W (power density was $0.15W/cm^2$), and the gas flow rates were 40sccm of CHF_3 and 10sccm of Ar. The chamber pressure was set to 60mTorr for 1 min at which the SiO_2 etch rate was 75Å/min, and then to 90mTorr for 0.5 min at which a continuous CF_x polymer film was deposited on the Si surface.

Post RIE Cleaning

Three approaches were studied and compared.

(1) UV-ozone & HF dip

The cleaning consists of a 10 min uv-ozone exposure followed by a 1 min dilute HF(10:1) dip. The procedure was repeated twice.

(2) Remote H-plasma Cleaning

The H-plasma was generated inductively by an rf field through a copper coil encircling a quartz tube[11,12]. The samples were positioned 40cm downstream relative to the center of the plasma tube, and were exposed to a H-plasma at 15mTorr and 450°C. The rf power was 20W or 100W.

(3) Remote H_2/SiH_4 plasma Cleaning

A dilute mixture of SiH_4 in H_2 was flowed into the plasma chamber. The gas flow rates were 86sccm of H_2, 10sccm of $H_2/1\%$ SiH_4, and the chamber pressure was 25mTorr. The samples were exposed to a 100W H_2/SiH_4 plasma at a substrate temperature of 450°C.

The plasma chamber was connected to a UHV surface analysis system which included AES and LEED[11]. Following the cleaning processes, the samples were transferred to the analysis system. The AES was used to investigate the surface chemical composition. The sample surfaces were characterized with atomic force microscopy(AFM) and cross-sectional transmission electron microscopy(TEM). The AFM used in this study was a Park Scientific SFM-BD2. The piezoscanner had a 10μm scanning range in both X and Y directions.

RESULTS AND DISCUSSION

Figure 1 represents a cross-sectional TEM micrograph of the reactive ion etched Si. Approximately 150Å of Ti was deposited in order to distinguish between the CF_x film and the epoxy glue used in the TEM specimen preparation. It is observed that the RIE resulted in the formation of ~150Å of a continuous CF_x passivation layer on the surface. In addition, a layer with a high density of defects is observed within several hundred Å of the surface.

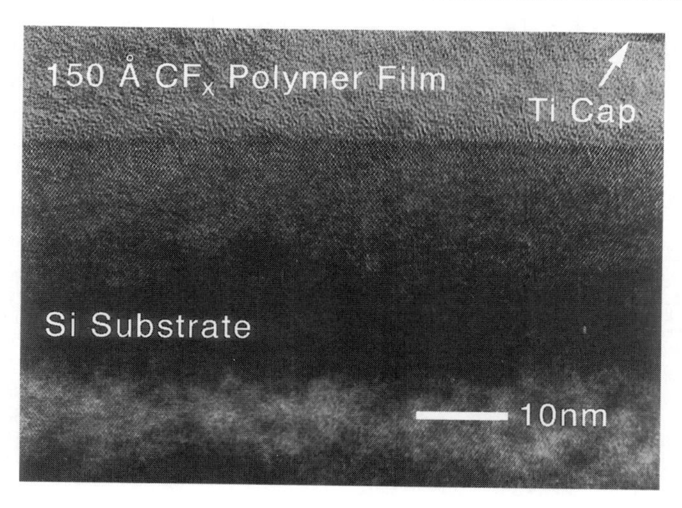

Fig. 1 Cross-sectional TEM micrograph of a Si wafer after the reactive ion etching process.

The AES spectra of the CF_x film removal by remote H-plasma is shown in Fig. 2. By exposing to 100W H-plasma at 15mTorr and 450°C, the CF_x polymer film was completely removed in less than 1 min. Longer exposure leads to a further decrease but not complete removal of the oxygen contamination. The oxide was apparently formed during the RIE process and might be buried in the defect sites at the near surface region. For comparison, the AES C/Si and O/Si peak-to-peak ratio are plotted versus the post-RIE-cleaning time in Figure 3 for the different cleaning approaches. In each case the F is below the AES detection limit. This agrees with the experimental observation that F is more effectively removed from the contamination layer than C[9]. The solid lines connecting the points are to correlate similar processes, and should not be viewed as representing the actual time dependence of the contaminant removal process. The data does establish the trends of lower surface contaminant with longer processing times. All three processes exhibit a significant reduction in surface contaminant with essentially submonolayer carbon and oxygen residuals. Both in situ techniques show no detectable C after a 10 min exposure, while the ex situ approach of uv-ozone & HF dip shows some C that could be attributed to the wafer transfer into the AES system. Both plasma processes show slight increases of oxygen after the removal of the RIE CFx passivation layer. It is probable that this oxygen formed during the RIE process as the passivation layer was formed. With the addition of a small amount of SiH_4, the remote H_2/SiH_4 plasma is more effective for oxygen removal[13]. In another study in this same volume it was shown that the very dilute H_2/SiH_4 plasma results in SiO_2 etching with no Si deposition on any oxide surfaces[13].

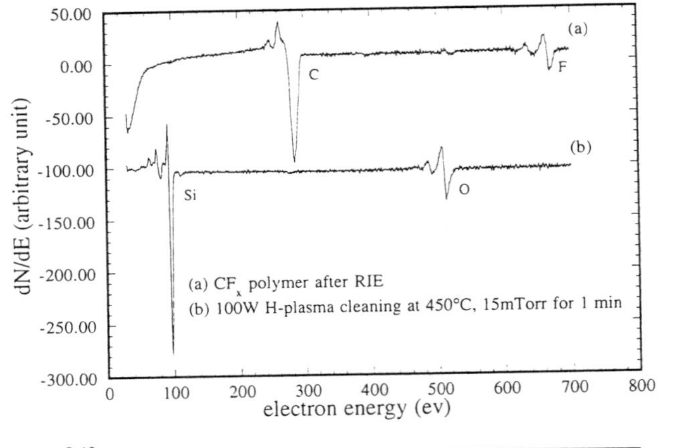

Fig.2 RIE passivation layer removal by remote H-plasma cleaning.

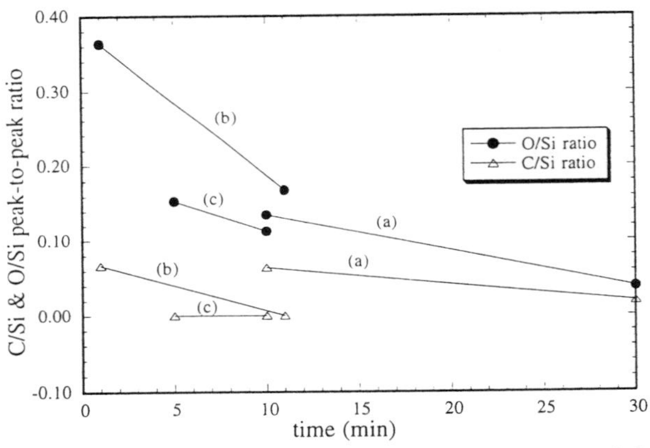

Fig. 3 AES C/Si and O/Si peak-to-peak ratio vs post-RIE-cleaning time.
(a) uv-ozone exposure followed by dilute HF dip
(b) H-plasma cleaning at 450°C, 15mTorr. The power was 20W for the first 1min and 100W after
(c) 100W H_2/SiH_4 plasma cleaning at 450°C, 25mTorr

Figure 4 shows representative AFM images of the surfaces. The RMS surface roughness increased from 3Å to 8Å after the H-plasma exposure for the conditions studied.

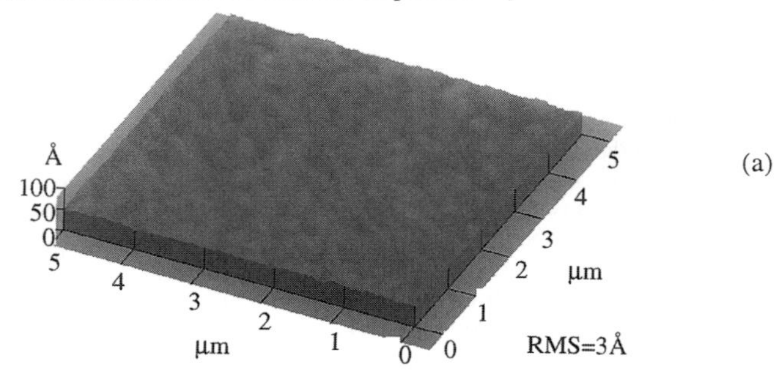

(a)

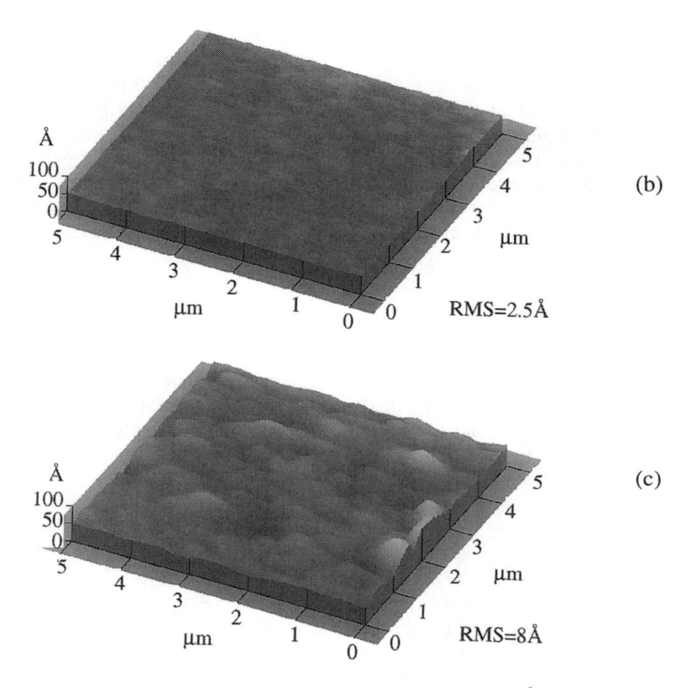

Fig. 4 AFM images of the samples (a) after RIE with ~150Å CF_x passivation layer
(b) after RIE and uv-ozone, HF dip
(c) after RIE and 100W H_2/SiH_4 plasma cleaning at 450°C, 15mTorr for 10 min

Although there is a small increase in roughness, low energy electron diffraction (LEED) measurements display the expected 2x1 reconstruction indicating that the surface is crystalline and not amorphized by either the RIE or the H-plasma clean. Future studies will explore whether the epi quality will be effected by the small increase in roughness.

It is also relevant to comment on the subsurface defect formation. The presence of subsurface defects were identified in the Si after the RIE. In addition it has been shown that H-plasma processing can result in the formation of platelet and other extended defects in the near surface region. In another study from this laboratory, it has been shown that H-plasma exposure results in substantial defect formation at wafer temperatures of 100 to 300°C, but for H-plasma exposure at 450°C no subsurface defects are observed in plan view or cross section TEM measurements[12]. We employed similar H-plasma processing conditions in the experiments described here, so it is anticipated that there should be no increase in subsurface defects due to the H-plasma. Future studies will explore issues related to whether the H-plasma produces defects in the RIE processed surfaces or whether the RIE damage can be affected by the process.

CONCLUSIONS

The results presented here display three processes that have potential to be employed as a post RIE clean after the formation of a CF_x passivation layer. The uv/ozone process followed by an HF dip showed the lowest oxygen residual, but a slightly increased C residual attributed to the exposures following the HF dip. The two plasma processes demonstrate effective and rapid removal of the passivation layer, and only slightly increased oxygen levels and rms surface roughness. The presence of a small amount of silane in the plasma results in a reduction in the oxygen concentration. The H-plasma exposure at a wafer temperature of 450°C has previously been shown to result in cleaning without defect formation, and we suggest this may be the case here also. Future studies will focus on issues related to epi growth and defect structures. The plasma processes described here could prove useful as cleans in single wafer clustered systems.

ACKNOWLEDGMENTS

This work is supported by the NSF Engineering Research Center for Advanced Electronic Materials Processing at North Carolina State University. The authors would specially like to thank K. Violette, M.C. Öztürk for the valuable discussions and R. Haymaker, J. Todd in the Microelectronics Lab of NCSU for their help in RIE processes.

REFERENCES

1. G.S. Oehrlein, Phys. Today. **39(10)**, 26 (1986)
2. J.C. Lou, W.G. Oldham, H. Kawayoshi, and P. Ling, J. Appl. Phys. **71**, 3225 (1992)
3. L.M. Ephrath and R.S. Bennett, J. Electrochem. Soc., **129**, 1822 (1982)
4. M.A. Jaso, and G.S. Oehrlein, J. Vac. Sci. Technol. **A6**, 1397 (1988)
5. G. E. Potter, G.H. Morrison, P.K. Charvat, and A.L. Ruoff, J. Vac. Sci. Technol. B10, 2398 (1992)
6. G.S. Oehrlein, G.J. Scilla, and S.J. Jeng, Appl. Phys. Lett. **52**, 907 (1988)
7. K. Reinhardt, B. Divincenzo, C.-L. Yang, P. Arleo, J. Marks, P. Mikulan, T. Gu, and S. Fonash, Mat. Res. Soc. Symp. Proc. vol 315, 267 (1993)
8. T. Gu, R.A. Ditizio, S.J. Fonash, O.O. Awadelkarim, J. Ruzyllo, R.W. Collins and H.J. Leary, J. Vac. Sci. Technol. **B12**, 567 (1994)
9. J.P. Simko and G.S. Oehrlein, J. Electrochem. Soc., **138**, 277 (1991)
10. H. Cerva, E.-G. Mohr, and H. Oppolzer, J. Vac. Sci. Technol. **B5**, 590 (1987)
11. T.P. Schneider, J. Cho, Y.L. Chen, D.M. Maher, and R.J. Nemanich, Mat. Res. Soc. Symp. Proc. vol 315, 197 (1993)
12. T.P. Schneider, J.S. Montgomery, H. Ying, J.P. Barnak, Y.L. Chen, D.M. Maher, and R.J. Nemanich, Electrochemical Soc. Proc. vol **PV94-7**, 329 (1994)
13. J. Barnak, H. Ying, Y.L. Chen, and R.J. Nemanich, Mat. Res. Soc. Symp. Proc. (this volume)

MONITORING OF THE SURFACE SPECIES ON SILICON AFTER CHEMICAL CLEANING BY FTIR SPECTROSCOPY

CHAN-HWA CHUNG, CHANG-KOO KIM, AND SANG HEUP MOON
Department of Chemical Engineering, Seoul National University, Seoul 151-742, Korea

ABSTRACT

For the analysis of as-cleaned surface, we have used a the unique IR method that uses a high-surface-area porous sample. We have observed by experiments that the oxide growth rate on silicon is reduced to a minimum when the surface is treated with a proper amount of HF vapor obtained from 1% HF solution. FTIR and XPS observations of the treated surface suggest that the oxide growth rate is closely related to the amount of the surface fluorides. In UV/O_2 cleaning process, we have observed experimentally that addition of water vapor to cleaning gas stream enhances the cleaning efficiency by as much as 1.3 times.

INTRODUCTION

When silicon is exposed to air or other oxidants, the surface is oxidized into a thin oxide layer that causes problems in IC fabrication and eventually degrades the device performance. Moreover, a great amount of organic and inorganic impurities adhere onto Si wafer surface easily. Accordingly, many efforts have been made to remove the oxide and organic impurities from or suppress them on the surface [1].

A common method for this purpose is to treat the sample in aqueous solutions such as buffered HF, NH_4OH/H_2O_2, and H_2SO_4/H_2O_2 solutions. After dipping in an HF solution, the surface is terminated by hydrides that retard the surface oxidation [2-4]. However, these wet methods give other problems such as waste disposal and the difficulty of controlling impurities during rinse-dry cycles. Dry cleaning method is therefore considered because the gaseous chemicals used in the method contain fewer amounts of impurities [5].

Recently, Haring *et al.* [6] have obtained the F-terminated surface using XeF_2 as a cleaning gas and found that the surface is relatively stable in oxygen. The result and those from other similar studies [7-9] suggest that an improved dry-cleaning method may be developed by a proper selection of the cleaning chemicals and conditions.

For removal of organic impurities, UV/O_2 treatment technique has been developed and recommended as an effective and easy cleaning method [10].

In this work, we have studied the silicon surface treated with HF vapor and have observed that the growth rates of the surface oxides are reduced by the treatment. The surface after the treatment has been characterized with Fourier transform infrared spectroscopy (Midac) and X-ray photoelectron spectroscopy (Perkin-Elmer), and the role of the surface fluorides in retarding the surface oxidation has been discussed based on the results.

We also have studied the mechanism of hydrocarbon (organic contaminant) removal from the silicon surface by vapor phase UV/O_2 cleaning with or without the additional water vapor.

Mat. Res. Soc. Symp. Proc. Vol. 386 © 1995 Materials Research Society

EXPERIMENTAL DETAILS

Experiments were made with 0.5x12x13 mm^3 p-type Si(100) whose surface had been anodized into a porous layer following the procedure of Bomchil *et al.* [11]. The area of the sample surface after the anodization was high, about 2 m^2 per cm^2 of the sample cross-section, enough to allow observation of the surface species by IR transmission. Furthermore, the porous Si obtained by the above method retained the crystallinity of the original monocrystalline Si wafer. The porous samples were then immersed in an HF:H_2O (1:10) solution for 4 minutes and dried in a nitrogen atmosphere before treatment with HF vapor. After the treatment, the IR results indicated a total removal of the native oxide.

HF vapor was obtained by flowing dry nitrogen through an evaporator containing an HF solution. Among the conditions of the sample treatment, we varied only the concentration of the HF solution to obtain different amounts of HF vapor, assuming that the latter increased with the solution concentration. The other conditions were fixed, i.e., the evaporator temperature to 30°C, the nitrogen flow rate to 150 sccm, and the treatment period to 20 minutes.

For the study of hydrocarbon removal, the samples with a porous surface layer were intentionally contaminated with vacuum oil. A low-pressure mercury lamp (Kumkang, Co.) was used as a UV light source. A synthetic quartz (Suprasil) window, transparent to the 185 nm and 254 nm lines, was placed between the lamp and substrate. High purity oxygen was introduced directly from the pressure tank to the reactor or bubbled through the water maintained at 70°C to add water vapor to the process stream. The oxygen flow rate and the treatment period were 9 sccm and 1 hr, respectively.

RESULTS AND DISCUSSION

Suppression of Oxide Growth

Figure 1 shows the temperature-programmed IR spectra of the surface species which change with temperatures. Figure 1 shows that the hydride peaks (910, 2089, and 2116 cm^{-1} [12]) disappear at the temperatures above 350°C, which is accompanied by appearance of a large broad peak at about 1000 cm^{-1} assigned to Si-O vibration [13]. This result confirms that the surface covered with the hydrides is protected from oxidation but one without the hydrides is easily oxidized [2-4].

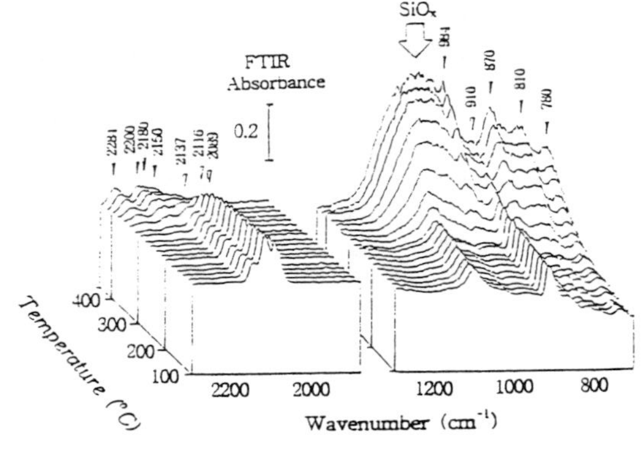

Figure 1. Changes in the IR spectrum with a sample heating at 10°C/min.

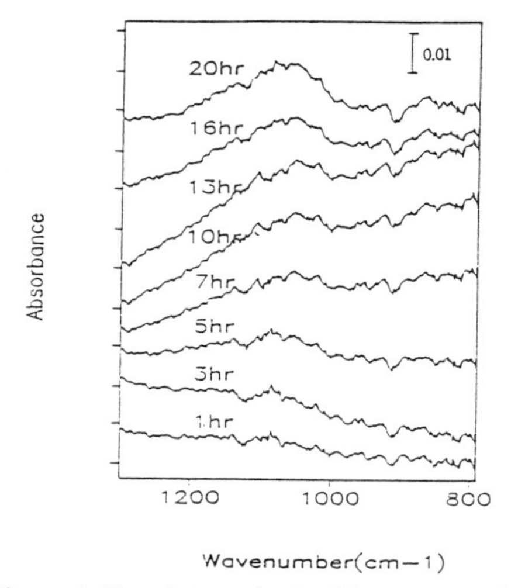

Figure 2. The changes in the IR spectrum of Si(100) treated with HF vapor during exposure to air.

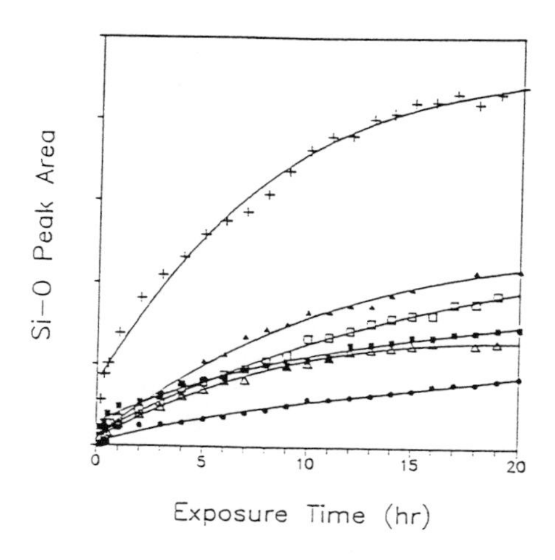

Figure 3. The growth of Si-O IR peak area on the HF-treated sample by exposure to air. (+; no HF treatment, ▲; HF vapor from 0.5%, ■; 0.7%, ●; 1%, △; 5%, □; 50% HF solution)

Figure 2 shows changes in the IR spectrum with time after the HF-treated silicon surfaces were exposed to air. Figure 2 represents the spectral differences from the reference one taken immediately after the HF treatment. Accordingly, a broad peak near 1000 cm^{-1} grows with time, indicating growth of the native oxide after the exposure to atmosphere.

The growth rate of the native oxide varies with the extent that the sample has been treated with HF vapor. In Figure 3, the oxide growth rates of several samples that have been treated with HF vapor from solutions of different concentrations are compared. Apparently, the sample without the HF treatment shows the highest growth rate, thus indicating that the HF-vapor treatment is effective for suppressing the oxide growth.

A close observation of Figure 3 reveals that the oxide growth rate is not reduced proportionally but goes through a minimum with the concentration of the HF solution used for the treatment. Namely, the sample treated with the vapor from a 1% HF solution shows the lowest growth rate compared to the other cases. The trend is shown more clearly in Figure 4, which is a plot of the oxide IR peak areas after 20-hour exposure versus the concentrations of the HF solution.

The above observations may be summarized into two as follows. One is that the HF treatment suppresses the native oxide growth on the wet-cleaned silicon surface, and the other is that an optimum amount of HF vapor is needed to reduce the growth rate to a minimum.

About the role of fluorides on the silicon surface, the following explanations have been proposed [3,14,15]. When the silicon substrate is simply wet-cleaned with the HF solution, the surface is terminated mostly by hydrides such as SiH, SiH_2, and SiH_3, which are relatively stable enough to protect the surface from oxidation. The surface, however, is subject to gradual oxidation in air because a few exceptionally reactive sites on the surface, usually the crystalline defects, initiate the oxidation [3,14].

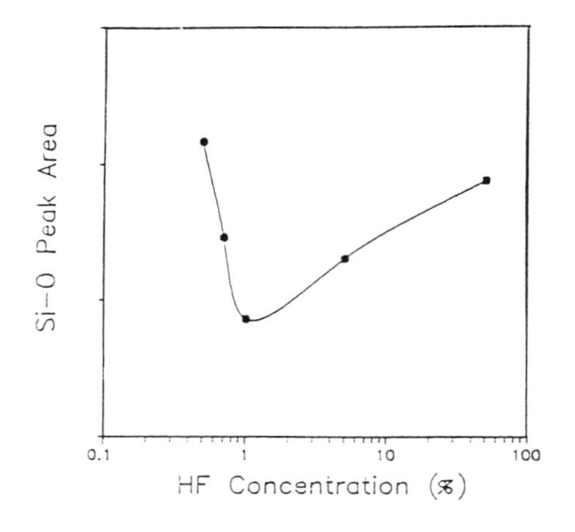

Figure 4. Dependence of the oxide IR peak area measured after 20 hr exposure to air on the concentration of the HF solution.

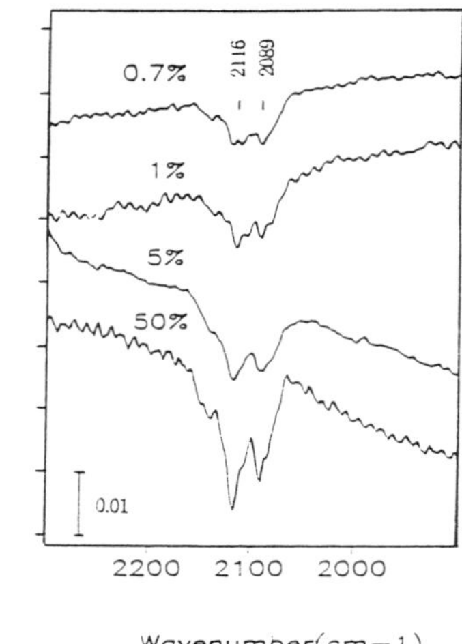

Figure 5. The changes in the IR spectrum of the HF-treated Si(100) with the concentration of the HF solution.

When the H-terminated surface is treated with HF vapor, fluorides adsorb preferentially on the reactive sites [15] and replace partially the surface hydrides[8]. Reduction of the surface hydrides with the HF treatment is indicated in Figure 5, which shows changes in the spectrum difference between the samples with and without the HF-vapor treatment. Since the spectrum difference has been obtained referring to the sample with no HF treatment, growing of the negative peaks at 2089 and 2116 cm^{-1} represents a decrease of the hydride species on the surface. The F-terminated surface is more stable to oxidation than the H-terminated one [15] and therefore the oxide growth is suppressed on the HF-treated surface.

Higashi *et al.* [16] have proposed that the surface Si-F bond is polarized due to a large electronegativity difference between the two atoms and consequently the adjacent Si-Si bond is relatively reactive. When the silicon surface is exposed to a large amount of HF vapor, the Si-Si bond is eventually broken and reacts with the excessive fluorides to produce volatile SiF_x species. Such a process leads to etching of the surface, which generates many dangling bonds of silicon without fluorides or hydrides. The dangling bonds with no surface protection become vulnerable to oxidation in air. In fact, the above explanation about the effect of excessive HF vapor agrees with the experimental observation by Helms [17] who studied the layer-by-layer oxidation of the HF-treated silicon surface.

We have attempted to observe the surface fluoride species on our samples by an *in situ* IR method, but have failed to do so because of the low IR intensity of the fluoride peak. Instead, we have observed the fluorides by XPS. Figure 6 shows the F_{1S} peak areas from XPS of the HF-treated samples. The fluoride content is the maximum when the sample is treated with the vapor from a 1% HF solution, in agreement with the above results of the minimum oxide growth rate.

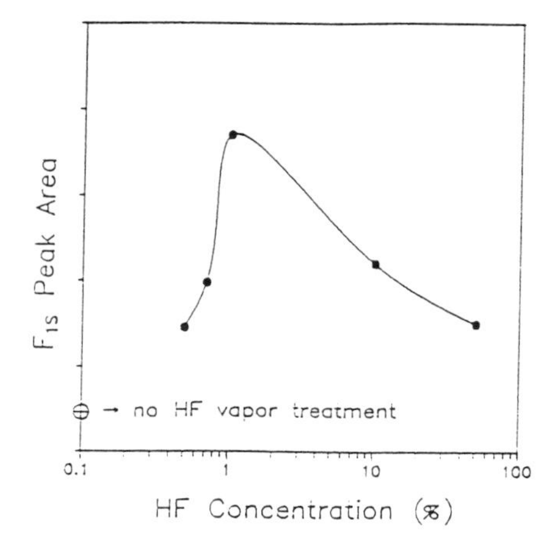

Figure 6. Dependence of F_{1S} XPS-peak area on the concentration of the HF solution.

Removal of Organic Impurities

Figure 7 shows differences in the FTIR spectrum of hydrocarbons after the vapor phase UV treatments. The water vapor without oxygen was not effective to remove the hydrocarbon contamination (Figure 7(c)). But, when the sample was exposed to UV together with pure and wet O_2, the hydrocarbons on silicon were drastically removed as shown in Figure 7(a) and 7(b).

Accordingly, oxygen in the presence of UV radiation is recommended as a strong cleaning agent suitable for the removal of hydrocarbons from silicon surface. The reaction mechanism of hydrocarbon removal is as follows. The reaction of the 185 nm line with oxygen leads to creation of ozone(O_3). The reaction of the 254 nm line with ozone results in generation of atomic oxygen, which is responsible for removal of hydrocarbon from the surface as CO_2 [10].

This process is enhanced by as much as 1.3 times water vapor is added to the ambient gas as shown in Figure 7(a). The reaction of water vapor with atomic oxygen results in the formation of hydroxyl radicals [10] that significantly enhance the hydrocarbon cleaning.

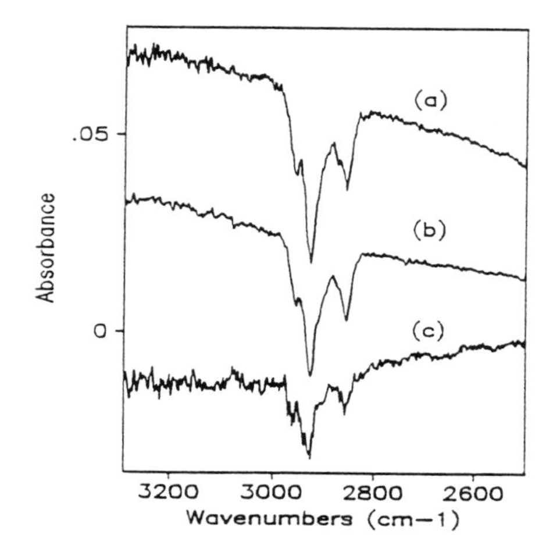

Figure 7. The UV cleaning efficiency of hydrocarbon removal under the (a) O_2 and H_2O, (b) O_2, and (c) H_2O gas stream.

SUMMARY

For analysis of contaminants on the silicon surface, we have used the FTIR transmission method. Generally, the transmission method can not observe a very small amount of contaminants, but this problem has been solved by using high-surface-area samples that allow relatively intense and sharp IR peaks.

We have shown by experiments that the native oxide growth on silicon is suppressed to a minimum by treating the surface with a proper amount of HF vapor and that the optimum condition is related to the fluoride amount on the surface.

When used under UV irradiation, oxygen is a strong cleaning agent suitable for removal of hydrocarbon contaminants from the silicon surface at room temperatures. UV/O_2 cleaning process is enhanced by addition of water vapor to the cleaning gas stream.

REFERENCES

1. J.P. Gambino, M.P. Monkowski, J.F. Shepard, and C.C. Parks, J. Electrochem. Soc. **137**, 976 (1990).
2. T. Takahagi, A. Ishitani, H. Kuroda, and Y. Nagasawa, J. Appl. Phys. **69**, 803 (1991).
3. S. Watanabe, N. Nakayama, and T. Ito, Appl. Phys. Lett. **59**, 1458 (1991).
4. Y.J. Chabal, G.S. Higashi, and K. Raghavachari, J. Vac. Sci. Technol. **A7**, 2104 (1989).
5. D. Burkman, C.A. Peterson, L.A. Zazzera, and R.J. Kopp, Semiconductor Processing Microcontamination **6**, 57 (1983).
6. R.A. Haring and M. Liehr, J. Vac. Sci. Technol. **A10**, 802 (1992).
7. P.A.M. van der Heide, M.J. Baan Hofman, and H.J. Ronde, J. Vac. Sci. Technol. **A7**, 1719 (1989).
8. M. Nakamura, T. Takahagi, and A. Ishitani, Jpn. J. Appl. Phys. **32**, 3125 (1993).
9. C.R. Helms and B.E. Deal, J. Vac. Sci. Technol. **A10**, 806 (1992).
10. J. Ruzyllo, G.T. Duranko, and A.M. Hoff, J. Electrochem. Soc. **134**, 2052 (1987).
11. G. Bomchil, R. Herino, K. Barla, and J.C. Pfister, J. Electrochem. Soc. **130**, 1611 (1983).
12. Y.J. Chabal and K. Raghavachari, Phys. Rev. Lett. **54**, 1055 (1985).
13. P.G. Pai, S.S. Chao, Y. Takagi, and G. Lucovsky, J. Vac. Sci. Technol. **A4**, 689 (1986).
14. T. Itoga, A. Hiraoka, F. Yano, J. Yugami, and M. Ohkura, in Extended Abstracts of the 1994 International Conference on Solid State Devices and Materials, (Jpn. Soc. Appl. Phys., Yokohama, Japan, 1994), pp.667-669.
15. T. Sunada, T. Yasaka, M. Takakura, T. Sugiyama, S. Miyazaki, and M. Hirose, Jpn. J. Appl. Phys. **29**, L2408 (1990).
16. G.S. Higashi, Y.J. Chabal, K. Raghavachari, R.S. Becker, M.P. Green, K.Hanson, T. Boone, J.H. Eisenberg, S.F. Shive, G.N. DiBello, and K.L. Fulford, in Proc. of the 4th International Symposium on ULSI Science and Technology, edited by G.K. Celler, E. Middlesworth, and K. Hoh (Electrochem. Soc. Proc. **93-13**, Pennington, NJ, 1993) pp.189-198.
17. C.R. Helms, J. Vac. Sci. Technol. **16**, 608 (1979).

EFFECTS OF UV/O_3 AND SC1 STEPS FOR THE HF LAST SILICON WAFER CLEANING

Hyeongtag Jeon[1], Hyungbok Choi[1], Y. S. Cho[2], K. K. Ryoo[3], and S. D. Jung[4].
[1]Department of Metallurgical Engineering, Hanyang Univ. Seoul, Korea 133-791
[2]Hyundai Electronics Industries Co., Ltd. Ichon, Korea
[3]Research Institute of Industrial Science and Technology, Pohang, Korea
[4]Electronics and Telecommunication Research Institute, Teajon, Korea.

ABSTRACT

As the size of integrated circuit is scaled down, the importance of Si wafer cleaning has been emphasized. Especially, in the ULSI level device, the cleaning has a great influence on device yields and reliabilities. In this study, we investigate the effects of UV/O_3 and SC1 steps in the HF last cleaning procedures. The UV/O_3 and SC1 cleaning steps are known to remove the organic contaminants. However, the combination of UV/O_3 cleaning step with HF wet chemical solution to remove the metallic impurities has not been studied extensively. We have applied the UV/O_3 and SC1 steps in the middle of the HF last cleaning procedures and have analyzed Si substrates with TXRF, AFM, I-V, etc. The cleaning splits we applied consist with 4 different types, split 1 (piranha + HF), split 2 (piranha + UV/O_3 + HF), split 3 (piranha + SC1 + HF) and split 4 (piranha + UV/O_3 + HF repeated 3 times). The contents of metallic impurities were measured with using TXRF and splits 2 and 4 showed low average values of metallic contents. The surface morphologies after each cleaning examined with AFM and the split 3 exhibited rough surface. The electric properties were measured after forming a MOS capacitor with oxide thickness of 250Å. The high leakage current and low breakdown voltage were observed at split 3.

INTRODUCTION

Contaminations such as the organics, particles and metallic impurities are common contaminants on Si surface to be removed. The organics can be remove by the piranha(H_2SO_4 : H_2O_2) and SC1 chemical solutions and the particles can be eliminated by SC1 solution(NH_4OH : H_2O_2 : DI water)[1, 2]. However, the metallic contaminants is very difficult to remove by wet chemical solution treatments. In semiconductor manufacturing fab. the SC2(HCl : H_2O_2 : DI water) chemical solution is utilized to remove the metallic impurities. But if the electronegativity values of metallic impurities exhibit the higher values than Si's, these metallic impurities such as Cu, Ag, etc are very difficult to eliminate on the Si substrate by this wet chemical process. In these day, the device manufacturing technology is moving toward deep submicron structures. Si substrate cleaning technology with the miniaturizing of Si device becomes to be considered as one of the most important processes in device manufacturer[1, 2, 3, 4]. The wet chemical cleaning and dry chemical cleaning methods are two major concepts for Si wafer cleaning. The wet cleaning method based on the RCA cleaning developed by W. Kern has been used for more than 20 years and is still dominant process to remove the contaminants on Si surface[1]. But the new concepts of Si substrate cleaning called dry cleaning method are under consideration to apply the device manufacturing process. HF vapor phase cleaning[5], H-plasma cleaning[6], and photoexcited

Mat. Res. Soc. Symp. Proc. Vol. 386 © 1995 Materials Research Society

cleaning (UV/Cl_2, UV/O_3)[7, 8] have been proposed and developed by some other scientists. The HF vapor phase cleaning process is now applied to the contact, and pre-silicide cleans. But this vapor phase cleaning system is still under consideration to apply in the pre-gateoxide cleaning. Plasma cleanings such as H-remote plasma or H-ECR plasma are now under development to apply the contact, via hole, and pre-gateoxide cleans. The important advantages of dry clean are low chemical consumption, improved process uniformity, and surface passivation to protect the Si surface out of the contamination. The photoexcited cleaning steps such as UV/Cl_2 and UV/O_3 are also suggested to eliminate the metallic and organic contaminants. But the wet chemical cleaning is still dominantly used in the actual device manufacturing process.

The main concept of this experiment is to combine the dry and wet chemical methods to eliminate the metallic impurities. The metallic contaminants are very difficult to remove in HF-last or HF-only cleaning procedures[9] and should be controlled below $10^{-10}/cm^2$ of each element. In this study we have combined the dry cleaning method and the wet cleaning method which are the UV/O_3 and HF chemical solution cleaning steps. This UV/O_3 is well known as organic contaminants remover due to the reactions of ozone and atomic oxygen with organics. But the atomic oxygen and/or ozone also oxidize the Si surface and metallic species on Si surface at the same time and form the silicon oxide and metallic oxides[7, 10]. These oxides can be removed by the next step of wet HF chemical treatment. The surface chemical analysis after each clean has been done with TXRF (total reflection x-ray fluorescence), surface roughness with AFM (atomic force microscope), and the electrical properties with I-V measurements.

EXPERIMENTAL

For this study, Si (100)-oriented substrates (4" dia) with resistivity of 15～25 Ω-cm (P-type B-doped) were used. The Si substrate cleaning was performed by the four different procedures such as split 1 (piranha + HF as a reference), split 2 (piranha + UV/O_3 + HF), split 3 (piranha + SC1 + HF), and split 4 (piranha + UV/O_3 + HF repeated 3 times). The conditions of piranha (1:4 of H_2SO_4 : H_2O_2) cleaning were at 120℃ and for 10 minutes, those of SC1 (1:1:5 of NH_4OH : H_2O_2 : DI water) were at 85℃ and for 10 minutes, those of HF (1:10 of HF : DI water) were at room temperature and for 10 seconds and those of UV/O_3 cleaning were at room temperature for 1 minute exposure. After cleaning, four Si wafers out of each cleaning split were transferred to oxidation furnace to grow the oxide. The thickness of oxide was 250Å. And then, Al dot, area of $7.85 \times 10^{-3} cm^2$, on gate oxide were deposited to form the MOS capacitors. The leakage current and breakdown voltage were measured. The metallic impurities on each cleaning split were examined with TXRF. The surface roughness of Si substrates were analyzed with using AFM. The I-V characteristic of MOS capacitors were measured with the semiconductor parameter analyzer (HP 4145B).

RESULTS AND DISCUSSIONS

In this study, we examined the metallic impurities and surface roughness of cleaned Si substrates and the electrical properties of silicon oxide. The amount of metallic contents were examined with using TXRF. Figure 1 is the data of bare Si substrates after each chemical solution dipping. The bare Si substrate as a reference was contaminated with Zn and some other impurities on the bare wafer such as Fe, K, and Cu were detected by TXRF. Figure 1 shows the results of the metallic impurities on a bare Si wafer and Si wafers with cleaned with chemicals of DI water, SC1, HF, and Piranha. This shows the piranha solution does not remove Zn effectively and also exhibits the large amount of sulfur (S) on Si surface. SC1 solution dipping also shows a small amount of Fe contamination. However, DI water rinse and HF dipping exhibit effective removal of metallic impurities.

Table I is the TXRF data of 4 different cleaning splits. Here split 1 which is established in ISRC(inter-university semiconductor research center) for a standard cleaning method is to be considered as a reference to compare each split. Splits of 2 and 4 show good removing efficiency of metallic impurities such as Cu, Zn, K and Fe compared to the split 1. The splits of 2 and 4 are the cleaning splits of the combination with the HF and UV/O_3. The UV/O_3 cleaning steps is known as the organic contamination removal step due to the generation of ozone and atomic oxygen. The organics on Si substrate under the UV/O_3 lamp will decompose into CO_2 and H_2O which are volatile[7, 11]. But at the same time the atomic oxygen will react with the metallic impurities on Si substrate and form the metallic oxide.

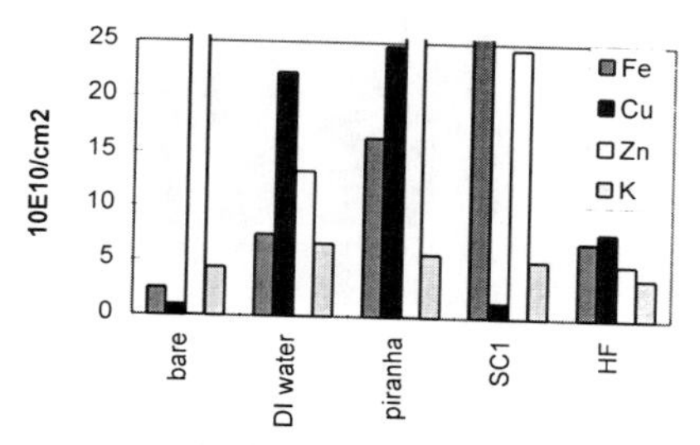

Fig. 1. Metallic impurities on a bare and the cleaned Si wafer with chemicals of DI water, SC1, HF, and Piranha ($\times 10^{10}/cm^2$)

Table I. Average values of metallic impurities after 4 different cleaning splits. ($\times 10^{10}/cm^2$)

	Bare	split 1	split 2	split 3	split 4
Fe	2.5	3.1	2.3	5.1	2.5
Cu	1.0	0.6	0.5	36.4	<DL
Zn	1534.4	7.9	1.4	102.7	<DL
K	4.4	5.6	20.5	13.4	1.0

These metallic oxides will be removed by next cleaning step which is HF chemical solution. These cleaning steps are very effective for metallic impurity removal. Especially the elements which exhibit the higher electronegativity values than Si can be removed effectively. The metallic impurities of Cu, Ni, and Fe should be removed in pre-gate oxide cleaning of the high quality oxide. But the element of Cu is very difficult to eliminate in the HF chemical cleaning. The chemical solution of HF with H_2O_2 is suggested by some other scientists[12, 13, 14, 15]. In this cleaning step, the H_2O_2 chemical solution oxidize the metallic impurities and HF chemicals will eliminate the these oxides. To apply this process into the current HF wet chemical procedure it is needed to modify the current cleaning procedures. But this UV/O_3 application to HF last clean can be applied as the same purpose of H_2O_2 in HF solution and shows the excellent removing efficiency. The removing efficiency of metallic impurity, especially the Cu, which is not removed by wet chemical treatment effectively is increased by the application of the UV/O_3 into HF cleaning steps (splits of 2 and 4). And this is very simple step (UV lamp exposure only) to apply current wet cleaning procedure.

The roughness of cleaned Si surfaces were measured by AFM and shown in Figure 2 and Figure 3. In this results we compared the results with a reference of split 1 which is standard cleaning split in ISRC. The split 3 which is combination with SC1 and HF shows

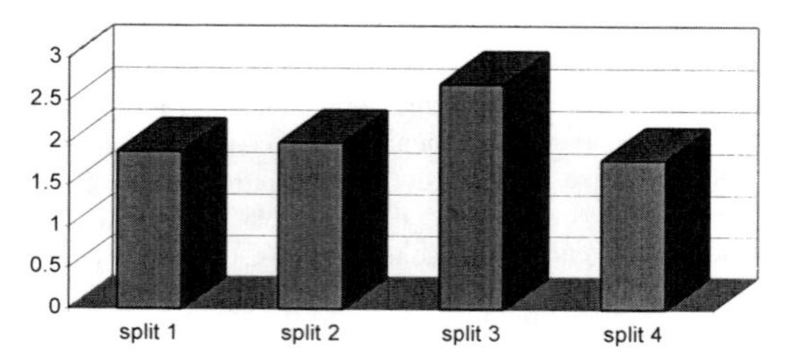

Fig. 2. RMS roughness values of Si surfaces after cleaning by each split (Å)

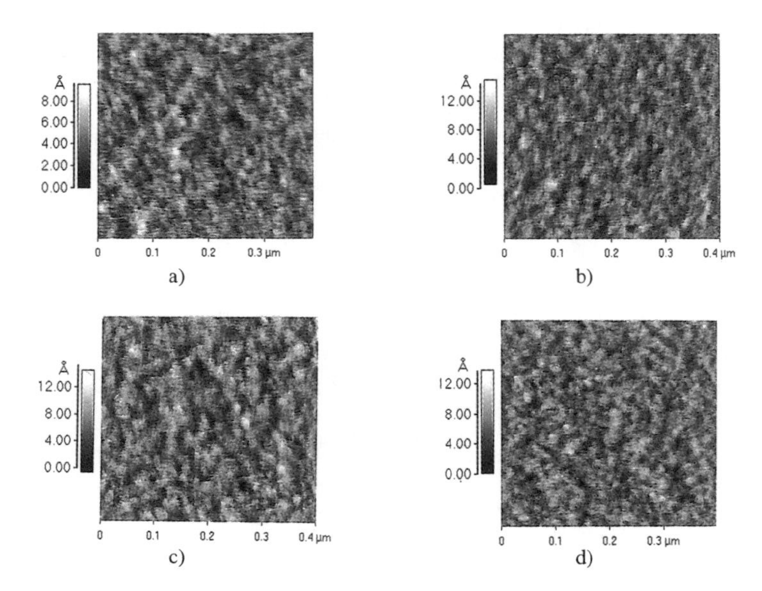

Fig. 3. Surface morphologies of cleaned Si surfaces after a) split 1, b) split 2, c) split 3, 4) split 4

much rougher than other splits. Splits of 1, 2 and 4 exhibits the root mean square roughness values in between 1.8 and 2.0 Å. The repeated results of split 4 which is UV/O_3 and HF repeated 3 times actually shows consistent smooth surface morphologies. These data tell us that the repeated UV/O_3 and HF cleaning step does not degrade the Si wafer surface. From the surface morphology and metallic impurity point of views, the split 4 exhibits the smooth surface and low metallic impurities after cleaning.

The electrical properties of oxide grown immediately after cleaning shown in Table II, III, and Figure 4. Table II and III are the average values of the leakage current and breakdown voltage of oxide and the Figure 4 is the typical I-V curves of MOS capacitor of

Table Ⅱ. Average leakage currents of 4 different cleaning splits

	split 1	split 2	split 3	split 4
nA/cm^2	9.2 ± 3.8	20.2 ± 6.4	175.8 ± 116.6	14.1 ± 8.3

Table Ⅲ. Average breakdown voltages of 4 different cleaning splits

	split 1	split 2	split 3	split 4
MV/cm	6.4 ± 0.1	7.0 ± 0.4	6.1 ± 1.1	7.7 ± 0.1

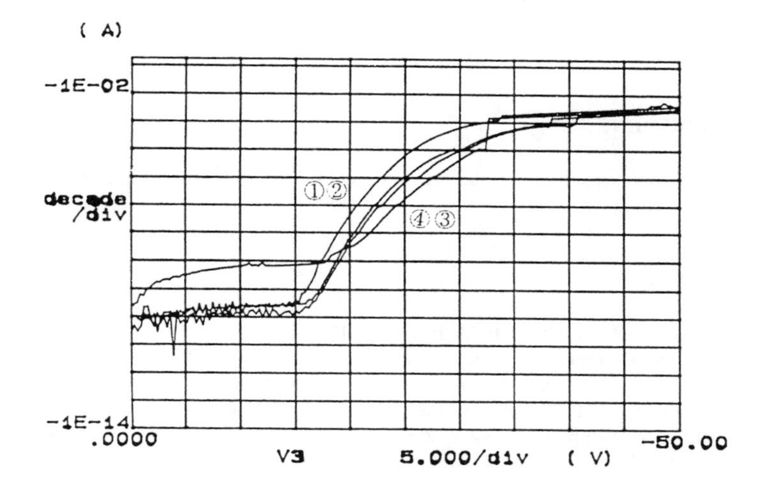

Fig. 4. Typical I-V characteristic curves of ① split 1, ② split 2, ③ split 3, and ④ split 4

each split. The split 3 shows the high leakage current and low breakdown voltage. These results of split 3 are considered due to the high metallic impurities and rough surface morphologies on cleaned Si substrate. The split 4 exhibits the high oxide breakdown voltage compared to other splits. The cleaning splits (splits of 2 and 4) with UV/O_3 and HF are analyzed to exhibit better physical and electrical properties than the splits of 1 and 3. In this study, the UV/O_3 application into HF wet chemical cleaning to remove metallic impurities has been examined and showed the good results to remove the metallic impurities. This combination of the dry and wet cleaning process which is UV/O_3 and HF shows the another possibility to be applied for current cleaning process due to its simplicity and effectiveness in metallic impurities removal.

SUMMARY

The quality of each chemical solution used in this experiment was verified before cleaning. Splits of 2 and 4 exhibit low metallic impurities on Si substrate and the high electrical breakdown voltage and low leakage current. Split 3 shows high metallic contamination and high leakage current. In this experiment, metallic impurities are removed effectively in splits of 2 and 4 which are the combination of the UV/O_3+HF. The surface roughness measured by AFM exhibit rough in split 3 and relatively smooth in splits of 1, 2, and 4. The repeated UV/O_3 and HF treatment (split 4) does not degrade the Si surface morphology. These results indicate the concentration of metallic impurities and the degree of surface roughness are considered to be related with electrical properties. The metallic impurities are effectively removed by the combination of the UV/O_3 and HF cleaning steps. In this experiment, we introduced the UV/O_3 step into HF-last and/or HF-only cleaning processes and found the relatively smooth surface and good effectiveness in removing the metallic impurities.

ACKNOWLEDGMENT

This study was supported by the Ministry of Education (Semiconductor Field) through grant ISRC-94-E-1105.

REFERENCES

[1] W. Kern and D. A. Puotinen, RCA Rev., **31**, 187 (1970)

[2] L. Mouche and F. Taradif, J. Electrochem. Soc. **141**, 1684(1994)

[3] G. Gould and E. A. Irene, J. Electrochem. Soc., **134**, 1031 (1987)

[4] J. S. Montgomery, J. P. Barnak, A. Bayoumi, J. R. Hauser, and R. J. Nemanich, Electrochem. Soc. Symp. Proc., **94-7**, 296 (1994)

[5] J. Ruzyllo, A. M. Hoff, D. C. Frystak, and S. D. Hossain, J. Electrochem. Soc. **136**, 1602(1989)

[6] J. Cho, T. P. Schneider, J. Vanderweide, H. Jeon, and R. J. Nemanich, Appl. Phys. Lett. **59**, 1995(1991)

[7] J. R. Vig, J. Vac. Sci. Technol. A **3**, 1027(1985)

[8] S. Watanabe, R. Sugino, T. Yamazaki, Y. Nara, and T. Ito, Jpn. J. Appl. Phys, **28**, 2167(1989)

[9] Hyeongtag Jeon, Y. L. Kang, Y. S. Cho, *Korean Journal of Materials Research,* **4(8)**, 921(1994)

[10] L. A. Zazzera and J. F. Moulder, J. Electrochem. Soc. **136**, 484(1989)

[11] R. R. Sowell, R. E. Cuthrell, D. M. Mattox, and R. D. Bland, J. Vac. Sci. Technol. **11**, 474(1974)

[12] C. R. Helms and H. Park, Mat. Res. Soc. Symp. Proc., **315**, 287(1993)

[13] T. Takahagi, I. Nagai, A. Ishitani, and H, Kuroda, J. Appl. Phys. **64**, 3516 (1988)

[14] Akihiro Miyauchi, Yousuke Inoue, and Takaya Suzuki, J. Appl. Phys. Lett. **57**, 676 (1990)

[15] T. Ohmi, T. Imaoka, I. Sugiyama, and T. Kezuka, J. Electrochem. Soc., 139, **11**, 3317(1992)

WAFER CLEANING INFLUENCE ON THE ROUGHNESS OF THE Si/SiO_2 INTERFACE.

A. MUNKHOLM*, S. BRENNAN*, AND JON P. GOODBREAD**
*Stanford Synchrotron Radiation Laboratory, Stanford, CA 94309
**Hewlett-Packard Co., Palo Alto, CA 94301

ABSTRACT

The roughness of the Si/SiO_2 interface has a great impact on the electrical properties of the gate-oxide in integrated circuits and consequently it is a large concern for the semiconductor industry. As the thickness of the oxide is decreased, the role of the roughness becomes more critical for the device. The nature of a buried interface prohibits the use of commonly used surface techniques. By the use of crystal truncation rod (CTR) x-ray scattering, it is possible to get information on the termination of the bulk silicon in a non-destructive fashion. The authors have investigated the influence of different cleanings on interfacial roughness using synchrotron radiation-based CTR-scattering. In particular, we looked at silicon(001) wafers both before and after the growth of a 1000Å thermal oxide. The results show that the use of HF during cleaning results in a smoother interface between silicon and its native oxide. Due to smoothing of the interface during the oxidation process, the difference between the various cleaning methods becomes less significant for these thick oxides.

INTRODUCTION

As device sizes decrease, the importance of interfacial roughness on the quality of gate oxides increases. With future generations of gate oxide thicknesses expected to be 35-50Å, both surface and interfacial roughness can dramatically effect device performance. Although the top surface roughness can be measured using probes such as scanning tunneling microscopy (STM) or atomic force microscopy (AFM), in order to measure the roughness of a buried interface, either the oxide layer must be stripped off or a probe must be used which can penetrate through the oxide layer.

Although specular reflectivity has long been used to measure surface and interfacial roughness, for materials whose index of refraction differ only slightly the deconvolution of the roughnesses from the two interfaces is problematic. Over the past decade x-ray scattering techniques have been developed to study the truncation of bulk order in crystalline materials[1,2]. By measuring only the termination of bulk order, the roughness of the Si/SiO_2 interface can be determined independently. The measurement is performed by studying the decay of x-ray scattered intensity away from a bulk diffraction peak in the direction of the surface. This fall-off in intensity is known as a Crystal Truncation Rod, or CTR. Many of the early studies of CTRs involved clean, reconstructed surfaces where the goal was to determine the rearrangement of the top layers of the bulk crystal. Recently, groups have started looking at interfacial roughness with CTRs, especially related to the technologically important Si/SiO_2 interface[3-5]. One of the advantages of using x-rays to measure interfacial roughness is the wide range of lateral distances over which the technique is sensitive. It ranges from the inverse of the length of the scattering vector ($\sim$ 2 Å) to the longitudinal coherence of the photons, which is related to the energy spread of the incident beam and the size of the detector slits ($\sim 1\mu$m).

In this study we have investigated the influence of different cleaning procedures on interfacial roughness using synchrotron radiation to perform CTR-scattering. To emphasize

Mat. Res. Soc. Symp. Proc. Vol. 386 © 1995 Materials Research Society

the utility of this technique for buried interfaces, we have looked at silicon(001) both before and after the growth of a 1000Å thermal oxide.

THEORY

Crystal truncation rods (CTRs) are intensity streaks in reciprocal space perpendicular to the surface which arise from the abrupt change in electron density from the termination of the bulk. The phenomenon can be thought of as a relaxation of the diffraction condition in the direction perpendicular to the surface, so that spots of intensity turn into rods. Because the diffracted intensity in reciprocal space is the fourier transform of the electron density in real space, the transformation of an infinite crystal (and the correspondingly tiny points in reciprocal space) into a semi-infinite crystal results in the points smearing out in the direction of the surface normal. The more abrupt the termination of the crystalline solid, the more fourier components in reciprocal space are needed to represent that transition, and the stronger the intensity of the CTRs. In contrast, a very rough interface can be represented by only a few fourier components and the scattered intensity decays more quickly in reciprocal space. In this paper we will be using a modified version of the theory by Andrews & Cowley[1] which only uses the scattering from a single reciprocal lattice point to describe the CTR scattering[6].

The termination of the bulk is a one dimensional function in the direction perpendicular to the surface. For a perfectly flat surface the electron density is of the form:

$$\begin{aligned} \rho(z) &= \rho_0 \qquad \text{for z} < 0 \\ \rho(z) &= 0 \qquad \text{for z} \geq 0 \end{aligned} \tag{1}$$

The intensity is accordingly given by the Fourier transform of the electron density:

$$I(\vec{q}_\perp) = \frac{|\rho_0(\vec{\tau})|^2}{q_\perp^2} \exp(-\sigma^2 q_\perp^2) \tag{2}$$

To account for crystallinity of the structure, the scattering vector in the above equation is a reduced scattering vector $\vec{q}_\perp$, which is the perpendicular momentum transfer relative to the reciprocal lattice $\vec{\tau}$ ($\vec{Q} = \vec{\tau} + \vec{q}$). In this model the roughness is described by the addition of an exponential decay similar to a thermal Debye term, but using the reduced scattering vector $\vec{q}_\perp$ rather than $\vec{Q}$.

EXPERIMENTAL

The data were collected on Beam line 6-2 at the Stanford Synchrotron Radiation Laboratory (SSRL) using a focused beam. The photon energy passed by the Si(111) double crystal monochromator was 10 keV. The wafers were mounted on a vacuum chuck and kept in a helium environment to reduce air scattering.

Both the incident and scattered beam slits were defined using the procedure of Specht and Walker[7]. This uses a large in-plane slit opening, which allows the entire diffracted beam to be collected by the detector simultaneously, so that the integrated intensity is directly measured. The incident slits were 0.25 mm horizontally by 1.0 mm vertically, which results in a beam size on the sample from 1.0 to 2.7 mm^2. The 4-circle diffractometer was operated in the ω=0 mode and the sample surface normal was aligned with the ϕ axis, which results in the incident and exit beam angles being equal. Rod intensities were obtained by scanning θ, which results in a trapezoidal rocking curve with the flat top of the curve resulting from diffraction of the entire rocking curve simultaneously. Thus the integrated intensity is the difference between that top and the background. The resolution along the rod is defined by the size of the scattered beam slits in the direction perpendicular to the scattering plane. This resolution also depends on the relationship of the rod to the horizontal plane, which changes as the rod is scanned. For these measurements the resolution varied between 0.01-0.03 reciprocal lattice units (rlu). The data were corrected for the background, the Lorentz-factor, the change in atomic form factor as a function of the magnitude of the scattering vector, the area of the sample illuminated by the beam and the resolution function of the rod.

The samples used are all well oriented silicon (001) wafers and consist of three sets of two wafers, a native and a 1000Å thermal oxide wafer both cleaned by the same methods. They are: an RCA clean only; HF dip followed by an RCA clean; and an RCA clean followed by HF dip. The RCA clean is a two step process performed at 70°C in which the wafers first are exposed to an SC1 (H_2O, H_2O_2 and NH_4OH in the ratio 8:1:1) followed by an SC2 clean (H_2O, H_2O_2 and HCl in the ratio 8:1:1). The HF clean was processed at 25°C with a water to HF ratio of 60:1. Note that the wafers cleaned with an HF dip had their original native oxide removed. The x-ray measurements were performed several weeks after these cleanings, so a native oxide reformed. Although there are several rods which can be measured, we focused on three, above and below the 202 reflection and below the 311 reflection.

RESULTS

The crystal truncation rod intensities from the 202 reflection are shown in Figure 1. The rod intensity data from the samples terminated by a native oxide are connected with lines showing the best fit of the modified Andrews & Cowley model. The symbol is specific to the cleaning procedure. For the data shown here and for the 311 reflection significant differences are observed amongst the roughnesses of the native oxide interfaces. After oxidation all three cleans result in smoother interfaces, seen in the figure as higher scattered intensity away from the bragg reflection.

The results of fits to these data using the modified Andrews & Cowley model are presented in Table I. These show that the use of HF after an RCA cleaning procedure results in the smoothest silicon-native oxide interface of the three recipes studied. The average roughness value for the three rods studied is 1.93 Å, which is roughly 0.7 Å smoother than the next best clean, that of HF before RCA. The RCA only clean has an average roughness of 3 Å, which is roughly 0.4 Å rougher than the RCA after HF clean. The error bars for the roughness measurements quoted here are ± 0.03 Å.

After thermal oxidation, the differences between the three cleans is much lower, although the RCA only clean is still roughest. The HF first clean has an average roughness after oxidation of 1.47 Å, while the HF after RCA clean has an average roughness of 1.51 Å. The RCA only clean has an average roughness of 1.56 Å, which, considering the error bars, is slightly rougher. There seems to be an asymptotic roughness value that the three are close to reaching. The reduction in roughness through oxidation ranges from 1.4 Å for the RCA clean to 0.4 Å for the HF after RCA clean.

CONCLUSIONS

There is clear evidence of a correlation between the roughness of the Si/SiO_2 interface and different wafer cleaning procedures. Use of HF at any point during the clean results in a smoother surface, and HF following an RCA clean results in the smoothest of the

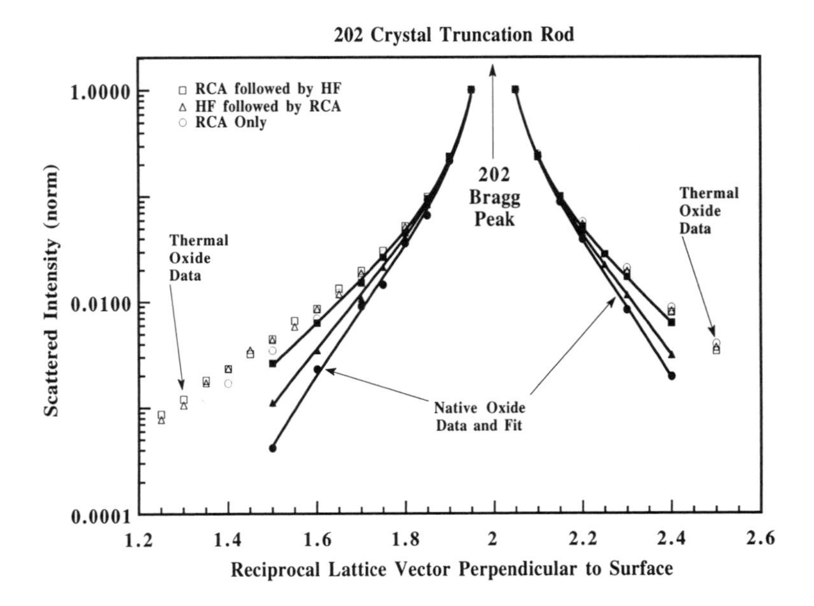

Figure 1: CTR scattered intensity around the 202 Bragg reflection for three different cleans on Si(001) wafers. The samples are terminated by a native oxide (filled symbols) and a 1000Å thermal oxide (empty symbols). The fits to the native oxide data are shown as solid lines.

	RCA only		HF first+RCA		RCA+HF last	
	native	thermal	native	thermal	native	thermal
	●	○	▲	△	■	□
202 low	3.01	1.50	2.56	1.45	1.98	1.45
202 high	3.18	1.69	2.79	1.76	2.09	1.79
311 low	2.78	1.48	2.43	1.20	1.71	1.30
Average	2.99	1.56	2.59	1.47	1.93	1.51

Table I: RMS roughness in Å.

three cleans studied. The growth of a thick thermal oxide reduces the initial roughness significantly. It would seem that the importance of the cleaning procedure decreases as the oxide thickness increases, as the thermal oxides cleaned by any of the three methods resulted in very similar roughness values.

We would emphasize that the roughness values quoted here are not affected by roughness at the oxide-air interface, because the measurements are only sensitive to the termination of the crystalline silicon.

A surprising result of this study is that the roughness values for different rods for the same sample have such different values, with deviations which are large compared to the estimated error of the roughness for any one rod. We do not have a good explanation for this variation and intend to explore this further. It should be clear, however, that with our present level of understanding x-ray scattering is a valuable tool for comparing interfacial roughness non-destructively.

ACKNOWLEDGMENTS

This work was performed at and supported by SSRL, which is supported by the Department of Energy through the Office of Basic Energy Sciences. The authors wish to thank A. Fischer-Colbrie for facilitating this collaboration and A. Bienenstock for many illuminating discussions.

REFERENCES

1. S.R. Andrews and R.A. Cowley. *J. Phys. C: Solid State Phys.* 18 (1985).
2. I.K. Robinson, *Phys. Rev. B* **33**, 6 (1986).
3. M.-T. Tang, K.W. Evans-Lutterodt, G.S. Higashi, and T. Boone, *Appl. Phys. Lett.* **62**, 24 (1993).
4. M.-T. Tang, K.W. Evans-Lutterodt, M.L. Green, D. Brasen, K. Krisch, L. Manchanda, G.S. Higashi, and T. Boone, *Appl. Phys. Lett.* **64**, 6 (1994).
5. M.L. Green, D. Brasen, K.W. Evans-Lutterodt, L.C. Feldman, K. Krisch, W. Lennard, H.-T. Tang, L. Manchanda and M.-T. Tang, *Appl. Phys. Lett.* **65**, 7 (1994).
6. For an explanation of why a single reflection is sufficient, see A. Munkholm, *et al.* in preparation.
7. E.D. Specht and F.J. Walker. *J. Appl. Crystallography* **26**, 2 (1993).

IN SITU AUGER SPECTROSCOPY INVESTIGATION OF InP SURFACES TREATED IN RF HYDROGEN AND HYDROGEN/METHANE/ARGON PLASMAS

J. E. PARMETER, R. J. SHUL, AND P. A. MILLER
Departments 1126, 1322, and 1128, Sandia National Laboratories, Albuquerque, NM, 87185-0601

ABSTRACT

We have used in situ Auger spectroscopic analysis to investigate the composition of InP surfaces cleaned in rf H_2 plasmas and etched in rf $H_2/CH_4/Ar$ plasmas. In general agreement with previous results, hydrogen plasma treatment is found to remove surface carbon and oxygen impurities but also leads to substantial surface phosphorus depletion if not carefully controlled. Low plasma exposure times and rf power settings minimize both phosphorus depletion and surface roughening. Surfaces etched in $H_2/CH_4/Ar$ plasmas can show severe phosphorus depletion in high density plasmas leading to etch rates of ~ 700 Å/min, but this effect is greatly reduced in lower density plasmas that produce etch rates of 30-400 Å/min.

INTRODUCTION

Plasma treatment of III-V semiconductor materials is commonly employed as a method of etching these materials, and, to a lesser degree, as a method of surface cleaning. Despite the common use of plasma etches, there is currently relatively little in situ data available on the effects of various plasma treatments on the stoichiometry of compound semiconductor surfaces. Most available data on surface stoichiometry have been obtained using ex situ Auger or X-ray photoelectron spectroscopy measurements, and, while these data have some utility, the degree to which exposure to atmosphere alters the composition of the plasma treated surface is always open to question. Since surface composition is likely to affect the properties of devices fabricated from semiconductor materials, systematic studies of the effects of various plasma treatments on surface composition are needed.

In this paper, we present an in situ Auger spectroscopic study of InP surfaces treated in rf H_2 and $H_2/CH_4/Ar$ plasmas. The studies have been performed in a newly constructed experimental system that interfaces a plasma processing chamber with an ultrahigh vacuum (UHV) surface analysis chamber. Hydrogen cleaning of InP using ECR plasmas or atomic hydrogen produced at hot filaments has been investigated previously, and since some previous in situ data exist this system provides a good starting point for comparing data obtained in our apparatus to data in the literature. The $H_2/CH_4/Ar$ gas mixture is a commonly used etch mixture for InP, and while ex situ Auger studies of etched surfaces exist there is currently little if any in situ compositional data. We therefore present some initial data on InP surfaces etched with this gas mixture under various plasma conditions.

Mat. Res. Soc. Symp. Proc. Vol. 386

EXPERIMENTAL

The UHV portion of the instrument used in this study has been described in detail elsewhere [1], and contains a PHI Auger spectrometer, a PRI reverse view LEED system, a UTI 100C mass spectrometer, and an LK technologies NGI3000 ion gun for sputter cleaning. The UHV chamber is pumped by a 450 l/s Balzers turbomolecular pump, and the base pressure is in the 10^{-10} Torr range. The UHV chamber is coupled to the plasma processing chamber via a differentially pumped penetration seal placed directly below a 2.75" gate valve. This chamber is pumped by a 210 l/s Balzers turbomolecular pump. At the opposite end of the plasma chamber, a copper coil facing a 6" viewport is used to couple rf power into the chamber and thus strike a plasma by inductive and/or capacitive coupling. Further experimental details will be given elsewhere [2].

The InP samples were cleaved wafer pieces that were ~ 1 cm^2. Two inch 3 x 10^{18} cm^{-3} S-doped InP wafers from Sumitomo were used. Details of sample mounting will be given elsewhere [2]. In the $H_2/CH_4/Ar$ studies, two samples were mounted side by side. One sample was unpatterned and was used to obtain Auger data, while the other was patterned with nickel ~ 3000 Å thick. The patterned samples were used for ex situ Dektak profilometry measurements to calculate etch rates. Ex situ atomic force microscopy (AFM) measurements of rms surface roughness were also performed on some unpatterned samples, using a Digital Instruments Dimension 3000 AFM operating in tapping mode. Samples were not deliberately heated. Samples were grounded during H_2 plasma treatment and were given a DC bias during $H_2/CH_4/Ar$ etching.

In some of the early hydrogen plasma experiments, including those represented in Figures 1 and 2, a 60 l/s turbomolecular pump was used on the plasma chamber, and the length of the rf cable between the rf generator and copper coil was longer than in subsequent experiments. As a result, plasma parameters for these data and subsequent hydrogen plasma data are not strictly comparable. However, these data can still be compared based on the ion saturation current measured at a wire probe placed ~ 2 cm from the sample during plasma treatment. To make data more comparable, both rf power and probe current are given in Figures 2 and 3, e.g. 10 W rf (16 μA).

RESULTS

A. Hydrogen Plasma Cleaning of InP

Figure 1(a) shows a typical Auger spectrum for an as-received InP wafer that has not been subjected to any cleaning. The P/In peak-to-peak Auger ratio for such surfaces is variable, falling between 0.2 and 0.6. The spectrum also shows large carbon and oxygen signals due to the presence of hydrocarbon and native oxide species from atmospheric contamination. The intensity of these signals also varies, but the values in the spectrum shown (C/In ~ 0.3 and O/In ~ 0.2) are typical.

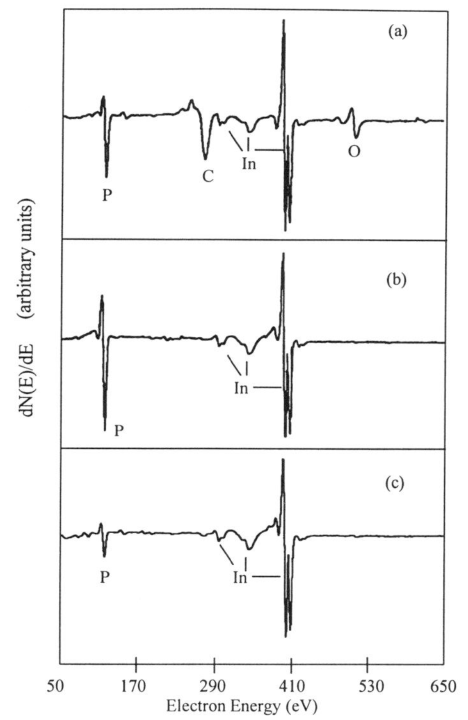

Figure 1. Auger Spectra of InP surfaces: (a) as-received (100) wafer, (b) sputter cleaned, and (c) cleaned 2 hours in H_2 plasma.

For comparative purposes, Figure 1(b) shows a typical InP surface cleaned by Ar^+ sputtering. Sputtering was performed on different samples using 0.5 or 1.5 kV Ar^+ at various incidence angles, and except for near grazing incidence the P/In Auger ratio was found to be nearly constant at 0.7 ± 0.1. It can be seen that the sputter treatment is completely effective in removing surface carbon and oxygen. Based on comparison to a pure indium sample mounted next to the sputtered InP, it is estimated that a stoichiometric InP surface would have P/In ~ 0.93 and that the average composition of a sputtered surface is $InP_{0.75}$. While there is some uncertainty in these numbers, they are in qualitative agreement with literature data that indicate a preferential removal of phosphorus during InP sputtering [3].

Figure 1(c) shows the Auger spectrum of an InP surface cleaned for two hours in an rf hydrogen plasma. The plasma conditions were pressure = 14 mTorr, flow rate = 6 sccm, and rf power = 50 W (15μA), but the significant changes in the Auger spectrum compared to an uncleaned sample are largely independent of the precise plasma parameters used. As in the case of sputtering, the hydrogen plasma is effective in removing the carbon and oxygen impurities from the InP surface. Compared to the sputter cleaned surface, the plasma treatment leads to a much greater phosphorus depletion of the near surface region, with an estimated average composition of ~ $InP_{0.2}$. Since there is no measurable etch rate for pure hydrogen plasmas, the P-depletion effect must be limited to the first few atomic layers.

Figure 2 shows the time evolution of the P/In Auger ratio for an InP surface treated in a hydrogen plasma identical to that used in Figure 1(c). In this case, the surface was first sputter cleaned to remove impurities and thus give the most accurate possible initial P/In value. The P/In value drops rapidly during the first 30 minutes, and eventually reaches a steady state value of ~ 0.25. The carbon Auger signal remains at a nearly constant low value throughout, while there is no detectable oxygen signal associated with these spectra.

Since the main change in P/In in Figure 2 occurs in the first 30 minutes, the initial changes in composition were looked at in more detail. The results are shown in Figure 3. An unsputtered InP sample is used so that changes in the carbon and oxygen signals could also be monitored. The figure shows that in the first few minutes of plasma treatment, the P/In ratio actually rises as the carbon and oxygen are removed. After 18 minutes of plasma exposure, P/In is ~ 0.6, while the oxygen and carbon signals have been reduced to the level of noise. This suggests that surface

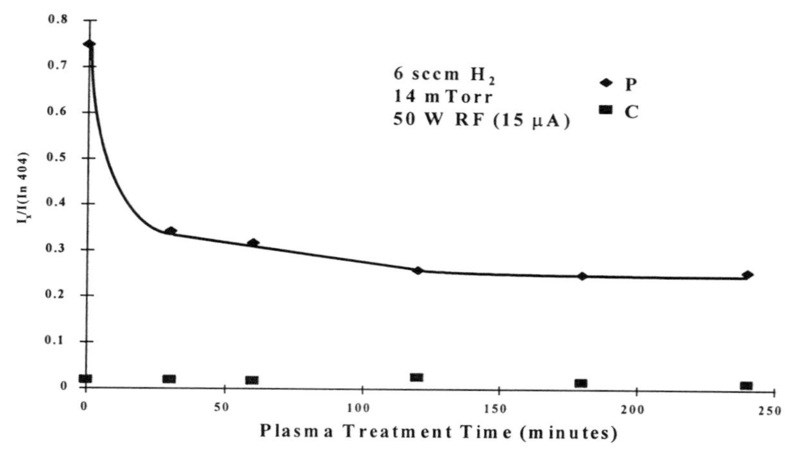

Figure 2. P/In Auger ratio as a function of time for an InP surface treated in a H_2 plasma.

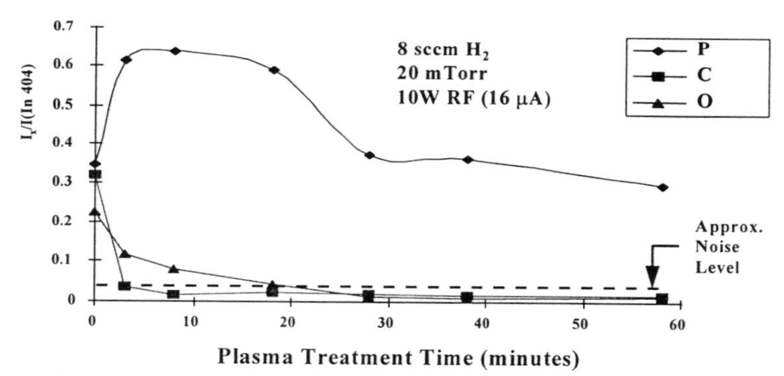

Figure 3. Evolution of the P, C, and O Auger signals on an InP surface during the early stages of H_2 plasma cleaning.

impurities tend to retard the phosphorus signal relative to the indium signal, and that removal of these impurities largely precedes surface phosphorus depletion. Note also that the carbon is removed much more rapidly than the oxygen; it takes 18 minutes for the oxygen signal to be reduced to a level that the carbon signal reaches in only 3 minutes. After 58 minutes of plasma treatment, P-depletion has occurred to a significant degree and the surface has a composition nearly equal to that in Figure 2 after one hour of treatment.

The effect of changing rf power was investigated by obtaining Auger and AFM data for InP surfaces treated in H_2 plasmas with identical conditions (20 mTorr, 8 sccm) apart from rf power, which was set at 3, 10, or 100 W. For one hour treatments, the P/In Auger ratio decreased from 0.58 to 0.13, while the rms roughness increased from 11 to 21 Å as the rf power increased. This indicates that surface roughening is correlated with phosphorus depletion.

B. Hydrogen/Methane/Argon Etching of InP

Figure 4 shows Auger spectra of two InP surfaces etched in $H_2/CH_4/Ar$ plasmas. The surfaces are not cleaned prior to plasma treatment, and in each case the flow rates (sccm) of $H_2/CH_4/Ar$ are 20/2/10. The two plasmas used are representative of two fundamentally different modes of operation attainable in our system. The plasma in (a) is a capacitively-coupled mode which appears dim and produces only 2 mA of current at the sample holder. The plasma in (b) is an inductively-coupled mode that produces a 7 mA current at the sample holder. In (b), the surface is etched at a rate of 700 Å/min and, as with extended hydrogen plasma treatment, there is a large surface P-depletion, with P/In = 0.29. There is also an impurity peak near 95 eV that probably results from silicon sputtered off of the 6" viewport. The surface in (a) shows an improved composition, with much less P-depletion (P/In = 0.66) and without the impurity peak. The etch rate for the surface represented in (a) is only 30 Å/min, but variation of the plasma parameters can produce etch rates of up to 400 Å/min with similar resulting compositions, as long as one operates in the lower density plasma mode. Both surfaces in Fig. 4 show trace oxygen and carbon impurities, but these are reduced dramatically compared to an unetched surface.

DISCUSSION AND CONCLUSIONS

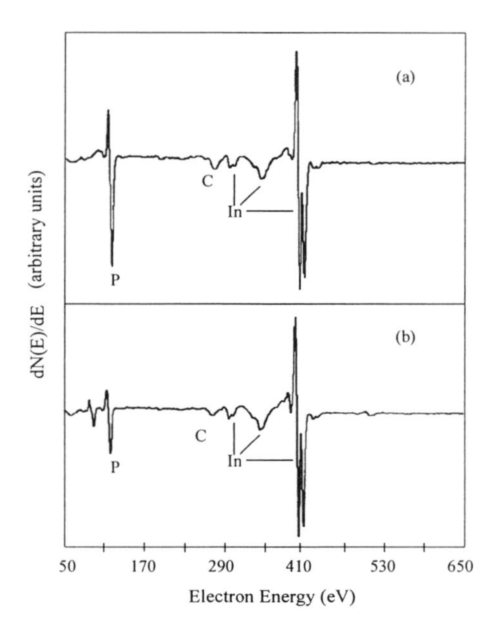

Figure 4. Auger spectra of two InP surfaces etched one hour in $H_2/CH_4/Ar$ plasmas: (a) "dim" mode, 10 mTorr, -200 V bias, 150 W rf. and (b) "bright" mode, 30 mTorr, -30 V bias, 150 W rf.

The results presented here for rf H_2 plasma cleaning of InP are generally in good agreement with previous studies of InP surface cleaning using ECR plasmas or hydrogen atoms produced at hot filaments [4-8]. The rf hydrogen plasma can completely remove surface carbon and oxygen impurities, but unless the process is very carefully controlled this occurs at the expense of surface P-depletion. This depletion results from the reaction of hydrogen atoms and phosphorus to form PH_3 [4,5, 9-14]; since the group III hydride is not volatile, indium is not removed and only the topmost atomic layers are affected. One recent study by Chun et al. reported that irradiating InP (100) at 350 °C for 30 minutes with H atoms led to an *increase* of 50 % in the P/In Auger ratio [7]. While this seems to contradict most other work in this area, Figure 3 in this paper suggests that this increase may be an artifact of impurity removal. Our observation that surface oxide removal is the rate-limiting step in surface cleaning is in agreement with a hydrogen atom beam study by Kikawa et al. [8]. Using XPS, these workers concluded that removal of $In(PO_3)_3$ was rate-limiting in obtaining a clean surface.

Addition of argon and methane to the hydrogen plasma initiates indium removal and hence InP etching [9-13]. Removal of In-containing species may be via sputtering (due to Ar) and/or indium alkyl formation (due to CH_4). Our in situ Auger results for InP surfaces etched in $H_2/CH_4/Ar$ mixtures show that surface P-depletion also occurs for these etches. However, the P-depletion can be minimized by using low density plasma conditions that produce low etch rates. In fact, it is possible that etched surfaces with P/In ~ 0.7 are stoichiometric but In-terminated, and thus have a lower P/In Auger ratio than the value of ~ 0.93 predicted for a surface with a 50/50 composition in each atomic layer. These results suggest that following a rapid etch with a slower etch under milder plasma conditions may be a method for restoring a better stoichiometry to a P-depleted, etched surface.

ACKNOWLEDGMENTS

The authors thank Arnold J. Howard for AFM data, Dennis Rieger for loaning a mass flow controller, Ray Hibray for InP wafers, and R. Miesch, P. Pochan, and P. L. Glarborg for technical assistance. This work was supported by the United States Department of Energy under Contract DE-AC04-94AL85000.

REFERENCES

[1] J. E. Parmeter, J. Phys. Chem. 97, 11530 (1993).
[2] J. E. Parmeter, R. J. Shul. P. A. Miller, and A. J. Howard, manuscript in preparation.
[3] J. B. Malherbe and W. O. Barnard, Surface Sci. 255, 309 (1991).
[4] P. G. Hofstra, D. A. Thompson, B. J. Robinson, and R. W. Streater, J. Vac. Sci. Technol. B 11, 985 (1993).
[5] D. Gallet, G. Hollinger, C. Santinelli, and L. Goldstein, J. Vac. Sci. Technol. B 10, 1267 (1992).
[6] E. J. Petit, F. Houzay, and J. M. Moison, Surface Sci. 269/270, 902 (1992).
[7] Y. J. Chun, T. Sugaya, Y. Okada, and M. Kawabe, Jpn. J. Appl. Phys. 32, L287 (1993).
[8] T. Kikawa, I. Ochiai, and S. Takatani, Surface Sci. 316, 238 (1994).
[9] S. J. Pearton, U. K. Chakrabarti, A. P. Kinsella, D. Johnson, and C. Constantine, Appl. Phys. Lett. 56, 1424 (1990).
[10] C. Constantine, D. Johnson, S. J. Pearton, U. K. Chakrabarti, A. B. Emerson, W. S. Hobson, and A. P. Kinsella, J. Vac. Sci. Technol. B 8, 596 (1990).
[11] S. J. Pearton, U. K. Chakrabarti, A. P. Perley, C. Constantine, and D. Johnson, Semicond. Sci. Technol. 6, 929 (1991).
[12] S. J. Pearton, U. K. Chakrabarti, A. Katz, A. P. Perley, W. S. Hobson, and C. Constantine, J. Vac. Sci. Technol. 9, 1421 (1991).
[13] S. J. Pearton, C. R. Abernathy, R. F. Kopf, F. Ren, and W. S. Hobson, J. Vac. Sci. Technol. B 12, 1333 (1994).

STUDY ON SULFUR PASSIVATION FOR $CuInSe_2$ POLYCRYSTALLINE THIN FILM WITH $(NH_4)_2S_X$ SOLUTION

Y. H. CHENG, B. H. TSENG*, J. J. LOFERSKI AND H. L. HWANG

Dept. of Electrical Engn., National Tsing Hua Univ., Hsinchu, Taiwan R.O.C.

*Inst. of Material Sci. & Engn., National Sun Yat Sen Univ., Kaohsiung, Taiwan, R.O.C.

ABSTRACT

In this work, we studied the removal of the native oxide on polycrystalline $CuInSe_2$ thin films by KCN and effect of subsequent chemical sulfurization with $(NH_4)_2S_x$ solution on these films. As a result of the treatment, a portion of the selenide film was transformed into $CuIn(S,Se)_2$. The Auger Electron Spectroscopy and X-ray photoelectron spectroscopy studies showed that KCN removed the oxygen and the sulfurization prevented regrowth of the oxides. The optical bandgap of the sulfurized films increased about 0.27 eV. From these experiments, we concluded that sulfur atoms are incorporated in the $CuInSe_2$ to form a stable and higher bandgap layer, $CuIn(S,Se)_2$ which may passivate the $CuInSe_2$ and improve the performance of polycrystalline photovoltaic cells made from it.

INTRODUCTION

Ternary chalcopyrite semiconductors $CuInSe_2$ (CISe) is one of the promising candidates for high efficiency, stable thin film photovoltaic (PV) cells for large scale solar energy conversion. Thin film solar cells based on this material have been reported to have solar energy conversion efficiency up to 16%[1,2]. The role of surface oxide layers on the performance of CISe PV cells has not been investigated to date even though a KCN treatment is sometimes used to improve CISe cell performance. The KCN is presumed to remove excess Cu along with the native oxide. In this investigation, we have followed the KCN etch with immersion in an $(NH_4)_2S_x$ solution which we hoped would sulfurize the surface thus preventing regrowth of oxides removed by the KCN. We expected that sulfur atoms would replace the oxygen on the surface and in the grain boundaries of the thin films. What we found is that the KCN treatment did indeed remove the oxygen and that the sulfur atoms entered into the CISe grains to transform the layers into alloy of CISe and copper indium sulfide, CISu.

Mat. Res. Soc. Symp. Proc. Vol. 386 © 1995 Materials Research Society

EXPERIMENTAL

$CuInSe_2$ thin films were prepared by the three-source evaporation method. Before chemical treatment, the films were cleaned in a conventional organic solvents sequence, namely trichlorethylene, acetone and methanol. It is well known that KCN is a good etchant for copper selenides and copper sulfides and that it can remove oxides very effectively. Such KCN etching has been successfully used to improve performance of solar cells based on $Cu(In,Ga)Se_2$[3] and $CuGaSe_2$[4]. The films were divided into two groups. One group was etched in an aqueous solution of 5% KCN for 5 minutes at room temperature, followed by immersion in an $(NH_4)2S_x$ solution containing about 5% excess sulfur for 1 hr and a deionized water rinse. After samples were taken out, the surface was blown dry with high purity nitrogen gas , and the samples were quickly loaded into the vacuum chamber of the surface analyzer. The second group of samples were left untreated and served as a control.

Differences between the surface and bulk composition were determined with Auger Electron Spectroscopy (AES) compositional depth profiling. X-ray photoelectron spectroscopy (XPS) measurements were performed with PHI Model 1600 ESCA system using Mg K_α for the determination of binding energies (E_B). Ion-beam sputter etching was performed with a 3KV Ar^+ ion beam (10^{-2} Pa Ar pressure). Previous studies have shown that the CISe films can be sputtered with Ar ions having energies up to 3KeV without changing the surface composition[5]. The optical bandgaps were determined from the Tauc plot using IR absorption spectrophotometry[6] in the wavelength region from 800 to 1150 nm.

RESULTS AND DISCUSSION

The qualitative AES compositional depth profiles of the $CuInSe_2$ thin film before and after KCN etching are shown in Figs. 1(a) and 1(b), respectively. No oxygen is observed in films subjected to KCN etching and sulfurization. The XPS measurement support the view that the native oxide on the CISe films is in the form of almost pure In_2O_3[7] . Analysis of the energy position of the O 1s before KCN etching reveals that the oxygen is present in a mix of carbonates and indium oxide ($E_B \cong 530.4$ eV)[8] as shown in Fig. 2a. After sputtering for 1 min, the carbonates decrease drastically as demostrated by the fading away of the C 1s peak and the oxygen is present mainly as indium oxide. The oxygen peak disappears after sputtering for 15 min, which agrees with with the AES depth profiles that show a large amount of oxygen present in the CISe films before KCN etching. After KCN etching and sulfurization, the O 1s peak is shifted to higher binding energy; this peak is removed by sputtering , i.e. oxygen is only present

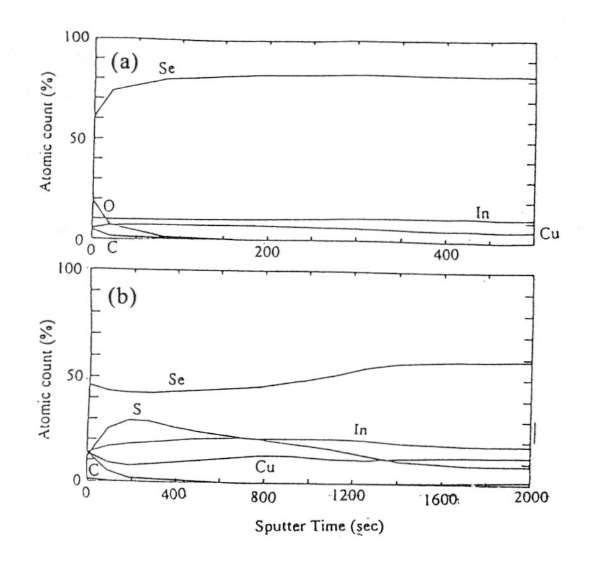

Fig. 1
AES compositional depth profile on $CuInSe_2$ thin film (a) before and (b) after etching in KCN solution and $(NH_4)_2S_x$ sulfurization.

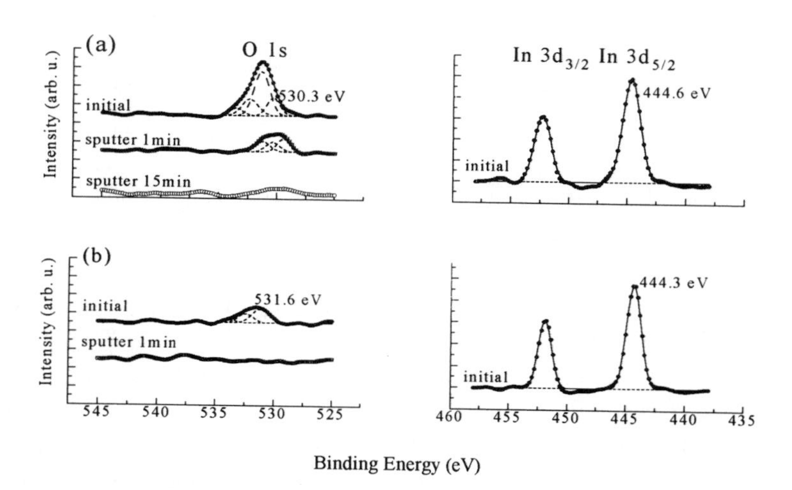

Fig. 2 Dependence of X-ray photoelectron spectra of O 1s and In $3d_{3/2}$, $3d_{5/2}$ on (a) before and (b) after etching in KCN solution and $(NH_4)_2S_x$ sulfurized $CuInSe_2$ thin film.

in a combination with carbon on the surface, as shown in Fig. 2b. Simultaneously, the In $3d_{5/2}$ peak is shifted to a lower binding energy characteristic of $CuInSe_2$[9]. Those results demostrate that KCN removes the native oxide and that the subsequent sulfurization can prevent the regrowth of the native oxide on the $CuInSe_2$.

The effect of immersion in an $(NH_4)2S_x$ solution is shown in the AES compositional depth profile (Fig. 1). A significant amount of sulfur atoms diffused deeply into the $CuInSe_2$ thin films. On the basis of XPS results, copper is oxidized and the S $2p_{3/2}$, $2p_{1/2}$ peaks make their appearance in the $(NH_4)2S_x$ treated samples as shown in Fig. 3. Analysis of the energetic position of the Cu $2p_{3/2}$ and S $2p_{3/2}$, $2p_{1/2}$ peaks leads to the conclusions that a copper sulfide layer is present in the sulfurized samples[10].

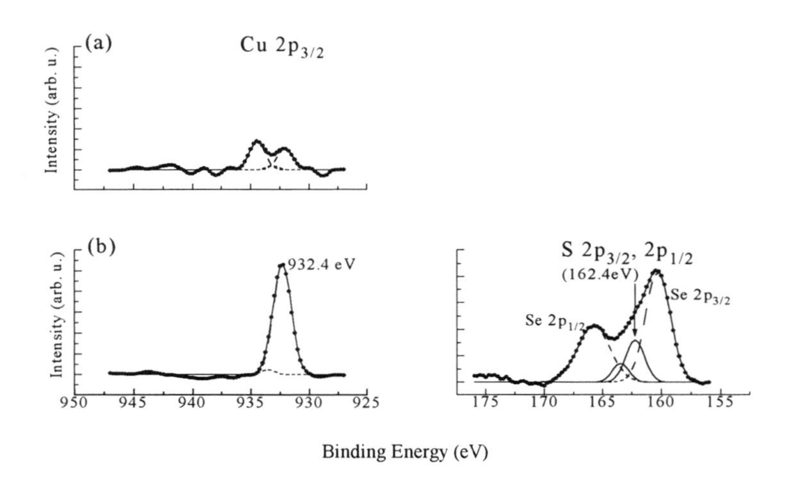

Fig. 3 Dependence of X-ray photoelectron spectra of Cu $2p_{3/2}$ and S $2p_{3/2}$ on (a) before and (b) after etching in KCN solution and $(NH_4)_2S_x$ sulfurized $CuInSe_2$ thin film before sputtering.

Based on the Cahen's defect chemical model[11], all group VI elements, O, S, Se and Te are Lewis acids. They will react readily with Lewis bases, e.g. V_{Se} and In_{Cu}. Thus, after removal of oxygen by the KCN treatment, sulfur atoms react with the Lewis bases on the films to form sulfides. We note that there are more sulfur atoms than oxygen atoms in the films after KCN etching and the Cu/In ratio is decreased by 17% (shown in Fig. 1). The defect chemical model might explain this observation. In addition to removal of native oxide, the KCN treatment may also remove Cu_xSe and/or excess copper phases on the Cu-rich $CuInSe_2$ films and consequently the Cu/In ratios are decreased[3,4,12], i.e. the concentration of Cu_{In} which can neutralize V_{Se} is decreased, and the films becomes more sensitive to sulfur. Thus a portion of the selenide film is transformed into an alloy of $CuInSe_2$ and $CuInS_2$. The optical bandgap as determined from a Tauc plot increases about 0.27 eV on the treated films. This supports the view that sulfur atoms are indeed incorporated in the $CuInSe_2$ layer, transforming it into a stable, higher bandgap $CuIn(S,Se)_2$ alloy.

CONCLUSIONS

Polycrystalline $CuInSe_2$ prepared by three-source evaporation were etched in a 5% KCN solution followed by sulfurization using an $(NH_4)_2S_x$ solution. This treatment results in the removal of the native oxide by KCN solution followed by sulfurization which prevents the regrowth of the native oxide. As a result of the $(NH_4)_2S_x$ treatment, a significant amount of sulfur atoms enter the $CuInSe_2$ thin films and replace Cu_xSe with Cu_xS. Based on Cahen's defect chemical model, the subustitution of S for Se may be the result of a higher sensitivity to sulfur caused by the KCN etching. The sulfur atoms are then incorporated into the $CuInSe_2$ film to form a stable and higher bandgap film, CuIn(S,Se)2. This hypothesis is supported by our observation that the optical bandgap increases by about 0.27 eV in the sulfurized films.

We conclude that a $CuIn(S,Se)_2$ layer with a higher bandgap is formed as a result of KCN etching and the inorganic sulfurization. The effect of formation of this layer on the reduction of the surface recombination losses and improvement of the electrical properties of the films will be investigated in the future.

ACKNOLEDGMENT

The authors thank Mr. S. B. Lin for the molecular beam depositional growth of the $CuInSe_2$ films. Support form the National Science Council is gratefully acknowledged.

REFERENCES

1. L. Stolt, J. Hedstrom, J. Lessler, M. Ruckh, K.O. Velthaus and H.W. Shock, Appl. Phys. Lett. **62**, 597 (1993)
2. L. Stolt, M. Bodegard, J. Hedstrom, J. Lessler, M. Ruckh, K.O. Velthaus and H.W. Shock, Proc. 11st E.C. Photovoltaic Solar Energy Conf., Montreus 1992
3. R. Klenk, R. Menner, D. Cahen and H.W. Schock, Proc. 21st IEEE Photovoltaic Specialist Conf., Orlando 1990

4. R. Klenk, R. Mauch, R. Schaffer, D. Schmid and H.W. Schock, Proc. 22nd IEEE Photovoltaic Specialist Conf., Las Vegas 1991
5. L.L. Kazmerski, N. Burnham, A.B. Swaitzlander, A.J. Nelson and S.E. Asher, Proc. 19th IEEE Photovoltaic Specialist Conf., New Orleans 1987
6. J. Tauc, R. Grigorovici and A. Vancu, Phys. Stat. Solid. **15**, 672 (1966)
7. O. Jamjouin, L.L. Kazmerski, D.L. Lichtman and K.J. Bachmann, Surf. Interface Anal. **4**, 227 (1982)
8. D.T. Clark, T. Fok, G.G. Roberts, R.W. Sykes, Thin Solid Films **70**, 4761 (1985)
9. D. Cahen, P.J. Ireland, L.L. Kazmerski, F.A. Thiel, J. Appl. Phys. **57**, 4761 (1985)
10. V.I. Nefedov, Y.V. Salyn, P.M. Solozhenkin and G.Y. Pulatov, Surf. Interface Anal. **2**, 171 (1980)
11. D. Cahen and R. Noufi, Appl. Phys. Lett. **54**, 558 (1989); Solar Cells **30**, 53 (1991)
12. Y. Ogawa, A. Jager-Waldau, T.H. Hua, Y. Hasimoto and K. Ito, presented at the 7th Int. Conf. on Solid Films and Surfaces, Hsinchu, 1994 (unpublished).

PRE-GATE OXIDATION CLEANING OF SILICON WAFER BY ELECTRIC ARC PLASMA JET TREATMENT

G.Ya. PAVLOV
Centre for Analysis of Substances, 9, Elektrodnaya St., 111524 Moscow, Russia

ABSTRACT

The effect of arc plasma jet treatment (APJT) of silicon surface used for pre-gate oxidation cleaning on the electrophysical parameters of MOS structures ($Si/SiO_2/Si^*/Al$) has been studied. We show that APJT etching cleaning considerably improves the constant current charge to breakdown of MOS structures in comparison with conventional wet chemical cleaning. We have analyzed the effect of plasma cleaning conditions on the quality of gate oxide and SiO_2/Si interface.

INTRODUCTION

Wet chemical cleaning remains promising for ULSI fabrication. However, the major trend of the research into surface cleaning turns to the vapor/gas/plasma processes [1]. They consume far less reagents and can easily be performed on cluster modules to satisfy the requirements of wafer cleaning. Etching in HF vapors (HF/H_2O, HF/CH_3OH, HCl, NO/HCl), H_2 annealing, and the use of UV excitation (UV/O_3, $UV/F_2/H_2$, UV/Cl_2), vacuum-plasma stimulation (O_2, H_2, NF_3, etc.), and Ar ion sputtering have already found general use in cleaning of silicon wafers from organic contaminants, metal atoms, and natural oxide. Dry processes are widely used for additional *in situ* cleaning of wafers before oxidation, metallization, epitaxy, etc., but they do not enable cleaning from mechanical particles.

A promising technique of dry surface cleaning is APJT [2]. A hydrodynamically continuous high-enthalpy ($> 10^4$ $J \cdot g^{-1}$) low-temperature ($< 10^4$ K) arc plasma jet affects the surface for a short time (0.01 to 0.1 s) at atmospheric pressure. The high jet velocity (100 m/s) in combination with the high density of the heat flow ($> 10^3$ $W \cdot cm^{-2}$) and a dense flow of chemically active and excited particles ($> 10^{18}$ $cm^{-2} \cdot s^{-1}$) at the plasma/surface boundary ensures surface cleaning. Addition of reactive gases to the jet leads to their thermal dissociation to chemically active particles, which produce volatile compounds with wafer material and enable an etching rate of up to 1 μm/s. The cleanliness of the APJT environment is maintained at atmospheric pressure using hydrodynamical shielding (HDS) [3] in the form of a circular laminar flow of purified gas (Ar, N_2, etc.) which envelops the plasma jet.

A most important procedure of the ULSI fabrication route is pre-gate oxidation cleaning of the wafer surface [4]. The electrophysical parameters of MOS structures largely depend on cleaning quality and may therefore serve as reliable criteria of process efficiency.

The aim of this work was to analyze the effect of APJT conditions on the electrophysical parameters of MOS structures, such as the flat band voltage (V_{fb}), transient capacity relaxation time (τ), breakdown field (E_{bd}), constant current charge to breakdown (Q_{bd}), density of states on the Si/SiO_2 interface in the middle of the band gap (N_{ss}), and the distribution of electrically active impurities in the superficial silicon layer (n_a).

Mat. Res. Soc. Symp. Proc. Vol. 386

EXPERIMENTAL

The 25 test samples, CZ *n*-Si(100) ∅100 mm wafers (P doping to 10^{15} cm^{-3}) in polyethylene packaging, were numbered and subdivided in two sets. Wafers 1-11 were subjected to wet chemical cleaning (RCA) before oxidation, wafers 12 and 13 were not cleaned, and the others were subjected to various modes of APJT.

Plasma-chemical cleaning of silicon wafers was performed on a modified DPO 100M unit. The arc plasma generator (Figure 1) produced a plasma funnel into which additional $CF_4/O_2/Ar$ flow with permanent total gas consumption was introduced. In some cleaning modes, the plasma jet was protected from the air environment using hydrodynamical shielding (HDS) in the form of a circular laminar ~1 m/s Ar flow. In other modes without HDS, air was injected into the plasma.

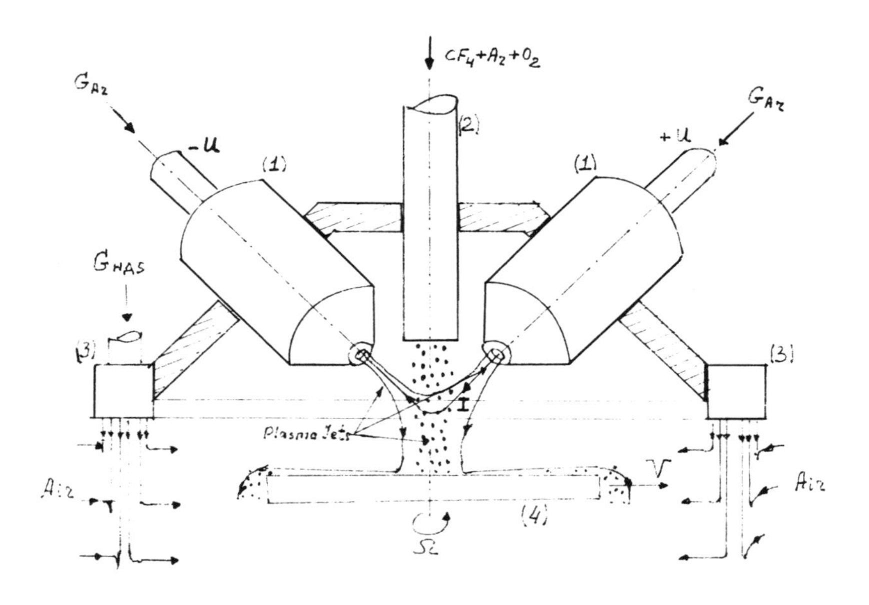

Figure 1. Shematic of APJT plasma generator: (1) double-jet electric arc plasmotrone, (2) reagent supply tube, (3) hydrodynamic shielding (HDS) system, (4) wsfer holder; (2U=100V, I=100A, G_{Ar} =1 l/min, G_{HDS} =from 10 to 20 l/min , V=1 m/s)

The wafers were in rotary-reciprocating motion to ensure homogeneous surface processing. The APJT modes were distinguished by the Ar plasma jet composition, which was determined by the reactive gas consumption and oxygen and HDS (Table 1).

Table 1. APJT conditions.

Wafer	$CF_4/(Ar+O_2+CF_4)$	$O_2/(Ar+O_2+CF_4)$	HDS (Ar)
14,15	0	0	no
17,18	0,03	0	no
19,20	0,07	0	no
21,22	0,15	0	no
23	0,15	0,03	no
24	0,07	0	yes
25	0,15	0,03	yes

After cleaning for 30-50 min the wafers were brought again into a single set and oxidized in an HCl-containing atmosphere to produce a 27 nm gate oxide layer. Then the wafers were coated with 0.4 μm polycrystalline silicon films and phosphosilicate glass (PSG) to dope Si* with P to < 15 Ω/□ Then PSG was etched off, an 1 μm Al + 3-5% Si film was deposited, and the Al and Si* layers were lithographed. Finally, Ti films were deposited onto the back sides of the wafers and then the wafers were subjected to anneal in N_2 at 470 °C for 30 min.

The thickness of the etched surface layer was proportional to the CF_4 consumption and was 10, 20, and 50 ±5 nm. MOS structures were fabricated in a 256 kbit - 1 Mbit IC module, their electrodes being 1, 0.5, 0.25, and 0.1 mm in diameter (0.0078 - 0.00078 cm^2).

The electrophysical parameters of the test MOS structures were estimated using RF (10^6 Hz) C-V characteristics, I-V characteristics, and the techniques of constant current charge to breakdown and DLTS. The measurements were performed on 100 chips for each wafer, and only one method was applied to one MOS structure in order to avoid interference. The results were statistically processed, and histograms of the electrophysical parameters were plotted.

RESULTS AND DISCUSSION

The electrophysical parameters were distributed over the surfaces quite homogeneously irrespective of the cleaning mode. The constant current charge to breakdown (Q_{bd}) measured at 100 $mA \cdot cm^{-2}$ had the greatest scatter because it is the most sensitive to oxide contamination and surface preparation (Fig. 2). Other MOS parameters were far less scattered.

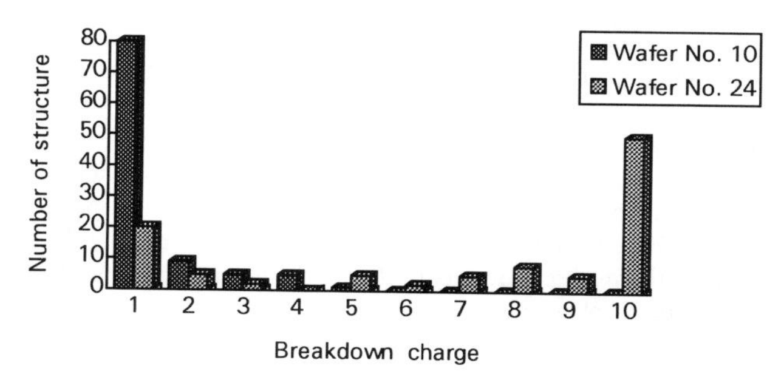

Figure 2. Typical histograms of MOS constant current charge to breakdown Q_{bd} ($C \cdot cm^{-2}$) after wet chemical cleaning (wafer 10) and plasma jet cleaning (wafer 24)

The scatter of the averaged electrophysical parameters of the MOS structures after wet chemical cleaning did not exceed 5%, i.e., one can reliably attribute the change in the parameters of wafers 14-25 solely to the effect of pre-gate oxidation plasma cleaning. Table 2 presents averaged E_{bd} measured at exponentially increased current, the maximum breakdown currents I_{bd}, and averaged constant current charge to breakdown Q_{bd} for various cleaning modes.

Table 2. Gate oxide stability in the MOS structures after wet (1-11) and plasma (14-25) cleaning.

Wafer	E_{bd}, MV/cm	I_{bd}, $mA \cdot cm^{-2}$	Q_{bd}, $C \cdot cm^{-2}$
1-11	7.7	5	0.7
12-13	7.0	1	0.6
14-15	9.6	1000	8.1
17-18	6.7	0.1	0.5
19-20	4.4	0.5	1.0
21-22	5.5	5	3.7
23	4.4	10	4.0
24	9.2	250	6.9
25	6.2	250	1.6

Wafers 14 and 15 had very high breakdown voltages and charge stability. Plasma-cleaned wafers had higher Q_{bd} but lower E_{bd} than after wet chemical cleaning. In that context, note the good parameters of wafer 24. We observed large-scale Q_{bd} inhomogeneities in some wafers after APJT which can be attributed either to a spatial distribution of impurities (mainly oxygen) during CZ-growth of silicon or to an unknown effect of cleaning.

Analysis of data (Table 2) suggests that after inert gas plasma cleaning, the gate oxide has the higher quality because this cleaning mode removes organic and sorption contamination and it is also likely to modify the structure of the natural oxide. The most technologically important result was obtained using oxygen-free cleaning in a plasma jet with Ar HDS (wafer 24). If oxygen was supplied to the plasma jet from the air or introduced with the reactive gas flow, E_{bd} was lower, but Q_{bd} higher in comparison with wet cleaning.

Pre-gate oxidation plasma cleaning slightly increases the positive charge in comparison with wet chemical cleaning, which can be seen from a change in V_{fb} by ~-0.1 V (Fig. 3). The relaxation time τ was quite short (1 to 5 s) because the defects in the wafers were not annealed in H_2 as it is usually done in IC technology, but it is correlated with the high breakdown voltages and charges of wafers 14, 15, and 24.

DLTS measurements of N_{ss} at 300 K proved the SiO_2/Si interface to be very good after any plasma treatment mode: N_{ss} was lower than $2 \cdot 10^{10}$ $cm^{-2} \cdot eV^{-1}$ (the sensitivity of experimental unit). For wet cleaning, the surface was more contaminated: N_{ss} was about 10^{11} $cm^{-2} \cdot eV^{-1}$; and for wafers 12 and 13, which were not cleaned, N_{ss} ranged from $2 \cdot 10^{11}$ to 10^{12} $cm^{-2} \cdot eV^{-1}$.

Despite the high thermal flow density of APJT, the reactive impurities were not found to redistribute in the superficial layer. The concentration of electrically active impurities increased in wafers 14 and 15 from 10^{15} to 10^{16} cm^{-3} after APJT.

An increase in Q_{bd} and a decrease in V_{bd} after CF_4/O_2/Ar plasma cleaning can be accounted for by a development of silicon surface morphology and the formation of a 0.5-1.5 nm nonstoichiometric plasma oxide layer. Indirect evidence for these processes is the fact that

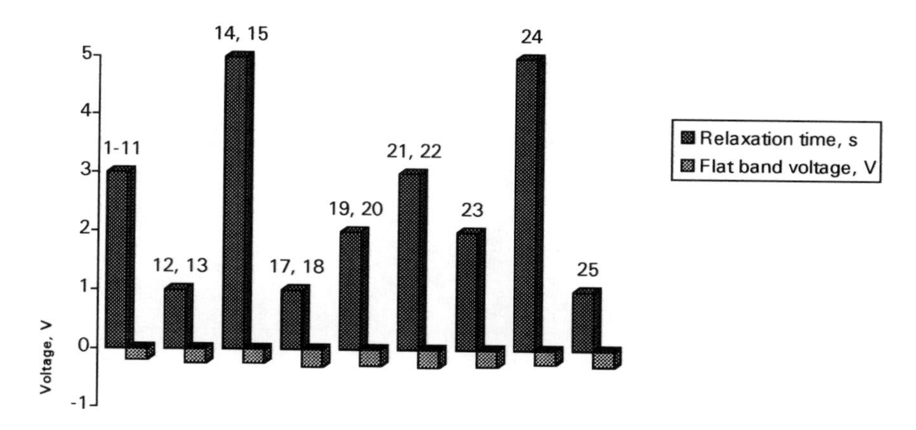

Figure 3. Histograms of relaxation time and flat band voltage for various pre-gate oxidation cleans.

cleaning with < 0.1% O_2 (HDS), which is unlikely to produce any change to the surface morphology, noticeably improved the electrophysical parameters. The most important result are the good MOS structure parameters after neutral plasma cleaning which removes sorption contaminants, in particular, OH-groups and produces no radiation defects. The removal of OH-groups is very essential for oxide quality.

CONCLUSIONS

• Arc plasma jet pre-gate oxidation cleaning of silicon surfaces considerably increases Q_{bd} in comparison with wet chemical cleaning and changes C-V characteristics of the MOS structures only slightly;

• Hydrodynamical shielding increases E_{bd} and Q_{bd} of the MOS structures;

• Argon-air plasma jet removes organic contamination and sorption OH-groups from Si surface and thus ensures the highest E_{bd} and Q_{bd};

• The technique of arc plasma jet cleaning should be developed because it is promising for IC technology.

REFERENCES

1. T. Hattori in Proceedings of the 2nd International Symposium on Ultra-clean Processing of Silicon Surface (UCPSS' 94), Brugge, 1994, pp. 13-18.
2. V.M. Maslovsky and G. Ya. Pavlov in Proceedings of the 2nd International Symposium on Ultra-clean Processing of Silicon Surface (UCPSS' 94), Brugge, 1994, pp. 83-86.
3. A.A. Paveliev, Models and Formation of Non-Equilibrium Turbulent Flows, in Invited Lectures of the 2nd European Fluid Mechanics Conference, 20-24 September 1994, Warsaw, Poland.
4. R. Schild, K. Locke, M. Kozak, and M.M. Heyns in Proceedings of the 2nd International Symposium on Ultra-clean Processing of Silicon Surface (UCPSS' 94), Brugge, 1994, pp. 31.

SURFACE MODIFICATION OF $Zn_xCd_{1-x}Te$ DUE TO LOW ENERGY ION SPUTTERING

S. VIJAYALAKSHMI*, K.-T. CHEN, M.A. GEORGE, A. BURGER and W.E. COLLINS,
Center For Photonic Materials And Devices, Department of Physics,
Fisk University, Nashville, TN-37208, U.S.A.

ABSTRACT

$Zn_xCd_{1-x}Te$ is a widely used substrate for the epitaxial growth of HgCdTe, which is used in infrared detectors. Results of the effect of sputtering of $Zn_xCd_{1-x}Te$ single crystals with low energy Ar^+ beam are reported in this paper. X-ray photoelectron spectroscopy (XPS) and photoluminescence (PL) techniques were used to measure the concentration of Zn in these crystals. Selective sputtering of Zn atoms has been observed from freshly cleaved crystals using XPS studies. Sputtering is a common method of cleaning $Zn_xCd_{1-x}Te$ crystals in their device preparation and our studies show that this method of cleaning alters the surface which may introduce lattice mismatch on the surface. Surface morphology before and after cleaving the crystals is studied using Atomic Force Microscopy (AFM).

INTRODUCTION

$Zn_xCd_{1-x}Te$ single crystal is an important substrate for the growth of HgCdTe or HgZnTe, used in infrared detectors[1] and in room temperature gamma ray detectors[2], because of the tunability of the lattice constant and band gap. The lattice constant ranges from 6.102 Å for ZnTe to 6.481Å for CdTe and can be lattice matched to the detector alloys[3]. The band gap varies between 2.3 to 1.6 eV. The usual surface preparation technique for $Zn_xCd_{1-x}Te$ crystals is mechanical polishing followed by chemical etching using HCl or bromine/methanol solution etc. For MBE growth of thin films on these substrates, the crystals are sputter cleaned using a low energy ion beam (1-4 keV). Chemical etching on these crystals typically involve bromine and an inorganic acid which leave the surface depleted of Cd[4] and contaminated by impurities such as bromine and chlorine. Further more, exposure to atmosphere before introducing the sample to ultra high vacuum results in contamination by carbon and oxygen. Heat treatment of ZnCdTe crystals to achieve stoichiometric surface is discussed in reference 3.

In an earlier paper, we have reported[5] the effect of chemical etching on ZnCdTe crystals using X-ray photoelectron spectroscopy (XPS) and atomic force microscopy (AFM) techniques. XPS data show an increase of the tellurium and a depletion of the cadmium concentration while AFM study shows tellurium precipitation upon bromine/methanol etching. It has also been shown that CdTe and ZnCdTe surfaces oxidize readily and the compounds most often found on ZnCdTe were oxides of Cd and Te ($CdTeO_3$ and $ZnTeO_3$), with higher relative concentration of tellurium oxides (TeO_2). Raman scattering studies show the formation of polycrystalline layer of tellurium of thickness 1.0 to 4.0 monolayers due to Br-MeOH etching. Low energy ion sputtering of single crystals is a field very well studied[6]. The ion beam removes surface atoms due to momentum transfer and rate of sputtering is determined by the ion beam parameters as well as the density of target crystal. Quantitative determination of sputtering rate for compounds is a difficult process since different constituent atoms will have different sputtering yields and bonding of certain atom in a given matrix is not very well known. Metal and insulator compounds have been studied extensively and the sputtering rate of CdTe is reported[6] to be 56 nm/min due to a 500 eV Ar^+ ion

Mat. Res. Soc. Symp. Proc. Vol. 386

beam incident at 90° (flux = 1 mA/cm^2). However, to the best of our knowledge, no work on the surface modification of ternary semiconductors due to low energy ion sputtering has been published. Zn concentration from various freshly cleaved (110 plane) $Zn_xCd_{1-x}Te$ single crystals using photoluminescence (PL) and XPS are reported in this paper. Introduction of a mask during ion sputtering of these crystals and further AFM study shows that the sputtering rate is approximately 15 nm/min for a 4 keV Ar^+ beam (flux = 55 mA/cm^2) incident at 45°. AFM study also shows the development of noncrystalline structure due to ion sputtering and small aggregates which may be identified as precipitates of some of the constituent atoms. However, both AFM and XPS studies are not capable of identifying these precipitates and further spectroscopic studies are in progress to identify these aggregates.

EXPERIMENTAL

$Zn_xCd_{1-x}Te$ crystals with varying values of x were grown using Physical Vapor Transport method. The crystals were cleaved and PL experiments were performed to determine the value of x. XPS spectra were obtained using a Kratos X-SAM 800 spectrometer with a dual anode and an aluminum monochromator. The AlK_α source on the dual anode having a characteristic X-ray energy of 1486.6 eV was employed for the surface analysis studies. The crystals were cleaved in air and were mounted on the sample holder and the system was evacuated immediately. The typical time for achieving ultrahigh vacuum was approximately one hour. A mini beam ion gun was used for sputter cleaning the crystals. XPS spectra were taken before and after ion bombardment. Experiments were performed on three different crystals with x values ranging from 0.12 to 0.14. A Digital Instruments Nanoscope E which had a piezoelectric tube scanner with an effective scan range from about 200 nm to 14 μm, was used to perform the AFM studies. For AFM studies, the samples were imaged in ambient air and the mapping was obtained in constant height mode.

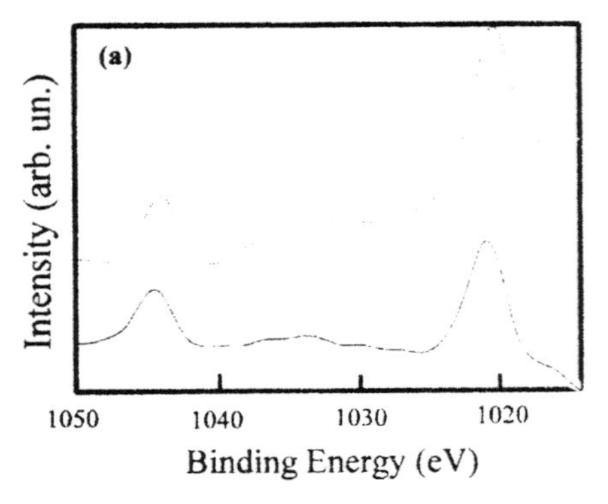

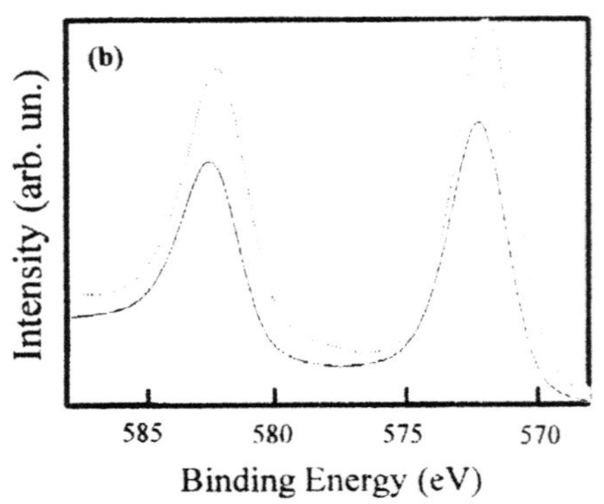

Figure 1: Photoelectron yield from a ZnCdTe surface for (a) Zn 2p, and (b) Cd 3d peaks. The solid and dotted lines indicate the spectrum before and after sputtering, respectively.

RESULTS AND DISCUSSION

Figure 1 shows the XPS spectra before and after ion beam sputtering from a sample which has a Zn concentration, x=0.13, as measured by PL experiments. Spectra for Zn, Cd, and Te are shown in panels (a), and (b), respectively. The solid lines represent the peaks obtained immediately after etching in air and introduction into the UHV chamber. The dotted lines represent the spectra obtained after two 3 min etches for a total of 6 min etch time. The XPS spectra indicate that the peak intensity increased after etching. In each case the peak intensity increases upon Ar^+ bombardment and this is directly attributed to a nominal decrease in the carbon and oxygen contamination from the cleavage in air. From the XPS data, shown in table 1, the ratio of Zn to Cd was calculated to determine the Zn concentration. Prior to etching, the value was 0.141. After the first three minute etch, the Zn:Cd ratio increased to 0.157 and after six minutes etch, it was 0.134, consistent with the data obtained from photoluminescence measurements. Also calculated from the XPS data was the total Zn and Cd to the Te concentration. In a stoichimetric crystal, this ratio would be expected to be 1:1, however, the as determined values shown in Table 1 were 0.80, 0.78 and 0.74 for the unetched, 3 min Ar^+ etch and 6 min Ar^+ etch respectively. The discrepancy may be attributed to species mixing induced by ion bombardment as well as the preferential ion etching of Zn and Cd. In addition to the variation in peak intensity, the binding energy shifts were examined and it was observed that the peaks, Zn 2p, Cd 3d and Te 3d, had decreased in their binding energy positions. These decreases in binding energies, shown in Table 2, are primarily attributed to the reduction of surface oxygen that occurred after Ar^+ ion sputtering.

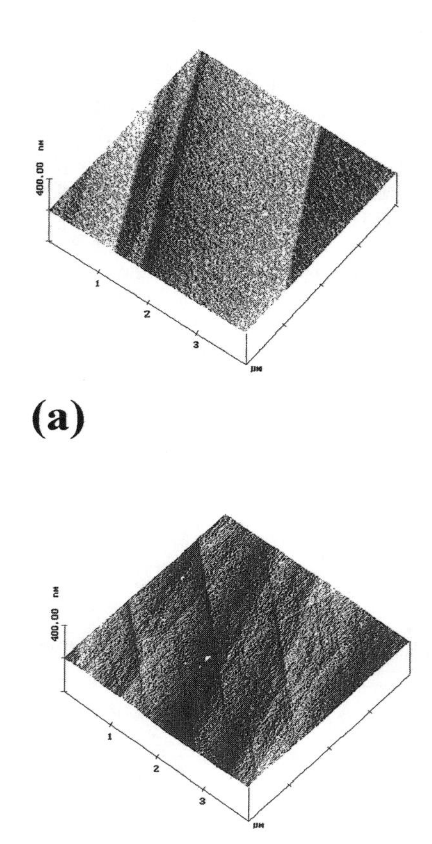

Figure 2: AFM images of the crystal (a) before and (b) after Ar^+ ion sputtering.

Figure 2 shows AFM images of the ZCT surface before and after ion etching. The morphology of the freshly cleaved crystal is shown in figure 2(a). Growth steps, commonly observed in II-VI semiconductors, are observed in the freshly cleaved samples. Calculations of the height of these visible steps as well as the surface roughness were performed and found to be 4.5 nm high and $R_a = 0.52$ nm. Figure 2(b) is an AFM micrograph of a freshly cleaved crystal after sputtering with 4 keV Ar^+ beam with an approximate flux of 55 mA/cm^2 for 6 minutes. Apparent in this image are the cleavage steps observed in 2(a), however, also seen are some additional line-like features that are a result from the ion etching process.

The surface roughness has increased to $R_a = 2.5$ nm over the freshly cleaved crystal in 2(a). In order to facilitate the determination of the sputtering rate for this crystal, the surface was masked in various regions. The masking was achieved by deliberately introducing small particles on the surface at selected positions. Figure 3 shows a sputtered crystal from which sputtering yield was estimated. The height difference between masked and unmasked region of the sputtered crystal is in the range of 70-90 nm. We believe that this range in sputtering is introduced due to the gaussian profile of the beam. As a conservative lower limit, we estimated the sputtering yield of ZnCdTe as 15 nm/min when sputtered by a 4.0 keV, Ar^+ beam incident at 45°.

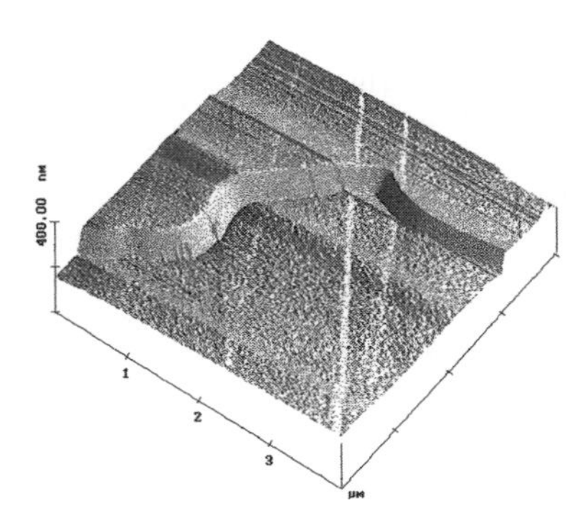

Figure 3: AFM image showing the masked (top) and unmasked (bottom) regions.

TABLE 1. The ratio of calculated mass % from the XPS data

TREATMENT	Zn/Cd	(Zn + Cd)/Te	(C1s+O1s)/ZnCdTe
Cleaved	0.141 ∓ 0.016	0.801 ∓ 0.088	0.342 ∓ 0.038
Ion etched 3 min	0.157 ∓ 0.017	0.778 ∓ 0.086	0.183 ∓ 0.020
Ion etched 6 min	0.134 ∓ 0.014	0.742 ∓ 0.082	0.165 ∓ 0.018

TABLE 2 : Peak shift after 6 min Ar^+ ion etch

XPS Peak	Peak shift (eV)
Zn 2p	-0.41
Cd 3d	-0.27
Te 3d	-0.32

ACKNOWLEGEMENT

This work was performed through funding provided by NASA through the Fisk Center for Photonic Materials and Devices, Grant No. NAGW-2925, by the NASA Lewis Research Center Grant No. NAG3-1430 and by the Office of Naval Research, Contract No. BMDO/ONR N00014-93-1-1344.

REFERENCES

1. A.A. Khan, W.P. Allred, B. Dean, S. Hooper, J.E. Hawkey, and C. J.Johnson, J. Electron. Mater. 15 (1986) 181
2. E. Reiskin and J. F. Butler, IEEE Trans. Nucl. Sci. (1988) 35
3 Y.S. Wu, C. R. Becker, A. Waag, R. N. Bicknell-Tassius, and G. Landwehr, Appl. Phys. Lett. 60 (15) (1992) 1878
4. A. Waag, Y. S. Wu, R. N. Bicknell-Tassius, and G. Landwehr, Appl. Phys. Lett. 54 (1989) 2662
5. M. A. George, M. Azoulay, H. N. Jayatirtha, A. Burger, W.E.Collins, and E. Silberman, Surf. Sci. 296 (1993) 231
6. R. J. McDonald and B. V. King in Ion Beams for Materials Analysis, edited by J. R. Bird and J.S. Williams, Academic Press, 1989

SURFACE PROPERTIES OF GaAs PASSIVATED WITH $(NH_4)_2S_x$ SOLUTION

Kyung-Soo SUH, Hyung-Ho PARK*, Jong-Lam LEE, Haechon KIM, Kyung-Ik CHO, and Kyung-Soo KIM
Semiconductor Technology Division, ETRI, Yusung P.O. Box 106, Taejon, 305-600, KOREA
*Dept. of Ceramic Engineering, Yonsei University 134, Shinchon-dong, Sudaemoon-ku, SEOUL, 120-749, KOREA

ABSTRACT

Surface properties of GaAs passivated with $(NH_4)_2S_x$ solution have been compared with HCl-treated GaAs using X-ray photoelectron spectroscopy. Sulfur treatment on GaAs surface results in the formation of S-Ga and S-As bonds, which remain after successive rinsing for 1 minute in DI water. The evolution of Ga 2p3 and As 3d peaks in the sulfidation treated GaAs was monitored with the exposing time to air. After 10 days exposure to air, the Ga-O and As-O bonds slightly increased, but maintained almost constant for further exposure. The increase of Ga-O and As-O bonds induces the partial decomposition of sulfur bonds. Decomposition and evaporation behaviors of sulfur and oxygen were observed through the heat treatment of sulfidation treated GaAs under ultra high vacuum (less than 1×10^{-9} torr). After anneal at 350 - 450 °C, slight decrease of sulfur and oxygen due to the decomposition of As-O bond were observed. No more sulfur was found after anneal at 550 - 650 °C, where the decomposition of Ga-O bond was completed.

INTRODUCTION

III - V compound semiconductors such as GaAs and InP are well adapted for microelectronic and optoelectronic device applications due to their unique properties. However, the extent of their applications is somewhat limited due to the high surface state density and difficult control of the Fermi level position.[1-3] According to a number of works done for improvement in the electronic properties, the sulfidation process has been known to be the most probable solution for the surface treatment.[4-6] However, due to the variety of conditions of substrate before treatment and the types of reactants such as Na_2S, $(NH_4)_2S_x$, and H_2S, various different experimental results for the formation of passivated surface and its bonding characteristics have been reported. Also, the existence of As-S bond has been known to depend on the rinsing time after the sulfidation treatment.

In this study, surface chemical state after the sulfidation treatment has been analyzed using X-ray photoelectron spectroscopy (XPS). The change of the bonding states due to the exposure to air has been also monitored for 30 days. Also, through in-situ anneal treatment under ultra high vacuum (UHV) condition in XPS chamber, the thermal behaviors of several chemical bonding states have been analyzed.

EXPERIMENTAL

The samples used in this study were p-type (100) GaAs substrates. Two samples were rinsed with methanol and dipped in HCl for 1 minute to remove native oxide and impurities. After water rinsing, one is delivered to XPS chamber within few minutes and the other was

Mat. Res. Soc. Symp. Proc. Vol. 386 © 1995 Materials Research Society

immersed in $(NH_4)_2S_x$ solution at 60 °C for 10 minutes and rinsed with distilled water for 1 minute. The above procedure has been used to fabricate MESFET with good electrical properties.[7] Within 1 minute after the treatment, the samples were introduced into XPS chamber to minimize the contamination from atmosphere.

The XPS analyses were performed in a V.G.Scientific ESCALAB 200R spectrometer with Al Kα (1486.6 eV) radiation operating at 300 W. Narrow scan spectra of all regions of interest were recorded with 20 eV of pass energy and 15 degree of take-off angle in order to quantify the surface composition and identify the elemental bonding states. Passivation effect was studied by exposure of the sulfidation treated sample to air and monitoring of the surface composition. A specially designed heatable stub was used for in-situ anneal of sulfidation treated sample. The sample was heated for 1 hour and cooled down to room temperature for analysis.

RESULTS AND DISCUSSION

Figure 1 represents a profile on the HCl treated GaAs with mild etching (etching rate ~ 20Å/min.) by XPS. Due to the large affinity of arsenic with oxygen, within 70 Å layer from the surface, the relative atomic concentration of arsenic has been revealed to be higher than that of gallium. In the narrow scan spectra for arsenic and gallium with the substrate composition obtained after etching for 70 Å, any undesirable effect as peak broadening or compositional variation due to the sputtering was not found. These spectra were exactly same as those obtained for the cleavaged GaAs under UHV.

Figure 2 shows the deconvoluted result for As 3d, Ga 2p3, and overlapped region of Ga 3s and As 2p plasmon loss ω_p from the surface of HCl-treated GaAs. Note that the

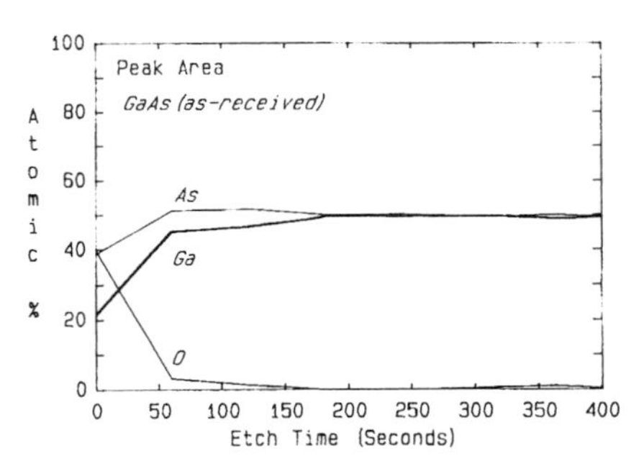

Fig. 1 Depth profile result on HCl-treated GaAs with mild etching (etching rate : ~20 Å/min.).

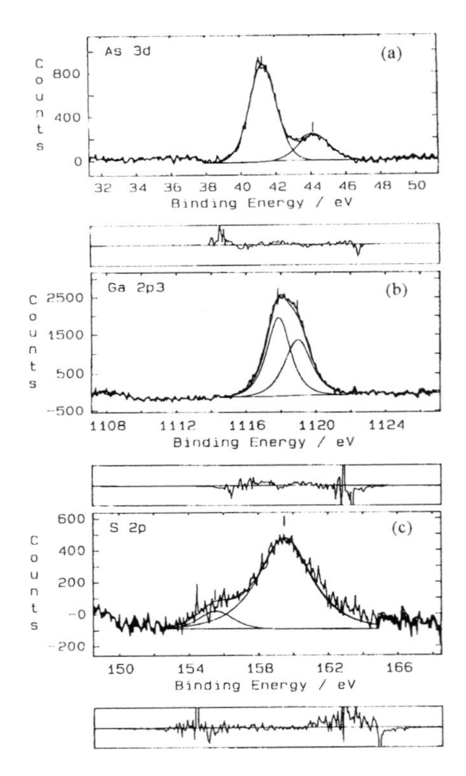

Fig.2. Deconvolution results on narrow scan XPS spectra from HCl-treated GaAs; (a) As 3d, (b) Ga 2p3, and (c) overlapped region of Ga 3s and As 2p plasmon loss ω_p.

overlapped region is monitored to examine the presence of sulfur after sulfidation treatment[8]. The As 3d spectrum could be resolved into As-Ga and As-O. The 44.1 eV binding energy for As-O bond corresponds to normal As-O bond in As_2O_3. The Ga 2p3 spectrum was also resolved into two bonding states such as Ga-As and Ga-O of Ga_2O_3 (1119 eV of peak binding energy). The overlapped region constituted Ga 3s of 159.5 eV binding energy and As 2p plasmon loss ω_p of 155.5 eV. The peak binding energies and full width at half maximum (FWHM) values obtained through the above analyses were used as fixed reference values for analyzing the chemical states after the treatments. Figure 2 clearly shows that GaAs surface is oxidized within few minutes.

Figure 3 shows the narrow scan spectra obtained on a sulfidation treated sample with $(NH_4)_2S_x$ and successive rinsing for 1 minute. When we assumed two chemical states for arsenic, i.e., As-Ga and As-O, the difference between experimental spectrum and simulated one was clearly seen. This peak deconvolution process was found in Fig. 3(a). The difference results from the existence of another chemical bond for arsenic of binding energy 42.6 eV, which corresponds to As-S bond. We can see that the exactly same spectrum is obtained when the peak was deconvoluted in three chemical states for As. For gallium, the same deconvolution process was applied and the result is given in Fig. 3(b). An additional peak of binding energy 1118.5 eV was found for Ga-S bond. The overlapped region in Fig. 3(c) was revealed to contain S 2p peak of binding energy 161.5 eV. Peak attributions, binding energy's, FWHM's, and percentages of each peak area are listed in Table 1.

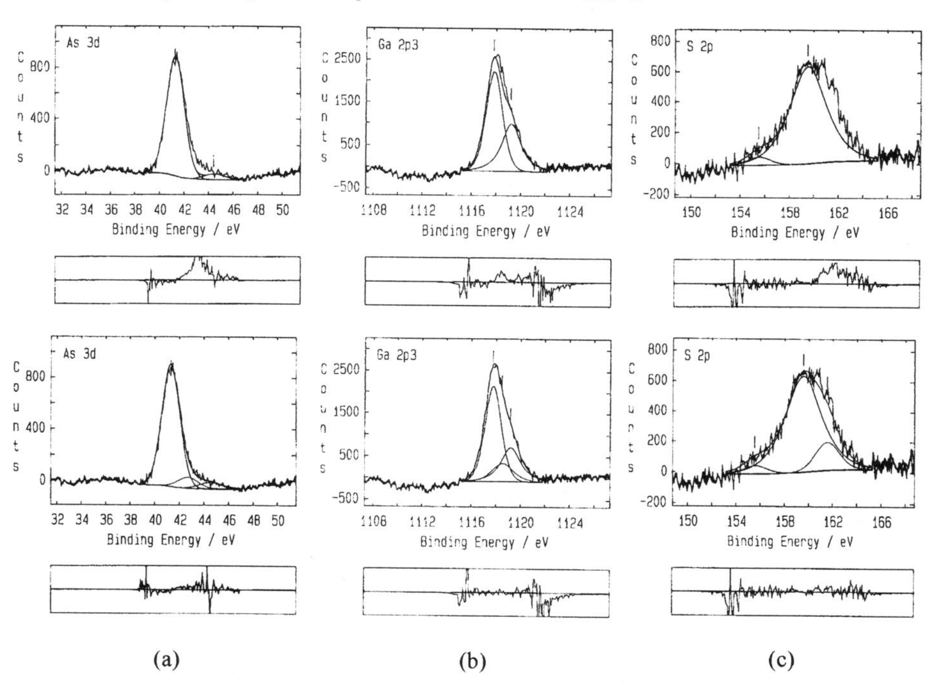

Fig.3. Deconvolution results of narrow scan XPS spectra from the sulfidation treated GaAs ; (a) As 3d, (b) Ga 2p3, and (c) overlapped region of Ga 3s and As 2p plasmon loss ω_p.

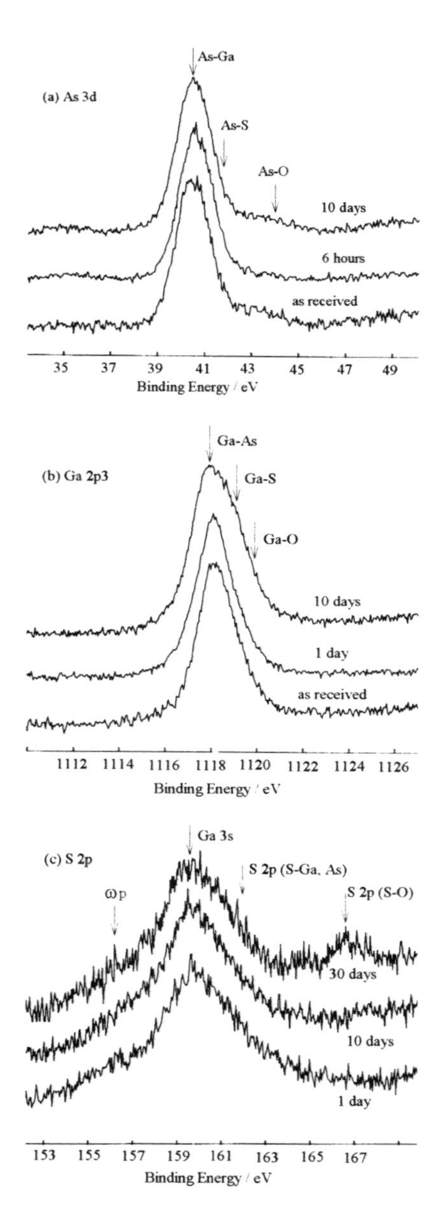

Fig. 4. Evolution of (a) As 3d, (b) Ga 2p3, and (c) S 2p peaks in the sulfidation treated sample according to the air exposing time.

Figure 4 shows peak evolution in the sulfidation treated sample according to air exposing time. Within 6 hours, no change in Ga 2p3 peak but decrease of As-O bond in As 3d peak was found. Up to 10 days, the increase of Ga-O and As-O bonds were also observed. They maintained almost constant within 30 days. The extent of the formation of Ga-O bonds was observed more than that of As-O bonds during the exposure to air. In the overlapped region in Fig. 4(c), S 2p peak decreased after 3 days of exposure in air. The amount of sulfur was observed to gradually decrease up to 10 days and maintain almost constant after 10 days. From the above results, it is suggested that the oxidation of As and Ga induced the decomposition of S-Ga(As) bonds and liberation of sulfur. Normally, the surface of sulfidation treated GaAs is completely covered with at least 1 monolayer of sulfur and some oxygen. Although the samples were exposed to air for 10 days, sulfur bonded with As and Ga was partially replaced by oxygen. In other words, GaAs surface is still passivated with sulfur atoms, comparing the results of sulfidation treated GaAs in Fig. 4 with those of HCl-treated one in Fig. 2.

For comparing the thermal stability of sulfur or oxygen bonds with As or Ga, sulfidation treated GaAs was in-situ annealed for 1 hour from 20 °C to 750 °C with 100 °C interval under UHV condition. Figure 5 shows absolute intensity changes of the elements. The quantity of sulfur was estimated by considering the relative intensity relationship between Ga 2p3 and Ga 3s, and between As 3d and As 3d plasmon loss w_p. The quantity of sulfur was slightly decreased at 350 - 450°C and completely disappeared at 550 - 650 °C. The thermal behavior of oxygen was

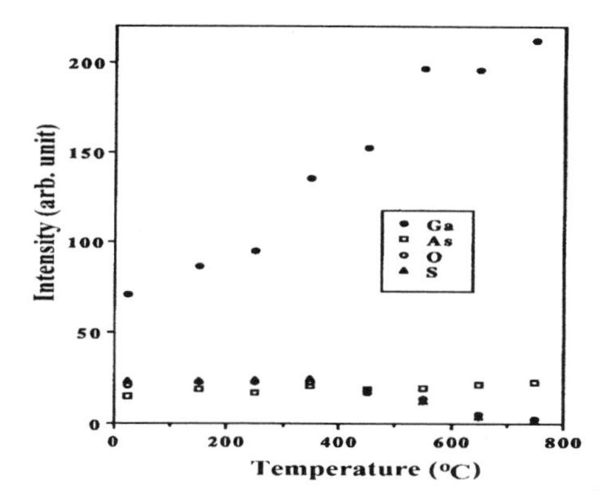

Fig. 5. Absolute intensity changes of sulfidation treated GaAs surface after anneal under UHV condition.

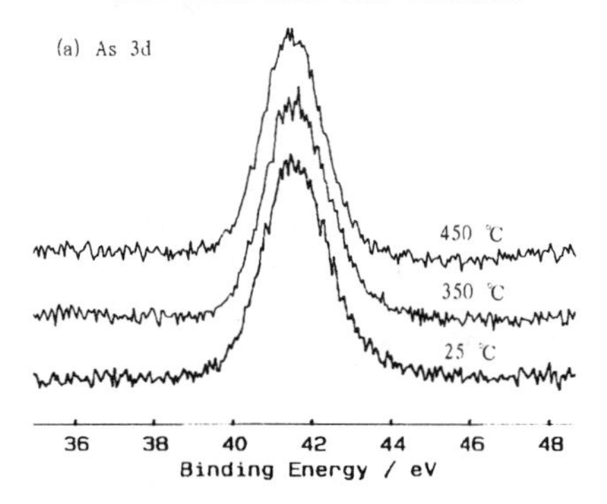

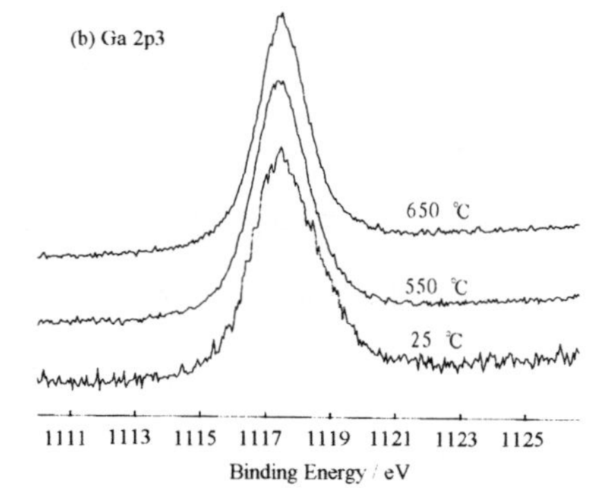

almost identical to that of sulfur.

Figure 6 represents the narrow scan spectra for As and Ga obtained from in-situ annealing. The decomposition of As-O occurred at 350 - 450 °C and Ga-O at 550 -650 °C. From above results, it is suggested that the slight decrease of oxygen composition at 350 - 450 °C corresponds to the decomposition of As-O and sharp decrease at 550 - 650 °C corresponds to the decomposition of Ga-O. The relatively large increase in absolute intensity of gallium might be a result from the high surface coverage of gallium due to the disappearance of oxygen and sulfur from the surface. This implies that the relative atomic concentration of gallium becomes larger than that of arsenic after the annealing, which is a consequence of the high volatile nature of arsenic compared to gallium during the heat treatment.

Fig. 6. Evolution of (a) As 3d and (b) Ga 2p3 peaks in the sulfidation treated sample according to the anneal under UHV condition.

Table 1. Deconvolution results of the Ga 2p3, As 3d, O 1s, and S 2p core level distribution.

	Peak attribution	Peak binding energy (eV)	FWHM (eV)	% of area	% of area*
Ga 2p3	Ga-As (GaAs)	1117.8	1.58	57	55
	Ga-O (Ga_2O_3)	1119.2	1.86	27	45
	Ga-S	1118.5	2.13	16	-
As 3d	As-Ga (GaAs)	41.3	1.74	85	76
	As-O(As_2O_3)	44.4	2.22	6	24
	As-S	42.6	1.97	9	-
O 1s	O-Ga, As	532.5	2.52	100	100
S 2p	S-Ga, As	161.5	2.42	100	-

* Values obtained from HCl-treated GaAs

CONCLUSIONS

Sulfidation treatment and successive rinse of GaAs for 1 minute left Ga-S and As-S bonds on the surface. The Ga-S and As-S bonds were deconvoluted into Ga 2p3 and As 3d peaks with peak binding energies of 1118.5 eV and 42.6 eV, and FWHM values of 2.13 eV and 1.97 eV, respectively. Sulfur was partially replaced by oxygen after the exposure to air for 10 days. Namely, the passivation effect by the sulfur treatment is maintained, comparing the results of HCl-treated surface. The thermal behaviors between S-Ga (As) bonds and O-Ga (As) bonds were observed identical. The decomposition of S-As and O-As bonds was observed after anneal at 350 - 450 °C, and S-Ga and O-Ga bonds completely disappeared after anneal at 550 - 650 °C. The evaporation of arsenic during the anneal was also observed.

REFERENCES

1. Y. Mada, K. Wada, and Y. Wada, Appl. Phys. Lett., **61**(25), 1992, p. 2993.
2. H. Sugahara, M. Oshima, H. Oigawa, and Y. Nannichi, J. Vac. Sci. Technol., **A11**(1), 1993, p. 52.
3. D. Landheer, G.H. Yousefi, J.B. Webb, R.W.M. Kwok, and W.M. Lau, J. Appl. Phys., **75**(7), 1994, p. 3516.
4. Y. Too, A. Yelon, E. Sacher, Z.H. Lu, and J. Graham, Appl. Phys. Lett., **60**(21), 1992, p. 2669.
5. Y. J. CHUN, T. Sugaya, Y. Okada, and M. Kawabe, Jpn. J. Appl. Phys., Vol. **32**, 1993, L287.
6. C. Debiemme-Chouvy, D. Ballutaud, J. C. Pesant, and A. Etcheberry, Appl. Phys. Lett., **62**(18), 1993, p. 2254.
7. J.-L.Lee, D. Kim, S. J. Maeng, H. H. Park, J. Y. Kang, and Y. T. Lee, J. Appl. Phys., **73**(7), 1993, p. 3539.
8. Z. H. Lu, M. J. Graham, X. H. Feng, and B. X. Yang, Appl. Phys. Lett., **62**(23), 1993, p. 2932.

EXTREMELY LOW TEMPERATURE SILICON LIQUID PHASE EPITAXY

M. KONUMA*, I. SILIER*, A. GUTJAHR*, E. BAUSER*, F. BANHART**, H. FREY***, AND N. NAGEL*+
*Max-Planck-Institut für Festkörperforschung, Heisenbergstrasse 1, 70569 Stuttgart, Germany
**Max-Planck-Institut für Metallforschung, Heisenbergstrasse 1, 70569 Stuttgart, Germany
***LSG GmbH, Mörikestrasse 2, 73773 Aichwald, Germany
+ present address, Sony Corp., Atsugi Technology Centre, 4-14-1, Asahi-cho, Atsugi-shi, 243 Japan

ABSTRACT

By liquid phase epitaxy (LPE) we have grown silicon layers on silicon and partially masked silicon at temperatures below 450 °C from Ga and Ga-In solutions. Oxidation of the cleaned silicon substrate surfaces before epitaxial growth has been prevented by a buffered hydrofluoric acid treatment. The epitaxial layers reached a thickness of 7 μm and were free of extended defects.

Low growth temperatures make it possible to grow silicon layers also on pre-treated glass substrates. The amorphous glass is first coated with a thin nano-crystalline silicon layer which is deposited by plasma processes from a mixture of SiH_4/H_2 gas. The grains in the silicon layers grown from Ga solution on glass have reached sizes up to 100 μm.

INTRODUCTION

Growth of high quality crystalline layers at very low temperatures is of interest because substrates can be used which do not withstand high temperatures. It is important also because the energy consumption is low for layers in mass production, such as solar cells. We have studied liquid phase epitaxy of silicon on silicon at extremely low temperature and we have likewise studied the deposition of silicon on dissimilar substrates.

Epitaxial growth of defect-free silicon requires silicon surfaces which are free from oxide and contamination. Chemical reduction of native oxides by having these oxides react with silicon bulk can not be expected at the applied low temperatures. We applied surface passivation techniques, however, that are helpful to protect Si surfaces against oxidation.

Surfaces of the silicon which were cleaned by various wet etching techniques have been investigated during the last several years. The urgent requirement for an ultra-clean Si surface which has micro-roughness arises from the application of Si ULSI technology. Higashi et al.[1] reported that atomically flat and ideally H-terminated Si (111) surfaces can be obtained by etching Si surfaces in pH-modified buffered HF (BHF) solutions with pH $\geq$5. Yasaka et al.[2] observed that the oxide thickness, which is determined from the chemically shifted Si 2p photoelectron spectrum, was below the detection limit after a BHF (pH=5.3) etched Si (111) surface was exposed for 300 min to clean room air at room temperature. They concluded that a BHF treated Si (111) surface is atomically flat and has no significant reactive site. Silicon (100) surfaces exposed to HF vapours are hydrogen-terminated and termination is stable in air for several tens of a minute, and in a vacuum for several hours[3]. More recently, Bender et al.[4] reported that Si (100) surfaces annealed at high temperature in an H_2 atmosphere are 2 x 1 reconstructed such that the step density is minimized and hydrogen-terminated. The surface

Mat. Res. Soc. Symp. Proc. Vol. 386

has been shown to be resistant against contamination and oxidation. The desorption peak maxima of hydrogen are 510 °C from Si (100) and 490 - 520 °C from (111) surfaces in a vacuum[5-8]. This hydrogen passivation technique has been applied in our substrate preparation. In view of the hydrogen desorption temperatures we have performed epitaxy below 450 °C even though our LPE is carried out at normal pressure in a pure hydrogen atmosphere .

EXPERIMENTAL

Silicon substrates were cleaned and their surfaces subsequentlly were passivated in a BHF solution which had a pH=9.0. The pH of a BHF solition was raised by adding ammonium hydroxide. The single crystal Si wafers used in this study were (111) oriented. For the studies of epitaxial lateral overgrowth (ELO), Si wafers were thermally oxidized. Photolithography and wet etching techniques serve to open rectangular seeding windows through the 60 nm thick thermal oxide. Liquid phase epitaxy was performed using conventional tipping boat systems. After a substrate and solvent metal were mounted in a graphite crucible, a reactor tube was evacuated to below 10^{-6} mbar. The tube was then filled with hydrogen purified in a palladium diffusion cell. During the LPE process hydrogen was made to flow through the reactor tube. As a solvent we chose Ga or Ga-In. The epitaxial layers were grown in the temperature interval between 450 and 50 °C.

In order to initiate epitaxial growth on glass substrates, nano-crystalline Si layers with a thickness of 2 - 3 µm were deposited on the substrates by a plasma technique. Borosilicate glass was chosen for Si layer growth because of the similar thermal expansion coefficients of both materials. The SiH_4/H_2 (1:10) plasma is generated by inductively coupled high frequency (2 MHz) at 0.5 Pa. We fixed the glass substrates on a water cooled substrate holder in the plasma chamber. This unique plasma technique is described in detail in reference[9]. The nano-crystalline Si-coated glass substrates were cleaned and BHF treated in the same way as described for single crystalline Si substrates.

RESULTS AND DISCUSSION

Low temperature LPE on single crystalline silicon substrates

The thickness of the epitaxial layers reaches 7 µm. Growth rates of between 0.4 and 0.5 µm/h are obtained. Due to a 0.3 ° misorientation of the substrates the layer surfaces are slightly terraced. Figure 1 shows a Nomarski differential interference contrast (NDIC) micrograph of the surface of a 2 µm thick layer. Investigations by scanning electron microscopy, SEM, and transmission electron microscopy, TEM, show that the layers are free from dislocations and other defects, and that the interface between substrate and epitaxial layer is almost free from solvent inclusions and defects.

Carrier concentration and mobility in the LPE silicon layers thus obtained are determined by Hall effect measurements at room temperature. Layers grown from Ga solution are of p-type conductivity with hole concentrations about 5×10^{18} cm^{-3}. Growth from Ga-In alloys leads to a lower concentration of free holes in the epitaxial silicon layer. A Ga/In ratio of, for example, 80/20 yields layers with hole concentrations of only 1×10^{18} cm^{-3}. The layers can, if necessary, be doped.

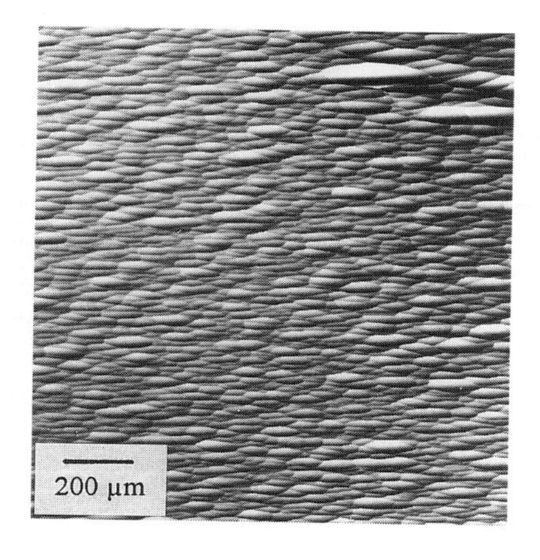

Fig. 1 Surface of a epitaxial Si layer grown on a (111) oriented mono-crystalline Si substrate in a temperature range between 450 and 50 °C. NDIC micrograph.

Low temperature silicon ELO on oxide masked silicon substrates

Defect-free semiconductor-on-insulator (SOI) layer structures are of considerable interest not only for application in electronic devices but also for basic studies of crystal growth. By the ELO technique we have earlier, at high temperatures, prepared defect-free SOI layers of high quality from In solutions[10-12]. We have also prepared large-area SOI layers by utilizing defect-free coalescence of ELO layers[13,14].

Recently we prepared defect-free SOI layers at very low temperatures by means of area selective epitaxial growth and ELO. Figure 2a gives a plan view of a SOI layer taken by NDIC microscopy using a high pressure Hg lamp with filters. Two SOI layers have grown laterally, starting from the long edges of a rectangular seeding area which is 20 µm wide and about 250 µm long. The contrast fringes in the SOI areas are due to interference of light (wavelength 545 nm). The sharpness of the fringes and their mutual distance suggest that an air slit exists between the bottom face of the ELO layer and a substrate oxide surface. The slit height increases with increasing distance from the seeding area. By surface profilometer measurements the top surface of the ELO layer is found to be extremely flat. The direction of substrate misorientation is aligned to <$11\underline{2}$> i.e. it is parallel to the long sides of the rectangular seeding areas. The microscopic growth mechanisms of low temperature SOI layers are described and discussed in separate papers[15,16]. The layer shown here is grown in the temperature interval 450 - 440 °C from Ga solution with a cooling rate of 6 K/h. The SOI layers on both sides of the seeding area have a maximum width of 50 µm. We prepared transmission electron micrographs of samples in cross-section. An example marked in Fig. 2b is shown in Fig. 2c. The micrograph Fig. 2c shows that the epitaxial SOI layer is about 1.7 µm thick and defect-free.

Low temperature growth of Si on glass substrates

Silicon deposition on glass is currently one of the foremost research topics in the field of

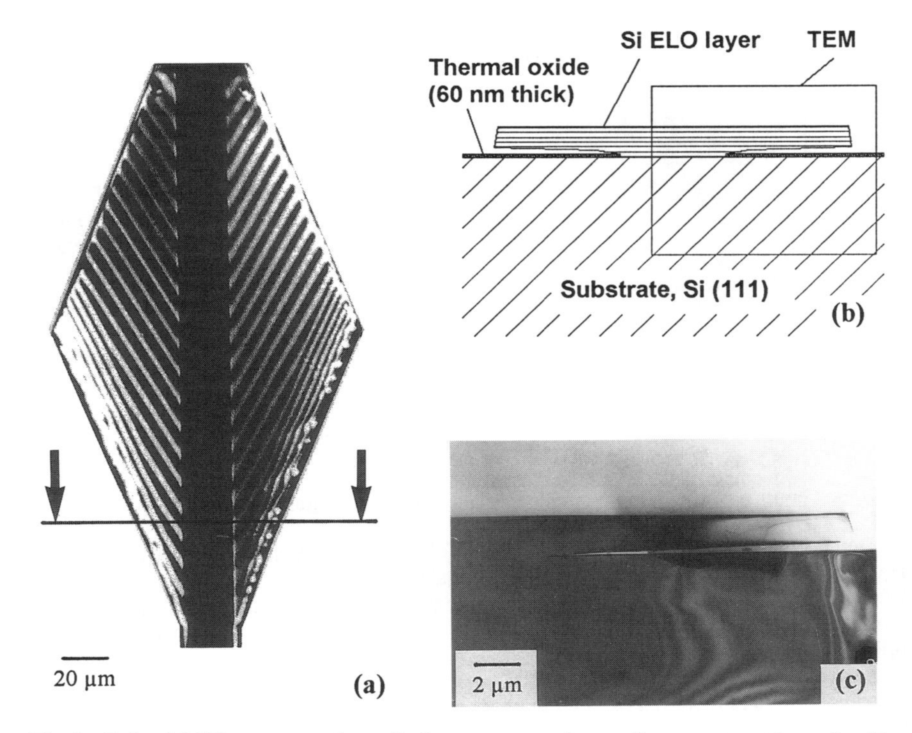

Fig. 2 Epitaxial Si layer grown laterally from a rectangular seeding area over thermal oxide. The ELO layers are grown in the temperature interval 450 - 440 °C from Ga solution. (a): plan-view, NDIC micrograph, (b): schematic of a cross section marked on (a), (c): cross sectional TEM.

crystal growth on dissimilar substrates. Applications for Si on glass include solar cells and thin film transistors (TFTs)[17-19].

We have obtained silicon layers on glass with the help of a silicon seeding layer. The seeding layer consists of a thin intermediate layer of amorphous Si next to the glass. This layer is covered by nano-crystalline Si. Grains in the nano-crystalline seeding layer reach sizes up to about 50 nm. X-ray diffraction shows that the nano-crystalline layers orientate preferentially with the (110) planes parallel to the substrate surface. On these seeding layers silicon can be deposited from Ga solutions at 350 °C. Apart from a few twins, the solution grown Si crystals are practically defect-free. They are sufficiently large, their sizes reach 100 µm, and grow close together, as shown in Fig. 3; yet they grow individually. Lateral connections between individual crystals with grain boundaries in-between hardly develop. The low-temperature silicon deposits lack, therefore, the character of a *layer*. In contrast, the individual crystals in silicon formed good lateral connections, when grown at 950 °C from In solution on quartz substrates covered with nano-crystalline Si seeding layers[20]. The individual Si crystals grew like *grains* in a polycrystalline layer at high temperature.

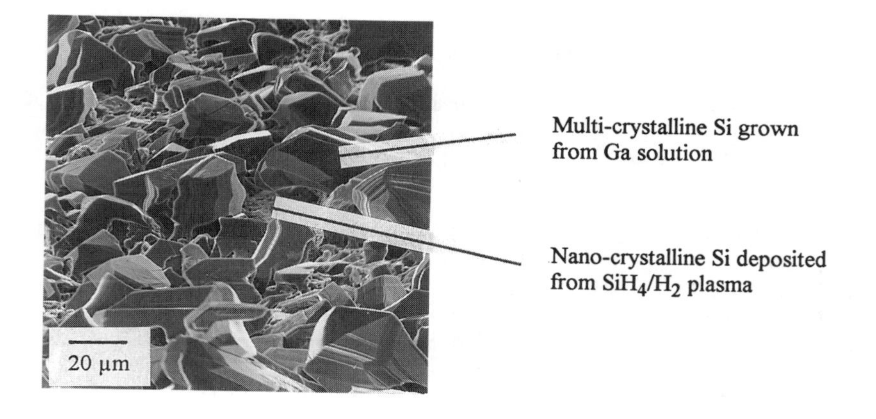

Fig. 3 Multi-crystalline Si grown on a borosilicate glass substrate with thin seeding layer which consists of amorphous and nano-crystalline Si. SEM photograph.

So far, it remains an ambitious and fascinating goal to achieve the desired layer character in low-temperature Si deposition on glass.

SUMMARY

We have applied the buffered-hydrofluoric acid (BHF) treatment for preparing not only mono-crystalline Si substrate but also nano-crystalline coated glass substrates. The BHF treatment yields clean surfaces with no oxygen and contamination and allows us to grow Si layers from metallic solution at very low temperatures. Liquid phase epitaxy (LPE) was performed to grow Si layer on mono-crystalline Si substrates below 450 °C from Ga solution. The thickness of the epitaxial layers reaches 7 μm. Semiconductor-on-insulator (SOI) structures can also be prepared by growing Si layers laterally over thermally grown thin oxide at 450-440 °C. When the epitaxial growth takes place on mono-crystalline silicon substrates, the layers are free from extended defects.

Low growth temperatures make it possible to grow silicon layers also on glass substrates. In order to initiate epitaxial growth, the amorphous glass surface is first coated with a thin nano-crystalline silicon layer which is deposited by plasma processes. The silicon layers grown on glass from Ga solution have grains whose sizes are in the hundred μm range.

ACKNOWLEDGEMENT

The authors are grateful to H.-J. Queisser for his continuous interest and encouragement. The authors thank M. Hafendörfer, J. Rudhard, A. Müller, K.S. Löchner, A. Weisshardt, and B. Fenk for technical assistance. We acknowledge financial support from the Bundesministerium für Forschung und Technology under contract 01M 2920 A.

REFERENCES

1. G.S. Higashi, Y.J. Chabal, G.W. Trucks, K. Raghavachari, Appl. Phys. Lett. **56**, 656 (1990).
2. T. Yasaka, K. Kanza, K. Sawara, S. Miyazaki, M. Hirose, Jap. J. Appl. Phys. **30**, 3567 (1991)
3. S.S. Iyer, M. Arienzo, E. de Frèsart, Appl. Phys. Lett. **57**, 893(1990).
4. H. Bender, S. Verhaverbeke, M. Caymax, O. Vatel, M.M. Heyns, J. Appl. Phys. **75**,1207 (1994).
5. S.H. Wolff, S. Wagner, J.C. Bean, R. Hull, J.M. Gibson, Appl. Phys. Lett. **55**, 2017 (1989).
6. N. Hirashita, M. Kinoshita, I. Aikawa, T. Ajioka, Appl. Phys. Lett. **56**, 451(1990).
7. M.C. Flowers, N.B.H. Jonathan, Y. Liu, A. Morris, J. Chem. Phys. **99**, 7038(1993).
8. Y. Morita, K. Miki, H. Tokumoto, Surf. Sci. **325**, 21(1995).
9. H. Frey, Appl. Phys. **A 47**, 193(1988).
10. R. Bergmann, E. Bauser, J.H. Werner, Appl. Phys. Lett. **57**, 351(1990).
11. R. Bergmann, J. Crystal Growth **110**, 823(1991).
12. R.P. Zingg, N. Nagel, R. Bergmann, E. Bauser, B. Höfflinger, H.J. Queisser, IEEE Electron Device Lett. **13**, 294(1992).
13. N. Nagel, F. Banhart, E. Czech, I. Silier, F. Phillipp, E. Bauser, Appl. Phys. **A 57**, 249 (1993).
14. F. Banhart, N. Nagel, F. Phillipp, E. Czech, I. Silier, E. Bauser, Appl. Phys. **A 57**, 441 (1993).
15. I. Silier, A. Gutjahr, N. Nagel, P.O. Hansson, E. Czech, M. Konuma, E. Bauser, F. Banhart, R. Köhler, H. Raidt, B. Jenichen, in Proc. 11th International Conference on Crystal Growth, to be published.
16. R. Köhler, H. Raidt, F. Banhart, M. Konuma, A. Gutjahr, I. Silier, E. Bauser, to be published.
17. J.B. McNeely, R.B. Hall, A.M. Barnet, W.A. Tiller, J. Cryst. Growth **70**, 420(1984).
18. Z. Shi, Materials Lett. **15**, 359(1993).
19. S.H. Lee, R. Bergmann, E. Bauser, H.-J. Queisser, Materials Lett. **19**, 1(1994).
20. I. Silier, M. Konuma, A. Gutjahr, F. Banhart, E. Bauser, H. Frey, to be published.

DEVICE QUALITY OF HYDROGEN PLASMA CLEANING FOR SILICON MOLECULAR BEAM EPITAXY

W. HANSCH*, I. EISELE*, H. KIBBEL** AND U. KÖNIG**
* Universität der Bundeswehr München, Fakultät Elektrotechnik, Institut für Physik D-85577 Neubiberg, Germany
** Daimler-Benz Forschungszentrum, D-89081 Ulm, Germany

ABSTRACT

Different substrate cleaning procedures were used before fabrication of pin diodes by silicon molecular beam epitaxy (MBE). We investigated the quality of these diodes in order to demonstrate the superior quality of a low energy plasma cleaning in an ultra-high vacuum (UHV). This plasma cleaning by hydrogen makes a wet-chemical cleaning or a high-temperature desorption step unnecessary. Moreover, the plasma-cleaned substrates are so strongly hydrogen passivated, that they can be transported through air and processed in another MBE chamber without any additional cleaning steps.

INTRODUCTION

Contaminations on the substrate surface strongly influence the quality of layers grown using silicon MBE. One of the standard methods of achieving a clean substrate surface is to do a wet-chemical treatment [1] outside the growth chamber followed by a high-temperature (~900°C) oxide desorption step [2] in the ultra-high vacuum (UHV) deposition chamber immediately before the growth process. The problem with this procedure is that it cannot desorb the residual carbon contaminations resulting from wet-chemical cleaning or transport through air. During high-temperature desorption step SiC precipitates are formed, which are the origin of strain in the epitaxial overgrowth. Under critical conditions, such as in high-level doping or high-temperature treatments, the induced strain relaxes via the formation of dislocations, which are fatal defects for devices. This restricts the quality of devices that involve highly doped multi-layer stacks, especially vertical devices or three-dimensional integration. Furthermore, the use of high-temperature post-growth processes like implantation or thermal oxidation is not possible.
To solve these problems encouraging results have been obtained using different kinds of hydrogen plasma to remove the native oxide and carbon contaminations from the silicon wafer surface [3,4]. With remote plasma processes, one must be careful to avoid surface damage and roughness on an atomic scale, the incorporation of gas ions, and the activation and contamination of atoms of the surrounding materials. These effects are influenced by the plasma parameters, especially the applied external and internal electromagnetic fields. In order to minimize substrate damage, chemically active species with low energies must be generated.
For this purpose, we used a newly developed plasma source from Balzers Ltd. to create a low energy argon/hydrogen gas discharge [5] in an UHV environment. The cleaning efficiency has already been described and the etch rates determined for SiO_2 and diamond-like carbon films [5,6]. The low-temperature growth of silicon MBE layers on these plasma cleaned substrates [6] and the possibility of cleaning patterned substrates [7,8] have also been shown. In order to demonstrate the device quality that results from this plasma cleaning, we fabricated locally grown triangular barrier diodes (TBDs) in a Modular UHV Multichamber System (MUM 545, Balzers Ltd.)[8]. For this local growth, micro-shadow masks were deposited on the substrate, which first required a wet-chemical cleaning. UHV plasma cleaning was then performed immediately before MBE growth. Although good layer and device quality were achieved after the plasma cleaning, the role of wet-chemically pre-cleaning was not clear.

Mat. Res. Soc. Symp. Proc. Vol. 386 © 1995 Materials Research Society

In this paper, we compare the efficiency of our developed plasma cleaning and surface protection via hydrogen passivation with various cleaning recipes, all applied before MBE growth.

EXPERIMENTAL

In order to fabricate pin-diodes we deposited p^+-i-n^+ and n^+-i-p^+layer sequences on various substrates with (100) surface orientation. MBE growth was done for half of the diodes in the MUM system without breaking UHV conditions between plasma cleaning and layer growth. The other half of the diodes were transported through air after plasma cleaning into another MBE system (single MBE chamber, SMC), which consists of a cassette station and the UHV growth chamber.
Various cleaning procedures were used for p^+(B, 0.02Ωcm) and n^+(As, 0.004Ωcm) substrates: wet-chemical cleaning, plasma cleaning, thermal desorption and combinations of these.

Wet-Chemical Cleaning (RCA clean)
For wet-chemical cleaning the widely used RCA clean [1], consisting of two steps, was used. The first step uses hydrogen peroxide (H_2O_2) to remove organic contaminations and several metals by oxidative dissolution. The use of not-stabilized H_2O_2 is very critical and must be kept within temperature and time limits; if H_2O_2 decomposes, metal and carbon recontamination takes place and causes surface roughening. The second step removes remaining metals and creates a thin protective layer of SiO_2, which is about 2nm thick.
Secondary ion mass spectrometry (SIMS) shows that our substrates usually have a carbon contamination of about $10^{19}cm^{-3}$ after going through this RCA-clean.

Thermal Desorption (TD)
At elevated temperatures above 850°C, the protective SiO_2 layer becomes desorbed in the ultra-high vacuum within a few minutes [2]. Depending on the amount of carbon remaining after RCA clean, several laboratories use an additional annealing step at medium temperatures (600°C for several hours) to desorb carbon without forming precipitates. But this procedure requires process time, which makes it impractical for industrial application.
For our thermal desorption (TD), we raise the temperature to 900°C within 2 min and have an annealing time of 5 min.

Plasma Cleaning
For wafer cleaning with activated hydrogen in the MUM system, we used the new plasma source (Balzers Ltd.), which is described in detail elsewhere [5]. The working principle is shown in Fig. 1.

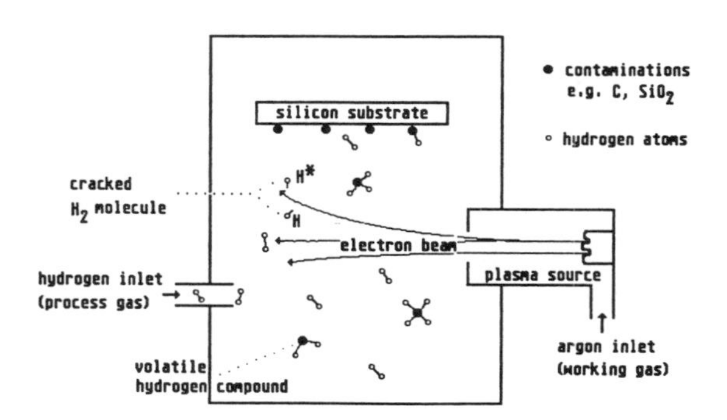

Fig. 1: Principle of plasma cleaning (description in the text).

Argon is fed into the plasma source and is used as the working gas for the discharge. This dc discharge generates high electron currents up to 100 A at low voltages (~25 eV) through the whole chamber, which acts as an anode on ground potential. The reactive gas (in this case, hydrogen for cleaning) is fed directly into the chamber. Depending on the electrical parameters of the plasma source (i. e., value of negative potential at the filament), the gas pressure within the chamber (which affects the mean free path for collisions) and the potential distribution in the chamber, the kinetic energy of the electrons varies between some electronvolts up to 70 eV. The shape of the electron beam can be influenced by external magnetic fields. We use two external coils fed by a dc current up to 40 A, which guide the electron beam directly onto the substrate or close beneath without touching it. Argon and hydrogen pressures each of about $1 \cdot 10^{-1}$ Pa, a discharge current of 30 A and a discharge voltage of about 25 V are standard parameters in our cleaning process. The whole cleaning procedure consists of exposing the substrate to the excited gas for some minutes. No external heating for the substrate is used during hydrogen cleaning.

Cleaning Sequences

The cleaning procedures we used are shown in the following table:

Id. No.	Cleaning process
C1	RCA + / + TD
C2	RCA + Plasma + Transport through air + TD
C3	/ + Plasma + Transport through air + /
C4	/ + Plasma + Transport through air + TD

Tab. I: Overview of cleaning sequences used

C1 was the reference cleaning without plasma treatment. The substrates were cleaned wet-chemically using RCA clean, then loaded into the cassette station of the single MBE chamber (SMC) within one hour. The deposition of the layer sequence for pin-diodes was carried out immediately after a high-temperature annealing step (TD, 900°C, 5min) to desorb the protective oxide.
C2 includes the same RCA clean, but the substrates were then loaded into the cassette station of the MUM system. The plasma cleaning described above was done in one of the UHV chambers. The cassette station containing the plasma-cleaned substrates was vented with nitrogen, and the substrates were loaded into the cassette station of the SMC system within one hour. After several days of storage, thermal desorption (TD) was performed and followed immediately by MBE growth.
C4 correspond to C2 without preceding RCA clean; in C3, both the RCA clean and the TD-step were left out. The substrates were loaded right out of their packages into the MUM system for plasma cleaning, and further processing was done in the SMC system.
The plasma-cleaned substrates were transported through air in order to investigate the effectiveness of surface passivation by plasma hydrogen. Also, before MBE growth, part of these transported substrates were stored for varying numbers of days in the cassette station of the SMC system at a pressure of about 10^{-6} Pa .

Layer Growth

The deposition of the epitaxial silicon layers was performed at a pressure of $1 \cdot 10^{-7}$ Pa with a flux rate of 0.1nm/s. P-type boron doping as well as n-type antimony doping have been achieved by coevaporation of the dopants with silicon, using effusion cells at a substrate temperature of 575°C for boron doping and 475°C for antimony doping. To avoid unintentional doping of the intermediate intrinsic layer, this layer is grown at 700°C. The incorporation of antimony at this temperature is less than $5 \cdot 10^{16}cm^{-3}$, which is the detection limit of our measurements with SIMS. The doping levels for the antimony and boron layers were $3 \cdot 10^{19}cm^{-3}$ and $5 \cdot 10^{18}cm^{-3}$,

respectively. All layers deposited had a thickness of about 300nm.
SiC precipitates cause dislocations to form in epitaxial layers. Because this process is enhanced under critical growth conditions such as high temperatures or high doping levels, a highly doped buffer layer was first grown on half of the substrates to investigate the formation of dislocations.

Device Fabrication

The diodes were fabricated in our process line using two mask steps. The diode diameters were defined in a first process . Free-standing mesas were formed in a CF_4/O_2 gas mixture using dry plasma etching with an etch rate of about 0.5nm/s. After an RCA clean, the diode mesas were thermally oxidized at 700°C for 4 hours. This resulted in a 30nm thick SiO_2-layer for electrical surface passivation.
In a second mask step, a window was opened in the oxide on top of the mesa for contact metallization. Titanium/tungsten and platinum were sputtered and patterned by a lift-off process of the photo resist.
On each substrate we processed 12 chips, each carrying about 40 diodes ranging in diameter from 20 to 520 microns.

Measurements

A Hewlett-Packard 4145B Parameter-Analyzer was used to make current and voltage measurements for the fabricated diodes. The top contact was made using a needle, the back contact was achieved by eutectic bonding of the substrate. All measurements were done at room temperature. The I(V)-characteristics were stored and evaluated. The ideality factor and reverse (leakage) current at -1V were determined .

RESULTS AND DISCUSSION

Diode characteristics were measured for all of the substrates cleaned using the various procedures. To estimate the quality of the diode, and thus of the cleaning procedure, we measured the reverse (leakage) current at -1V. The ideality factor was extracted from the slope of the logarithmic plot of current vs. voltage in the forward direction. Figure 2 shows an example of diode characteristics; the right vertical axis and "lin" curve show the linear behavior, the left axis and "log" curve show the logarithmic behavior of this particular diode.

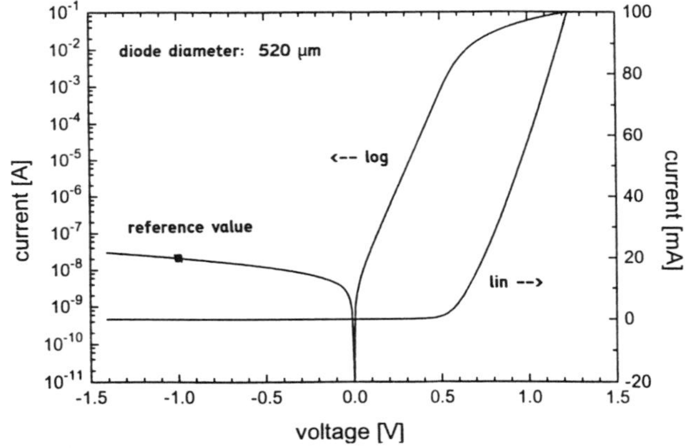

Fig. 2: I(V)-characteristics of a pin-diode fabricated on a plasma-cleaned substrate without any additional cleaning. The substrate was then transported through air and the layer sequence grown in another UHV-chamber.

For this diode the n^+-substrate was cleaned in sequence C3 (see table 1). The substrate was taken from the package and loaded directly into the MUM system. After a 15 min plasma cleaning, the substrate was moved back into the cassette station; vented; transported through air into the SMC-system; and stored in the SMC's cassette station for 4 days at 10^{-6}Pa. The substrate was then moved into the growth chamber and heated up to 475°C. After 15 minutes, deposition of an n^+-buffer layer was begun. The intrinsic layer was grown on this buffer layer at 700°C, and the boron doped p^+-layer at 575°C.
The performance of this diode is good, but it is not the best of the diodes we fabricated. The ideality factor is about 1.5 ± 0.08, the reverse current at -1V is about $2 \cdot 10^{-8}$ A, which corresponds to a current density of 10^{-13} A/µm at -1V in the reverse direction. All diodes fabricated with this layer sequence have almost the same values as above for ideality factor and reverse current density. The lower diode quality lies not in the chosen cleaning procedure, but in the growth of the n^+-buffer layer. Diodes with and without the n^+-buffer were fabricated in all of the cleaning sequences, and in each case the diodes without buffer layer showed better performance. For example, the same diode without the n^+-buffer yields values of about 1.05 ± 0.05 for the ideality factor and about $2 \cdot 10^{-10}$ A for the reverse current at -1V.
For diodes with a buffer layer, a plot of measured reverse currents at -1V versus diode area yields a line of slope 1. The reverse current is therefore caused by defects that are proportional to the diode area. On the other hand, diodes that had no buffer layer and were treated with an effective cleaning procedure yielded a line of slope 0.5. This shows, in connection with the very low values of reverse current, that these diodes are limited by our surface passivation. A surface current is proportional to the diode circumference, which is proportional to the square root of the diode area; thus the 0.5 slope on the logarithmic plot.
The diodes fabricated in the MUM system without breaking UHV conditions had similar reverse currents that are also limited by surface passivation. This demonstrates, that on substrates, only cleaned by low energy hydrogen plasma, high quality diodes can be fabricated.
A comparision of the different cleaning procedures (C1 - C4) is shown in Fig. 3.

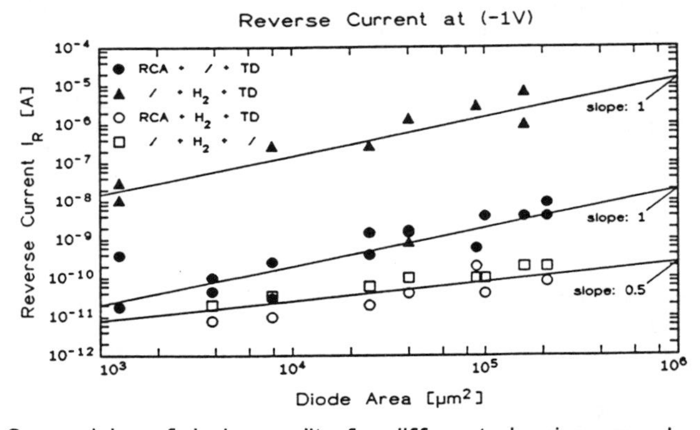

Fig. 3: Comparision of device quality for different cleaning procedures

All of the diodes shown in Fig. 3 were fabricated on n^+-substrates <u>without</u> a buffer layer, in order to allow a judgement of the cleaning procedures themselves. The closed circles (•) are diodes treated with our former standard clean C1. The corresponding line of slope 1 shows that the reverse current is limited by area defects. Open squares (□) are diodes that were only plasma cleaned (C3). Their values are best fitted by a line of slope 0.5, which could be seen more clearly in an ex-

panded plot; the reverse current for these plasma-cleaned diodes is therefore limited by our surface passivation. Similiar values are obtained for the standard clean with intermediate plasma treatment (C2), represented by the open circles (o) in Fig. 3. The values for the cleaning procedure C4 (▲), which had no previous RCA clean, are limited by defects again.
The differing behavior can by explained by the formation of SiC precipitates as follows. After our RCA-clean and transport to the UHV camber, a carbon contamination of about $10^{19}cm^{-3}$ can be detected. If a high-temperature step (TD) is used to desorb the native oxide layer, the atomically distributed carbon starts to diffuse and form SiC precipitates. Defects and dislocations are induced in the epitaxial layer that is then grown on the substrate.
With hydrogen-plasma cleaning, on the other hand, the native oxide layer is etched at room-temperature. Depending on the amount of carbon contamination, the plasma parameters, and the etching time, the carbon contamination is reduced or completely removed. Even when a certain amount of carbon remains, such as after transport through air, it is still atomically distributed. When the substrate temperature is raised for epitaxial growth, the hydrogen passivation desorbs and still leaves the carbon contamination atomically distributed. Depending on the carbon contamination that remains, a optional high-temperature step may (as in the case of C4, filled triangles) or may not (as in C2, open circles) result in SiC formation and device degradation.
It is thus clear that the plasma cleaning introduced above permits the removal of the protective SiO_2 layer at room temperature without the formation of SiC precipitates. This reduces the cleaning time for in situ applications to only a few minutes. The advantage of strong hydrogen passivation can be applied to other applications as well.

CONCLUSIONS

We have demonstrated that high quality diodes can be fabricated on plasma-cleaned substrates, where the leakage current is limited by our device passivation. There is no longer a need for wet-chemical cleaning or for an in situ high-temperature desorption step. Moreover, the plasma-cleaned substrates can be transported through air and be processed in another UHV-chamber without any additional cleaning steps.
An increase in device performance has been demonstrated with UHV plasma cleaning as compared to common cleaning procedures. A costly wet-chemical cleaning can be omitted, and the reduction in cleaning time opens up new possibilities for the use of MBE-machines in the fabrication of industrial devices.

REFERENCES

[1] e.g. RCA clean: W. Kern and D. A. Puotinen, RCA Rev. **31**, 187(1970)
[2] M. Tabe, Jpn. J. Appl. Phys. **21**, 534(1982)
[3] R. P. H. Chang, C. C. Chang and S. Darack, J. Vac. Sci. Technol. **20**, 45(1982)
[4] I. Suemune, Y. Kunitsugu, Y. Tanaka, Y. Kan and M. Yamanishi, Appl. Phys. Lett. **53**, 2173 (1988)
[5] J. Ramm, E. Beck and A. Zueger, Mat. Res. Soc. Symp. Proc. **220**, 15(1991)
[6] J. Ramm, E. Beck, A. Züger, A. Dommann and R. E. Pixley, Thin Solid Films **222**, 126(1992)
[7] J. Ramm, E. Beck, I. Eisele, W. Hansch, B.-U. Klepser and H. Senn, Mat. Res. Soc. Symp. Proc. **315**, 91 (1993)
[8] W. Hansch, E. Hammerl, W. Kiunke, I. Eisele, J. Ramm and E. Beck, Jpn. J. Appl. Phys. **33**, 2263(1994)

REMOVAL OF SiO_2 FROM Si (100) BY REMOTE H_2/SiH_4 PLASMA PRIOR TO EPITAXIAL GROWTH

J.P. BARNAK, H. YING, Y.L. CHEN, J. MONTGOMERY, AND R.J. NEMANICH,
Department of Materials Science and Engineering, Department of Electrical and Computer Engineering, and Department of Physics
North Carolina State University, Raleigh, NC 27695-8202

ABSTRACT

This study demonstrates the cleaning of Si(100) surfaces with a remote H_2/SiH_4 plasma. The surfaces were prepared with a chemical oxide the remains after an RCA clean. The plasma cleaning process was designed to remove contaminants such as C, F, and SiO_2. The key to successful removal of the oxide is to have the plasma chemistry in a neutral deposition regime. The neutral deposition process regime is a balance between the deposition of Si by SiH_4 and the etching of the deposited Si by atomic H. During the neutral deposition mode the SiO_2 was removed without deposition of Si on the SiO_2 surface. Once the SiO_2 layer is removed, the underlying Si surface is exposed to the H_2/SiH_4 plasma and a thin epitaxial film may be deposited. The final Si surface configuration after plasma cleaning is a 2x1 hydrogen terminated surface. The characterization of the interface and epitaxial film were investigated using Auger electron spectroscopy (AES) and transmission electron microscopy (TEM).

INTRODUCTION

Various methods have been investigated for performing *in situ* cleaning prior to Si epitaxial growth. These cleaning methods include: thermally assisted cleaning [1-2], photon assisted cleaning [3], plasma assisted cleaning [4], and vapor plasma cleaning [5]. These methods have varying degrees of success in removing organic and inorganic contaminants. Some of these cleaning techniques have been developed into low temperature cleaning processes (less than 800°C). In this study our goal is to lower the temperature at which SiO_2 is removed from a Si surface by adding SiH_4 to a remote H-plasma.

The remote H_2/SiH_4 plasma process is being developed as a precleaning step prior to Si epitaxy. Previous studies have shown that a remote H-plasma is effective in removing C, O and F from the Si (100) surface [6-8] whereas high temperature cleans (> 800°C) using SiH_4 or Si_2H_6 are effective in removing SiO_2 from Si [9]. The remote rf H-plasma generates atomic H which is the primary reactive species for removing these contaminants. However, atomic H is not effective in removing SiO_2.

In this study a low temperature remote H_2/SiH_4 plasma process is described which has two main processing regimes: blanket silicon deposition and selective SiO_2 etching. In the blanket deposition regime Si is deposited on both the Si and SiO_2 regions of the wafer. In the etching regime SiO_2 is selectively etched with respect to Si. At the exposed Si regions of the surface the deposition and etching rates of Si are approximately equal (i.e., a neutral deposition process), however, in the SiO_2 regions etching of the oxide dominates. The SiO_2, as well as any Si deposited on the SiO_2, are etched off during processing in the neutral deposition regime.

EXPERIMENTAL PROCEDURES

The substrates used in this study were p-type, boron doped, 25 mm diameter Si (100) wafers with a resistivity of 0.8-1.2 Ω-cm. The Si wafers were RCA cleaned by the supplier. Prior to processing, the wafers were dipped in a 10:1 HF solution for 10 seconds to strip the oxide and hydrogen terminate the Si surface. An additional RCA clean was used to terminate the surface with a thin SiO_2 layer. Each wafer was mounted onto a molybdenum sample holder.

Mat. Res. Soc. Symp. Proc. Vol. 386 © 1995 Materials Research Society

The holder was placed into a load-lock before transferring the sample into the UHV transfer-line. The transfer-line acts as a linear wafer handler used in cluster tools. The analysis and plasma cleaning chambers are attached to the transfer-line and each chamber is isolated from the transfer-line by a gate valve. Currently there are eight stainless steel UHV chambers along the 35 ft. long transfer-line. Once a sample is placed onto the sample cart, which travels the length of the transfer-line, the sample can be moved from chamber to chamber *in situ*. The base pressure is maintained at 1×10^{-9} Torr through the use of Cryopumps and turbomolecular pumps.

The remote plasma cleaning chamber has been described previously [10]. The process gases flow through a quartz tube located on top of the chamber. An inductively coupled plasma is generated using a rf power supply (13.56 MHz) and rf matching network attached to a Cu coil placed around the quartz tube. The sample is located 40 cm from the center of the rf coil. The sample is heated by a tungsten coil facing the backside of the wafer. The base pressure of the plasma cleaning chamber is 8×10^{-9} Torr. A point-of-use purifier/filter was located downstream of the H_2 and SiH_4 mass flow controllers. The SiH_4 gas is delivered as a mixture of 1% SiH_4 in H_2. Depending on the chamber pressure and rf power, the plasma can be maintained in the quartz tube or extended down toward the sample region. For the experiments in this study, the plasma extends down to the sample region although the plasma was excited remotely.

The samples were ramped from 150°C to 450°C at 100°C/min. The H_2 (93.9 sccm) and SiH_4 (0.1 sccm) flow rates as well as the process pressure (25 mTorr) were established prior to ramping up the temperature. The pressure was controlled by the H_2 and SiH_4 flow rates. The wafers were exposed to a 20, 100, or 400 Watt rf plasma, and the plasma was struck once the set point was reached. After the plasma exposure, the plasma, the gases, and the heater were simultaneously turned off. Once a plasma cleaning experiment is complete, a low temperature (450°C) PECVD or a (800°C) CVD epitaxial Si film was deposited in the H-plasma chamber to facilitate TEM analysis.

Auger analysis was done using a Perkin-Elmer static probe Auger system which utilizes a cylindrical mirror analyzer and a spot size less than 1 mm. The differentiated Auger spectra were normalized using the $Si_{(LMM)}$ peak at 92 eV (H_2/SiH_4 plasma cleaned wafers) or normalized using the $O_{(KLL)}$ peak at 503 eV (as-loaded RCA cleaned wafers). The LEED diffraction patterns were taken at an accelerating voltage set at 60.8 V.

TEM lattice images were obtained to determine the quality of the interface and the surface roughness. The TEM was also used to investigate the crystalline quality and the growth rate of the PECVD deposited Si film.

RESULTS AND DISCUSSION

Neutral Deposition Processes

The Auger $O_{(KLL)}/Si_{(LMM)}$ peak-to-peak ratio and TEM images were used to determine the experimental conditions that lead to a neutral deposition. The balance that is to be achieved is that Si etching from the atomic H is offset by deposition from the silane containing species. If neutral deposition is achieved on the Si surfaces then Si will not deposit on the oxide surfaces and etching can potentially be achieved. Figure 1 displays the ratio of the $O_{(KLL)}/Si_{(LMM)}$ Auger peaks versus plasma exposure for conditions of different plasma power and silane ratio. The conditions employed for the studies here were as follows: rf power varied from 20 to 400 Watts, SiH_4:H_2 ratio equal of 0.0010 and 0.0017, and a substrate temperature of 450°C. For rf powers of 100W and 400W, the underlying Si substrate is exposed in approximately 10 min. Since the RCA oxide thickness is ~20 Å, the SiO_2 etch rate is deduced to be ~2 Å/min. For the 20 Watt plasma exposure shown in Figure 1, the $O_{(KLL)}/Si_{(LMM)}$ dependence indicates that the oxide removal rate is significantly reduced, and the same low level was not achieved for the exposure times employed.

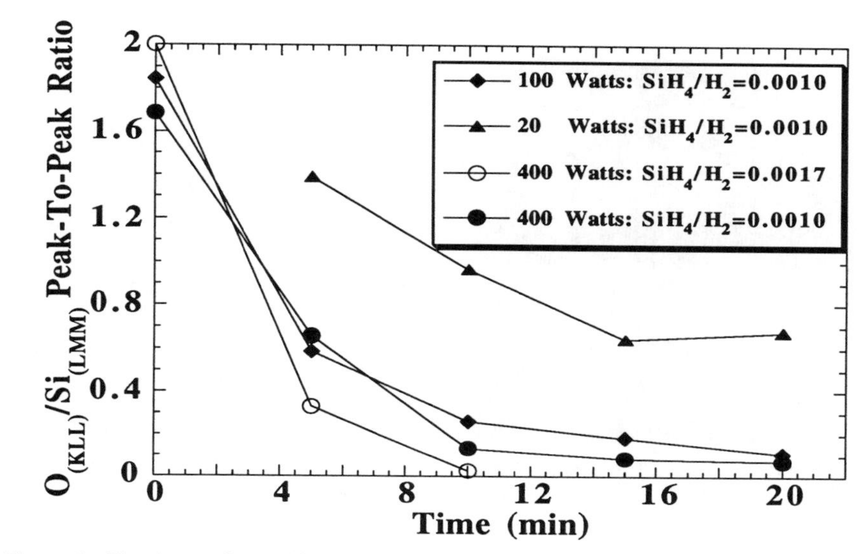

Figure 1. The Auger $O_{(KLL)}/Si_{(LMM)}$ peak-to-peak ratios of RCA cleaned Si(100) wafer after various remote plasma exposures. The results display the effect of rf power and SiH_4/H_2 ratio on the etch rate of the SiO_2 layer. The substrate temperature was 450° C.

To more closely examine the process, TEM analysis was carried out for the wafer exposed to a 400 Watt plasma, $SiH_4:H_2$ ratio of 0.0010, wafer temperature of 450°C, and 15 min total exposure. A high resolution cross-section image is shown in Fig. 2. The image shows that a thin epitaxial layer has been deposited on the surface, and no defects were observed at the interface. The micrograph confirms that the oxide layer has been removed. We presume that the thin epitaxial layer was grown after the surface was oxide free. If we assume that the 50Å epi layer was deposited in the last 5 min of the process, then we deduce a rate of ~10Å per min. The slow growth of the epitaxial Si shows that the system was in a near neutral deposition mode.

When the SiH_4/H_2 ratio was increased from 0.0010 to 0.0017 for a 400 Watt plasma exposure, the $O_{(KLL)}/Si_{(LMM)}$ ratio approaches zero in a time similar to that of the more dilute processes (Figure 1). Cross-sectional TEM analysis indicated, however, that the oxide film was only partially removed and that the remaining oxide film was capped by a heavily defective crystalline Si film. This suggests that the reduction of the oxygen AES signal was a combination of oxide removal and Si deposition.

Two possible mechanisms for etching SiO_2 are ion bombardment and neutral radicals of SiH_4. To explore these effects several plasma exposures were completed with different sample bias. Experiments were carried out at a substrate bias of -25 volts (attracting ions), and in these measurements no significant change in the SiO_2 etch rate was observed. Previously, the plasma system has been characterized for H_2 gas. The atomic H concentration is 3 orders of magnitude greater than the H ion concentration [10]. Even though the plasma is remote, atomic H can react with SiH_4 in the gas phase to form SiH_x radicals [12]. We suggest that neutral SiH_x radicals in combination with the atomic H are the primary species for etching SiO_2 and not the ions. The etching mechanism of deposited Si on the SiO_2 film is by atomic H to form SiH_4. The etching of Si from the surface of the SiO_2 prevents stable Si nuclei formation which inhibits film growth on the SiO_2. A high etch rate for Si clusters on SiO_2 is likely because these deposits would be amorphous. The chemistry for the neutral deposition process is controlled by having a sufficient flux of atomic H to etch off any excess Si on the SiO_2 surface to maintain Si deposition selectivity and etching of the SiO_2.

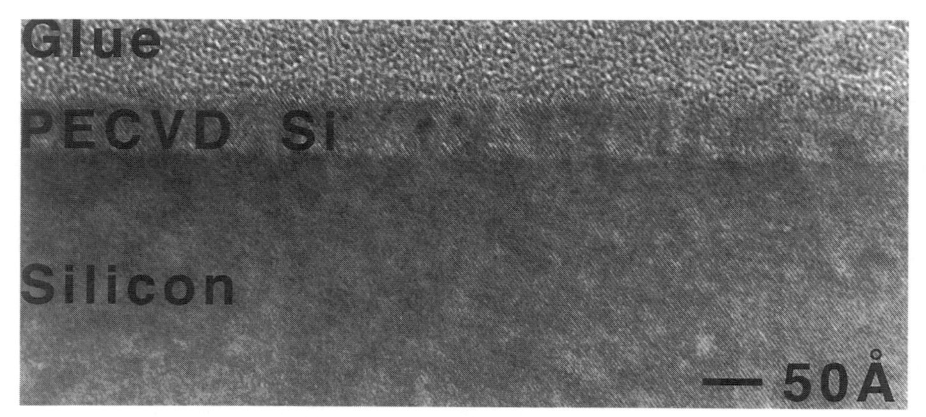

Figure 2. High resolution TEM image of a Si surface after H_2/SiH_4 plasma cleaning. The surface was exposed with a wafer temperature of 450°C, plasma power of 400W and SiH_4/H_2 ratio of 0.0010.

The chemical processes for the etching of the SiO_2 are not established at this time. Previous studies have indicated that the remote H-plasma did not etch the SiO_2. However, oxide has been removed from Si by exposure to a Si flux in MBE systems at temperatures of ~750°C. In these experiments it was presumed that the excess Si allowed the formation of SiO which is volatile at these temperatures. We suggest that the combination of SiH_x and excess atomic H allows the formation of molecules or radicals that are volatile at the lower temperatures employed in this study.

Surface Characterization

The Auger analysis was also used to monitor the level of oxygen and carbon contamination after each process step as shown in Figure 3. The AES scans in Figure 3 depict the carbon remaining after the RCA clean (as-loaded condition) and after removal of the SiO_2 film by the H_2/SiH_4 plasma. After an RCA clean some carbon is detected on the SiO_2 surface (Figure 3a). The carbon peak is reduced below the detection limits of AES after H_2/SiH_4 plasma cleaning. Previous studies of H-plasma cleaning have indicated that the carbon is removed from the wafer surface through the interactions of atomic H with the carbon contaminants.

LEED was used to investigate the surface structures from the as-loaded RCA cleaned surface to a PECVD Si surface due to the H_2/SiH_4 plasma exposure. There was no LEED pattern for the as-loaded wafer due to the amorphous SiO_2 film (15-20 Å)., and this is consistent with the AES scan (also indicating oxide) for the wafer in Figure 3a. In contrast, the 400 Watt plasma cleaned surface exhibits a 2x1 reconstructed surface which is similar to the LEED pattern for H-plasma cleaning of Si above 300°C [10]. The streaked 2x1 LEED pattern of the H_2/SiH_4 plasma cleaned surface is consistent with a hydrogen terminated monohydride configuration.

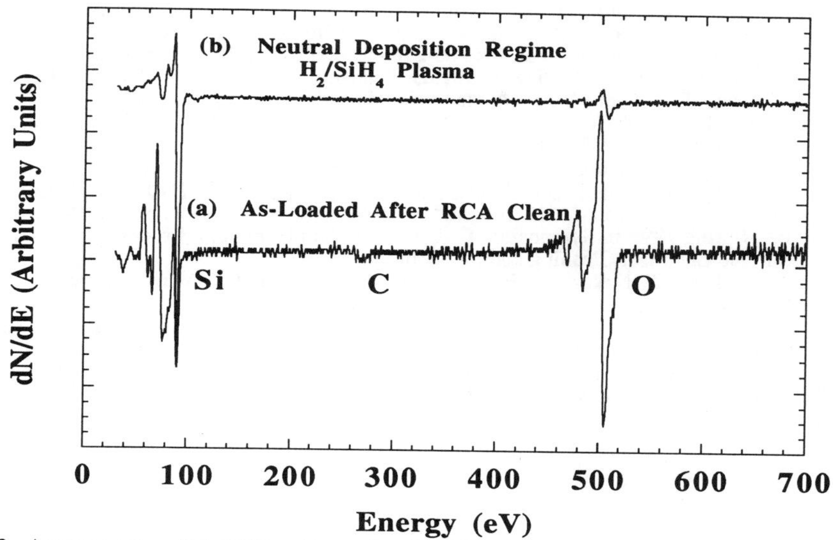

Figure 3. Auger spectra of Si (100) surfaces after (a) RCA clean and (b) H_2/SiH_4 plasma clean.

SUMMARY

This study has established that the remote H_2/SiH_4 plasma removes an SiO_2 film from a Si surface at an etch rate of approximately 2 Å/min. The trace amount of SiO_2 that remained after the plasma clean was insufficient to prevent a high quality PECVD Si epitaxial film from being deposited. For the neutral deposition process regime, the experimental conditions for removing SiO_2 was 450°C, 25 mTorr, 100-400 Watts of rf power, and a SiH_4:H_2 flow ratio equal to or less than 0.0010. Future studies will explore methods to enhance the etch rate without damaging the surface.

ACKNOWLEDGMENTS

This work has been supported by the NSF Engineering Research Centers Program through the Center for Advanced Electronic Materials Processing (Grant CDR 8721505).

REFERENCES

1. Tatsuya Yamaraki, N. Miyata, T. Aoyama, and T. Ito, J. Electrochem. Soc., Vol. 139, No.4, April 1175-1180 (1992)
2. T.Y. Hsieh, K.H. Jung, D.L. Kwong, T.H. Koschmieder, and J.C. Thompson, J. Electrochem. Soc., Vol. 139, No.7, July 1971-1978 (1992)
3. T. Aoyama, T. Yamazaki, and T. Ito, J. Electrochem. Soc., Vol. 140, No.2, February 366-371 (1993)
4. R.P.H. Chang, C.C.C Chang, and S. Darack, J.Vac. Sci. Technol., Vol. 20, No.1 January 45-50 (1982).

5. A. Izumi, T. Matsuka, T. Takeuchi, and A. Yamano, in Cleaning Technology in Semiconductor Device Manufacturing, J. Ruzyllo and R. Novak, Eds., The Electrochem. Soc., Inc. Princeton, NJ, Vol. 12: 260-276 (1992).(Vapor Phase Cleaning)
6. T. Hsu, B. Anthony, R. Qian, J. Irby, D. Kinosky, A. Mahajan, S. Banerjee, C. Magee, and A. Tasch, Journal of Electronic Materials, Vol. 21, No. 1, 1992, 65-74.
7. Jaewon Cho, T. P. Schneider, J. VanderWeide, Hyeongtag Jeon, and R. J. Nemanich; Appl. Phys. Lett. 59: 1995-97, (1991)
8. J.P. Barnak, S. King, J. Montgomery, Ja-Hum Ku, and R.J. Nemanich, to appear in Removal of Fluorine From a Si (100) Surface by a Remote RF Hydrogen Plasma, edited by M. Liehr, M. Heyns, M. Hirose, and H. Parks (Mater. Res. Soc. Symp. Proc., Pittsburgh, PA, Spring 1995 in Symposium O: Ultraclean Semiconductor Processing Technology and Surface Chemical Cleaning and Passivation).
9. A. Tsukune, T. Nakazawa, F. Mieno, Y. Furumura, and K. Wada, Abstracts, Vol. 89-2, Hollywood, FL, Oct. 15-20, (1989).
10. T. P. Schneider, D. A. Aldrich, Jaewon Cho, and R. J. Nemanich, in Mat. Res. Soc. Symp. Proc., Spring 1991 in Symposium B: Silicon Molecular Beam Epitaxy.
11. S.V. Hattangady, J.B.Posthill, G.G. Fountain, R.A. Rudder, M.J. Mantini, and R.J. Markunas, Appl. Phys. Lett. 59 (3), 15 July 339-341 (1991).
12. N.M. Johnson, J. Walker, and K.S. Stevens, J. Appl. Phys. 69 (4), 15 February 2631-34 (1991).

REMOVAL OF FLUORINE FROM A Si (100) SURFACE BY A REMOTE RF HYDROGEN PLASMA

J.P. BARNAK, S. KING, J. MONTGOMERY, JA-HUM KU, AND R.J. NEMANICH
Department of Materials Science and Engineering, and Department of Physics
North Carolina State University, Raleigh, NC 27695-8202

ABSTRACT

Fluorine contamination was removed from a Si(100) surface by an atomic H flux. The surface was intentionally contaminated to approximate the residual fluorine concentration remaining after a concentrated HF last process. By dipping the wafers in concentrated HF the thin oxide was removed and replaced with a hydrogen and fluorine terminated surface. This surface was then either vacuum annealed or exposed to a 20 Watt rf excited H-plasma at 50 mTorr, in order to achieve an atomically clean surface. The substrate temperature during the H-plasma exposure and vacuum anneal was 450°C. The surface chemistry was characterized with x-ray photoemission spectroscopy (XPS), auger electron spectroscopy (AES), and angle-resolved UV photoemission spectroscopy (ARUPS). The surface symmetry was characterized with low energy electron diffraction (LEED). Before the H-plasma exposure, the XPS spectra indicated Si-F bonding, and a 1x1 LEED diffraction pattern was observed. Immediately following the H-plasma exposure, the fluorine concentration was reduced below detection limits of XPS, and the surface showed a 2x1 reconstruction. A mechanism is proposed by which molecular HF results from atomic hydrogen interactions with fluorine on the surface.

INTRODUCTION

The minimum feature size for device structures is continuing to shrink well into the deep submicron realm. As this continues and wafer sizes increase, it may become more economical for critical process steps to be done using single wafer cluster tools. To achieve this it is necessary to develop various cleaning processes which are vacuum compatible with cluster tools. Wet cleans may still be used in the overall process flow for device fabrication, but probably not as modules on cluster tools since wet cleaning tools are not vacuum compatible. The processes being developed to address the dry cleaning issue are plasma cleaning, UV gas phase cleaning, and vapor phase cleaning. These techniques are vacuum compatible and are potentially adaptable to be used as cluster modules.

One of the most promising techniques is HF vapor phase cleaning which can be used to remove native or thin oxide layers. This process will, in general, result in residual fluorine on a nearly oxide free Si surface. We set out to develop a process which removes the fluorine contaminant from the Si surface that might occur after an HF vapor phase cleaning step. We examined Si surfaces exposed to a concentrated HF dip because this surface would be similar to that from an HF vapor phase clean. A concentrated HF dip was chosen as the HF last step since it has been shown that as the HF concentration in solution increases, the amount of residual fluorine remaining on the surface also increases therefore maximizing the amount of residual fluorine on the Si (100) surface [3]. Other HF vapor phase cleaning processes rely on water or alcohol to ionize the HF [4]. After removing the SiO_2 by the HF vapor phase cleaning technique residual contaminants of carbon and fluorine may remain on the Si surface.

In this study, a mechanism will be proposed for fluorine removal by atomic H generated from a remote rf H-plasma. The results of this study are compared to previously reported results on the removal of halides from Si (100) surfaces by a hot filament atomic H source [6-7].

Mat. Res. Soc. Symp. Proc. Vol. 386 © 1995 Materials Research Society

EXPERIMENTAL PROCEDURES

The wafers used in this study were p-type boron doped 25 mm diameter Si (100) wafers with a resistivity of 0.8-1.2 Ω-cm. The Si wafers were RCA cleaned by the manufacturer. Prior to the HF dip, the wafers were placed in a UV ozone cleaning chamber for 5 min to remove hydrocarbon molecules from the surface. The wafers were then dipped in a concentrated 49% HF solution for 10 seconds to strip off the oxide and leave a F and H terminated surface. The wafer was then mounted onto a molybdenum sample holder. The holder was placed into the load-lock before transferring the sample into the transfer-line. The analysis and plasma cleaning chambers are attached to the transfer-line and each chamber is isolated from the transfer-line through the use of a gate valve. Currently there are eight stainless steel UHV chambers along the 35 feet long transfer-line. The transfer-line is a linear wafer-handler for our cluster tool. Once a sample is placed onto the sample cart, which travels the length of the transfer-line, the sample can be moved from chamber to chamber *in situ*. The base pressure was maintained at $1x10^{-9}$ Torr through the use of cryogenic and turbo molecular pumps.

The remote plasma cleaning system has been previously described [5]. The process gas flows through a quartz tube located on top of the chamber. An inductively coupled plasma was generated using a rf power supply (13.56 MHz) and rf matching network attached to a Cu coil placed around the quartz tube. The base pressure of the H-plasma cleaning chamber is $8x10^{-9}$Torr. The sample is located 40 cm from the center of the rf coil. A point-of-use purifier/filter is located downstream of the hydrogen mass flow controller. The sample was heated by a tungsten coil facing the backside of the wafer. This heating coil is bent in the shape of a ring, and the control thermocouple is positioned at the center of the heater. Depending on the chamber pressure and rf power, the plasma can be maintained in the quartz tube or extended down toward the sample region. At 50 mTorr and 20 Watts the plasma does not extend down to the sample, i.e. a remote plasma. If the pressure is reduced below 20 mTorr, the plasma envelopes the sample region. In the experiments described here the pressure was controlled by the hydrogen flow rate.

The surface of each sample was characterized using XPS, AES, and LEED after each of the following steps: (1)10 s HF dip (as loaded condition for all samples), (2) 450°C vacuum anneal for 5 min, (3) 450°C H-plasma for 0.1 min, and (4) 450°C H-plasma for 1 min. The samples processed at 450°C were ramped from 150°C to 450°C at 100°C/min. For the H-plasma cleaning experiments, the hydrogen flow was established prior to ramping up the temperature. The plasma was struck once the 450°C set point was reached. After the plasma exposure, the plasma, hydrogen gas, and heater were simultaneously turned off. The sample cooled in the plasma chamber before it was transferred to the analysis chambers. In the H-plasma cleaning experiments, the wafers were exposed to a remote H-plasma with the conditions of 50 mTorr and 20 Watts. The flow rate of hydrogen was 140 sccm.

All XPS spectra were taken using Al K alpha radiation (1486.6 eV) with a 20 mA emission current and a 12 kV anode potential. Calibration of the binding energy scale for all scans was achieved by periodically taking scans of the Au 4f7/2, Cu 2p3/2, Cu LMM, and Cu 3p peaks from in-house standards and correcting for the discrepancies between the measured and known values of these peaks.

Auger analysis was done using a Perkin-Elmer static probe Auger system which utilizes a cylindrical mirror analyzer and a spot size less than 1 mm. The spectra were taken by averaging over 3 scans with a step size of 0.5 eV/step and a dwell time of 25 ms/step. The differentiated Auger spectra were normalized using the $Si_{(LMM)}$ peak at 92 eV.

The LEED diffraction patterns were taken at an accelerating voltage set to 60.8 V.

RESULTS AND DISCUSSION

Auger surface analysis is an established probe of the surface O and C contamination. The carbon peak is reduced below the detection limits of AES by comparing the as loaded wafer (Figure 1a) to the 1 min H-plasma cleaned wafer (Figure 1c) . The oxygen peak is only slightly reduced at best, by comparing the as-loaded AES scan to the H-plasma clean AES scan in Figure 1. By comparing the as-loaded wafer to the vacuum annealed wafer (Figure 1b) the AES analysis revealed that the carbon and oxygen concentrations increased after vacuum annealing. Also note that the fluorine is below the detection limit of AES for all wafers examined in this study.

It has been previously reported that oxygen and carbon contaminants are removed below the detection limits of AES for Si surfaces exposed to the H-plasma [8]. To achieve this the surfaces were prepared with a dilute HF etch. However, when the wafers are dipped in the concentrated HF solution, the AES scan after H-plasma cleaning shows that the oxygen peak is only slightly reduced. The remaining oxygen is most likely chemisorbed in the form of SiO_2. The increase in the carbon and oxygen concentrations after vacuum annealing is a direct result of hydrogen evolution and of the wafer being in vacuum for several hours in order to complete the surface analysis and annealing experiments.

The LEED patterns also reflect aspects of the surface structure. The LEED pattern for the as-loaded wafer is 1x1. The surface structure after H-plasma cleaning is a 2x1 reconstruction [5]. The 2x1 LEED pattern is not as sharp as a high temperature reconstructed surface. The diffuse cross pattern of this low temperature plasma assisted reconstructed surface has a small domain size, which leads to this type of feature in the LEED pattern [9]. The H-plasma cleaned 2x1 Si surface is hydrogen terminated in a monohydride configuration whereas the 1x1 Si surface exhibits mostly dihydride termination.

The surface termination was determine using ARUPS by scanning the silicon surface after H-plasma exposure and then rescanning the wafer after a thermal desorption step [8]. The hydrogen induced surface states in the spectra are eliminated after thermal desorption of hydrogen from the Si surface.

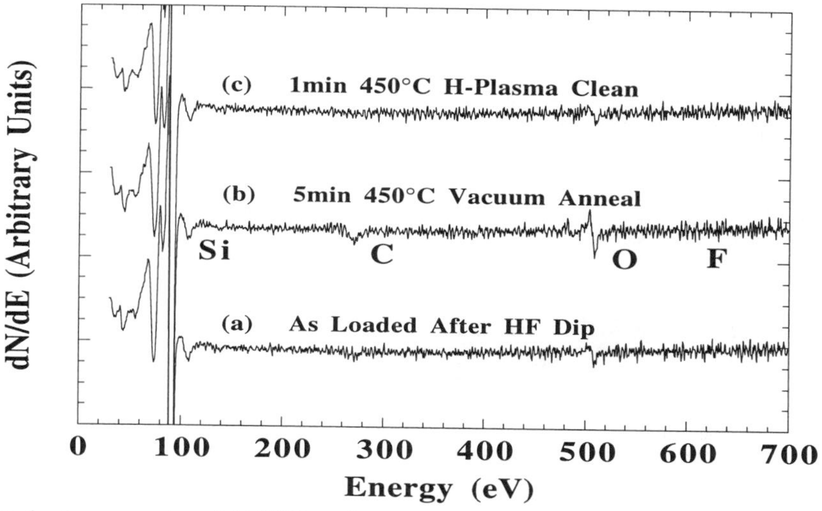

Figure 1. Auger spectra of Si (100) surfaces (a) as-loaded after an HF dip, (b) after a 450°C 5 min vacuum anneal , and (c) after a 450°C 1 min H-plasma clean.

The residual surface concentration of fluorine on a Si (100) wafer after dipping in concentrated HF was measured by XPS (Figure 2a) since the fluorine concentration was below the detection limit of AES (Figure 1a). Since the F 1s peak was so weak, long narrow energy scans were required in order to resolve the peak. Following an HF dip the F 1s peak is centered at 685.9 eV (Figure 2a) and the surface concentration is in the range of 0.01-0.1 monolayers. The XPS spectrum of the F 1s peak after annealing is shown in Figure 2b. Figure 2b indicates that the majority of fluorine still remained on the silicon surface after annealing, however the position of the F 1s peak has shifted to a higher binding energy and is now positioned at 686.2 eV. Figure 2c shows the XPS spectrum of the F 1s peak after a 0.1 min exposure to the remote H-plasma. As can clearly be seen, fluorine is removed below the detection limit of XPS.

It may be argued that the F 1s peak in Figure 2a is actually a convolution of two peaks. Since the Si 2p peak position did not shift either before or after annealing, the peak shift cannot be due to a Fermi level shift or other spurious effects. One of the peaks is associated with a weakly bound state of fluorine, which forms the tail on the left side of the peak (Figure 2a), and the other is due to a strongly bound chemisorbed state of fluorine (i.e. Si-F bonds). Annealing at 450°C results in the desorption of the more weakly bound fluorine leaving behind only the chemisorbed fluorine with the peak centered at 686.2 eV. This is indicative of Si-F bonding, and the binding energies agree with previously reported results [6-7].

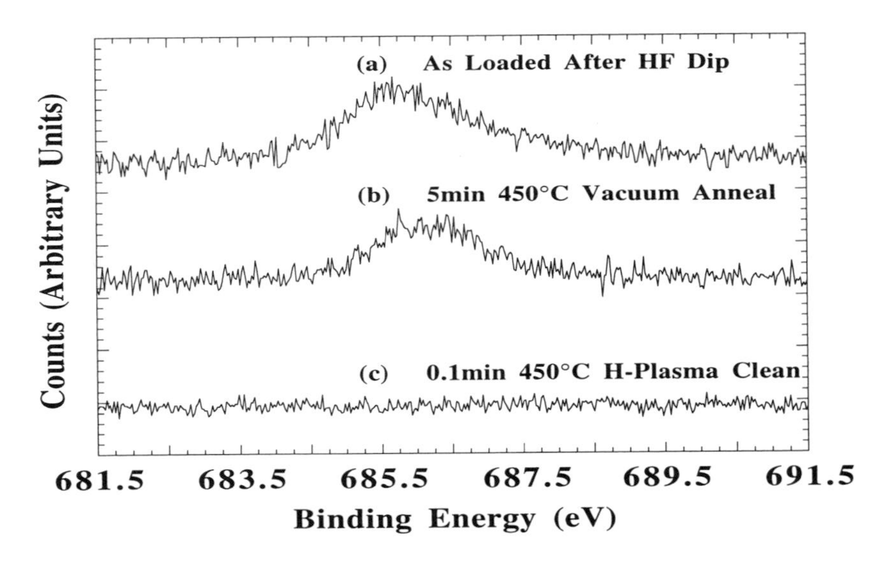

Figure 2. XPS spectra of Si (100) surfaces depicting the F 1s peak (a) as-loaded after an HF dip, (b) after a 5 min vacuum anneal, and (c) after a 0.1 min H-plasma clean.

The ionic and atomic hydrogen fluxes in the sample region were measured to be 10^{13} and 10^{16} atoms/cm^{2}s, respectively[8]. The ionic and atomic H fluxes during the 0.1 min H-plasma exposure were calculated to be approximately $6x10^{13}$ and $6x10^{16}$ atoms/cm^2, respectively. The number of silicon atoms per cm^2 is $7x10^{14}$. For the 0.1 min H-plasma exposure the ionic flux is an order of magnitude less than the number of silicon atoms per cm^2 whereas the atomic hydrogen flux is two orders of magnitude greater than the number of silicon atoms per cm^2. The fluorine coverage is less than 0.1 monolayers, therefore the surface concentration will be less than $7x10^{13}$ per cm^2. In order for the fluorine to be removed, the flux of the reactive species must be equal to or greater than the number of Si atoms per cm^2. The only flux in sufficient quantity to remove the fluorine is the atomic H flux.

It is also of interest to examine if it is energetically more favorable for the fluorine to remain bonded to the Si surface or for the atomic H to react with the fluorine, removing it from the Si surface. If the bond strength between the Si and the adsorbate species is greater than the atomic bond between the adsorbate and the reactive species, then there is no driving force (i.e. lowering the systems energy) for the reactive species to break the Si-adsorbate bond. In this study the reactive species is atomic H and the adsorbate is fluorine. The bond energies of H-F and Si-F are 5.91 eV and 5.73 eV, respectively. Since the H-F bond energy is greater than the Si-F bond energy, the energy of the system would be lowered by atomic H breaking the Si-F bond and forming HF. The proposed mechanism is that the atomic H flux from the gas phase reacts with the physisorbed and chemisorbed fluorine on the Si surface to form HF. The HF then desorbs from the Si surface. Thermal desorption was not considered the primary reaction in removing fluorine from the Si surface due to the low process temperature in this study and the small reduction observed in the F 1s peak after vacuum annealing.

It has been previously established that the reaction rate for the removal of halide species from silicon increases as either the temperature or the atomic hydrogen flux increases [6-7]. Additionally, the etch rate of Si by atomic H for the test conditions in this study is negligible [5]. By reducing the etch rate of Si and increasing the etch rate of the halide contaminant, a process regime was established where the halide is removed below the detection limit of XPS without roughening the surface or causing subsurface damage.

Other studies involving removal of halide species (I, Br, and Cl) from Si (100) by atomic H show that the reaction follows an Eley-Rideal process (a pick-off reaction) [6-7]. If the removal of fluorine from Si by atomic H follows an Eley-Rideal process in the same manner that other halides are removed from Si by atomic H, then the atomic hydrogen flux is sufficient to remove fluorine in 0.1 min. The results from this study are insufficient to conclude that the removal of fluorine by atomic H is an Eley-Rideal process. One method for determining if a reaction follows an Eley-Rideal process is to measure the activation energy of the reaction [6-7]. If the activation energy for the reaction approaches zero then the reaction follows an Eley-Rideal process.

CONCLUSIONS

The results presented here establish that the remote hydrogen plasma removes the residual fluorine below the detection limit of XPS from the Si surface in less than 0.1 min at 450°C. The fluorine is replaced by a hydrogen terminated surface. Annealing the fluorine terminated sample at the same temperature only desorbs the physisorbed fluorine leaving the majority of the fluorine on the surface. The proposed mechanism is that atomic hydrogen and fluorine react to form HF which either desorbs or the molecule is removed in a pick-off type reaction (i.e. Eley-Rideal). An additional benefit of the plasma clean is that it removes carbon and oxygen to the extent that the oxygen is not in the form of SiO_2.

ACKNOWLEDGMENTS

This work has been supported by the NSF Engineering Research Centers Program through the Center for Advanced Electronic Materials Processing (Grant CDR 8721505).

REFERENCES

1. C.C. Cheng, S.R. Lucas, H. Gutleben, W.J. Choyke, and J.T.Yates, Jr.; J. Am. Chem. Soc. 114: 1249-1252 (1992).
2. D.D. Koleske and S.M. Gates; J. Chem. Phys. 98 (6), 15 March: 5091, (1993).
3. T. Takahagi, I. Nagai, A. Ishitani, and H. Kuroda; J. Appl. Phys. 64(7), 1 October 1988, 3516.
4. A. Izumi, T. Matsuka, T. Takeuchi, and A. Yamano, in Cleaning Technology in Semiconductor Device Manufacturing, J. Ruzyllo and R. Novak, Eds., The Electrochem. Soc., Inc. Princeton, NJ, Vol. 12: 260-276 (1992).
5. T.P. Schneider, D.A. Aldrich, Jaewon Cho, and R.J. Nemanich, in Mat. Res. Soc. Symp. Proc., Spring 1991 in Symposium B: Silicon Molecular Beam Epitaxy.
6. T.J. Chang, J. Appl. Phys. 51: 2614, (1980).
7. C.D. Stinespring and A. Freedman, Appl. Phys. Lett. 48: 718, (1986).
8. Jaewon Cho, T.P. Schneider, J. VanderWeide, Hyeongtag Jeon, and R.J. Nemanich; Appl. Phys. Lett. 59: 1995-97, (1991).
9 D.H. Lee, J. Joannopoulos, and A. Berker, J.Vac. Sci. Technol. B1(3), July-Sept.: 705(1983).

Part VI

Characterization of Cleaned Surfaces

PREFERENTIAL ETCHING OF Si(111) and Si(001) IN DILUTE NH_4F SOLUTIONS: AS PROBED BY IN SITU STM

KINGO ITAYA*, SHUEH-LIN YAU**, AND KAZUTOSHI KAJI**
*Department of Applied Chemistry, Faculty of Engineering, Tohoku University, Sendai 982, Japan
**Itaya Electrochemiscopy Project, JRDC/ERATO, Yagiyama-minami 2-1-1, Taihaku-ku, Sendai 982, Japan

Abstract

In situ scanning tunneling microscopy (STM) was used to examine the etching process of n-Si(111) and Si(001) electrodes in dilute NH_4F under cathodic potential control. For Si(111), time-dependent STM images have revealed the pronounced effect of the microscopic structure of surface Si atoms on their dissolution rates. The multiple hydrogen bonded Si atoms at the kink sites and dihydride steps eroded faster than the monohydride terminated Si. Presumably, the higher polarity at these defect sites is responsible for the difference. Steric consideration further favors the higher activity at the more open kink sites. The monohydride terminated Si(111) surface represents the most stable surface structure, which guides the dissolution of the Si(001) surface to the formation of {111} facets. The initial stage {111} facet formation on a Si(001) surface was revealed by in situ STM.

I. INTRODUCTION

Wet chemical etching of Si with buffered HF solutions is known to be an effective method for preparing ideal hydrogen-terminated Si(111) surfaces [1,2]. This simple etching process, which can produce an oxidation resistive Si surface with low density of surface states [3], is of importance to the semiconductor processing technology. Early spectroscopic studies employing surface sensitive infrared [1,2] and electron energy loss specotroscopies [4], have elucidated the atomically flat morphology of the etched Si surfaces. It is shown that alkaline etching solutions of NaOH or NH_4F indeed lead to atomically flat Si(111)surfaces. This is further supported by STM [5,6] and AFM [7] studies, which provide topography and atomic views of the etched Si(111) surfaces in UHV as well as in solutions [8].

The electrochemical potential, which controls the surface charge concentration of the Si electrode, plays a main role in guiding the Si etching process in electrochemical environment [6]. This potential effect was recently demonstrated in 1 M NaOH [9]. Interestingly, chemical etching, which dissolved Si in a layer by layer manner, was revealed in the cathodic region, in contrast to the pit formation at anodic potentials. The latter observation seems to contradict the previous model [10], which attributed the dark current to the local electron injection into the conduction band at some surface defects. In fact, in situ STM has revealed anodic potential-induced random nucleation of pits, arising from the oxidation of chemisorbed hydrogen and the subsequent dissolution at terrace sites [9]. The role of water molecules in implementing chemical etching has been established in the absence of F^-, while the effect of F^- on the etching of Si has been unclear. We have recently examined the etching process of an Si(100) surface in NH_4F and found that preferential etching also prevailed in NH_4F solution as was evidenced by the fact that long range ordered {111} microfacets were revealed by the STM [11].

In this study, in situ STM was used to examine the etching process of n-type silicon (111) and (001) electrodes in 0.27 M NH_4F. Weak cathodic polarization of the Si electrode resulted in the preferential etching at multiple H-bonded defects, which is tentatively attributed to the nucleophilic attack of Si at the defects by H_2O and F^-. It was found that chemical dissolution of Si proceeded through the initial nucleation and growth of pits, followed by lateral erosion of the uppermost Si layer. Furthermore, preferential etching of multiple H

Mat. Res. Soc. Symp. Proc. Vol. 386

bonded Si species such as the kinks and the dihydride steps was found. This result is tentatively attributed to the H-adsorption induced polarization of the Si atoms and the steric effect in the nucleophilic reaction of H_2O and Si. Time sequenced STM images were acquired to evaluate the orientation dependent lateral and vertical etching rates. The thermodynamic stability of the monohydride terminated (111) structure was also responsible for the formation of {111} microfacets on Si(001) electrodes. In situ STM also revealed the initial stage of the facet formation.

II. EXPERIMENTAL

The samples used were n-type silicon (111) (Osaka Titanium Co. Ltd., Tokyo, Japan) with 5-14 Ω cm resistivity. Our sample pretreatment was somewhat different from the conventional methods [1] in which the Si(111) samples were first oxidized in 12 M nitric acid at 350 K for 10 min. The oxide layer was removed by 0.59 M HF at room temperature, followed by etching in 11 M NH_4F for 3 min at room temperature. Millipore water was used to rinse the Si sample throughout the etching process. The Si(111) surface appeared hydrophobic after the etching process. This method resulted in long range, atomically flat terraces on the Si samples. Electronic grade NH_4F (40% by weight, Morita Chemical Co., Tokyo, Japan) were used to prepare dilute NH_4F solutions with Millipore water. The cyclic voltammograms were measured in a Teflon electrochemical cell with a saturated calomel reference electrode (SCE) and a Pt counter electrode. The STM was a NanoScope III (Digital Instruments Inc., Santa Barbara, CA, USA), and an electrochemically polished W wire was used as the tip. They were further insulated with nail polish to decrease the residual current of the tip to less than 50 pA.

III. RESULTS AND DISCUSSION

A. Cyclic Voltammogram of n-Si(111) in 0.27 M NH_4F.

Fig. 1 shows current-potential curves of an n-Si(111) electrode in nitrogen-saturated 0.27 M NH_4F in the dark (solid line) and under illumination (dashed line). Cathodic potential scan was firstly made from the open-circuit potential (-0.5 V) to -1.1 V. The corresponding cathodic current is attributed to the discharge of protons and subsequent hydrogen evolution at the Si electrode. The anodic current peaked to 12 $\mu A/cm^2$ at -0.2 V and decreased slowly to 8 $\mu A/cm^2$ at 0.2 V. The dark current has been explained by the electron injection from chemisorbed H atoms (Si-H) to the conduction band of the n-Si(111) electrode, leading to local surface oxidation and dissolution. These processes were proposed to occur at defect sites [9]. Illumination of this n-Si(111) electrode led to the emerging of photocurrent at -0.4 V, peaking to 350 $\mu A/cm^2$ at -0.1 V, and followed by an exponential decay to 120 $\mu A/cm^2$. This current decay implies the passivation (growth of an insulating oxide film) of the Si electrode, which also resulted in the decrease of cathodic current in the subsequent negative going potential scan. The Si(001) samples yielded essentially identical CV results.

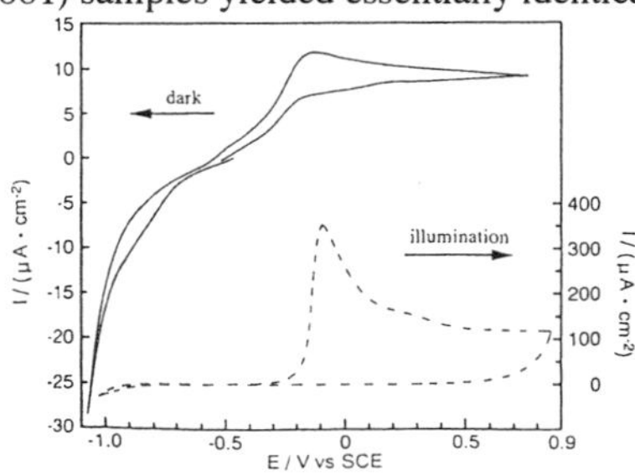

FIG. 1. The CV curves for n-Si(111) in a nitrogen saturated 0.27 M NH_4F under dark (solid line) and illumination (dotted line). The scan rate is 50 mV/s.

B. Time Dependent STM Imaging of Si(111) Etching.

Detailed preferential etching of Si can be illustrated by the time-dependent STM images shown in **Fig. 2(a)** to **2(d)** [12].

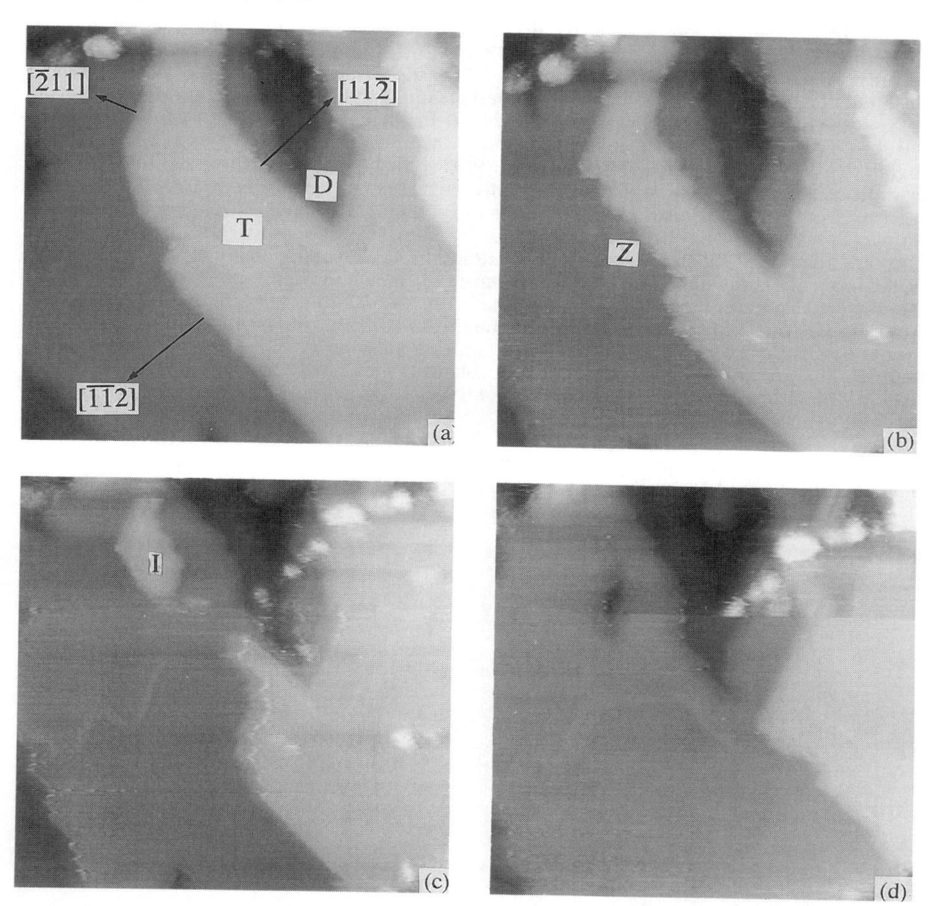

FIG. 2 Time-sequenced STM images (50×50 nm) showing the continuous etching process of Si(111) in 0.27 M NH_4F. The images were acquired continuously at a time interval of 12.8 s. The Si and tip potentials were held at -1.04 and -0.1 V, respectively [12].

The initial surface morphology in Fig. 2(a) which dominated the etching process will be described first. The well-defined step-and-terrace structures can immediately be identified. The terraces, spanning more than 50 nm, are stacked up in the [111] direction. Their brightness in the STM image reflects their height. The internal atomic structure of the terrace (marked T) could be discerned in the higher resolution STM scan. The well-ordered hexagonal pattern obtained, (not shown) having a 0.38 nm interatomic spacing, is in good agreement with the ideal Si(111)-(1x1):H structure. In addition to these atomically flat (111) terraces, line defects of steps and kinks were also revealed by the STM. The step orientations, as defined by their outward normals, were in the $[\bar{2}11]$, $[11\bar{2}]$, and $[\bar{1}\bar{1}2]$ directions. The former two steps are monohydride-terminated, while the latter has a dihydride structure. All the steps are bilayer in depth (with a step height of 0.31 nm).

The 13 s etching of Si, from Fig. 2(a) to 2(b), decreased the upper portion of the terrace T from 16 to 8 nm, as compared to the decrease from 18 to 12.5 nm for the lower portion of the terrace T. These numbers translate into 38 and 26 nm/min etching rates for the upper and lower portions of the terrace *T*, respectively. While the terrace T remains atomically flat, the different etching rates at different parts of a terrace are then likely to be due to the dissolution of Si at the defects of kinks and steps. The role of the defects in controlling the etching of Si will be be discussed further below. Specifically, the upper portion of the terrace T contains a higher density of kinks at the $[\bar{1}\bar{1}2]$ dihydride step so that Si dissolves more rapidly. This fast erosion of the upper portion of the $[\bar{1}\bar{1}2]$ dihydride step leads to a rather jagged step ledge, marked Z in Fig. 2(b). This ragged step ledge then quickly dissolved Si to remove most of the terrace T, leaving a small isolated island I in Fig. 2(c). It is surprising to find that the lower portion dihydride $[\bar{1}\bar{1}2]$ step is well-defined (straight). As Si etching continues, ca. 30 nm^2 area of the terrace T at the uppermost Si layer is removed, shown in Fig. 2(d). The $[\bar{2}11]$ and $[11\bar{2}]$ monohydride steps at the lower portion of the depression D do not exhibit noticeable change from Fig. 2(a) to 2(c). More interestingly, the location where the protruding island I in Fig. 2(c) is now replaced by a bilayer deep (0.31 nm) depression in Fig. 2(d). The shape of the depression is roughly identical to that of the island I. This newly formed pit is apparently a reality, not an imaging artifact, because it continues to grow and coalesces with an adjacent $[11\bar{2}]$ step (not shown).

The presence of kinks apparently plays a main role in controlling the etching rate of Si. This point can be seen more clearly in the two time sequenced (10 s difference in time) STM images (**Fig. 3(a)** and **3(b)**). These two images were obtained at the (111) facets on Si(001).

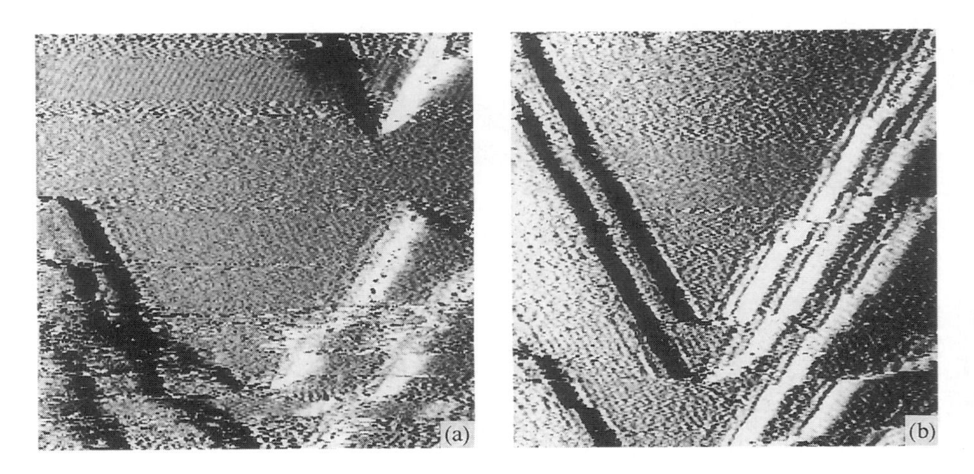

FIG. 3. Time-dependent in situ STM images (17×17 nm) showing the etching process on Si(111) in 0.1 M NH_4F. The time difference between them is 10 s [12].

The local surface includes well-ordered (111) facets, outlined by well defined monohydride steps. Although the internal structure of the facets was not perfectly resolved, a portion of Fig. 3(a) indeed reveals the Si(111)-(1×1) structure. In addition, the 60° angle enclosed by the step ledges are indicative of the (111) facets. These features are particularly clear in the second image. The first image was scanned upward, and the fuzzy appearance at the corner (kink site) of the facet suggests its higher reactivity than the step sites. In situ STM imaging of an ongoing chemical process always introduces instability to the scanning probe and results in fuzzy images. Nevertheless, the two monohydride steps have roughly retracted 2.5 nm or 8 atomic rows along the $[\bar{2}11]$ and $[11\bar{2}]$ directions. Only the Si atom at the corner of the (111) facet (intersection of two monohydride steps) has a dihydride configuration, and therefore it is preferentially removed to expose more dihydride Si atoms. Preferential etching would blunt the originally sharp corner because of the steric advantage of the kink. It allows easy access by the H_2O and F^- etchants. More kink sites were likely exposed to the solution species and the etching proceeded along the step ledges (in the <110> directions). This process can be compared to those results from Fig. 2(a) to Fig. 2(d) which include two different steps in the $[\bar{1}\bar{1}2]$ and $[\bar{2}11]$ directions. In this case, the etching front discriminated step ledges for their atomic structures so that the $[\bar{1}\bar{1}2]$ dihydride step rapidly retracted while the $[\bar{2}11]$ monohydride step remained mostly unchanged.

C. {111} microfacet formation on Si(001) surface.

The ideal H-terminated (111) structure is likely to be the most stable surface arrangement for the uppermost Si layer. This is the driving force for the etching process. This is further demonstrated by the in situ STM results obtained with Si(001) in dilute NH_4F solution. Firstly, in situ STM imaging revealed the development of a staircase morphology which descends from the left to the right in the image in **Fig. 4 (a)** [11].

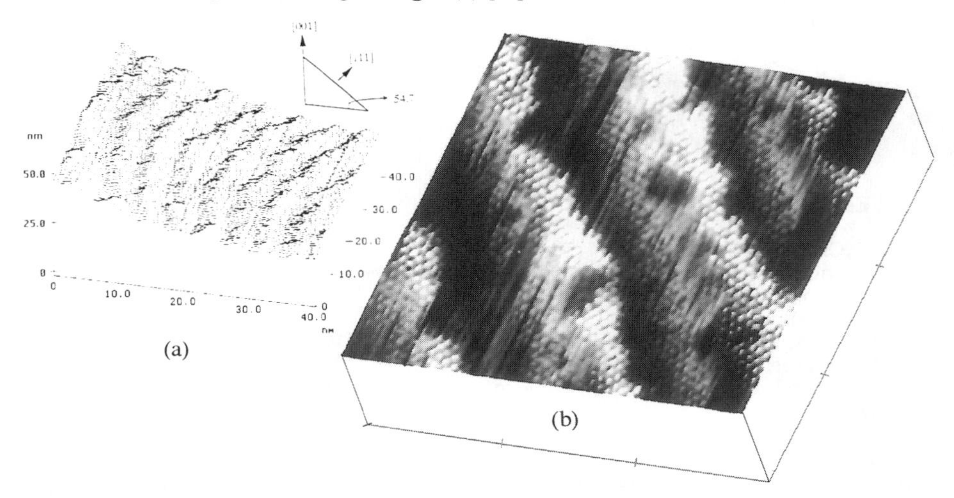

FIG. 4. (a) Topographic line scans (40×40 nm) showing the (111) facets formation after 1 hr etching of Si(001) in NH_4F solution. Inset in the figure shows the tilting geometric relationship between the (111) and (001) planes. (b) High resolution STM scan (15×15 nm) showing the atomic features on the {111} facets [11].

All stairs are only 4 nm in width but extend over more than 40 nm in length. This staircase-like feature is attributed to the formation of {111} microfacets. The height of each step is ca. 1.2 nm. The inset in Fig. 4(a) shows an angle of 54.7° between the (111) and (001) planes. The angle observed in the image shown in Fig. 4(a) seems to be consistent with the expected value. Apparently, step bunching occurs on the extensively etched Si(001) surface, producing the appearance of stairs with step heights greater than the height of a monolayer. It is remarkable that the in situ STM of the smaller scan area (15×15 nm) shown in **Fig. 4(b)** clearly reveals atomic features on the tilted {111} microfacets. Hexagonal arrays with an interatomic spacing of 0.39 nm are clearly resolved on rather narrow atomically flat planes. The steps separating these {111} microfacets are most certainly double layers. Many different types of kink sites can be clearly seen along the step lines. These poorly defined step edges were obtained owing to multiple hydrogen bonded Si atoms, although monohydride Si atoms existed along the [$\bar{1}$10] direction as shown in Fig. 4(b). The pits appeared on the narrow terraces indicate that the etching also occurs by removal of monohydride Si on the {111} microfacets. The formation of pits on Si(111) was previously explained by a random dissolution of Si atoms on the (111) terraces. This is in contrast to other microfacets where double layer steps form in parallel to the <110> direction. Incidentally, we obtained time-dependent STM atomic images at the {111} facets, which highlight the selective etching initiated at kink sites.

IV. SUMMARY AND CONCLUSION

In situ STM has revealed preferential etching of Si(111) at the kinks and steps in dilute NH_4F under cathodic potential control. The dihydride steps, having a higher density of kinks than monohydride steps, eroded faster. The thermodynamically stable (111) monohydride structure drives the etching process of Si(001) to the formation of {111} microfact. The time-dependent STM images allow estimation of the etching rate of 16-38 and 0-15 nm for the dihydride and monohydride steps, depending on the specific morphology of the steps.

ACKNOWLEDGMENT

This work is supported by the JRDC/ERATO, Japan.

REFERENCES

1. Y.J. Chabal, Surf. Sci. **168**, 594 (1986).
2. P. Jacob and Y.J. Chabal, J. Chem. Phys., **95**, 2897(1991).
3. E. Yablonovitch, D.L. Allara, C.C. Chang, T. Gmitter, T.B. Bright, Phys. Rev. Lett. **57**, 249 (1986).
4. P. Dumas, Y.J. Chabal and P. Jakob, Surf. Sci. **269**, 867 (1992).
5. R.S. Becker, G.S. Higashi, Y.J. Chabal and A.J. Becker, Phys. Rev. Lett.**65**, 1917 (1990).
6. H.E. Hessel, A. Feltz, M. Reiter, U. Memmert, and R.J. Behm, Chem. Phys. Lett. **186**, 275 (1991).
7. Y. Kim, and C.M. Lieber, J. Am. Chem. Soc. **113**, 2335 (1991).
8. K. Itaya, R. Sugawara, Y. Morita and H. Tokumoto, Appl. Phys. Lett. **60**, 2534 (1992).
9. P. Allongue, V. Costa-Kieling, and H. Gerischer, J. Electrochem. Soc.**140**, 1009, 1018 (1993).
10. H. Gerischer, and M. Lubke, Ber. Bunsenges. Phys. Chem. **91**,394 (1987).
11. S. L. Yau, K. Kaji, and K. Itaya, App. Phys. Lett. **66**, 766(1995).
12. K. Kaji, S. L. Yau, and K. Itaya, J. Appl. Phys. submitted (1995).

CONDUCTING AFM: APPLICATIONS TO SEMICONDUCTOR SURFACES

MARTIN P. MURRELL*, SEAN J. O'SHEA†, JACK BARNES†,
MARK E. WELLAND† and CARL J. SOFIELD‡
*East Coast Scientific Ltd, 14 Bishop's Road, Cambridge, England, CB2 2NH
†Department of Engineering, University of Cambridge, Cambridge, UK. CB2 1PZ;
‡AEA Technology, Oxford, England OX11 0RA.

ABSTRACT

The use of Conducting Probe Atomic Force Microscopy to give nm scale electronic characterisation of surfaces is reviewed. Local conductance, Kelvin Probe work function measurements, Fowler-Nordheim tunnelling and local C-V characterisation techniques are outlined. The principle results of these and their applications to the semiconductor surface and thin film characterisation are discussed. We present tunnelling data from silicon through varying oxide thickness using conducting AFM and scanning Kelvin Probe measurements from sub micron MOS capacitors. The F-N tunnelling technique has also been used on epitaxial silicon surfaces with atomically flat topography.

The inherent problems associated with quantitative, reproducible measurements are outlined, and the potential applications of the measurements to surface and thin film technology are discussed.

INTRODUCTION

VLSI AND SPM

In the past 15 years VLSI geometry has shrunk below conventional optical microscopy limits and commercial devices now have 100 nm geometry. This demands ever tighter control over the initial semiconductor materials, the fabrication processes and the final device integrity. To provide diagnostics for individual devices and fabrication process steps, both physical and electronic characterisation is required at an ever higher resolution and sensitivity. It is perhaps fortunate that while device dimensions have been shrinking Scanning Probe Microscopy (SPM) has developed into a mature technology, well placed to analyse the next generation of VLSI devices. Limitations to conventional SPM exist however: STM is only sensitive to the electronic structure of conducting materials, and conventional AFM provides only topographic information. In the first case the convolution between the surface topography and electronic structure prevents the unambiguous interpretation of features in certain instances, and in the second no localised electronic information is provided.

In this paper we shall show how Conducting Atomic Force Microscopy provides a unique capacity to investigate the topographic and electronic structure of devices, surfaces and

Mat. Res. Soc. Symp. Proc. Vol. 386

thin films. We present data using the technique to study and image local conductivity, dielectric strength, charge to breakdown, static local charge and contact potential on a lateral scale suitable for 0.1 micron geometries.

Instrumentation

In all the experimental studies detailed below the instrument used was an optical deflection sensing AFM, equipped with both contact and attractive mode control and a maximum scan area of 50 microns. The microscope was modified with carefully screened electrical connections to the sample and tip. Where electrical currents were measured, an STM type current amplifier was attached to either the sample or tip, with a gain of 10^{10} V/A and less than 1pA of noise. To protect the tip from excessive currents, a series resistor was usually incorporated. The AFM control was achieved using a digital feedback loop[1] and the sample bias and current measurement were controlled and recorded using additional I/O channels. In the contact potential measurements two lock-in amplifiers were used in addition, to separate the signals at the fundamental and first harmonic driving frequency of the oscillating tip-sample bias.

The cantilever and tip combinations used for these studies included: etched gold wire, aluminium coated silicon nitride levers and tips. silicon microfabricated levers and silicon levers coated with diamond or platinum/iridium. It is worth emphasising that with the addition of careful electrical screening and low noise current amplification, most conventional AFM systems could be used for these studies if equipped with the appropriate conducting tips and levers.

TECHNIQUES

In conducting AFM the measured current between a tip and surface may be described as :

$$I = \frac{V}{R} + C\frac{dV}{dt} + V\frac{dC}{dt} + \alpha A V^2 e^{-\beta / V} \tag{1}$$

where R is the series resistance, C(t) the tip-sample capacitance, V(t) the applied voltage, A the effective area of the capacitor (assumed constant) and α and β the coefficients for FN tunnelling. The techniques reviewed below rely on the various terms of this equation to measure specific electronic properties of devices, films and surfaces.

Local Conductance

If a Conducting AFM image is taken in contact mode with a tip bias applied, the ohmic current flowing from tip to sample may be correlated with the tip position, producing a local conductance image. In the application of this technique to silicon surfaces however, the resistivity of the oxide is > 10^{16} Ω cm and a detectable current only flows when the tip bias is sufficiently large that tunnelling occurs, or the contact force is sufficient to allow the probe to "push through" the insulator. In either case the dielectric is rapidly degraded and the technique no longer probes the properties of the original film. This technique may be used to image localised defects which provide a conduction path through the oxide. Such defects tend to be extremely rare in modern VLSI technology but they may be observed in UHV processed gate oxides and other research based fabrication processes[2].

Dielectric Breakdown

The maximum electronic stress a gate oxide will withstand prior to breakdown is an useful measure of the quality of the dielectric and indicative of final device reliability. Low field breakdown may be connected with metal related defects combined with non ideal processing conditions[3]. Conventional MOS capacitor test structures are used to measure either the breakdown field or the charge injected before breakdown, however the spatial resolution is restricted to the dimensions of the test capacitors. Conducting AFM, in contact mode, may provide analogous measurements over much smaller areas, by monitoring the tip-sample current as the bias is increased. The current recorded will show a rapid transition to saturation as the oxide irreversibly breaks down.

Initial AFM studies of the breakdown field of SiO_2 have indicated a proportionality between the breakdown voltage and the film thickness[4]. It has also been observed that the degradation of the oxide at breakdown causes large-scale topographic features. This destruction of the film over micron sized regions limits the spatial resolution of the technique if operated in an imaging mode.

Fowler Nordheim tunnelling

To assess the dielectric strength of the oxide film with improved reproducibility and better spatial resolution than simple breakdown field measurements described above, we have developed a current limited technique of Fowler Nordheim (FN) emission measurement. Using high speed control of the tip bias it is possible to monitor the tip-sample current and remove the bias when the current reaches a pre-determined threshold value. Fig 1 shows two current limited I/V spectra taken from 7 and 12 nm thick gate oxides using a gold tip. The current threshold is set at 40 pA, when the tip current reaches this value the bias is set to zero (within 5 μS). If the voltage ramp continues, or if the injected current damages the oxide excessively, catastrophic breakdown occurs. If, however, breakdown is avoided excellent agreement is observed between the theoretical FN current and the measured emission from the silicon interface. Emission from the AFM tip appears more difficult to understand, due to uncertainties in the nature of the tip-surface contact. With etched gold tips we observe fluctuations in the I/V spectra, presumably due to changes in the effective work function of the gold surface or the precise geometry of the AFM tip[5]. It has been reported elsewhere[6] that measurements of the work function of metal surfaces in air are extremely sensitive to mobile charges and external water films, and we believe that such contamination may also play a considerable part in the discrepancies between the observed and predicted behaviour of current injection from the AFM tip.

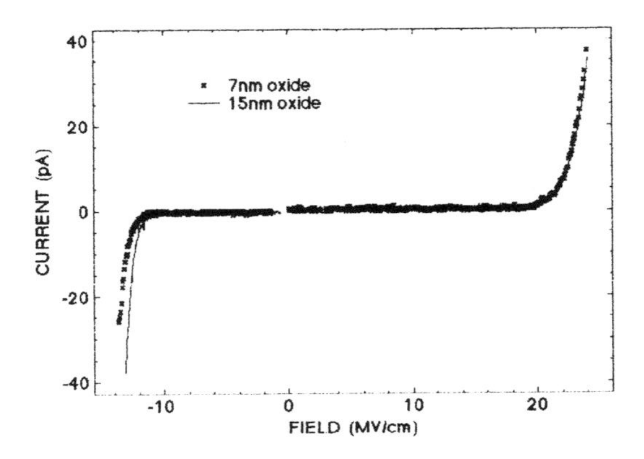

Fig 1. The Current limited I/V spectra from 7 and 12 nm gate oxides, using an etched Gold AFM lever and tip.

After repeated electronic stressing, eventually the oxide will undergo catastrophic breakdown. Values for the injected charge before breakdown vary, but we have measured between 2 and 20 C/cm^2 from AFM tips on gate oxides or using the AFM tip to contact to contact directly to the gate electrode of 0.4 micron MOS capacitors. Such values appear to be consistent with conventional measurements of large scale devices[7].

In addition to the electronic stressing of the oxide layers, field enhanced growth of material under the tip may occur during the measurement. It has been shown that the application of even relatively low voltages to AFM tips on silicon surfaces in air can produce locally enhanced oxidation[8] and this will certainly affect repeat measurements on silicon surface with thin oxides. Figure 2 shows an attractive mode topographic AFM image of an epitaxial grown silicon surface after two current limited I/V measurements of opposite polarities. The image shows the individual atomic terraces of the silicon substrate, and two raised features generated by the measurements. The heights of the features are 1.2 and 1.8 nm for the tip and substrate injection respectively. This is comparable to the thickness of the native oxide and demonstrates the problems involved with acquiring multiple spectra from single areas of oxide. It has been reported that oxide growth occurs for positive sample bias relative to the tip, the speculated oxidation mechanism being field enhanced migration of oxygen from the tip to the silicon interface[8]. We observe a build-up of insulating material below the tip with either sample bias provided a current flow occurs. This is inconsistent with the field enhanced diffusion

Fig 2. Attractive mode AFM topography of Epitaxial Si (100)with native oxide, after Current Limited I/V spectroscopy. Position A: Tip negative, position B: Tip Positive.

of negative oxygen species alone and the composition of the grown material is currently being studied in an attempt to clarify the formation mechanism in more detail.

Current limited I/V spectra may be taken at single points on the oxide surface, as demonstrated above. We have extended this process, to record the bias required to generate a pre-determined threshold current at each pixel of a scanned image[9]. Specifically at each imaging point the AFM feedback loop allows the tip to image the topography at constant force as in conventional contact AFM. At this point the feedback loop is frozen and the tip bias increased until the threshold current has been reached. The bias is returned rapidly to zero and the topography, force offset, tip bias at threshold and the corresponding current are recorded. The feedback loop is enabled and the AFM moves

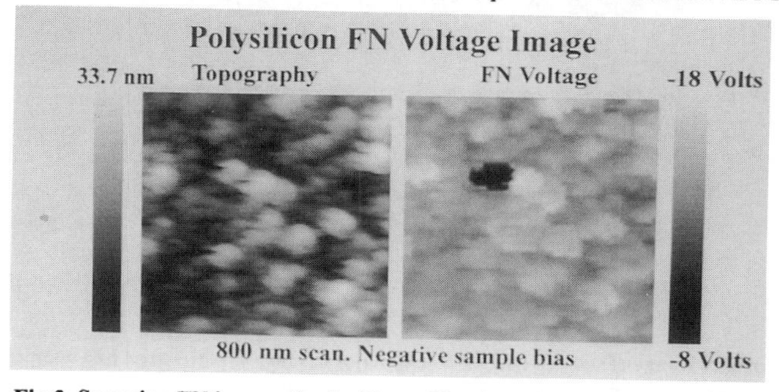

Fig 3. Scanning FN image of polysilicon. The threshold current was 25 pA

the sample to the next pixel. Fig 3 shows a polysilicon surface imaged in this manner using a diamond coated silicon tip. The individual grains are imaged in the topography, and the corresponding FN voltage map shows a contrast between individual grains, presumably arising from changes in the oxide thickness or barrier heights from grain to grain.

To illustrate this imaging technique on submicron devices, Fig 4 shows an array of 0.4 micron MOS capacitors with a 7 nm thick gate oxide. The corresponding voltage map has a number of features of note. In the square in the top right corner, the capacitors have a consistently higher threshold voltage than on the other side of the image. These capacitors have been previously stressed with repeat threshold current measurements, and subjected to approximately 2C/cm^2 of injected charge. Also present in the voltage image is a slight increase in the threshold voltage applied to the silicon surface in between the MOS devices, in comparison to the surface of the unscanned area. This is due to the growth of insulating material on the surface during the electrical measurement. Lastly, a single capacitor may be seen at position (A) which has a threshold voltage substantially lower than the surrounding devices. This capacitor has undergone a low field breakdown and close analysis of the device topography indicates a slightly abnormal physical structure.

Capacitive techniques:

From equation 1. for high quality dielectric films, as described above, the V/R term may be neglected, as may the FN component for low fields (below 5 MV/cm in our measurements the FN term is immeasurably small). We are left with capacitive coupled current terms, both of which may be measured with high spatial resolution at low bias and without the problems associated with current flow through dielectrics. In particular the displacement current caused by capacitive variations can provide information analogues to CV measurements in conventional systems, and the electrostatic forces on the tip which arise from these interactions provides information on the local contact potential of the surface.

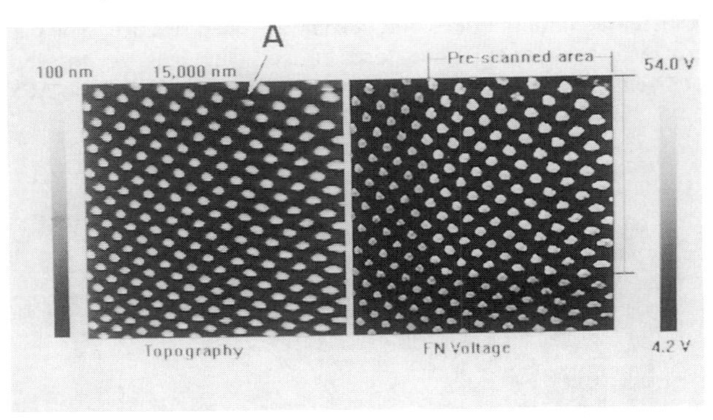

Fig 4, Scanning FN Image of 0.4 micron MOS capacitors. The top right area has been scanned previously, and each capacitor received ~2C/cm^2 of injected charge

Scanning Capacitance probes.

These techniques are based on modulating the tip/sample bias at high frequency and measuring the displacement current, in precisely the same way as CV measurement of MOS devices.

$$I_{tip} = C\ dV/dT. \qquad (2)$$

In Conducting AFM, the tip provides an upper capacitor electrode with an effective electrode area as small as 10 nm in diameter. Thus on a native oxide with 2 nm of separation between the tip and the silicon interface the tip-surface capacitance is about 10^{-17} F. However, for a conventional AFM cantilever (200 microns long, 20 microns wide and 5 microns above the substrate surface), the capacitance between the lever and substrate will be of the order of 10^{-14} F. The parasitic capacitive coupling between the surface and cantilever is thus considerably greater than that from the apex of the tip to the surface and measurements of the capacitance between the entire AFM probe and the surface will lack spatial resolution.

It is possible to screen the AFM probe with an earthed shield, only allowing the probe end to be unscreened from the surface. The small spatial dimensions of the AFM tip tend to make this fabrication difficult, but scanning capacitance microscope systems based on this technique have been successfully demonstrated, with quoted sensitivities of 10^{-21}F and lateral resolutions of 10 nm[10].

An alternative approach is to measure the entire capacitance, using sensitive C/V measurement electronics and modulate the depletion layer of the silicon under the tip with a slowly varying applied bias. The oscillation of the depletion layer provides a differential capacitance signal arising from depletion under the AFM tip, but it is relatively insensitive to the constant capacitance between the cantilever and sample surface. These measurements require conventional conducting AFM probes, and highly sensitive modulated CV measurement electronics. Spatially resolved quantitative CV measurements may be obtained, and lateral dopant profiles extracted using this technique with a resolution of 100 nm or better[11,12].

Kelvin Probe

Conventional (macroscopic) Kelvin probe techniques measure work function and local charge densities by detecting the displacement current induced on an oscillating probe placed in close proximity to a surface[13]. In this case

$$I_{kel} = V\ dC/dT.$$

In conducting AFM, the Kelvin probe measurements may be implemented with a cantilever oscillating above the surface.

Conventional attractive mode AFM imaging controls the tip-surface separation by controlling on the amplitude of oscillation of a modulated cantilever in the attractive force gradient above the surface. If a DC bias is applied between the tip and surface a displacement current may be observed from the tip. For a conventional, coated silicon nitride cantilever, we expect about 1 pA of current per volt of tip bias, at a driving frequency of 100kHz. As with most non contact techniques, both the lateral resolution and the sensitivity improve with decreasing tip-surface separation, but a practical limit is reached due to operating stability when the tip touches the surface.

As an alternative to the detection of displacement current, an improved sensitivity may be achieved by modulating the AFM tip with an oscillating sample bias and separately detecting the various electronic forces on the cantilever[6,14-19]. This technique is an elegant solution to the problems of high sensitivity and high bandwidth measurements of local charge and work function. A conducting tip is held close to the surface to be analysed, and an oscillating bias applied between the two with a frequency f. The electrostatic interactions between the tip and surface apply two forces to the AFM tip at frequencies f and 2f. The fundamental frequency component arises from the electrostatic force between charges on the AFM tip and the surface, this includes differences in the work function of the materials and the local charge of the sample, collectively referred to as the Contact Potential Difference (CPD). The 2f component derives from the force between two effective capacitor plates: the sample surface and the tip. Using lock-in amplification it is possible to detect the two frequency components independently, providing contact potential and tip-sample separation signals respectively. To fully separate the signals it is usual to apply a nulling bias, such that the contact potential signal is zeroed. We achieve this by applying a feedback voltage to the tip which is controlled from the "f detecting" lock-in amplifier. This ensures the quantitative measurement of the CPD voltage directly from the applied nulling bias. The AFM topography control is achieved using the digital feedback loop to hold the tip-surface separation constant, using the "2f" signal as a reference for the tip-surface separation. Topographic resolution of the order of 0.1 nm vertically and 20 nm laterally is achievable, with a contact potential sensitivity better than 5 mV. We have applied this technique to the measurement of contact potential difference on cross-sectioned multiple layer III-V compound structures, and the direct imaging of charge on sub micron MOS structures.

Fig 5 shows the topography and CPD images acquired simultaneously from four submicron sized MOS capacitors on a 7 nm gate oxide. Prior to this image being acquired, the capacitor in the top right of the image has been charged to +2 volts with the AFM probe and the top left capacitor charged to -2V. The topography of all four devices is essentially identical and shown in

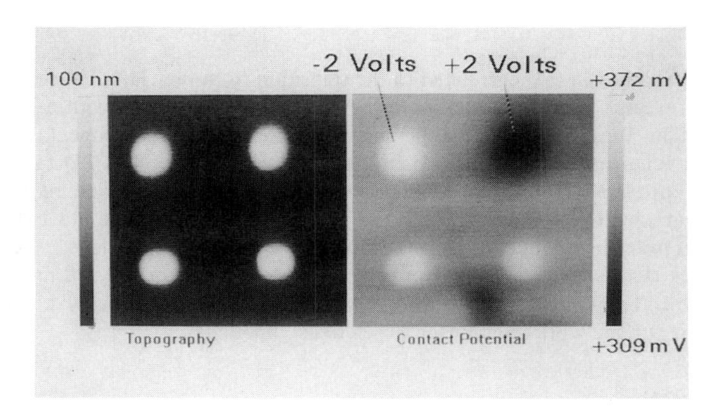

Fig 5. Scanning CPD image of 0.8 micron MOS capacitors, previously charged with the AFM probe. The voltages applied to the individual capacitors are indicated

the left hand image. The two uncharged capacitors appear to have a CPD signal slightly higher than the silicon substrate surface, the positively charged capacitor has a very distinct CPD image, appearing as a dark patch some 28 mV less positive than the silicon surface potential. The negatively charged capacitor appears 15mV more positive than remaining capacitors. This charge image appeared approximately constant over a period of 1 hour, indicating leakage from the capacitors in this case was small.

Fig 6 shows the topography and CPD images from a cleaved GaAs/AlAs epitaxial multiple layer structure. The bulk substrate was undoped, with an initial 500 nm layer of GaAs grown on it, followed by 50 alternating 100 nm layers of GaAs and AlAs, all grown layers were doped n^+ with 10^{18} Si atoms cm^{-3}. When these samples are cleaved in air, the AlAs layers relax slightly, resulting in a corrugated topography, with 1 nm height variations. The bulk crystal and the doped initial GaAs layers appear flat. The CPD image shows a change

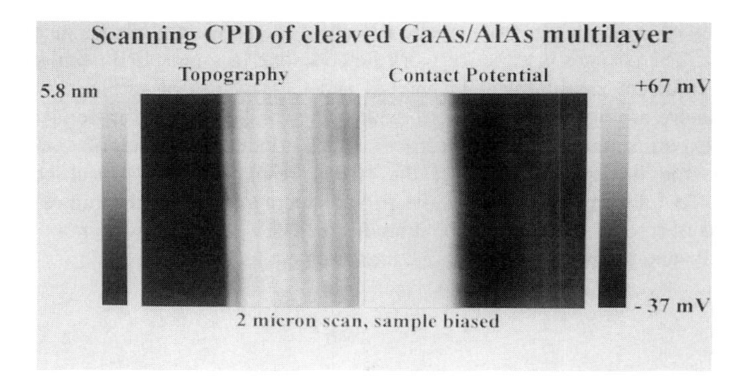

Fig 6. Scanning CPD image from cleaved GaAs/AlAs multiple layered structure.

in voltage by nearly 100 mV correlated to the transition from the GaAs to the layers. The CPD image reveals ~5 mV of structure in the CPD image correlated with each alternate layer of the film. We have confirmed the absence of topographic components in the CPD images by coating the cleaved surface with gold and re-imaging. No CPD features were seen which correlated to the individual layers, despite the preservation of the overall surface topography. Of particular interest in Fig 6 is the lack of change in CPD signal on the transition from the undoped GaAs substrate to the doped 500 nm layer. We attribute this to surface electronic states, which effectively mask both the doped and undoped bulk band structure. The contrast in the AlAs/GaAs CPD image is probably a function of changes in the surface work functions between the compounds.

CONCLUSION.

We have described a range of conducting AFM techniques applied to semiconductor surfaces and device characterisation. Conduction techniques may be used to detect defective, conducting regions of gate oxide, but such features are rare in high quality processes. Scanning F/N techniques have been used to image the relative tunnelling properties of the oxide layers on polysilicon grains and the uniformity of 0.3 micron MOS capacitors. These techniques are semi-reproducable, and provide high spatial resolution relative measurements of dielectric properties.

Using scanning Contact Potential Difference techniques we can image the charge stored on individual sub micron MOS capacitors and the changes in effective work function of epitaxial layered structures in cross-section. In all of these cases, the high spatial resolution of the AFM allows the topographic structure to be imaged, and the integration of conducting cantilever and associated electronics provides highly localised electronic measurements to be made, correlated with the topography. This combination represents a unique diagnostic for the miniaturisation of semiconductor devices in the future.

The presence of atmospheric contamination in all of the above techniques must be taken into account. The changes in effective work function due to atmospheric exposure affects both the FN current measurement techniques and CPD values of surfaces. Oxygen and surface moisture are likely candidates to explain the local growth of material during FN tunnelling experiments on thin dielectrics, and conducting films of contamination surrounding the tip apex may affect the lateral resolution of local conductance techniques. To fully exploit the potential for conducting AFM in the future, it will be necessary to overcome these effects, and studies in UHV are underway in an attempt to produce truly quantitative, non destructive measurement techniques.

Acknowledgements:

We are indebted to S.J. Brown and Dr. G.A.C. Jones for the supply of the III-V layered material and to P. Niedermann for the diamond coated levers. We wish to thank Mark Heyns, Michel Depas and Antonio Rotondaro and colleagues for the fabrication of the MOS structures and assistance with the interpretation of the data. This work was funded in part by the ESPRIT "PRONANO" research project, number 8523.

REFERENCES:

1) T.M.H. Wong and M.E. Welland, Meas. Sci. Technol. 4 (1993).

2) C.J. Sofield, M.P. Murrell, S.Sugden, M. Heyns, S. Verhaverbeke, M.E. Welland, B. Golan and J. Barnes. MRS Spring Meeting Proc. **259** 3, San Francisco CA, 1992

3) S. Verhaverbeke, M. Meuris, M.M. Heyns, R.F. De Keersmaecker, M.P. Murrell and C.J. Sofield, Electrochemical Soc. Proc **92** 187 1991

4) Y. Fukano, Y. Sugawara, S. Morita, Y. Yamanishi and T. Oasa, Extended abstracts 1992 SSDM Conference, p117-119, Tsukuba, Japan.

5) S.J. O'Shea, R.M. Atta, M.P. Murrell and M.E. Welland, Submitted to J.Vac. Sci. Tech.

6)N. Nonnenmacher, M. P. O'Boyle and H. K. Wickramasinghe. Appl. Phys. Lett. **58** (25) 1991.

7) D.J. Di Maria, Extended abstracts 1993 SSDM Conference,, Makuhari, Japan.

8) M. Yasutake, Y. Ejiri and T. Hattori, Jpn. J. Appl. Phys. Let. 32 (7B).

9) M.P. Murrell, M.E. Welland, S.J. O'Shea, T.M.H. Wong, J.R. Barnes, A.W. McKinnon, M. Heynes and S. Verhaverbeke, Appl. Phys. Lett. **62** (7) 1993.

10) S. Lanyi, J. Torok and P. Rehurek, Rev. Sci. Instrum. **65** (7) 1994

11) Y. Huang, C.C. Williams and J. Slinkman. Appl. Phys. Lett. 66 (3) 1995

12) Y. Huang, C.C. Williams, J. Vac. Sci. Technol. B 12 (1) 1994

13) Ian D. Baike PhD. Thesis, University of Twente 1988.

14) B.D. Terris, J.E. Stern, D. Rugar and H.J. Mamin, J. Vac. Sci. Technol. **A8** (1) 1990

15) B.D. Terris, J.E. Stern, D. Rugar and H.J. Mamin, Phy. Rev. Lett. **63** (24) 1989.

16) J.E. Stern, B.D. Terris, H.J. Mamin and D. Rugar, Appl. Phys. Lett. 53 (26) 1988

17) C. Schonenberger and S.F. Alvarado Phy. Rev. Lett. **65** (25) 1990.

18) Y. Martin, D.W. Abraham and H.K. Wickramasinghe, Appl. Phys. Lett. **52** (13) 1988

19) A.K. Henning, T. Hochwitz, J. Slinkman, J. Never, S. Hoffmann, P. Kaszuba and C. Daghlian, J. Appl. Phys. 77 (5) 1995.

ANALYZING ATOMIC FORCE MICROGRAPHS USING SPECTRAL METHODS

SAMEER D. HALEPETE, H. C. LIN, SIMON J. FANG AND C. R. HELMS,
Stanford University, CA 94305

ABSTRACT

Microroughness is a critical parameter in ULSI device interface reliability and has been shown to effect several critical MOS electrical properties. The atomic force microscope (AFM) has become the instrument of choice for silicon surface microroughness analysis. The parameters usually specified to characterize roughness are average and root mean square roughness. However, these parameters are spatial averages and can have the same value for two significantly different surfaces. Spectral analysis using the Fast Fourier Transform (FFT) has been applied as a powerful tool to analyze AFM data by looking at roughness as a function of spatial wavelength. The Fast Hartley Transform, being a real transform, is faster than the FFT and is better suited for this analysis. It has been used here to derive spectral information from the AFM height data. Before evaluating the transform, cancellation of any tilt or warp in the AFM data is done to remove frequency components which interfere with other spectral information. A PC-based computer program to determine the transform and its magnitude will be described. The application of this method to analyze data from Si and SiO_2 surfaces as a function of pre-oxidation cleaning chemistry will be presented. Significantly better insight into the spatial distribution of roughness is obtained, when compared to previous implementations.

INTRODUCTION

The trend of decreasing gate oxide thickness in MOS devices puts new restrictions on Si/SiO_2 interfacial microroughness due to its effect on the electrical properties of thin films[1,2]. The dielectric breakdown electric field intensity (E_{BD}) for oxide has been shown to increase with a decrease of surface microroughness[3]. The charge-to-breakdown (Q_{BD}) increases rapidly with a decrease in interfacial roughness[3]. The channel mobility (μ_{eff}), can be maintained at a very high level by suppressing the interface microroughness, since this prevents the carriers from being scattered[4,5].

Several methods have been used to characterize microroughness on the silicon surface[6]. The Atomic Force Microscope however is fast becoming the instrument of choice for silicon surface metrology. Improvements in AFM technology have led to better accuracy in the measurements and resolutions of up to a few hundredths of a nanometer have been reported. State-of-the-art AFM's have a lateral resolution of approximately 10 nm to 10 μm and a noise floor of around 0.1 nm. AFM measurements performed in ambient have been shown to measure the true silicon surface roughness, in spite of the presence of native oxide[7]. The standard parameters specified for microroughness from the data obtained using AFM are average roughness (R_a) and RMS roughness which are defined as follows

$$R_a = (1/N)\sum_{i=1}^{N}\left|z_i - Z_{av}\right| \tag{1}$$

$$RMS = \left[(1/N)\sum_{i=1}^{N}\left(z_i - Z_{av}\right)^2\right]^{1/2} \tag{2}$$

where z_i is the height of the i'th data point and Z_{av} is the average value of z_i over all i's. However, both these methods have information about height and no spatial information. Therefore they could have different values depending upon the area chosen to do the measurement. Also, the

Mat. Res. Soc. Symp. Proc. Vol. 386 © 1995 Materials Research Society

values of these parameters could be same for two very different surfaces. For instance, the two types of surfaces shown below in Fig.1, one where a particle is modeled in one dimension and another a surface with a wavy nature, could have the exact same R_a. However, the two types of surfaces have different properties and must be distinguished clearly by the measure of roughness we use. Therefore, it is difficult to make any conclusions about microroughness by looking at either R_a or RMS roughness.

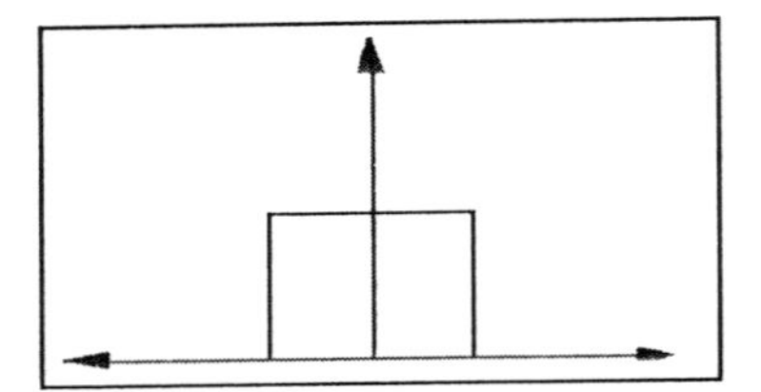

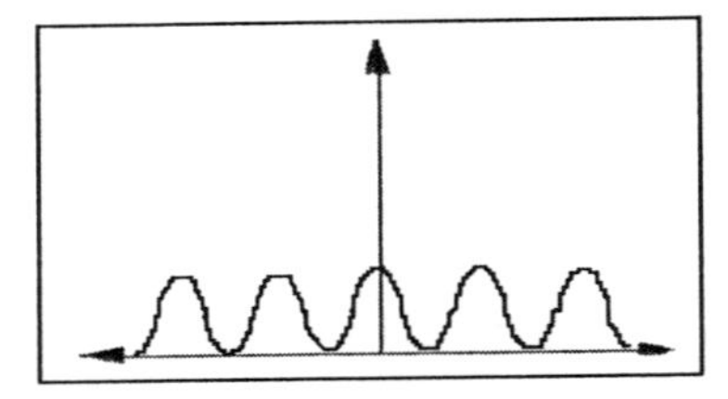

Fig.1 RMS roughness for a "particle" and a "wavy surface" could be same

The use of these measures of microroughness have lead to difficulties in correlating microroughness with both chemical parameters of cleaning chemistries and electrical properties of MOS structures.

SPECTRAL ANALYSIS OF ROUGHNESS

The problems with using the simple statistical descriptions for characterizing microroughness have lead to the introduction of more sophisticated methods of roughness analysis, spectral analysis being the most promising one. In this technique, a one-dimensional representation of RMS roughness of the surface is plotted against spatial frequency. This technique was proposed independently by Y.Strausser et al.[8] and P.Dumas et al.[9]. Both these groups use the FFT to do the spectral analysis. We believe that the Fast Hartley Transform is better suited for this application as explained below.

The Fast Hartley Transform

The two-dimensional Fast Hartley Transform (FHT) was proposed by Bracewell[10] in 1986. It is a real transform, i.e. real data gets transformed into real data in the Hartley domain, unlike the Fourier transform, where the complex exponential causes real data to get transformed into a real and an imaginary part in the Fourier domain. The FHT is twice as fast as the FFT. There is a simple transformation that relates the FHT to the FFT and can be used if the FFT is required. The Hartley transform is defined by:

$$H(u,v) = \sum_{x=0}^{M-1}\sum_{y=0}^{N-1} z(x,y) cas\left[2\pi\left(ux/M + vy/N\right)\right] \quad (3)$$

where $cas\theta = \sin\theta + \cos\theta$.

The 2D FHT gives us a 2D transformed version of the AFM data. It is difficult to interpret roughness features from this 2D data. Therefore, it is more meaningful to derive another representation of the same data, from which it is easier to extract roughness parameters. The standard method of doing this is to evaluate the one-dimensional magnitude of the transform (MHT). This can be done in a number of different ways. The usual method used to do this is to find the angular average of the 2D power spectrum, with the radius representative of the spatial frequency. This is represented in the discrete domain in the form of a summation as follows

$$MHT(k_r) = \left(\frac{2\pi}{L^3} \sum_{n=0}^{N_r} \left| H(k_x, k_y) \right|^2 \right) \tag{4}$$

$$k_r^2 = k_x^2 + k_y^2$$

This process is shown schematically in Fig.2(a). The center of the 2D transform is chosen as the reference. For every circle of radius k_r ('distance' in the frequency space is spatial frequency) in the frequency domain, points that lie on that circle are chosen. The square of the transform at these points is then averaged over all these points and this average is plotted against the spatial frequency k_r. The plot of this average versus spatial frequency is the MHT obtained using radial averaging. We also use two other methods to calculate the power spectrum: averaging along the x-axis and averaging along the y-axis. Averaging along x-axis is also shown schematically in Fig.2(b). The equation for these calculations are very similar to those for the radial average case. The center of the 2D transform is again chosen as the reference. The averaging is performed over all the points that lie k_x away to either side of the y-axis. This becomes the MHT of at the spatial frequency k_x. By looking at the three together, we can obtain the orientation of roughness features. Care must be taken to include the physical units in the transform. Both the FHT and MHT have units of length^3.

CANCELLATION OF WARP AND TILT

A precaution that needs to be taken while doing spectral analysis of roughness is to cancel out any warp, tilt or DC shift that might be present in the data. The MHT in the presence of any of these would not give us any extra information about the microroughness but add a lot of spectral components that interfere with those we are interested in. It is therefore important to cancel these out before evaluating the MHT. To do this, we make an assumption that it is enough to cancel the warp up to second-order. Corrections to the fourth-order are common, but they might remove real information[11]. We fit a quadratic surface of the form shown below, to the height data using the method of least squares and then subtract it from the original data.

$$z_{fit}(x, y) = ax^2 + by^2 + cxy + dx + ey + f \tag{5}$$

The corrected data would have much less effect of warp and tilt on the MHT.

COMPUTER PROGRAM FOR MICROROUGHNESS ANALYSIS

It is clear from the description of the analysis method that it involves significant data processing. We have developed a computer program written in Borland C++ 4.0 to do this analysis. It is designed to run under the Windows operating system on a PC. The program has a graphical user interface (GUI) that helps the user display the AFM image, analyze data, view results and store data. It loads the AFM data from a binary file format. It cancels any warp or tilt by the quadratic fit as described before. It evaluates the FHT and the MHT using the three averaging methods described earlier.

APPLICATIONS

The computer program described above can be conveniently used to demonstrate the power of the spectral analysis method. The MHT clearly distinguishes between two rough surfaces having different types of roughness. We have used it to look at various surfaces[11], some of which we will discuss. Consider the example of a wavy surface and a particle shown in Fig.1, which have the same R_a. Fig.3 shows how the roughness looks in the frequency domain. The MHT of the wavy surface has a peak at the frequency of the undulations. The "particle" however, has a $sinc^2$ type of an MHT, which is very different. The radial averaging method has been used here to evaluate the MHT. In both cases, the data is "artificial", i.e. it has been created numerically. We now present experimental results for two different cases: BOE etched samples and oxidized silicon samples.

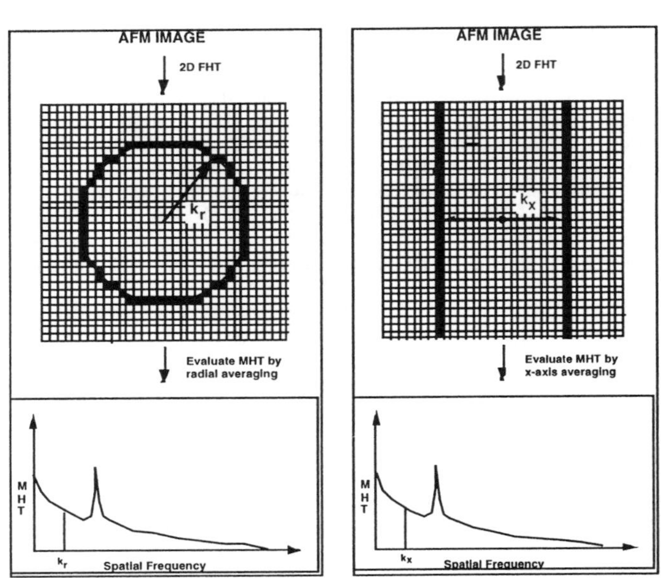

Fig.2 Evaluation of MHT using (a) radial and (b) x-axis averaging

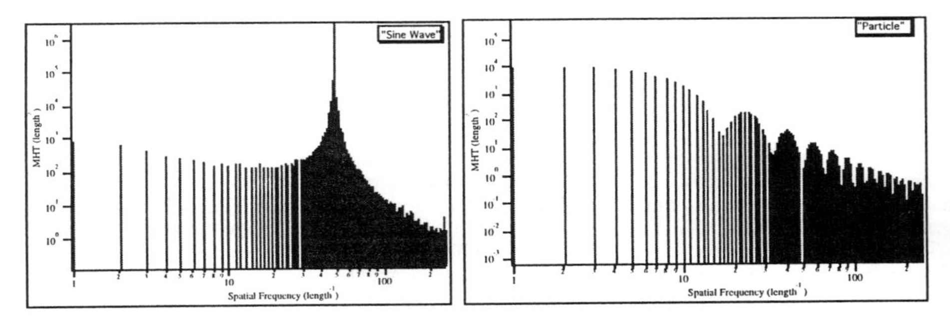

Fig.3 MHT of a wavy surface and a particle on a surface

BOE Etched Silicon

AFM scans of 0.5μ by 0.5μ and 5μ by 5μ scan sizes were performed on unprocessed silicon and silicon cleaned using 20:1 NH_4F:HF for 15 minutes. MHT spectra of the above mentioned scans are shown in Fig.4. We see the 15 minute BOE clean increases the roughness at almost all spatial frequencies. An important observation here is that the MHTs for the two scans lie on top of each other for most of the region of overlap. By performing scans of different scan-sizes, we can be confident about the data in the regions of overlap.

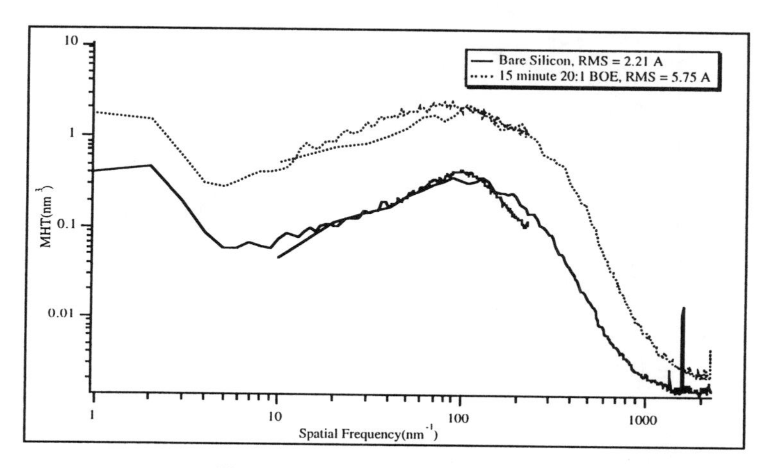

Fig.4 MHTs for BOE etched silicon

Oxidized Silicon

The effect of three different cleans: HF last (50:1 H_2O:HF), SC2 last (5:1:1 H_2O:HCl:H_2O_2, 70°C) and BOE (20:1 NH_4HF:HF, 15 minutes), on the microroughness of the oxide formed on the cleaned silicon was studied in this experiment. AFM scans (1μm by 1μm) of 6nm thick oxidized silicon were performed for each of the cleans. The MHTs of these three scans are shown in Fig.5. We see that the SC2 last clean produces the smoothest oxide surface of the three. The HF-last is rougher than the SC2 last in the lower spatial frequencies. The BOE oxide surface is the roughest and is uniformly rougher over most spatial frequency regions.

CONCLUSIONS

We have developed a methodology for silicon surface microroughness analysis, demonstrating the advantages of using the spectral technique of roughness analysis. We have shown the effectiveness of this technique for actual AFM data for silicon. With this analysis tool, a better understanding of the interrelationship between microroughness and cleaning chemistries and electrical parameters can be obtained.

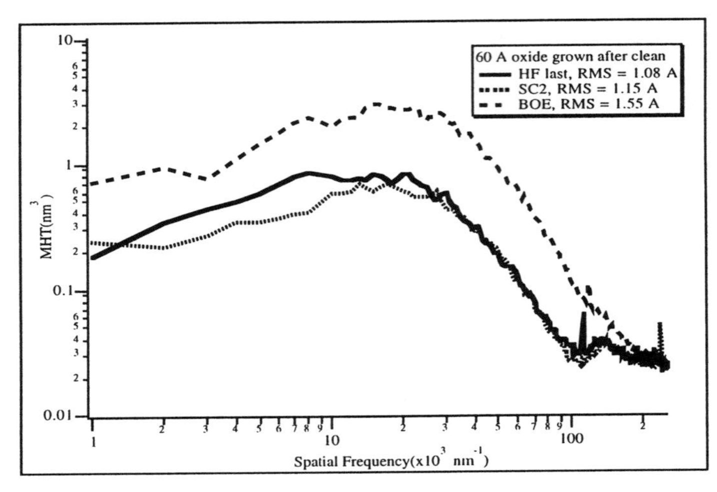

Fig.5 MHTs for oxidized silicon surface with different cleans

ACKNOWLEDGMENTS

This work was supported by SRC contract SJ-350. We would like to thank the Center for Materials Research, Stanford University for use of the Nanoscope III AFM and Prof. Bracewell of Stanford University for his help with the Fast Hartley Transform.

References

1. B. B. Triplett, Proceedings of the Seventh International Symposium on Silicon Material Science and Technology : Semiconductor Silicon 94, 333, 1994.
2. T. Ohmi, Proceedings of the IEEE, **81**, 716, 1993.
3. T. Ohmi, M. Miyashita, M. Itano, T. Imaoka, and I. Kawanabe, IEEE Trans. on Electron Devices, **39**, 537, 1992.
4. T. Ohmi, K. Kotani, A. Teramoto, and M. Miyashita, IEEE Electron Device Lett., **12**, 652, 1991.
5. J. Koga, S. Tagaki and A. Toriumi, Technical Digest of the IEDM meeting, San Francisco, CA, 18.6.1, 1994.
6. A. C. Diebold and B. Doris, Surface and Interface Analysis, **23**, 127, 1993.
7. Mark R. Rodgers, Yale E. Strausser and Kevin J. Kjoller, Presented at the Particles, Haze and Microroughness on Silicon Wafers, San Jose, CA, 1994.
8. Y. E. Strausser, B. Doris, A. C. Diebold, and H. R. Huff, presented at the 185th meeting of the Electrochem. Soc., San Francisco, CA, 1994.
9. P Dumas, B. Bouffakhreddine, C. Amra, O. Vatel, E. Andre, R. Galindo, and F. Salvan, Europhys. Lett., **22**, 717, 1993.
10. R. N. Bracewell, O. Buneman, H. Hao, and J. Villasenor, Proceedings of the IEEE, **74**, 1283, 1986.
11. Sameer D.Halepete and C.R.Helms, to be published.

IR AND MW ABSORPTION TECHNIQUES FOR BULK AND SURFACE RECOMBINATION CONTROL IN HIGH-QUALITY SILICON

A. KANIAVA[*,**], U. MENCZIGAR[***], J. VANHELLEMONT[**], J. POORTMANS[**], A.L.P. ROTONDARO[**], E. GAUBAS[*], J. VAITKUS[*], L. KÖSTER[****], D. GRÄF[****]
[*]Vilnius University, Sauletekio 10, 2054 Vilnius, Lithuania
[**]IMEC, Kapeldreef 75, B-3001 Leuven, Belgium
[***]Institute of Semiconductor Physics, Walter-Korsing Straße 2, 15230 Frankfurt (Oder), Germany
[****]Wacker-Chemitronic GmbH, P.O. Box 1140, D-84479 Burghausen, Germany

ABSTRACT

The carrier recombination rate in high-quality FZ and Cz silicon substrates is studied by contactless infrared and microwave absorption techniques. Different surface treatments covering a wide range of surface recombination velocity have been used for the separation of bulk and surface recombination components and evaluating of the efficiency of passivation. Limitations of effective lifetime approach are analyzed specific for low and high injection level. Sensitivity limits of the techniques for iron contamination are discussed.

INTRODUCTION

Nondestructive pump-probe techniques for recombination lifetime measurement have been approved as a powerful tool for the characterization of starting material and wafer processing as the carrier lifetime is one of the most sensitive parameter for material contamination. Microwave (MW) absorption/reflection and infrared (IR) absorption are widely used methods for the contactless detection of photoconductance (PCD) transients and nonequilibrium carrier concentration, respectively. The most conventional approach for the extraction of the recombination parameters is based on the analysis of the asymptotic part of excess carrier decay and its fitting by the exponential function with time constant, so called "effective lifetime". However, the interpretation of this effective lifetime is not straightforward as it is always composed of two components, namely the bulk and surface lifetime. The bulk lifetime τ_b is often the quantity of interest as it is sensitive to bulk defects and impurities. The surface lifetime τ_s or, actually, the surface recombination velocity (SRV) is a measure of the quality and cleanliness of the wafer surface. The in-line control of both surface and bulk quality is essential for the ultraclean silicon processing [1].

Thermal oxidation or immersion in passivating chemicals is often used to reduce the surface effects. In high-quality material even with passivated surfaces SRV is not negligible, and the simultaneous determination of the complete set of recombination parameters is required. The real situation, unfortunately, becomes even more complicated as different recombination mechanisms are possible depending on the material quality, doping and injection level. In that case only the combination of proper analysis of the signal shape and modeling of the carrier decay involving various recombination mechanisms can lead to the extraction of the recombination parameters.

Mat. Res. Soc. Symp. Proc. Vol. 386 © 1995 Materials Research Society

In this paper we address some problems related with recombination lifetime measurement in high-quality material, effectiveness of the surface passivation treatments and sensitivity limits of IR and MW techniques for the detection of low level iron contamination.

THEORETICAL CONSIDERATIONS

The theoretical analysis of PCD transients is based on the solution of the continuity equation for excess carriers with appropriate boundary and initial conditions. The most comprehensive analysis of the linear recombination model was developed by Luke and Cheng [2]. The decay of photoexcited carriers is governed by a multiexponential function with the characteristic decay constants τ_n given by

$$\tau_n^{-1} = \tau_b^{-1} + \alpha_n^2 D , \tag{1}$$

D is the carrier diffusion constant and α_n are the roots of the transcendental equation:

$$ctg(d\alpha) = \frac{D\alpha}{2s} - \frac{s}{2D\alpha} . \tag{2}$$

The higher order modes (n>1) decay much faster with time than the fundamental mode (n=1) and in the asymptotic the decay of carriers is dominated by the fundamental mode $\tau_1^{-1} \equiv \tau_{eff}^{-1} = \tau_b^{-1} + \tau_s^{-1}$. Thus, the measured effective lifetime can be significantly different from the bulk lifetime, especially when the diffusion length of carriers reaches or even exceeds the thickness of wafer. The surface lifetime τ_s depends on the surface recombination velocity s, the sample thickness d and the diffusion constant D. Two limiting cases are usually considered: firstly, s is small ($s \ll 2D/d$), and therefore $\tau_s = d/2s$ and, secondly, large s ($s \gg 2D/d$) with $\tau_s = d^2/\pi^2 D$. In the case of identical SRV on both surfaces of the wafer $s_1 = s_2 = s$, due to the symmetry of the carrier diffusion flows, τ_{eff} can be approximated by

$$\frac{1}{\tau_{eff}} = \frac{1}{\tau_b} + \left(\frac{d}{2s} + \frac{d^2}{\pi^2 D} \right)^{-1} . \tag{3}$$

The accuracy of this approximation is better than 5% for the complete range of s values [3].

At high excitation level the nonlinear Auger recombination mechanism starts to dominate the initial carrier decay and the assumption on the linearity of carrier recombination is no more valid. A method for the determination of recombination parameters has been developed based on the fitting of the experimental and theoretical decays. In this case there is no analytical solution and the nonlinear bipolar continuity equation is solved numerically:

$$\frac{\partial \Delta n}{\partial t} = D \frac{\partial^2 \Delta n}{\partial x^2} - \frac{\Delta n}{\tau_b} - C \Delta n^3 , \tag{4}$$

where $C = C_n + C_p$ - coefficient of band-to-band Auger recombination. The volume integrated carrier density $\langle \Delta n \rangle_d$ is then used to model the IR PCD transient for the respective injection level.

EXPERIMENTAL

High-quality FZ 4 inch wafers both n and p-type have been used for the determination of the measurement regimes and effectiveness of the surface passivation. One set consists of n-type FZ (111) both side lapped wafers with the thickness varying in the range of 200 to 685 μm. Another set are p-FZ (100), 7 Ωcm, 4 inch, 200 and 400 μm thick double side polished wafers. The measurements of τ_{eff} are carried out on the as-received wafers, oxidized in a dry O_2 ambient (1000°C, 2h) and afterwards annealed in forming gas (450°C, 25 min.) to improve the quality of oxide, after charging of both surfaces of the oxidized wafer by the corona-charging method [4], after the etching of the oxide and passivation in the 10% HF solution just before measurement and by immersion into an iodine-methanol 0.08 mol/l solution for in-situ passivation studies.

The recombination lifetime studies at high excitation level are performed on the p-FZ, 12 Ωcm, 4 inch double side polished wafers having different thickness in the range of 0.46 to 5 mm. To study the sensitivity of the techniques to iron contamination both at low and high excitation level p-Cz (100), 24-36 Ωcm, 5 inch single side polished 680 μm thick wafers were uniformly contaminated by iron from Fe spiked NH_4OH:H_2O_2:H_2O solutions. The resulting Fe surface concentration in the range of $8x10^9$ to $2x10^{12}$ cm^{-2} was driven into the bulk of the wafers by the annealing in dry O_2 (900°C, 30 min.). This treatment also provided good surface passivation by the SiO_2 layers.

The low level (LL) injection carrier lifetime with the density of photogenerated carriers of 10^{14} cm^{-3} is measured by MW reflection technique using 22 GHz Phoenicon GmbH instrument. For in-situ surface passivation in iodine-methanol solution a stainless steel cell with a glass cover plate is used. PCD transients at high excitation level (HL) with $\Delta n=10^{17}$ cm^{-3} are measured by the light-induced IR absorption technique [5]. In this case a decay of an additional absorption by free carriers of cw He-Ne laser probe beam with the wavelength λ=1.15 μm is detected. For both the MW and IR techniques a uniform generation of excess carriers within the bulk of the wafers is obtained by the illumination of the sample with 10 ns pulses of a Q-switched Nd:YAG laser. At each experiment 50-500 of transients are averaged depending on the signal/noise ratio. The effective lifetime is determined by the exponential function fitting of the asymptotic part of the decay.

RESULTS AND DISCUSSION

The lapped surfaces are characterized by a large SRV with $s > 10^5$ cm/s due to the rough surface. Even if the bulk lifetime in the material is expected to be high, τ_{eff} in the lapped wafers is completely governed by the surface recombination (SR), i.e. $\tau_{eff} \approx \tau_s$. From the plot of the average τ_{eff} measured by the MW technique as a function of the wafer thickness d (Fig. 1), one observes that $\tau_{eff} \sim d^2$. This indicates that for the lapped surfaces the SR is diffusion limited. Thus, the expression of $\tau_s = d^2/\pi^2 D$ can be used to extract the diffusion constant D of the minority carriers. The value D_p = 12.8 cm^2/s determined from the slope of τ_{eff} is in a good agreement with diffusion constant of holes in n-Si [6].

Although lapped wafers are useful when the minority carrier diffusion constant in the material is unknown, it is obvious that there is no possibility to get information on τ_b due to the prevalence of SR. To reduce the surface effects two pairs of double side polished p - FZ 200 and

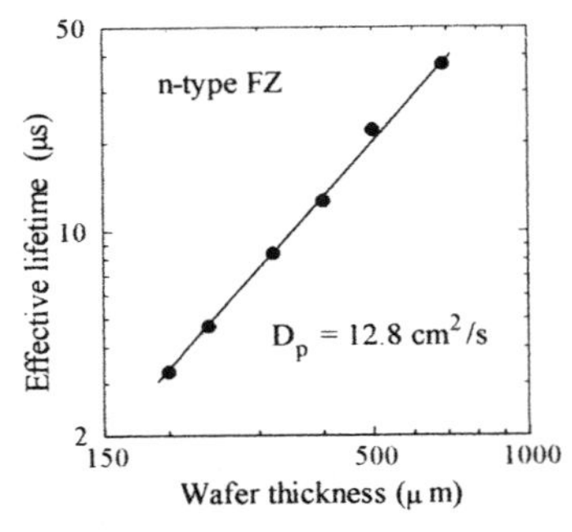

Fig. 1. Average τ_{eff} measured in n-type lapped wafers of different thickness

Table I. Average effective lifetime of as-received and passivated 200 and 400 μm thick wafers measured by MW reflection technique with bias light

Treatment	τ_{eff}^1, μs 200 μm	τ_{eff}^2, μs 400 μm	s, cm/s
as-received	1.9	6.3	$8x10^3$
oxidized	72	136	131
annealed oxide	133	181	40
HF dip	41	140	345
HF dip (no bias light)	3050	1530	-

400 μm thick wafers were subjected to different surface passivation treatments. The advantage of double side polished wafers is that polishing induced damage which degrades the carrier lifetime in the near surface region has been removed. The analysis of τ_{eff} measured on the different thickness wafers with symmetrical SR on both surfaces allows to evaluate s and τ_b using approximation (3).

The recombination rate in the as-received wafers without surface passivation is also purely dominated by the SR, s = $8x10^3$ cm/s (Table I). Surface passivation after thermal oxidation provided much higher τ_{eff} values, and s decreases to about 130 cm/s. The lowest SRV is achieved when the oxidized wafers are annealed in the forming gas. In this case s = 40 cm/s and bulk lifetime of $\tau_b = 280 \pm 30$ μs is obtained from the expression (3) with reasonable accuracy. As can be seen from Table I, removal of the oxide and passivation in diluted HF solution for 10 min. just before measurement remarkably deteriorates the surface quality (s = 345 cm/s).

Another pair of the thermally oxidized 200 and 400 μm thick wafers was used to investigate the influence of the surface potential on the variation of τ_{eff}. The surface potential is varied by the gradual increasing of negative charge in the oxide by corona charging. In the Fig. 2 τ_{eff} as a function of the surface potential measured by Kelvin probe is presented. The fixed oxide charge is removed by the immersion of the wafers into boiling ethanol for a few minutes. This procedure results in the strong decrease of τ_{eff} from 77 μs to 1.4 μs and from 102 μs to 5μs for the 200 and 400 μm thick wafers, respectively. Effective lifetime increases with surface potential and saturates at about -45 V for both wafers. The saturation values are quite reproducible after several charging/discharging cycles. One can conclude from the behaviour of τ_{eff} that the charge density in the oxide rather than thermal oxide itself determines the efficiency of the surface passivation. This assumption is further supported by the correlation which is observed between the distribution of the surface potential across the wafer and lateral variation of τ_{eff}. After etching the oxide, a 400 μm thick wafer was immersed into an iodine-methanol solution. The average τ_{eff} during in-situ passivation reaches 300 μs which is very close to the τ_b value obtained for the thermally oxidized wafers.

Care has to be taken for the correct interpretation of τ_{eff} measured at low injection level when carrier trapping occurs. Due to this effect the PCD transient breaks into two distinct parts, the slower of which is in the range of few milliseconds (Table I, HF dip, without bias light). The carrier trapping can be efficiently suppressed by background illumination of the sample with low

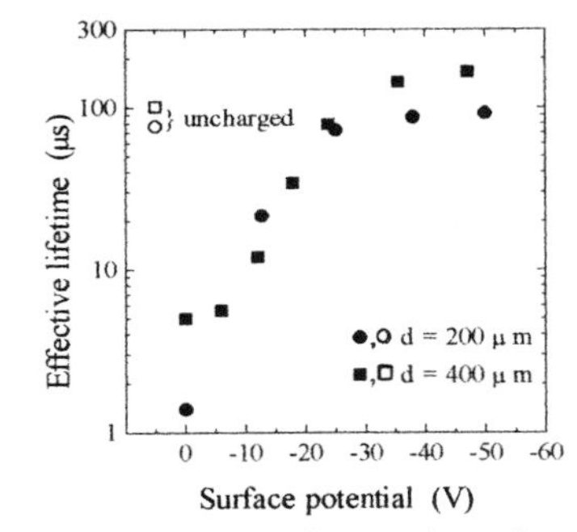

Fig. 2. Dependence of τ_{eff} on the surface potential after charging both oxide layers negatively

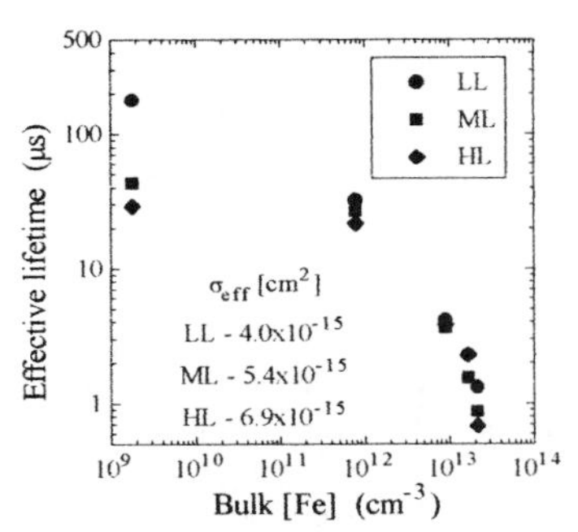

Fig. 3. Dependence of τ_{eff} on iron bulk concentration

intensity bias-light during the measurement. It can be eliminated also by increasing the excitation level when the density of excess carriers is large enough to saturate the carrier transitions to and from the trapping centers. However, at the high excitation level the influence of the nonlinear recombination mechanisms can not be avoided. The analysis of PCD transients as a function of the injection level reveals Auger recombination as the dominant mechanism at densities of photogenerated carriers Δn larger than 10^{17} cm^{-3}. The following recombination parameters have been determined by fitting the experimental IR PCD transients and theoretical decays taking into account Auger recombination: $\tau_b = 70$ μs, $s = 10^4$ cm/s, $D = 18$ cm^2/s and $C = 2\times10^{-30}$ cm^6/s. The obtained τ_b value is in good agreement with the upper limit of carrier lifetime reported by Häcker and Hangleiter [7] for carrier density of 10^{17} cm^{-3}.

Investigations of the recombination activity of iron-related defects in p-Cz wafers are performed to derive the sensitivity limits of MW and IR technique for low level Fe contamination at different excitation levels. Surface photovoltage (SPV) and deep level transient spectroscopy (DLTS) are used to address Fe content in the bulk and identify the deep levels introduced by iron related defects. DLTS analysis reveals a shallow donor trap at $E_v + 0.10$ eV related with the Fe_i-B_s pair. A clear correlation between the inverse trap density and the LL effective lifetime was found in the range of Fe concentration from 8×10^{11} to 2×10^{13} cm^{-3} [8]. In Fig. 3 τ_{eff} as a function of Fe bulk concentration (N_{Fe}) is plotted for low (10^{14} cm^{-3}), moderate (5×10^{15} cm^{-3}) and high (10^{17} cm^{-3}) level of injection. A linear decay of τ_{eff} with increasing N_{Fe} can be observed starting from 10^{12} Fe/cm^3. In this linear part the recombination activity of the Fe_i-B_s center can be characterized by an effective carrier capture cross-section σ_{eff}:

$$\sigma_{eff} = \left(\tau_b v_{th} N_{Fe}\right)^{-1} , \tag{5}$$

where τ_b is assumed to be equal to τ_{eff}, and v_{th} is the carrier thermal velocity. As clearly seen from the values of σ_{eff} given in Fig. 3, the recombination activity of Fe_i-B_s pairs increases with injection level. The linearity between the τ_{eff} and N_{Fe} is suppressed for lower contamination levels due to the dominance of surface recombination over the bulk recombination. The saturation value of τ_{eff} is limited by SRV which itself is the excitation level dependent. It can be estimated from (5) using $\sigma_{Fe\text{-}B} = 5\times10^{-15}$ cm^2 that Fe concentration of 10^{10} cm^{-3} corresponds with a τ_b of about 2 ms. To be able to extract such large bulk lifetimes from τ_{eff} in the wafers of standard thickness of 680 μm one should satisfy the requirement $\tau_b \leq \tau_s = d/2s$, and thus $s \leq 20$ cm/s. Alternatively, for the detection of surface contamination lower than 10^9 cm^{-2}, the carrier bulk lifetime should be larger than 1 ms. Both conditions put stringent demands on the purity of the starting material and on efficient surface passivation.

SUMMARY

The efficiency of various surface passivation treatments is analyzed by evaluation of the surface recombination rate from the effective lifetime measurements on wafers with different thickness. Although passivating with SiO_2 layers ensures low surface recombination velocity, the charge density in the oxide should be carefully controlled to get reliable lifetime values. Surface charging seems to be the most suitable technique to stabilize the oxide state. A method for the determination of the recombination parameters at high excitation level ($\Delta n \approx 10^{17}$ cm^{-3}) has been proposed by taking into account also nonlinear Auger recombination. To extend the range of the linear correlation between carrier lifetime and iron concentration up to 10^{10} Fe/cm^3 high purity material with $\tau_b > 2$ ms and perfect surface passivation with $s \leq 20$ cm/s is required.

References

1. W. Bergholz, G. Zoth, F. Gelsdorf, and B. Kolbesen in Defects in Silicon II, edited by W.M. Bullis, U. Gösele, and F. Shimura (The Electrochem. Soc. Proc. **91-9**, Pennington, NJ, 1991) pp. 21-39.
2. K. Luke and L. Cheng, J. Appl. Phys. **61**, 2282 (1987).
3. A.B. Sproul, J. Appl. Phys. **76**, 2851 (1994).
4. M. Schöfthaler, U. Rau, G. Langguth, M. Hirsh, R. Brendel, and J.H. Werner in Proc. 12th Europ. Photovolt. Solar Energy Conf. (Amsterdam, the Netherlands, 1994) pp. 533-536.
5. J. Vaitkus, E. Gaubas, K. Jarasiunas, and M. Petrauskas, Semicond. Sci. Technol. **7**, A131 (1992).
6. J. Linnros and V. Grivickas, Phys. Rev. B, **50**, 16 943 (1994).
7. R. Häcker and A. Hangleiter, J. Appl. Phys. **75**, 7570 (1994).
8. A. Kaniava, A.L.P. Rotondaro, J. Vanhellemont, E. Simoen, E. Gaubas, J. Vaitkus, T.Q. Hurd, P.W. Mertens, C. Claeys, D. Gräf in Proc. Ultra-Clean Processing of Silicon Surfaces (UCPSS '94) edited by M. Heyns (Acco Leuven, Amersfoort, 1994) pp. 197-200.

SILICON SURFACE CHEMISTRY BY IR SPECTROSCOPY IN THE MID- TO FAR-IR REGION: H_2O AND ETHANOL ON Si(100)

L. M. Struck[1,2,3], J. Eng Jr.[1], B. E. Bent[1], Y. J. Chabal[2], G. P. Williams[4], A. E. White[2], S. Christman[2], E. E. Chaban[2], K. Raghavachari[2], G. W. Flynn[1], K. Radermacher[5], and S. Mantl[5]

[1]Columbia University, Department of Chemistry, New York, NY; [2]AT&T Bell Laboratories, Murray Hill, NJ; [3]new address: National Institute of Standards and Technology, Gaithersburg, MD; [4]Brookhaven National Laboratory, Upton, NY; [5]Institut für Schicht-und Ionentechnik, Forschungszentrum Jülich, Jülich, Germany.

ABSTRACT

The technique of external reflection infrared (IR) spectroscopy is used to study silicon surface chemistry. External reflection is enhanced by implanting a buried cobalt silicide layer in silicon to act as an infrared reflector. The preparation of clean well-ordered surfaces from the ion implanted substrates is demonstrated. The reactions of water and ethanol with Si(100) are investigated.

INTRODUCTION

The understanding of chemical reactions on silicon surfaces is important to semiconductor processing technology. Infrared (IR) spectroscopy has contributed to the characterization of hydrogen-terminated silicon surfaces, including the identification of the surface hydride structures and chemical reactions important in the cleaning and passivation of silicon surfaces. However, the most commonly utilized surface IR technique, multiple internal reflection spectroscopy (MIRS), is limited to frequency regions above 1500 cm^{-1} due to silicon phonon absorptions, restricting observation to hydrogen and deuterium containing surface species. Vibrations of Si-O, Si-C, Si-N, and heavier atoms cannot be studied by MIRS; therefore, the technique used in the present experiments utilizes a buried layer of cobalt disilicide in single-crystal silicon making single external reflection IR spectroscopy of the silicon surface possible, as previously demonstrated.[1,2] The relatively short path length through the thin overlayer of silicon (20-50 nm) minimizes the Si phonon absorption, opening the accessible range to all frequencies. This technique has several advantages over other surface vibrational spectroscopies: (1) the entire spectral range can be accessed using various detectors and sources, (2) the sensitivity using the $CoSi_2$ reflector is higher than standard transmission through or external reflection from pure silicon samples, (3) the spectral range is competitive with broadband techniques such as High Resolution Electron Energy Loss Spectroscopy (HREELS) but the resolution (< 1 cm^{-1}) is better, (4) the versatility is greater because it does not require high vacuum as do electron spectroscopies, and (5) the sensitivity makes possible studies of species on single-crystal surfaces, avoiding use of porous material which has less well-defined surfaces. Therefore, this technique can be a useful tool in the study of silicon surfaces in systems at higher

Mat. Res. Soc. Symp. Proc. Vol. 386

pressures, such as CVD film growth on silicon, silicon interfaces with various materials, or oxidation of silicon surfaces in vacuum or by wet chemical treatments in air.

EXPERIMENTAL

Infrared spectra of silicon surfaces are acquired by external reflection from a buried layer of $CoSi_2$.[1,2] A schematic representation of the geometry is shown in Figure 1. The buried cobalt silicide layer (~50 nm) acts as an infrared reflector and the silicon overlayer (20-50 nm) is sufficiently thin for the electric field (p-polarization) at the silicon-vacuum interface to remain large (angle of incidence ≈ 85°). The strong normal electric field and high reflection intensity are key factors in the good signal-to-noise ratio attainable.[1-3]

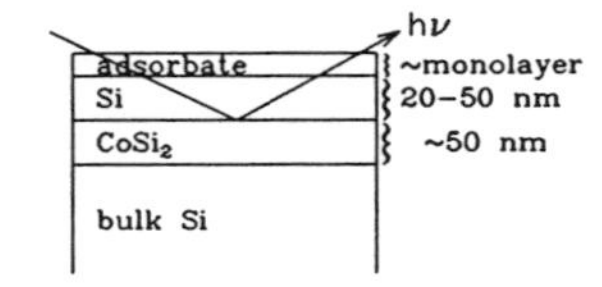

Figure 1. Schematic representation of geometry to obtain infrared spectra of silicon surfaces via single external reflection from a buried $CoSi_2$ layer.

The $CoSi_2$ buried layer is prepared[2,4] by 100-200 keV Co^+ ion implantation into either Si(100) or Si(111) with an ion beam density of approximately 10 μA/cm^2 and a dose of $\geq 1\times10^{17}$ cm^{-2}. Subsequent heating to ~1270 K forms a continuous single layer of $CoSi_2$ roughly 50 nm below the silicon surface and anneals the defects in the silicon overlayer.

The silicon substrates are degreased in trichloroethylene, acetone, and methanol, and then dried with nitrogen. The samples are oxidized/etched in an H_2O:H_2O_2:NH_4OH solution (4:1:1 by volume) at about 350 K for ≤5 minutes. Samples are rinsed thoroughly in deionized water after each step. The oxide is etched in 40% HF for 20 seconds. The silicon is oxidized again in a 4:1:1 H_2O:H_2O_2:HCl solution for 10-20 minutes at 350 K. The oxidizing and etching steps in the HCl and HF solutions, respectively, are repeated until a hydrophobic surface is observed after the HF etch. To test the quality of the resulting silicon surface, an oxidized Si(111) surface is etched in a pH-buffered HF solution to produce an ideally H-terminated surface.[5] The spectrum of the resulting surface, Figure 2a, shows the full-width-at-half-maximum (FWHM) linewidth of the Si-H stretch is about 1 cm^{-1}, similar to what is achieved for wafers without Co^+ implantation.[5] The same linewidth is observed on a sample where both external reflection from the $CoSi_2$ region and MIRS in the region with no Co^+ implantation could be performed. Since the infrared linewidth is a measure of the homogeneity of the surface, we conclude that the Co^+ implantation/annealing sequence does not disrupt the surface in any way to alter the chemical process or surface homogeneity.

Clean well-ordered Si surfaces can also be prepared in vacuum without sputtering, which roughens the surface and erodes the sample quickly. The samples are chemically

Figure 2. Infrared spectrum of (a) hydrogen terminated Si(111) prepared by wet chemical methods, spectra taken in air (0.5 cm^{-1} resolution); (b) hydrogen terminated Si(100) prepared in UHV (1.0 cm^{-1} resolution).

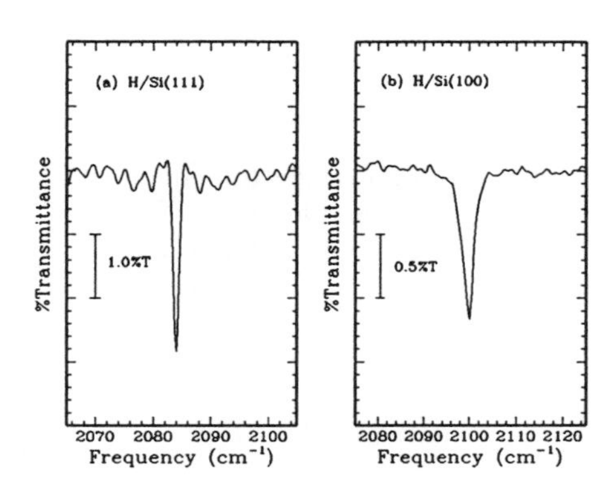

oxidized before mounting in the UHV chamber. The substrate is annealed to ~1170 K while maintaining the pressure below 1.3×10^{-7} Pa (1×10^{-9} Torr), resulting in a clean oxygen- and carbon-free surface. The monohydride on Si(100) is prepared in vacuum by exposing the clean surface to atomic hydrogen (from H_2 exposed to a tungsten filament at 1900 K) at a sample temperature of 570 K. Monohydride surfaces prepared here exhibit a FWHM of order 3 cm^{-1} (Figure 2b), tending to be more homogeneous than samples cleaned by typical sputtering and annealing.[6]

Experiments are performed in UHV chambers with base pressures $\leq1.3\times10^{-8}$ Pa ($\leq1\times10^{-10}$ Torr) equipped with Low Energy Electron Diffraction (LEED), Auger Electron Spectroscopy (AES), Temperature Programmed Desorption (TPD) and CsI windows for IR spectroscopy. Exposures of water and ethanol are made through a tube directed at the sample, and measured by an ionization gauge. Infrared spectra are collected using an interferometer. Broadband experiments (180-3000 cm^{-1}) are performed using Cu-doped Ge and bolometer detectors at the U4IR beamline[7] at the National Synchrotron Light Source, Brookhaven National Laboratory. Additional work focusing on the higher frequency region (>700 cm^{-1}) is done using HgCdTe and/or InSb detectors and a conventional black body source.

RESULTS AND DISCUSSION

Water on Si(100)

The interaction of water with silicon is important in silicon oxidation as well as silicon processing since water is present in most operating environments. The reaction of water with silicon has been the subject of numerous investigations such as surface studies of this reaction performed more than a decade ago.[8,9] H_2O on Si(100) is found to dissociate into Si-H and Si-OH surface species,[8,9] characterized by fundamental vibrations: the Si-H stretch and bend, the SiO-H stretch, Si-OH stretch, and the SiOH bend. These modes and the corresponding modes for SiOD species after D_2O exposure were previously observed with EELS.[8] The infrared spectrum of Si(100) exposed to D_2O using external reflection

Figure 3. (a) Infrared spectrum of clean Si(100) exposed to D_2O; (b) Infrared spectrum of clean Si(100) exposed to H_2O.

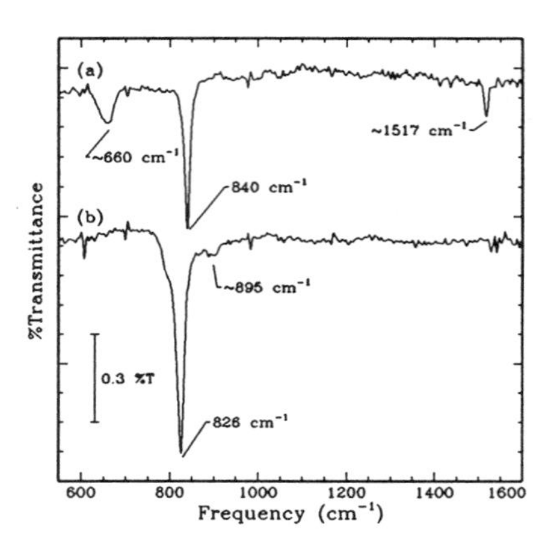

of a $CoSi_2$ buried layer is shown in Figure 3a (sample temperature at 318 K). The absorption at ~660 cm^{-1} is assigned to the SiOD bend, 840 cm^{-1} to the Si-OD stretch, and 1517 cm^{-1} to the Si-D stretch. The SiO-D stretch occurs at 2700 cm^{-1} as previously reported.[8,9] The infrared spectrum of Si(100) exposed to H_2O is shown in Figure 3b (sample temperature at 266 K). In contrast to the EELS work[8] which reported one strong feature at 820 cm^{-1}, the improved resolution of IR allows the detection of a weak feature at 895 cm^{-1}. Before the observation of this additional weak mode, the apparent frequency shift to *higher* frequency of the Si-OH stretch (820 cm^{-1}) to the deuterated Si-OD stretch (840 cm^{-1}) was unexplained. Based on a complete analysis,[10] the 826 cm^{-1} feature is assigned to predominantly an Si-OH stretch, while the 895 cm^{-1} feature has significant SiOH bending character. The Si-OH stretch and the SiOH bend are mechanically coupled, leading to frequency shifts and intensity changes, which will be discussed in more detail.[10]

This study has shown that the high resolution of IR spectroscopy makes it possible to resolve spectral features and characterize subtle shifts to provide a more complete picture of the surface chemistry. The knowledge of the precise Si-O stretch frequencies will be very useful in the studies of initial states of oxidation as well as surface studies of vibrations of heavier atoms than H or D which were previously inaccessible by infrared spectroscopy.

Ethanol on Si(100)

Ethanol on SiO_2 have been used as a model to understand the growth of SiO_2 by tetraethoxysilane (TEOS).[11] TEOS, $Si(OC_2H_5)_4$, has been used in industry to deposit SiO_2 films in applications where reduced deposition temperatures are desirable. Adsorbed ethanol mimics the siloxanes by formation of an ethoxide ($SiOC_2H_5$) structure.[11] An important step in the reaction mechanism of TEOS is that the hydrocarbon desorbs and does not become a primary constituent of the resulting film. Similar structures and decomposition mechanisms have been reported for clean silicon surfaces. Clean Si(100)[2a] and Si(111)7×7[12] exposed to methanol form surface methoxy species; methanol dissociates into $-OCH_3$ and -H surface groups. At higher temperatures, the methoxy species on Si(111)7×7 decomposes on the surface into oxygen, carbon, and hydrogen;[12b] however, the methoxy species on Si(100) is found to desorb with no surface decomposition.[2a] Isopropanol on silicon results in a similarly adsorbed isopropoxy group, $-OCH(CH_3)_2$.[1b] The decomposition of ethanol on SiO_2 is similar to that of the siloxanes; both involve the desorption of ethylene.[11] Ethanol on clean Si(111)7×7 forms a surface ethoxy species; however, the ethoxy decomposes on the surface at higher temperatures.[13]

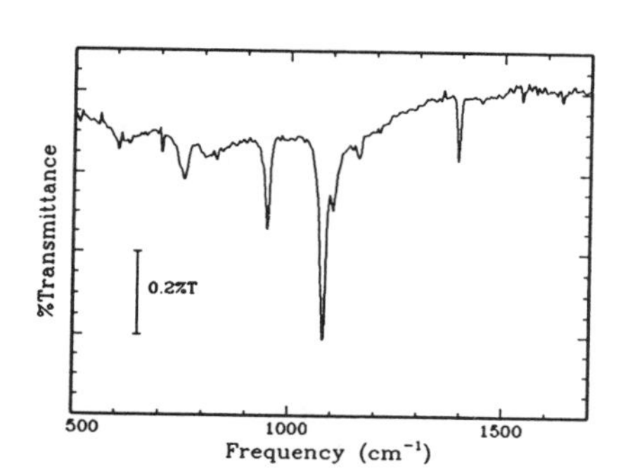

Figure 4. Infrared spectrum of clean Si(100) exposed to ethanol.

In this work, we investigate the exposure of C_2H_5OH to clean Si(100) and its subsequent decomposition mechanism. Clean Si(100) at 220 K is exposed to a nominal 1 Langmuir (1.3×10^{-4} Pa-s) dose of ethanol. The resulting infrared spectrum is shown in Figure 4. The absorption at 1394 cm^{-1} is primarily due to a CH_3 deformation mode.[1b,12-14] The 1081 cm^{-1} feature is tentatively assigned to the asymmetric C-C-O stretch, the feature at 950 cm^{-1} to the symmetric C-C-O stretch, and the 756 cm^{-1} peak to the Si-O stretch, however, a more detailed spectral analysis is underway.[15] The appearance of the Si-H stretch at about 2090 cm^{-1} (not shown) also supports the $Si-O-C_2H_5$ structure, since the dissociation of the O-H bond in ethanol results in the formation of a surface ethoxy species as well as addition of hydrogen to the surface. After heating this surface, the IR peaks disappear and additional carbon is not detected by AES measurements, indicating that surface decomposition does not occur.

The broad spectral range accessible in this technique allows the observation of the Si-H stretching mode as well as lower frequency modes of ethanol on Si(100) which are needed to assign the surface structure and understand the surface chemistry.

CONCLUSION

Single external reflection from a $CoSi_2$ buried layer in single-crystal silicon has been used to obtain IR spectra of silicon surface species. Clean well-ordered surfaces of Co^+

implanted Si samples can be prepared in air and in UHV as determined by the relatively narrow IR linewidth of the Si-H stretching vibration. We have used this technique to investigate water and ethanol adsorption on Si(100), and thus have demonstrated several advantages of this technique including its broad spectral range, high resolution, and surface sensitivity. In the future, this approach could be used for a wide range of interesting systems including surface studies of As/Si or Cl/Si; interface studies such as the GaAs/Si interface; oxidation of silicon surfaces; or surface studies in systems at higher pressures, such as CVD film growth on silicon.

ACKNOWLEDGMENTS

Financial support from the National Science Foundation Division of Materials Research, Joint Services Electronics Program through the Columbia Radiation Laboratory (grant number DAAH04-94-G-0057), and DOE (contract number DE-AC02-76CH00016) are gratefully acknowledged.

REFERENCES

[1]V. M. Bermudez, J. Vac. Sci. Technol. A **10**, 152 (1992); (b) M. McGonigal, V. M. Bermudez, and J. E. Butler, J. Elec. Spec. Rel. Phen. **54/55**, 1033 (1990).
[2]W. Erley, R. Butz, and S. Mantl, Surf. Sci **248**, 193 (1991); (b) S. Mantl and H. L. Bay, Appl. Phys. Lett. **61**, 267 (1992); The $CoSi_2$ layer is prepared in this paper by a deposition process.
[3]Y. J. Chabal, M. A. Hines, and D. Feijóo, J. Vac. Sci. Technol. A **13**, to be published.
[4]A. E. White, K. T. Short, R. C. Dynes, J. P. Garno, and J. M. Gibson, Appl. Phys. Lett. **50**, 95 (1987); (b) R. Hull, A. E. White, K. T. Short, and J. M. Bonar, J. Appl. Phys. **68**, 1629 (1990).
[5]G. S. Higashi, Y. J. Chabal, G. W. Trucks, and K. Raghavachari, Appl. Phys. Lett **56**, 656 (1990).
[6]Y. J. Chabal, Surf. Sci. **168** 594 (1986).
[7]G. P. Williams, Int. J. Infrared Millimeter Waves **5**, 529 (1984); (b) G. P. Williams, Nucl. Instrum. Methods A **291**, 8 (1990); (c) G. P. Williams, Rev. Sci. Instrum. **63**, 1535 (1992).
[8]H. Ibach, H. Wagner, and D. Bruchmann, Solid State Comm. **42**, 457 (1982); (b) F. Stucki, J. Anderson, G. J. Lapeyre, and H. H. Farrell, Surf. Sci. **143**, 84 (1984).
[9]Y. J. Chabal and S. B. Christman, Phys. Rev. B **29**, 6974 (1984).
[10]L. M. Struck, *et al.*, manuscript in preparation.
[11]L. L. Tedder, J. E. Crowell, and M. A. Logan, J. Vac. Sci. Technol. A **9**, 1002 (1991); (b) L. L. Tedder, G. Lu, and J. E. Crowell, J. Appl. Phys. **69**, 7037 (1991).
[12]K. Edamoto, Y. Kubota, M. Onchi, and M. Nishijima, Surf. Sci. **146**, L533 (1984); (b) J. A. Stroscio, S. R. Bare, and W. Ho, Surf. Sci. **154**, 35 (1985).
[13]Z. Ying and W. Ho, Surf. Sci. **198**, 473 (1988).
[14]T. Shimanouchi, Tables of Molecular Vibrational Frequencies, Consolidated Volume I, Nat. Stand. Ref. Data Ser., Nat. Bur. Stand. (U.S.) **39** (1972) pp.104-105; (b) The Sadtler Handbook of Infrared Spectra, (Philadelphia: Sadtler Research Laboratories, 1978) p. 464.
[15]J. Eng Jr., *et al.*, manuscript in preparation.

MONITORING OF HF/H_2O TREATED SILICON SURFACES USING NON-CONTACT SURFACE CHARGE MEASUREMENTS

P. ROMAN, D. HWANG, K. TOREK, J. RUZYLLO AND E. KAMIENIECKI*
Electronic Materials and Processing Research Laboratory, The Pennsylvania State University, University Park, PA 16802
*QC Solutions, Inc., Woburn, MA 01801

ABSTRACT

In this work, a new commercial system allowing non-contact measurement of the surface charge is used to monitor the condition of the silicon surface following HF/water etch. Results obtained demonstrate that by monitoring changes of surface charge using this system, a truly non-invasive, instant and easy to carry out characterization of Si surfaces after HF/water etch can be accomplished. The results show that HF/water exposure adds positive charge to the silicon surface. Change in the surface charge, considered to be indicative of the change in the electro-chemical condition of the surface, appears to precede initiation of the oxide etching process, and is proposed to be a factor in initiating etching reactions that involve mainly negatively charged species.

INTRODUCTION

The HF/water etch of native or chemical oxides is a common step in silicon surface cleaning and conditioning operations. This step not only removes oxide, but also affects the chemical state of the silicon surface resulting from etching, and in this way has a significant bearing on the outcome of the subsequent processing steps performed on the silicon wafer.

The specific role of HF/water etch varies depending on the application. Prior to epi deposition, poly-emitter formation or metallization, its role is to remove residual oxide and to leave the silicon surface mainly hydrogen terminated to prevent subsequent regrowth of the oxide. The role of the HF/water step in the surface cleaning sequence is less straightforward. In particular, the location of the HF/water etch in the pre-gate oxidation cleaning procedures is a matter of on going considerations concerning disadvantages, or advantages of the HF-last step (e.g.[1,2]). This discussion exemplifies still incomplete understanding of the very complex interactions between silicon and the species in the HF/water solution and the effect they may have on the subsequent process. One problem is a multiplicity of factors that may affect these interactions while the other is a lack of methods that would allow non-invasive, instant monitoring of the continuously changing condition of the silicon surface following HF/water etch.

In this experiment, the HF/water treated silicon surfaces are studied using a commercial Surface Charge Profiler (SCP) [3] that allows non-contact measurement of the electric charge on the etched silicon surfaces as well as surface recombination lifetime. Surface charge is very sensitive to the physical and chemical conditions of the silicon surface, and hence, its measurement can provide valuable information in process monitoring applications. The measurement of the surface charge can be performed in

Mat. Res. Soc. Symp. Proc. Vol. 386

various ways. Typically, surface charge measurement methods require either physical contact to the surface, or are contactless but use high voltage or high intensity illumination altering in each case the condition of the surface during the measurement. In contrast, the non-contact SCP method applied in this experiment uses no bias and very low intensity illumination, and hence, is truly non-invasive.

The SCP method has been proven very useful in the monitoring of wafer cleaning operations [4]. In this study, this method is used specifically to obtain new information concerning HF/water etching of silicon surfaces.

EXPERIMENTAL

The SCP method used in this study is based on the generation of the Surface Photovoltage (SPV) by illumination of the silicon surface with a beam of chopped low intensity light, having a photon energy higher than the silicon bandgap [5]. The value of the SPV is related to the density of electric charge accumulated on the illuminated surface. Because low intensity illumination is used in this method, the perturbance of the surface potential barrier during measurement is kept well below room temperature thermal energy, and hence, the surface charge measurement in this case is truly non-invasive. Besides surface charge, this method allows simultaneous measurement of the surface recombination lifetime. A schematic diagram of the measurement setup is shown in Fig. 1.

An experimental procedure in this study involved immersion of both P- and N-type, (100) silicon substrates in HF/water solution followed by D.I. water rinse and nitrogen blow drying. Fresh wafers as supplied by the wafer manufacturers were used. The composition of the HF/water solution used was 1:100 HF(49%):H_2O. Measurement of surface charge and surface recombination lifetime was performed within a few minutes following HF/water treatment. Additional information concerning the condition of the silicon surface following the HF/water etch was obtained from XPS analysis and contact angle measurements.

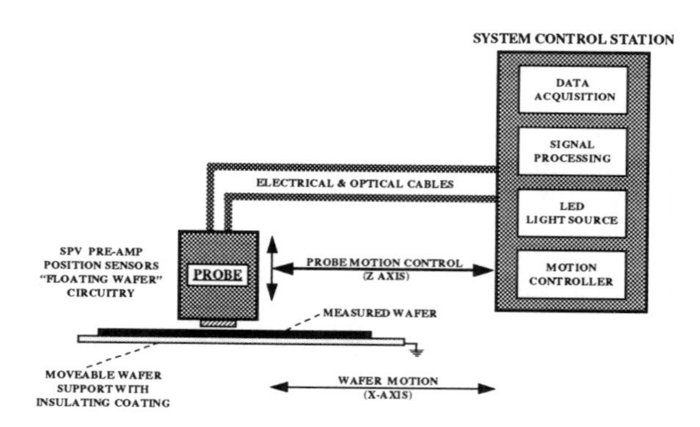

Fig. 1 Surface Charge Profiler experimental setup

RESULTS AND DISCUSSION

In the following discussion, it is assumed that the nature of surface charge measured in air following HF:H_2O immersion is probably different from the nature of charge on the same surface while in the solution. This is not considered to be a limitation of the introduced methodology for two reasons. First, what counts in this application is the charge on the surface after HF:H_2O treatment, and not the charge during the actual immersion. Second, regardless of the exact nature of surface charge in the solution and in air, the variations in the density of the latter were shown in this experiment to respond to the variations in the composition of the HF:H_2O mixture and time of immersion, which are also likely to affect surface charge in solution. Hence, one may draw the conclusion that the densities of these charges are related.

In the case of both N- and P-type wafers used in this study, surface charge on the wafers "out-of-the-box" was negative. Subsequent HF/water immersion always added positive charge to the surface (Fig. 2). For P-type wafers this added positive charge was making the Si surface highly positively charged with inversion of the near-surface region resulting. In the case of N-type wafers, the surface following HF/water immersion remained negatively charged, although the value of the negative charge was always substantially reduced as compared to the initial surface charge (Fig. 2).

The effect of the silicon surface acquiring positive charge during HF:H_2O treatment was found to be strongly dependent on the variety of process parameters. In order to follow changes in the surface charge density during HF:H_2O etching, fresh silicon wafers were exposed to an HF(1):H_2O(100) solution for periods of time varied from 1 second to 120 seconds. Subsequently, wafers were rinsed for 1 minute, blown dry, and then, in each case exactly five minutes later, measurement of the surface charge and surface recombination lifetime was performed. The results, shown in Fig. 3, indicate that

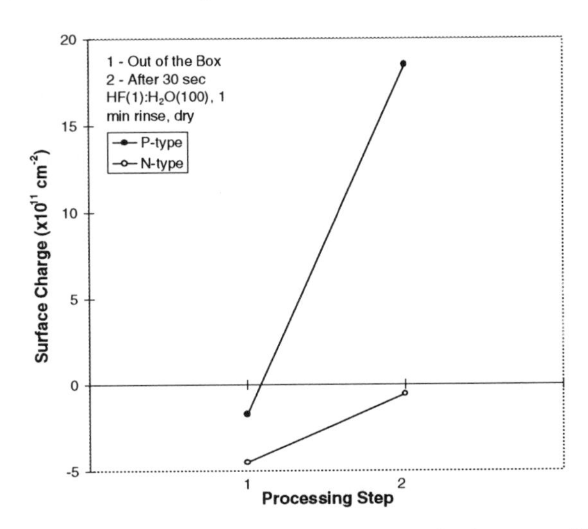

Fig. 2 Surface charge for P and N-type Si out of the box and following 30 sec HF(1):H_2O(100) immersion

positive charge is added to the silicon surface immediately after immersion in the solution, although, for etching times less than 10 seconds instability of the surface charge during measurements was observed.

The nature of the relationship illustrated in Fig. 3 suggests a connection between the behavior of surface charge and the oxide etching process. Based on such an assumption, as well as on the observation that surface charge reaches saturation as a result of very brief exposure to the HF/water mixture, one may draw a conclusion that the native oxide is completely etched off in the HF(1):H_2O(100) solution in the course of just a few seconds long immersion. However, the value of the oxide etch rate for this particular solution does not seem to confirm this notion. Moreover, the changes of the surface recombination lifetime shown in Fig. 3 indicate more gradual changes of the surface features that are more likely to reflect gradual dissolution of the silicon oxide in HF/water solution. If this were the case, then one might conclude that the recombination centers are associated with native/chemical oxide and their density decreases during dissolution of the oxide.

To explain the above effects, the extent of the oxide etching during very short immersions had to be determined independently using other methods. For this purpose, oxide wetting angles were measured on wafers prepared together with the wafers for surface charge measurements. The results shown in Fig. 4 clearly indicate that the transformation of the silicon surface from initially hydrophilic to fully hydrophobic requires times of immersion in HF:H_2O solution longer than the time needed to complete addition of the positive charge to the surface. Moreover, the values of wetting angles in Fig. 4 suggest that the process of oxide etching, in contrast to rapid changes of the surface charge, proceeds gradually, and the oxide is etched off completely only after approximately a 60 seconds long immersion. In fact, these changes appear to be accurately reflected by the behavior of the surface recombination lifetime shown in Fig. 3.

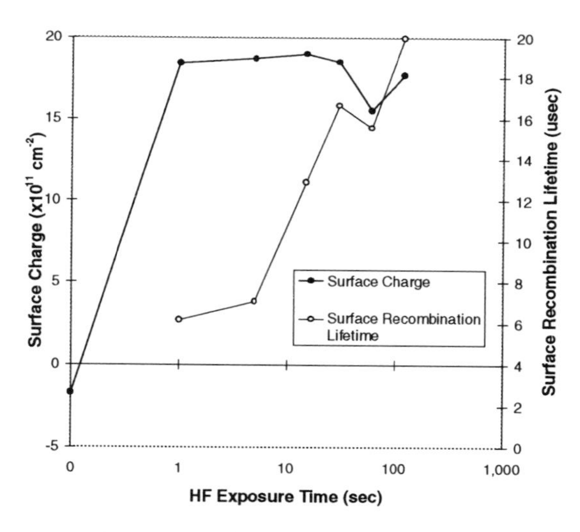

Fig. 3 Surface charge and surface recombination lifetime vs. HF(1):H_2O(100) exposure time for P-type Si

In order to further investigate the above phenomena, the XPS analysis of Si surfaces etched in the HF:H_2O solution for various times was carried out. Figure 4 illustrates changes of the O_{1s} peak as a function of the etch time and shows again that the oxide etching proceeds gradually, just as indicated by the variations of the wetting angle (Fig. 4), and surface recombination lifetime (Fig. 3).

Based on the results obtained, one may speculate that the process of oxide etching in HF/water solution is a two-step process in which dissolution of the oxide is preceded by the electrochemical reactions at the oxide surface resulting in the addition of the positive charge to this surface. These changes were not observed as a result of wafer immersion in D.I. water without any HF added, and hence, are considered instrumental in initiating etching in HF/water solution. As seen in Fig. 2, the oxide surface prior to HF/water immersion is negatively charged. Under such conditions, interactions of the surface with negatively charged etching species F^- and HF_2^- is highly unlikely. Therefore, to initiate etching, positive charges must first be formed on the oxide surface. It is known that the SiO_2/water interface contains neutral hydroxylated surface groups Si-OH which can protonate to give positively charged sites Si-OH_2^+. Once the surface becomes positively charged, the negative etching species in the solution begin to dissolve the silica. These processes strongly depend on several process parameters, including pH of the solution, but in general should be considered to be drivers of silica etching in HF/water solutions.

CONCLUSIONS

In this work, a new commercial system allowing a truly non-invasive measurement of the electric charge and surface recombination lifetime on the bare silicon surface was used to monitor the condition of the silicon surface following HF/water etch. By using this system, an experimental evidence that the etching of native silicon oxide in HF/water

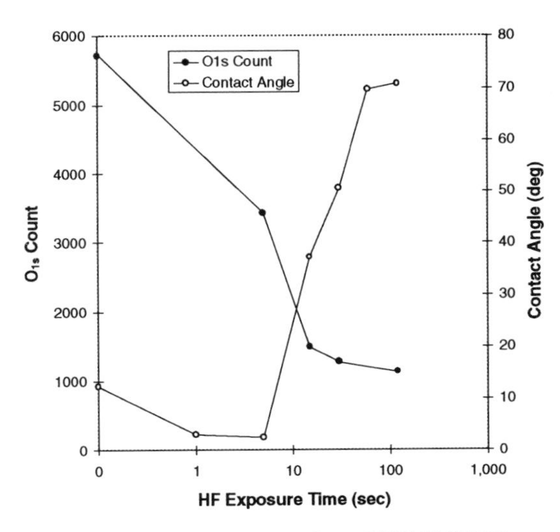

Fig. 4 XPS Oxygen count and contact angle vs. HF(1):H_2O(100) exposure time

solution is a two-step process was established. In the case of both N- and P-type wafers, surface charge on the wafers "out of the box" was negative and remained negative as a result of the D.I. water rinse. However, the HF/water immersion always added positive charge to the surface. This positive charging of the oxide surface is necessary to initiate oxide etching by negative species F^- and HF_2^-.

The results obtained indicate that the tool used in this study can be very effective in in-line monitoring of $HF:H_2O$ etching of oxide on P-type wafers due to the fact that the surface recombination lifetime measured accurately reflects the process of native oxide etching.

ACKNOWLEDGMENT

This study was partially supported by a grant from QC Solutions, Inc. Fruitful discussions with Dr. K. Osseo-Assare are gratefully acknowledged.

REFERENCES

1. A. Philipossian, J. Electrochem. Soc., **139**, 2956 (1992).

2. T. Ohmi in Proc. Third Intern. Symp. on Wafer Cleaning Technol. in Semicond. Device Manufacturing, edited by J. Ruzyllo and R. Novak (The Electrochem. Soc. Proc. **94-7**, Pennington, NJ, 1994) pp. 3-13.

3. Technical Information, SCP-110, QC Solutions, Inc., Woburn, MA 01801.

4. E. Kamieniecki, P. Roman, D. Hwang, and J. Ruzyllo in Proc. Second Intern. Symp. on Ultra-Clean Processing of Silicon Surfaces UCPSS '94, edited by M. Heyns (Acco, Leuven, 1994) pp. 189-192.

5. E. Kamieniecki and G.J. Foggiato, in Handbook of Semicond. Wafer Cleaning Technol., edited by W. Kern (Noyes Publications, 1993) pp. 497-536.

AUTHOR INDEX

SUBJECT INDEX